Fundamentals of Combustion

To my wife Ruby

Fundamentals of Combustion

ROGER A. STREHLOW

Professor of Aeronautical and Astronautical Engineering
University of Illinois, Urbana

ROBERT E. KRIEGER PUBLISHING COMPANY
MALABAR, FLORIDA

Original Edition 1968
Reprint 1979, with corrections, 1983

Printed and Published by
ROBERT E. KRIEGER PUBLISHING COMPANY, INC.
KRIEGER DRIVE
MALABAR, FLORIDA 32950

Printed in the United States of America

Library of Congress Cataloging in Publication Data

Strehlow, Roger A.
Fundamentals of combustion.

Reprint, with corrections, of the edition published by International Textbook Co., Scranton, Pa., in series: International Textbooks in mechanical engineering.
Includes bibliographies.
1. Combustion. I. Title.
[QD516.S89 1979] 541'.361 77-5368
ISBN 0-88275-539-0

Preface

This book is intended as an introductory text in combustion for the advanced undergraduate or beginning graduate student in engineering. As such, it presupposes a knowledge of compressible gas dynamics, the fundamentals of thermodynamics, and the student's ability to handle mathematics through differential equations and advanced calculus. However, because the text is intended for the engineering student, it was written on the assumption that the reader has had only a minimum of preparation in the fields of physical chemistry or reactive-gas dynamics.

The book is divided into two parts. The first part (Chapters 1 through 5) deals with fundamentals important to the combustion field, while the second part (Chapters 6 through 10) discusses the nature of combustion phenomena. The treatment is not intended to be exhaustive but, instead, introductory. It is hoped that the contents of this book will categorize the subject matter in the combustion field and prepare a foundation for the student who has not previously studied combustion phenomena. Thus, detailed discussions of very complex combustion phenomena are not included in this text. However, a thorough understanding of the material in this text should allow the student to think about more complex combustion processes in a constructive manner.

The first portion of the text contains chapters or sections on the particulate nature of matter, the kinetic theory of gases, advanced thermodynamics, the theory of transport phenomena, and the principles of chemical reaction kinetics, chemical equilibrium, and reactive-gas dynamics. Except for the treatments of advanced thermodynamics, chemical equilibrium, and reactive-gas dynamics, the approach is based on the concept that molecules exist and that their aggregate behavior is responsible for combustion phenomena. This part of the book was designed to include the majority of the material that the ordinary junior-senior engineering student does not have in his background. Individual instructors may choose to delete chapters or sections of this portion of the text, after evaluating the background of their students.

The second portion of the text deals with combustion phenomena. Chapter 6 "General Combustion Phenomena," introduces combustion processes which are not distinctly propagating wave phenomena; while the remainder of the chapters discuss the simpler combustion phenomena which do have a distinctly wave nature. Premixed gas flames are discussed in Chapters 7 and 8. Chapter 7 is used to introduce laminar premixed flames and discuss the properties of such flames. Chapter 8 discusses the aerodynamics of flames as a separate and distinct subject and in this respect it represents a departure from previous treatments. Chapter 9 discusses detonation in the light of some of the newer discoveries concerning the structure of the wave front and also mentions the subject of high-explosive detonation. Chapter 10 contains material on solid and liquid propellant combustion in a rocket engine. In this Chapter, solid propellant combustion is discussed rather extensively and liquid propellant combustion is mentioned only briefly. This treatment reflects a philosophy which is prevalent throughout the text. In general only those combustion phenomena which are relatively well understood are discussed in detail. Phenomena such as the aerodynamics of diffusion flames, turbulent combustion, and liquid propellant combustion in engines are not adequately understood and, therefore, in my estimation, should not be discussed in an elementary text on combustion. Let the student discover the difficulties himself, in the literature, after obtaining a good foundation in generally accepted concepts!

This text has evolved from a course in the fundamentals of combustion which I have taught in the Aeronautical and Astronautical Engineering Department for the past six years. As such, I owe a debt of gratitude to the many students who have contributed either directly or indirectly to the contents and organization of the text. I would also like to thank Professor H. O. Barthel for many helpful discussions concerning Chapters 1 and 3, Professor L. H. Sentman for reading and commenting on Chapters 1 and 2, and Professor L. D. Savage, Jr. of the Mechanical Engineering Department for our many discussions concerning the contents of Chapter 4. I would also like to acknowledge, with thanks, the work of Miss Dorothy Nugent and her secretarial staff in typing the manuscript as course notes.

ROGER A. STREHLOW

Urbana, Illinois

Contents

1

Atomic and Molecular Structure

1-1. INTRODUCTION

In this chapter we wish briefly to introduce and discuss some of the physical evidence for present-day concepts of atomic and molecular structure. In doing so we will touch lightly on many fields which are in themselves subjects for complete volumes. However, we are interested here in presenting only a few examples (hopefully the more important ones) of evidence for the particulate nature of material things. Nevertheless, the conclusion that particles exist and that their behavior controls the macroscopic properties of ordinary substances will be used throughout the text. The reader is referred to the bibliography at the end of this chapter for more complete works on our understanding of atomic and molecular structure.

1-2. EARLY SPECTROSCOPY

By far the most important tool available to the study of atomic and molecular structure is spectroscopy. Newton (1642–1727) is credited with discovering the fact that white light from the sun can be separated into colors with a prism. This discovery led to an investigation of optical dispersion in various glasses, and finally to the development of the diffraction grating and to the study of dark lines in the solar spectrum by Fraunhofer early in the nineteenth century. Fraunhofer's discoveries were very important to the development of spectral analysis and our concept of atomic structure. The existence of dark lines in the solar spectrum led various investigators to become inquisitive about other light sources. It was found that most light sources emitted light at discrete wavelengths. Futhermore, the invention of the diffraction grating allowed the early investigators to measure accurately the wavelength of the spectral lines and thereby make quantitative comparisons. The reason for this can be seen if one assumes that light is a wave phenomenon. Figure 1-1 shows that a fine grating effectively becomes a set of line sources when light passes through it. Each of these sources is in phase, however, and the light that is one wavelength behind the light from a neighboring slit will produce constructive interference at an angle θ_1 given by $\lambda = d \sin \theta_1$, where λ is the wavelength of the light and d is the slit spacing. In a similar manner, light two wavelengths behind that from a neighboring slit will produce constructive interference at an angle $2\lambda = d \sin \theta_2$ and in general, constructive interference and therefore a spectra will be produced at angles defined by $n\lambda = d \sin \theta_n$, where n is an integer equal to the order of the spectra. Since accurate gratings with known spacings can be constructed (albeit with some difficulty), the wavelength of particular spectral lines can be measured precisely.

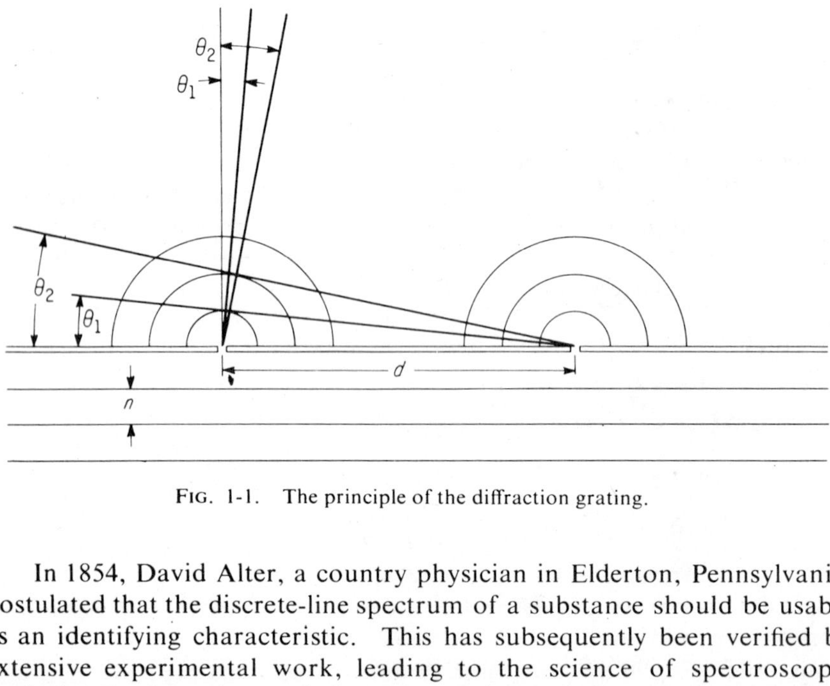

FIG. 1-1. The principle of the diffraction grating.

In 1854, David Alter, a country physician in Elderton, Pennsylvania, postulated that the discrete-line spectrum of a substance should be usable as an identifying characteristic. This has subsequently been verified by extensive experimental work, leading to the science of spectroscopic analysis. In time, the profusion of experimental data produced by this area of research (i.e., tabulated wavelengths of spectral lines) raised many unanswerable questions concerning the origin of discrete spectra and led to investigations of the relationship between the different spectral lines emitted by a single substance.

In 1885 Johann Balmer discovered the first of a series of simple empirical relationships between the wavelengths of the spectral lines of a single substance. The Balmer formula for the wavelength of hydrogen emission lines in the visible region is

$$\nu' = R\left(\frac{1}{2^2} - \frac{1}{n^2}\right) \qquad n = 3, 4, 5, \ldots$$

This equation predicts the wave number $\nu'(\text{cm}^{-1})$ or reciprocal wavelength $1/\lambda$ (where λ = cm) for lines appearing in the visible spectra of hydrogen with an internal consistency of 1 part in 10^7. The constant R is called the *Rydberg constant* and has the value $R = 109{,}677.6\ \text{cm}^{-1}$. This and other similar empirical relations attracted much interest in the late nineteenth century but unfortunately they could not be used to construct a model of the atom. Too much information was still missing from other areas of physical science.

1-3. THE ELECTRON AND THE NUCLEUS

In 1899 J. J. Thomson proved that cathode rays were particles with the same charge-to-mass ratio (e/m) irrespective of the cathode material. Electrons were thus discovered as an elementary particle presumably present in all atoms. This quickly led to the simple but unproven concept that an atom was a kind of "plum pudding" of positive charge with sufficient electrons immersed in it to make it neutral. This tentative concept was modified in 1911 when Rutherford in England showed that alpha-particle scattering data could be explained quantitatively if one assumed that the scattering material consisted of very small positively charged particles (diameter about 10^{-12} cm) separated by large distances. Rutherford's observation gave no information about the electrons, but the theory of a small positively charged nucleus predicted the experimental scattering results so well that it was almost immediately accepted. At this point it was realized that classical mechanics could not find a satisfactory position for the electrons because an orbiting electron is constantly accelerating, and electromagnetic theory predicts that a charged particle which is accelerating should radiate energy continuously. However, Rutherford's model is very close to our present understanding of atomic structure and stimulated a rapid growth of the concept of the quantum mechanical atomic model.

1-4. THE DUAL NATURE OF LIGHT

As indicated in Sec. 1-2, diffraction is easily explained if one assumes that light is a propagating wave. Other experimental evidence, such as the measurement of the velocity of light in water, led to the general acceptance of the wave theory of light about 1850 and to a rather complete electromagnetic radiation theory by the end of the nineteenth century. During the years 1900 to 1913, however, three relationships were discovered which, taken together, showed overwhelmingly that light exhibits a corpuscular nature in certain experimental situations. The first of these, the postulate of Plank concerning blackbody radiation, simply demonstrated that the dependence of intensity on frequency for radiation from a blackbody is explained if one assumes that the light is emitted in discrete energy bundles (called quanta) whose individual energies are related to the frequency of the light by the expression $e = h\nu$, where ν is the frequency in cycles per second. In addition, Plank determined that a value of $h = 6.55 \times 10^{-27}$ erg sec yielded the best fit between the experimental data and the theoretical curve for blackbody radiation.

In 1905 the second relationship was found when Einstein, using Plank's results, predicted that the maximum energy of a photoelectron emitted from a substance should be proportional to the frequency of the light causing the emission, and that the proportionality constant would be Plank's constant h. Einstein's equation is

$$E_m = h\nu - h\nu_0$$

where ν_0 is the threshold frequency for any substance and is related to the energy required to remove an electron from the crystal lattice. This relation was quickly verified and in 1916 Millikin's precise experimental technique yielded a value of $h = 6.56 \times 10^{-27}$ erg sec for the photoelectric constant.

The third confirmation of the corpuscular nature of light came from Bohr's original explanation of the discrete spectra of an atom containing a nucleus and a single electron. Bohr postulated that the electron could only have orbits whose angular momenta $I\omega$ are integral multiples of Plank's constant divided by 2π where I is the moment of inertia of the electron around the nucleus and ω its angular velocity. Emission or absorption was postulated to occur when a transition from one to another energy level (or orbit) occurred. A conformation of the corpuscular nature of light occurs because this model allows one to calculate Plank's constant h from the Rydberg constant in Balmer's empirical formula. Needless to say, the value is the same as that predicted by either Plank's blackbody radiation theory or the theory of the photoelectric effect.

1-5. THE DUAL NATURE OF MATTER

In 1924 Louis de Broglie postulated that since light has both a wave and corpuscular nature, free particles (particles not in a potential field) should have a wave nature. He further predicted that the "wavelength" of a free particle should be related to its momentum by the equation

$$\lambda = \frac{h}{p}$$

where h is Plank's constant in erg sec, λ is the wavelength in centimeters, and $p = mv$, the particle's momentum in gram centimeters per second. This relationship was first confirmed quantitatively by Davisson and Germer in 1927 when they observed diffraction of a monoenergetic beam of electrons in a single crystal of nickel. In this case the regular spacing of the atoms in the crystal lattice acts as a miniature diffraction grating. It is significant that the observed diffraction can be explained only if the free electron beam has a wave nature associated with it. Following that experiment the diffraction of beams of various particles such as hydrogen, mercury, or arsenic atoms have all confirmed de Broglie's equation for matter waves.

1-6. QUANTUM MECHANICS

De Broglie's relationship concerning free particles led Schrödinger to postulate that a bound particle (i.e., a particle trapped in a potential field) will also exhibit wave properties. The equation he postulated for the wave behavior of a single particle in a one-dimensional potential field is called *the Schrödinger wave equation for one dimension* and is written

$$-\frac{h^2}{8\pi^2 m}\frac{\partial^2\Psi(x,t)}{\partial x^2} + V(x)\Psi(x,t) = -\frac{h}{2\pi i}\frac{\partial\Psi(x,t)}{\partial t} \qquad (1\text{-}1)$$

where $\Psi(x,t)$ is the wave function (in general, complex), m the mass of the particle, h = Plank's constant, $V(x)$ the classical potential function for the particle, and $i = \sqrt{-1}$. The wave function in this equation has no true physical interpretation; however, the magnitude of $\Psi(x,t)$ can be interpreted as the probability of finding the particle in the region dx. Since $\Psi(x,t)$ is in general complex, its magnitude is given by its product with its complex conjugate $\Psi^*(x,t)\cdot\Psi(x,t)$, where this product must be independent of time and normalized $\left(\text{i.e., } \int_{-\infty}^{\infty} \Psi^*(x,t)\cdot\Psi(x,t)\,dx = 1\right)$ in order for a time-independent probability (or stationary state) to exist. This, of course, implies that Ψ be finite everywhere and approach zero at $x = \pm\infty$. In addition $\partial\,\Psi(x,t)$ and $\partial\,\Psi(x,t)/\delta t$ are assumed to be single-valued and continuous between $x = \pm\infty$. It is interesting to note that these (essentially mathematical) limitations on the behavior of the function $\Psi(x,t)$ are all that are necessary to predict many of the observed properties of bound particles. Before an example is discussed it should be mentioned that the statement that the quantity $\Psi^*(x,t)\cdot\Psi(x,t)\,dx$ represents the probability of finding the particle in the distance dx implies that a classical description cannot be made with exactness in the case of microscopic particles. This uncertainty about the location of a particle was first stated by Werner Heisenberg in 1927. He reasoned that the error in measuring the particle's momentum Δp times the error in measuring its position Δx must always be greater than Plank's constant divided by 2π, i.e.,

$$\Delta x \cdot \Delta p > h/2\pi$$

This is indeed reasonable if energy is quantized in units of $h/2\pi$. Notice that for ordinary substances in the macroscopic world this represents no real limitation to our classical statements concerning the motion of the body. However, for atomic sizes this represents a very real limit to the ultimate accuracy of observation. Remember that the first Bohr radius was chosen such that

$$mv \cdot r = h/2\pi$$

In this case the magnitude of a momentum-distance product is just equal to the constant used in determining the minimum uncertainty in a measurement. We would expect, therefore, that it would be impossible experimentally to verify any physical picture of atomic structure. We are left instead with a probability distribution for the particle or particles. We should also add in passing that quantum mechanics must always predict the correct classical behavior for macroscopic systems. In most cases it is found that the probability function essentially predicts a finite but small probability of our observing a behavior which does not follow the ordinary classical description. However, for ordinary macroscopic phenomena this probability is so small as to be entirely negligible for any ordinary observation time or for any ordinary number of observations.

1-7. THE HARMONIC OSCILLATOR

The utility of quantum mechanics can best be illustrated through the use of examples. We first choose to examine the harmonic oscillator. In classical mechanics the motion of a harmonic oscillator is represented by the equation

$$m \frac{d^2x}{dt^2} = -kx \tag{1-2}$$

where m is the mass of a particle and the quantity $-kx$ is the restoring force acting on the particle. The solution to this equation (which may be verified by direct substitution) is

$$x = A \sin (2\pi\nu_0 t + B)$$

where ν_0 is the frequency of the oscillation. The restoring-force constant is given by the expression

$$k = 4\pi^2\nu_0^2 m$$

where the potential energy is

$$V(x) = \tfrac{1}{2}kx^2 = 2\pi^2\nu_0^2 mx^2$$

In this conservative classical system the restoring-force constant and the mass determine the frequency of oscillation. In addition the arbitrary constant A is equal to the maximum displacement and is related to the total energy in the system by the equation

$$E = 2\pi^2\nu_0^2 mA^2$$

where E can take on any value and thereby determine the extent of the oscillation.

In quantum mechanics the behavior of the harmonic oscillator is represented by the one-dimensional Schrödinger amplitude equation with $V = 2\pi^2\nu_0^2 mx^2$

$$\frac{d^2\psi(x)}{dx^2} + \frac{8\pi^2 m}{h}(W - 2\pi^2 m\nu_0^2 x^2)\psi(x) = 0 \tag{1-3}$$

Here W represents the total energy of the conservative mechanical system which is being modeled and the wave function amplitude $\psi(x)$ is related to $\Psi(x,t)$ through the equation $\Psi(x,t) = \psi(x)\cdot\phi(t)$. The solution of this equation[1] with Ψ restricted to behave in the manner described in Sec. 1-6 yields the following interesting results. First, the total energy of the system is found to be restricted to a series of discrete states represented by eigenvalues of Eq. 1-3. For the harmonic oscillator potential these values are given by the equation

$$W_n = (n + \tfrac{1}{2})h\nu_0 \qquad n = 0, 1, 2, 3, \ldots$$

where n is called the *quantum number* of the particular state of the system. Notice that this equation predicts that the quantized harmonic oscillator has a minimum energy of $W_0 = \tfrac{1}{2}h\nu_0$ (called the *zero-point energy*) and that the energy difference between quantum mechanically allowable vibrational states (or vibrational energy levels) is constant and equal to $h\nu_0$. Second, the evaluations of ψ and $\psi^*\psi$ for each of these levels predict that there is a small but finite probability of finding the mass center of the harmonic oscillator in a position outside the range predicted by classical mechanics. Third, it is found that the quantum mechanical probability distribution asymptotically approaches the classical distribution for large quantum numbers. These last two results are illustrated in Figs. 1-2 through 1-4. The coordinate in these figures is a dimensionless displacement coordinate defined with the equation

$$\zeta = \frac{4\pi^2 m\nu_0}{h}x$$

Figures 1-2 and 1-4 also show the classical prediction for the probability of finding the mass center at any position for the energies $W = \tfrac{1}{2}h\nu_0$ and

[1] See, for example, Linus Pauling and E. B. Wilson, Jr., *Introduction to Quantum Mechanics*, McGraw-Hill, New York, 1935, pp. 67–82.

$W = 10.5h\nu_0$, respectively. We note from these figures that for low quantum numbers the classical and quantum probability distributions are quite different, whereas for a high quantum number the average height of the quantum probability appears to be approaching the classical probability distribution. This is typical of a quantum description: in systems which have sufficient stored energy, quantum and classical mechanics

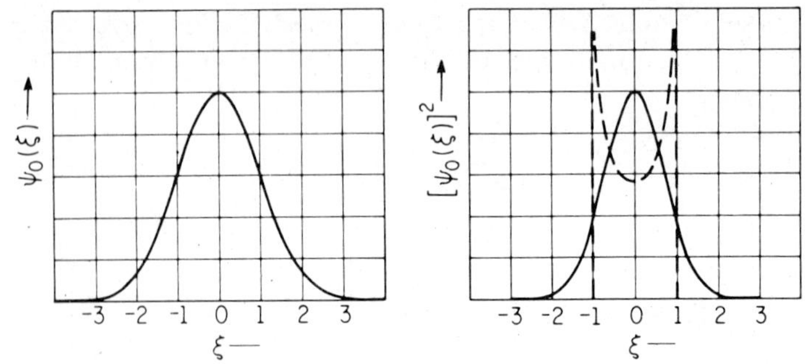

FIG. 1-2. The wave function $\psi_0(\xi)$ for the normal state of the harmonic oscillator (left), and the corresponding probability distribution function $[\psi_0(\xi)]^2$ (right). The classical distribution function for an oscillator with the same total energy is shown by the dashed curve.

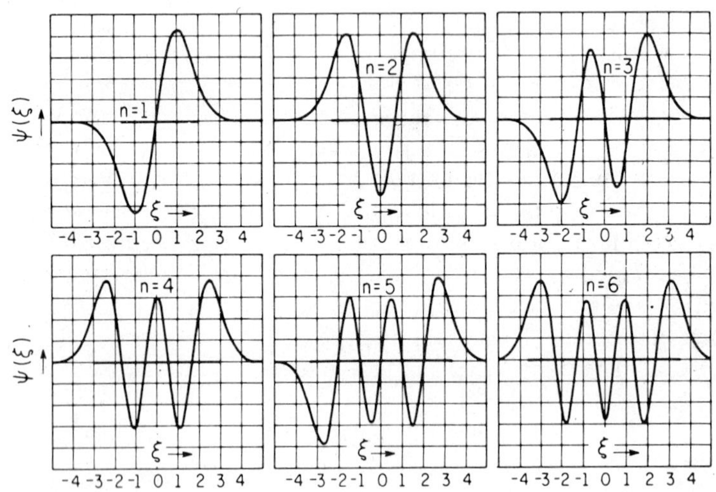

FIG. 1-3. The wave functions $\psi_n(\xi)$, $n = 1$ to 6, for the harmonic oscillator. For each case the heavy horizontal line indicates the region traversed by the classical harmonic oscillator with the same total energy.

Figs. 1-2 and 1-3 with permission from L. Pauling and E. B. Wilson, *Introduction to Quantum Mechanics*, New York, McGraw-Hill, 1935.

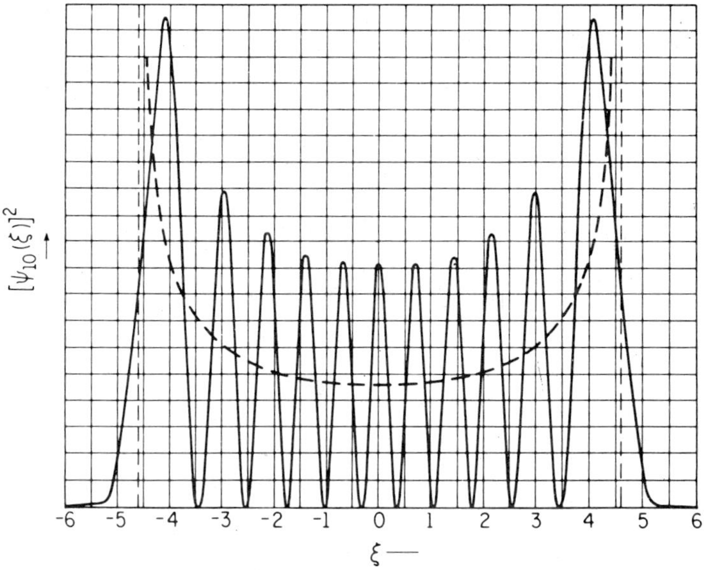

FIG. 1-4. The probability distribution function $[\psi_{10}(\xi)]^2$ for the state $n = 10$ of the harmonic oscillator. Note how closely the function approximates in its average value the probability distribution function for the classical harmonic oscillator with the same total energy, represented by the dashed curve.
With permission from L. Pauling and E. B. Wilson, *Introduction to Quantum Mechanics*, New York, McGraw-Hill, 1935.

predict essentially the same behavior but the descriptions are different when the energy storage is a few multiples of $h/2\pi$.

Notice that our discussion of the Schrödinger equation has not as yet led to any compelling reason for accepting the quantum description. The harmonic oscillator was chosen as a first point of discussion, however, because it is a rather simple system and because it corresponds rather well to the behavior of the stretched bond in a diatomic molecule. We will later use these results to indicate how the quantum-mechanical approach, coupled with spectroscopic measurements, allows the calculation of thermodynamic properties which are important to the combustion field. We will now briefly shift our attention to quantum predictions for the more complex systems representing the electronic and molecular structure of atoms.

1-8. THE HYDROGEN ATOM

By the time quantum mechanics was first postulated the hydrogen atom was known to be the simplest atom, consisting only of a heavy

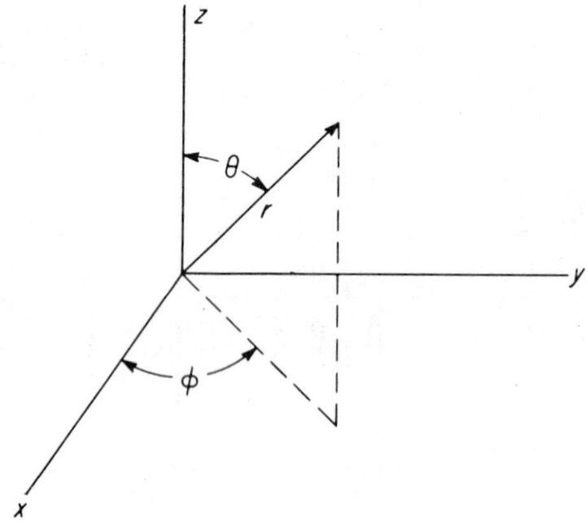

FIG. 1-5. Spherical coordinates.

proton (+ charge) and a light electron (– charge). The allowable energy levels (values of W) for this atom may be calculated by using the three-dimensional Schrödinger equation and by replacing the two-body problem with an electron of reduced mass $m_1 m_2/(m_1 + m_2)$ in a fixed central field. In this case spherical coordinates are simpler to use and the Schrödinger equation written in the coordinates r, ϕ, and θ is solved with the condition that the wave function be separable in these coordinates and the restraints that $\psi \rightarrow 0$ as $r \rightarrow \infty$, that $\psi_{\phi=0} = \psi_{\phi=2\pi}$ and that $\psi(\theta)$ be well behaved and have an integrable square in the interval $0 \leq \theta \leq \pi$. This set of restrictions yields a unique set of three quantum numbers, all of which must be specified to determine the state of the electron–proton configuration. These three quantum numbers and their (mathematically) allowable values are called

1. The *principle quantum number* $n = 1, 2, 3, 4, \ldots,$ (associated with the condition that $\psi = 0$ as $r \rightarrow \infty$).
2. The *orbital quantum number* $l = 0, 1, \ldots, (n - 1)$ [associated with the limitation on $\psi(\theta)$].
3. The *magnetic-moment quantum number* $m = -l, \ldots, -1, 0 + 1, \ldots, +l$ (associated with the $\phi_0 = \phi_{2\pi}$ condition).

In this system the energy associated with the various quantum-number combinations $W_{n,l,m}$ is not a simple function of the quantum number as it was for the harmonic oscillator. Furthermore, degenerate states (i.e., quantum states with the same value of W but a different configuration or wave function) appear. The quantum states for the single electron atom are listed in order of their respective quantum numbers in Table 1-1. The degeneracy of each state and its spectroscopic label or name are also listed. In addition, the last column of this table lists the relative position of these states if they were placed in order of increasing energy. This table of states for a single electron atom has been adequately verified by

TABLE 1-1
ENERGY STATES OF THE SINGLE ELECTRON ATOM*

Quantum Number			Degeneracy	Description	Order of Increasing Energy	Atomic Number of Element Listed in Table 1-2
n	l	m				
1	0	0	1	1*s*	1	1,2
2	0	0	1	2*s*	2	3,4
2	1	−1, 0, +1	3	2*p*	3	5, . . . , 10
3	0	0	1	3*s*	4	11,12
3	1	−1, 0, +1	3	3*p*	5	13, . . . , 18
3	2	−2, −1, 0, +1, +2	5	3*d*	7	21, . . . , 30
4	0	0	1	4*s*	6	19,20
4	1	−1, 0, +1	3	4*p*	8	
4	2	−2, −1, 0, +1, +2	5	4*d*	10	
4	3	−3, . . . , +3	7	4*f*	14	
5	0	0	1	5*s*	9	
5	1	−1, 0, +1	3	5*p*	11	
5	2	−2, −1, 0, +1, +2	5	5*d*	13	
5	3	−3, . . . , +3	7	5*f*	18	
5	4	−4, . . . , +4	9	5*g*	>18	
6	0	0	1	6*s*	12	
6	1	−1, 0, +1	3	6*p*	15	
6	2	−2, −1, 0, +1, +2	5	6*d*	17	
6	3	−3, . . . , +3	7	6*f*	>18	
6	4	−4, . . . , +4	9	6*g*	>18	
6	5	−5, . . . , +5	11	6*h*	>18	
7	0	0	1	7*s*	16	

*Adapted from Penner, *Chemistry Problems in Jet Propulsion*, Pergamon, 1957.

spectroscopic measurements. For example, transitions between states for $l = 0$ corresponds to the Balmer series when the transition occurs from a state with a high value of n to the state with $n = 2$. Furthermore, transitions from high n values to $n = 1$ yields the Lyman series in the ultraviolet and to $n = 3$, the Paschen series in the infrared region. In addition, this model allows one to calculate the ionization energy (the energy required to completely remove the electron from the proton) with great exactness. In fact, the calculated result appears to have a higher inherent accuracy than the experimental result. One additional feature of the quantum-mechanical description for a single electron system should be mentioned. The states of the system do not all correspond to equal radial probability for the electron distribution. In particular, the value l, the orbital quantum number, is of primary importance in determining the radial distribution. Typical distributions are shown in Fig. 1-6. These distributions as drawn are symmetric about the vertical axis and should be interpreted by relating the probability of finding the electron along any particular radius to the length of a line from the origin to the curve in that direction.

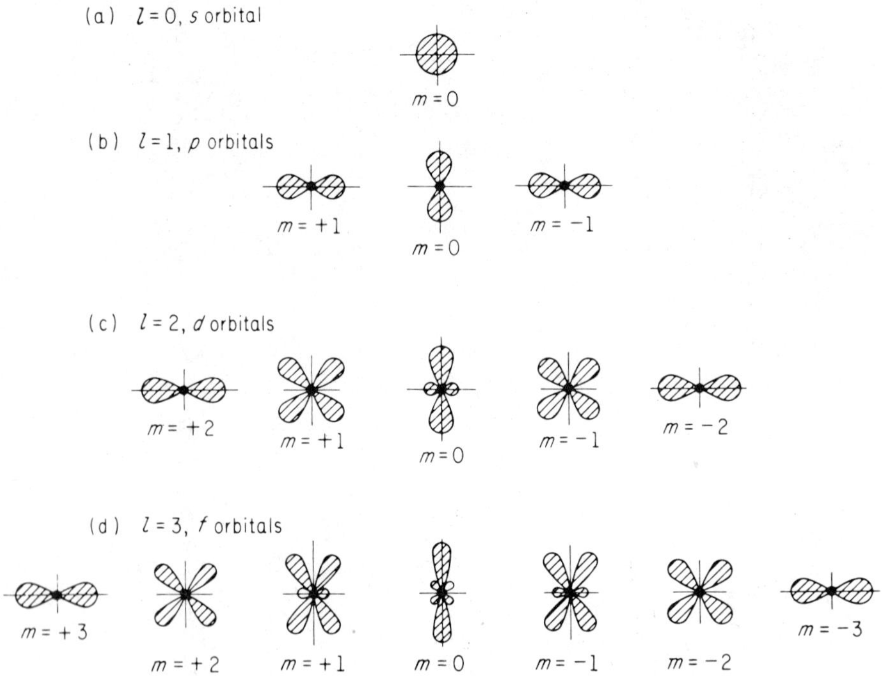

FIG. 1-6. Angular distribution functions for different electronic states of hydrogenlike atoms. (By permission from A. K. Syrkin and M. E. Dyatkina, "Structure of Molecules and the Chemical Bond." New York: Interscience Publishers (1950).

1-9. ATOMIC STRUCTURE

The first really significant generalization concerning atomic structure was made by Dmitri Mendeleev in 1863 when he constructed his periodic table of the elements from the various physical and chemical characteristics of the elements and their compounds. His most important discovery was that very striking regularities in behavior appear when the elements are arranged in order of increasing atomic weight. This empirical relationship was useful and led, for example, to the prediction of the properties of missing (undiscovered) elements. Unfortunately, as with most empirical relationships, the justification was lacking. Both the use of atomic weight as an indicator of relative position and the octet rule for constructing the length of the periods were without theoretical support.

The first real confirmation for the periodic arrangement of the elements came in 1913 when Henry Mosley discovered that the X-ray spectra of various elements were simply related to their relative position in the periodic table and when F. W. Aston in 1912 discovered isotopes (atomic species with the same chemical properties but different atomic weight) using the mass spectrometer. Mosley produced X-rays by electron bombardment of various elements and showed that the spectra contained two principal lines (now known to be due to nuclear transitions) whose frequencies are simply related to the position of the element in the periodic table. Aston showed, by using an apparatus similar to Thomson's apparatus for determining e/m of an electron, that the value of e/m, of a positively charged particle, depended markedly upon the source of the particles, and furthermore that one element could contain elementary particles (atoms) with two or more relatively close values of e/m. This work led to the present-day concept of the gross structure of the nucleus as a collection of positively charged protons and neutral bodies of about the same intrinsic mass called neutrons. In this model the protons determine the number of electrons needed to make the atom electrically neutral, the chemical properties of the atom and the characteristic X-ray spectra. The number of neutrons simply determines the total mass of an individual atom. The relative fraction of different isotopes present in macroscopic samples of the element determines its molecular weight on the scale in which the isotope of Carbon ${}_6C^{12} \equiv 12.00000$.[2] Further justification for the details of atomic structure had to wait for the development of quantum mechanics and the quantum-mechanical description of the single electron atom.

It would seem at this point that the solution of the wave equation for any particular atom should yield results significant to the theory of atomic structure. Unfortunately, the mathematics of this problem are extremely difficult and this approach has yet to yield any sweeping generalizations. Fortunately, however, it appears that the electrons do not materially interfere with each other in a multielectron atom, and in fact a very reasonable explanation of many physical and chemical phenomena can be obtained by assuming that the electrons occupy hydrogenlike "orbits" (or distributions) in a multielectron atom. However, before we look at this in detail we must introduce a new quantum number for the electron—its spin quantum number. The assumption that the electron is spinning and has a magnetic field associated with it allows for the correct explanation of many observed phenomena which would otherwise be unexplainable. For example, the doublet nature of alkali metal spectra, the existence of two forms of hydrogen at low temperatures, and the observed fact that electrons exhibit stable configurations when appearing as pairs all justify the concept of electron spin. The spin quantum number has a value of either $+\frac{1}{2}$ or $-\frac{1}{2}$ and depends upon the orientation of the electrons spin momentum vector. If in addition to postulating the existence of the spin quantum number $s = \pm\frac{1}{2}$, we postulate that no

[2] This scale was adopted by international agreement in 1961 and replaces the separate physicists' and chemists' ̣ale based on oxygen.

two electrons in any atom may simultaneously have the same four quantum numbers (the Pauli exclusion principle) we may construct a periodic table of elements. Recall that adding a proton to the nucleus (and its attendant electron in an orbit) changes the physical and chemical properties of the atom and increases the atomic weight by 1. We neglect

TABLE 1-2

ELECTRONIC CONFIGURATION OF THE ELEMENTS*

By Laurence S. Foster

References

1. Key furnished with "Periodic Chart of the Atoms" (1947) Edited by Henry D. Hubbard and William F. Meggers, W. M. Welch Sci. Co., Chicago.
2. William F. Meggers, "Electron Configurations of 'Rare Earth' Elements," *Science* (May 16, 1947), pp. 514–516.

Atomic No.	Element	*K*	*L*		*M*			*N*			
		1	2		3			4			
		s	*s*	*p*	*s*	*p*	*d*	*s*	*p*	*d*	*f*
1	H	1									
2	He	2									
3	Li	2	1								
4	Be	2	2								
5	B	2	2	1							
6	C	2	2	2							
7	N	2	2	3							
8	O	2	2	4							
9	F	2	2	5							
10	Ne	2	2	6							
11	Na	2	2	6	1						
12	Mg	2	2	6	2						
13	Al	2	2	6	2	1					
14	Si	2	2	6	2	2					
15	P	2	2	6	2	3					
16	S	2	2	6	2	4					
17	Cl	2	2	6	2	5					
18	A	2	2	6	2	6					
19	K	2	2	6	2	6	–	1			
20	Ca	2	2	6	2	6	–	2			
21	Sc	2	2	6	2	6	1	2			
22	Ti	2	2	6	2	6	2	2			
23	V	2	2	6	2	6	3	2			
24	Cr	2	2	6	2	6	5†	1			
25	Mn	2	2	6	2	6	5	2			

*With permission from *Handbook of Chemistry and Physics*, 46th ed., Chemical Rubber Company, Cleveland, 1965. (Only the first 25 elements are included for comparison with Table 1-1.)

†Note irregularity.

the neutrons in the nucleus temporarily because they do not markedly affect the gross properties of the atom. The electron which must be added to neutralize the additional nuclear charge is added to the state representing the lowest available (i.e., unoccupied) energy level in the quantum-mechanical description for the single-electron atom. Table 1-2 is a list of the observed internal structure of the first 25 atoms in the periodic table. The relationship of this table to Table 1-1 and its ability to allow prediction of the physical and chemical properties of the elements is striking, to say the least. Every element that has just the s and p levels of one shell filled is an inert gas (the principle quantum number n designates a shell). Every element that has five electrons in the p level of its outer shell is a halogen. Every element that has one electron in an outer shell (in its s level) and all its inner subshells filled is an alkali metal, and so on. Thus quantum theory is successful in providing a semi-quantitative description of atomic structure and chemical reactivity if one assumes that the electrons in a multielectron atom act rather independently of each other.

1-10. BOND FORMATION AND VALENCE

We are, of course, familiar with the fact that most elements exist in the form of chemical compounds which have properties quite different from the properties of the uncombined elements. In the above description of atomic structure, the s and p electrons in the outer shell are thought to contribute most to the formation of the bonds which lead to stable compounds and to determining the physical properties of the element and its compounds. In addition, bond formation is related to the strong tendency of the electrons to form a completed outer shell in the atom. There are two primary ways by which this condition may be met. In the first case the atom may completely lose or gain enough electrons to complete a shell. In this case a stable ion is formed and the chemical bond is said to be ionic (i.e., it consists of an electrostatic attraction between two charged particles). For example, the alkali metals have only one electron in the outer shell and losing this electron gives them a stable configuration corresponding to the structure of the next lowest completed shell. These metals all ionize rather easily and tend to form ionic compounds by acting as positively charged particles. On the other end of the behavior scale, the halogens can form a stable shell by adding one electron, thereby completing their octet. These elements have very strong electron affinities and tend to form ionic compounds with a valence of -1. Sodium chloride, for example, is well known to conduct electricity in solution and to have a crystalline form whose lattice points are occupied alternatively by positive sodium and negative chlorine ions.

There are, however, many chemical compounds which contain bonds that do not require complete ionization for bond formation. In these covalent bonds a number of the electrons from each atom are shared by both atoms, thereby satisfying the octet requirement. The simplest example of this type of bond is the H_2^+ ion which consists of a single electron shared by two protons. A quantum-mechanical description of this simple compound may be formulated and yields an interatomic distance of 1.06 Å (1Å = 10^{-8} cm) with stability arising because the electron can resonate between the two protons. Incidentally, this is an example of a molecule which contains an unshared electron and is therefore called a radical. Radicals are usually very reactive.

The neutral hydrogen molecule has the next simplest structure. It has four possible elementary or limit configurations:

(1) $H_A^1 \quad H_B^2$
(2) $H_A^2 \quad H_B^1$
(3) $H\!:_A^- \quad H_B^+$
(4) $H_A^+ \quad :\!H_B^-$

where the individual nuclei and electrons are labeled *A*, *B*, and 1, 2 respectively. A quantum-mechanical description of this molecule yields the correct dissociation energy and predicts that the molecule resonates between the four configurations and also can exist in more complicated configurations. The probabilities are found to be 80% for states 1 and 2 and 5% for states 3 and 4, with the remaining 15% representing the more complicated configurations. We would say that the bond is covalent with very little ionic character.

In some instances bond hybridization is observed. A particularly important example of bond hybridization occurs in the case of carbon. Carbon is known to form four covalent bonds which are essentially indistinguishable. However, the Pauli exclusion principle predicts that the outer shell of carbon should have two *s* electrons and two *p* electrons. Experiment has shown that in reality an individual carbon atom contains four unshared electrons and therefore only one *s* electron. This would still not lead to the equivalence of the four valence bonds. However, a quantum-mechanical description shows that the energy for four equal hybridized bond configurations is more stable than for one *s* and three *p* bonds, and carbon is said to form four sp^3 hybrid bonds in place of one *s* and three *p* bonds. Other types of stable molecular bonds are also observed in nature, but in the case of gases they are not very important and so will not be discussed here.

1-11. MOLECULAR STRUCTURE

In the case of gases most molecular species exist with a primarily covalent bond structure. The term *molecule* may therefore be defined in the context of this book as an aggregate of atoms which is electrically neutral and has sufficient stability to have measurable properties. Any free molecule (i.e., one which is observed in the absence of an external potential field) can store energy internally and thereby exist in a series of quantized energy levels which may be conveniently divided into rotational, vibrational, and electronic levels. The electronic levels of a molecule represent energy storage levels, each with a different electron configuration. Electronically excited states can represent a significant contribution to molecular properties at very high temperatures. Electronic transitions (even in complex molecules) are observed spectroscopically, and considerable experimental information has been gathered on excited molecular states. Notice that the calculation of these states (using quantum mechanics) must perforce contain a description of the electronic levels and be formulated to allow electronic transitions to occur. This subject and the process of ionization is considered to be outside the scope of this book.

Let us therefore consider a molecule in its ground electronic state. In this case the electronic configuration is not perturbed sufficiently during quantum transitions to require that electronic transitions be included in the description, and we can reduce the problem to one of discussing the energy of a system of n point masses (with mutual attractive and repulsive forces) as a function of the intermass distances and motion. Therefore we must specify $3n$ coordinates to completely describe the ground-level structure of an n-atom molecule. Since the molecule is considered to be a stable aggregate, it will have a center of gravity whose coordinates may be specified, and in addition it will have an equilibrium configuration in a molecule-oriented coordinate system. That is, one should be able to measure equilibrium bond distances. Furthermore, molecules in which $n \geq 3$ will exhibit equilibrium bond angles.

Thus we find that we can describe an n-atom molecule by listing the three cartesian coordinates of a center of mass and (if it is linear in its equilibrium configuration) the two coordinates for its rotation, or (if nonlinear) the three coordinates for its rotation. This leaves $(3n - 5)$ coordinates (linear molecule) or $(3n - 6)$ coordinates (nonlinear molecule) to be specified for a complete mechanical description. These remaining coordinates can be defined by assigning this number of normal (i.e., linearly independent) vibrational modes to the molecule. This separation of internal motions into $(3n - 3)$ separate normal motions is of course not exact, but represents a first approximation to the true behavior of the molecule in its ground state. However, for the lower quantum levels this linearly independent representation is relatively accurate and extremely useful.

In fact, even stronger simplifications, i.e., the simplification that all the vibrational modes have constant restoring forces (harmonic oscillator approximation) and the simplification that rotation will not stretch the molecule out of shape (rigid rotor approximation) yield quite good agreement with the experimental evidence. Thus we may assume, to a first approximation, that the molecule has an equilibrium configuration and that it can store energy internally in a series of independent modes which consist of three (or two for a linear molecule) rotational degrees of freedom represented by its principal moments of inertia in its lowest energy state and $(3n - 6)$ or $(3n - 5)$ (in a linear molecule) normal vibrational modes which can be represented as harmonic oscillator vibrations. We notice that these approximations should not hold at higher energy levels and we would expect to find various types of cross-term interactions appearing at the higher energy levels.

Since we have assumed that the molecule has an equilibrium configuration which represents the lowest energy of the molecule, we can also determine how much energy would be required to completely remove one of the atoms from the molecule. This would correspond to separating the molecule into the atom and a new stable configuration of the remaining $(n - 1)$ atoms. The energy required for this removal is called the *dissociation energy* when $n = 2$ or the *bond strength* when $n \geq 3$.

We should add that the harmonic oscillator approximation is not valid at the higher vibrational energy levels because the restoring force is not truly constant (since the harmonic oscillator does not allow dissociation). In addition, in a complicated molecule there is an interaction between the various vibrational degrees of freedom that allows an internal transfer of energy between the various vibrational modes which is not possible in the framework of the simple linear model. Furthermore, the simple rigid-rotor approximation for the rotational degrees of freedom is not always sufficient, because bond stretching at high rotational energies changes both the moment of inertia and the vibrational bond strength. However, these effects are important only insofar as one wishes either to obtain a quantitatively correct description of the molecular energy levels or the spectra of the molecule, or to calculate the exact thermodynamic functions of the molecule. Conceptionally, they represent a perturbation to the harmonic-oscillator rigid-rotor approximation of molecule structure. One more complication occasionally arises. In some molecules the second electronic level is close enough to the lowest electronic level to be partially populated at ordinary temperatures. This leads to quantitatively different thermodynamic properties for these molecules (oxygen and nitric oxide have been shown to exhibit this effect) which will not be discussed further in this text.

1-12. DIATOMIC MOLECULES

Diatomic molecules, the simplest molecules, are of two types: those in which both atoms are the same (homonuclear) and those containing two different atoms (heteronuclear). In both types the simplest model for the internal energy is the harmonic-oscillator rigid-rotor approximation. Since atomic masses are small the energy is quantized and can be written as a sum of two terms:

$$E_{\text{int}} = \left(n + \frac{1}{2}\right)h\nu_0 + \frac{J(J+1)h^2}{8\pi^2 I} \tag{1-4}$$

where E_{int} is the energy in ergs/molecule, J the rotational quantum number ($J = 1, 2, 3, \ldots$), n the vibrational quantum number ($n = 0, 1, 2, 3, \ldots$), h = Plank's constant, ν_0 is the fundamental oscillator frequency for the harmonic oscillator, and I the moment of inertia of the rigid-rotor molecule. The harmonic-oscillator energy was derived in Sec. 1-8 and the rotational part as represented by the last term is obtained from the Schrödinger equation for a rigid rotor. The proper values of ν_0 and I for each diatomic molecule are determined experimentally.

This model is reasonably accurate for lower vibrational-energy levels but it suffers from the fact that it does not predict dissociation. The model also predicts that the energy between neighboring vibrational levels should be a constant, and therefore that vibrational-energy transitions should all emit or absorb at the same frequency. This in fact does not happen, and the vibrational spectra contains a series of lines each slightly displaced from the next. This experimental evidence, when coupled with an experimental determination of the dissociation energy, allows one to predict the true shape of the potential-energy curve for the vibrational degree of freedom. This shape is quite different from the harmonic oscillator potential except at very low energy levels and a more reasonable analytic form for the "shape of the well" is desirable. The Morse potential is such an expression:

$$E_{\text{vib}}(r - r_e) = D_e[1 - e^{-2\beta(r-r_e)}]^2 \tag{1-5}$$

where $\beta = \omega_e \sqrt{\dfrac{\pi^2 c\mu}{2hD_e}}$ and $\omega_e = \nu_0/c\ \text{cm}^{-1}$, c = velocity of light, μ = reduced mass of the two atoms $\dfrac{m_1 m_2}{m_1 + m_2}$ in grams, D_e and E are energy in units of cm^{-1}, and r and r_e are respectively the separation distance and equilibrium separation distance in centimeters. Figure 1-7 shows the shape of the potential well of the hydrogen molecule for the Morse potential, the harmonic oscillator potential, and the experimental potential, as well as the quantized energy levels for the Morse and harmonic oscillator potentials. The value of D_0 in this figure is the usual thermodynamic dissociation energy at 0°K and is related to the Morse potential by the equation

$$D_0 + \frac{1}{2}\frac{h\nu_0}{c} = D_e$$

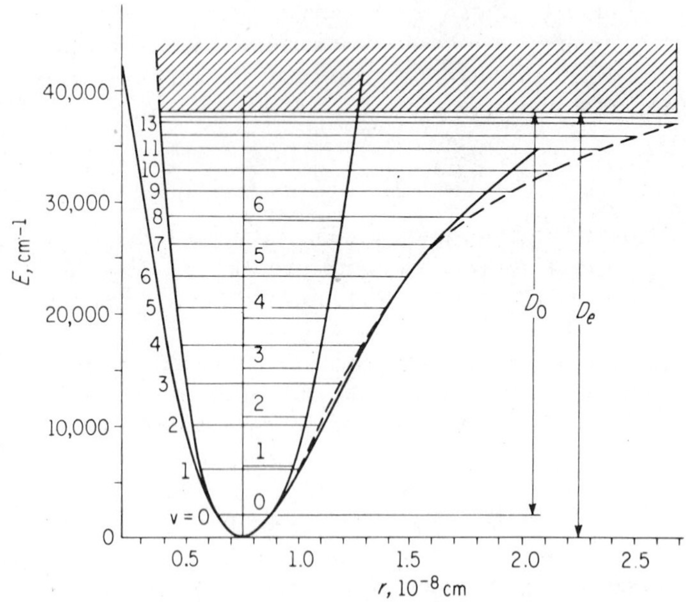

FIG. 1-7. Potential curve of the H_2 ground state with vibrational levels and continuous term spectrum. The broken curve is the Morse curve. The harmonic oscillator potential is also shown. (Adapted with permission from Herzberg, *Spectra of Diatomic Molecules*, 2d Ed., Van Nostrand Co., Princeton (1950).

The energy stored per molecule has been given in two forms in Eqs. 1-4 and 1-5. These energy units are interchangeable with each other and with some other commonly used energy units for molecular or quantized processes. Appendix B contains conversion factors for some of the more commonly used energy units.

Appendix C1 lists the equilibrium separation distance r_e, the dissociation energy D_0 and the characteristic vibrational and rotational constants for a series of diatomic molecules. The characteristic constants represent an energy per molecule in units of cm^{-1} and are defined by the equations

$$\omega_e = \frac{\nu_0}{c}$$

and

$$B_e = \frac{h^2}{8\pi^2 cI}$$

where c is the velocity of light.

3. F. K. Richtmeyer and E. H. Kennard, *Introduction to Modern Physics*, 4th ed. (New York: McGraw-Hill, 1947.)

4. Physics Staff of the University of Pittsburgh, *Atomic Physics*, 2d ed. (New York: Wiley, 1937.)

5. B. Hoffmann, *The Strange Story of the Quantum.* (New York: Harper, 1947.)

6. L. Pauling and E. B. Wilson, Jr., *Introduction to Quantum Mechanics*, 1st ed. (New York: McGraw-Hill, 1935.)

7. L. Pauling, *Nature of the Chemical Bond.* (Ithaca: Cornell, 1939.)

8. C. W. Sherwin, *Introduction to Quantum Mechanics.* (New York: Holt, 1959.)

9. S. S. Penner, *Chemistry Problems in Jet Propulsion.* (New York: Pergamon, 1957.)

10. G. M. Barrow, *Molecular Spectroscopy.* (New York: McGraw-Hill, 1962.)

11. A. G. Gaydon, *Dissociation Energies.* (New York: Dover, 1950.)

PROBLEMS

1-1. Calculate the de Broglie wavelength of (a) a golf ball (mass = 50 gm) traveling at 150 mph, (b) an electron traveling at 100 ft/sec, (c) a 100-volt electron.

1-2. Calculate the zero-point energy of a mass of one gm restrained to move in a straight line with a restoring force of 1,000 dynes/cm. Determine the quantum number of this system when its energy is kT, where k is the Boltzmann constant and $T = 298°K$. This corresponds to thermodynamic equilibrium for the particle as a whole at room temperature.

1-3. The Morse potential for a diatomic molecule is a three-parameter potential. Show that the distance of closest approach of the two neutral atoms of the diatomic molecule (with zero initial kinetic energy) is given by the expression

$$\frac{r}{r_e} = 1 - \frac{\ln 2}{2\beta r_e}$$

and is therefore a function of all the three parameters in the potential.

1-4. Use Eq. 1-4 to calculate the energy levels for $n = 0, 1, 2, 3, 4$ and $J = 1, 2, 3, 4$ for a nitrogen molecule and for a hydrogen molecule. Compare results. Notice that, in general, the vibrational and rotational constants are not simply related to each other or to the dissociation energy of the molecule.

2

Kinetic Theory and Statistical Thermodynamics of a Dilute Gas

2-1. AVAGADRO'S NUMBER

We have seen in the previous chapter that a great deal of experimental and collaborating theoretical evidence indicates that all matter consists of indivisible[1] particles which have distinguishable and distinct properties such as weight, structure, reactivity, etc. Thus atomic theory supports the concept of all substances having a specific molecular weight. This concept was originally derived before the development of atomic theory from the observation that substances formed new chemical compounds by combining in fixed-weight ratios (i.e., 1.008 gm of hydrogen will combine with 8 gm of oxygen to produce exactly 9.008 gm of water, a new chemical compound). Atomic theory and the law of combining weights leads one to the conclusion that a gram-molecular weight of a substance must contain a fixed number of molecules of that substance. This number, which we give the symbol $\mathfrak{N}$, is called Avagadro's number and is equal to $(6.0225 \pm 0.0011) \times 10^{23}$ molecules per gm molecular weight. It has been measured with varying accuracy using many independent techniques. For example:

I. *Radioactive decay.* Alpha particles emitted by radioactive processes have been shown to be helium atoms with a double positive charge. A measurement of the rate of alpha-particle production coupled with a measurement of the quantity of helium collected in a known time allows a direct calculation of Avagadro's number.

II. *Electrolysis.* The charge on a single electron has been measured and the quantity of electricity in one gram-molecular weight of electrons can be measured using conventional electrolysis techniques. The ratio of these quantities yields a value of Avagadro's number.

[1] In the sense that further division would destroy their identifying properties.

III. *Crystal structure.* Gratings can be ruled to measure the wavelength of X-rays, and crystal lattices can be used as gratings. Therefore the interatomic spacing in crystals can be measured and compared to the bulk density of the crystal to calculate an independent value of Avagadro's number. These first three techniques yield values with high intrinsic accuracy and show excellent agreement.

IV. *Brownian motion of small particles.* Small particles have been observed to have a random erratic motion in a liquid. If this is ascribed to collisions with individual molecules in the liquid, one can experimentally determine a value for Avagadro's number. While this determination is much less precise than the first three it yields a value that is quite acceptable.

V. *The kinetic theory of gases.* The kinetic theory of gases allows one to discuss the occurrence of density fluctuations at low density. If fluctuations are present random gradients will also be present and light passing through the gas should be scattered. A measurement of the intensity of the sunlight scattered by the upper atmosphere (i.e., the blue sky) yields an imprecise value of Avagadro's number which is nevertheless in good agreement with the other values.

Thus many independent techniques may be used to determine this fundamental quantity, and we use the fact that all the techniques yield the same answer to reaffirm the existence of atoms and molecules in nature.

2-2. THE IDEAL GAS STATE

The gaseous state under ordinary conditions (i.e., the air we breathe) is extremely dilute. If we assume that the molecules are touching in a solid, we find, by measuring the relative densities, that the molecules are separated by extremely large distances compared to their diameters in the gaseous state. Furthermore, we know from adequate experimental evidence that in the limit of very low density all gases obey the ideal gas law,

$$P = \rho \mathcal{R} T \tag{2-1}$$

where $\mathcal{R}$ has the units of dyne cm/gm °K and where the experimentally determined value of $\mathcal{R}$ is dependent upon the composition of the gas in question. It is also known from ample experimentation that at constant temperature and in the limit as P approaches zero the ratio P/ρ for various pure gases is equal to the inverse molecular weight ratio for these substances, i.e.,

$$\frac{\left(\dfrac{P}{\rho}\right)_A}{\left(\dfrac{P}{\rho}\right)_B} = \frac{\mathcal{R}_A}{\mathcal{R}_B} = \frac{M_B}{M_A}$$

We may therefore write

$$\mathcal{R}_A M_A = \mathcal{R}_B M_B = R$$

where R is a universal constant for all substances. The ideal gas law then becomes

$$PV = RT \tag{2-2}$$

where $V = M/\rho$, the volume of one gram-molecular weight of the gas. The volume of one gram-molecular weight of an ideal gas at $P = 1$ atmosphere and $T = 273.16°\text{K}$ (0°C) has been found to be 22.414 liters. This equation is a universal equation of state for all gases (if the density is low enough) and its existence leads one to the conclusion that at constant temperature and pressure equal volumes of different (ideal) gases contain an equal number of molecules. In addition, this universal behavior allows us to define a convenient temperature scale, the ideal gas temperature scale, by using Eq. 2-2 with the added definition that the temperature of melting ice is 273.16°K or 459.6°R. It is convenient at this point to recall from thermodynamics that the integrating factor needed to change

$$\delta Q_{\text{rev}}$$

into a perfect differential is $1/T$, where T is defined as the thermodynamic temperature and is equal to the ideal gas temperature defined above.

2-3. PROBABILITY AND THE LAWS OF LARGE NUMBERS

From our statements about Avagadro's number and the ideal gas law we can see that even a very small quantity of gas contains an extremely large number of individual molecules. While it is possible to think of the individual molecules each with its own coordinates and velocity components, it is obviously impractical to attempt any description of a gas based on a detailed description of the motion of each constituent molecule. Instead we base our analysis on the fact that a gas does contain a large number of molecules and that for this case the laws of probability can be used to obtain meaningful results. However, before we discuss the properties of a gas using this approach, we must first review some of the concepts of probability.

The same probability laws which are useful to the gambler will be useful in discussing the behavior of a gas. Therefore let us first consider the repeated throw of a pair of perfect dice. If the dice are identified by two distinctive colors (say red and green) which do not "load" the individual dice, we know that the probability of a red 6 appearing in a large number of independent throws is simply 1/6. Similarly the probability of (R_2, G_3) appearing is 1/36 and is equal to the probability of any (R_i, G_j) appearing (where $i = 1, \ldots, 6; j = 1, \ldots, 6$). We can paraphrase this statement by saying that each of the possible results of a throw (R_i, G_j) have equal *a priori* probability of appearing. We may also describe this situation with an equivalent statement by letting the (R_i, G_j) represent 36 cells of equal height and area and writing the probability in the form

$$Pr(R_i, G_j) = \frac{A(R_i, G_j)}{\sum_{x=1}^{6} \sum_{j=1}^{6} A(R_i, G_j)} \tag{2-3}$$

since the value of h cancels when all heights are the same (see Fig. 2-1). Now let us consider loaded dice. The fact that certain combinations

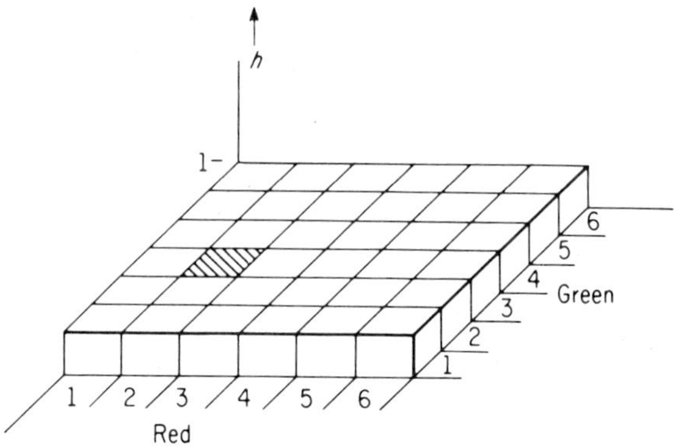

Fig. 2-1. The probability of a number appearing on throwing a pair of unloaded dice. The shaded cell represents a (R_2, G_3) throw and has a volume equal to 1/36 of the total volume.

(R_i, G_j) appear more frequently in this case may be stated mathematically by allowing a variable cell height h in Eq. 2-3.

$$P_r(R_i, G_j) = \frac{h(R_i, G_j)\, A(R_i, G_j)}{\sum_{i=1}^{6} \sum_{j=1}^{6} h(R_i, G_j)\, A(R_i, G_j)} \tag{2-4}$$

where $h(R_i, G_j)$ is not a simple constant and therefore does not cancel out. For example, in Fig. 2-2b we illustrate the probability of any number appearing in a dice game played with loaded dice. Notice that the example assumes that the value of h for R_4 or G_3 is twice what it is for $R_i, i \neq 4$ and $G_j, j \neq 3$. This yields a very high weighting factor for the occurrence of the throw (R_4, G_3). Equation 2-4 still yields a true probability, however, because the sum of the $P_r(R_i, G_j)$ over all i and j is just 1. In other words, the weighted probability that we have shown in Fig. 2-2b is still properly normalized in Eq. 2-4.

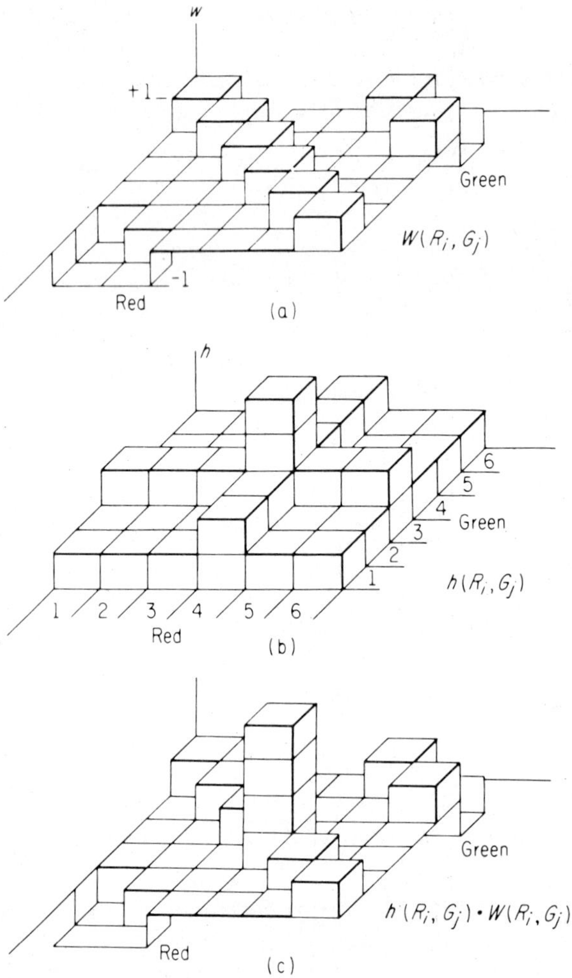

Fig. 2-2 The throw of a pair of loaded dice.

We may also consider the case where in addition to a weighting factor each cell has attached to it an intrinsic functional value $W(R_i, G_j)$. For example, in gambling, plus one for a win, minus one for a loss and zero for no effect on the outcome of the bet. This is illustrated for the first throw in craps in Fig. 2-2a, where the functional value is taken as $+1$ for a win, -1 for a loss and 0 for no effect. In this case the average value of this function for a large number of first throws is simply

$$\overline{W} = \sum_{i=1}^{6} \sum_{j=1}^{6} Pr(R_i, G_j)\, W(R_i, G_j)$$

which may be written

$$\overline{W} = \frac{\sum_{i=1}^{6}\sum_{j=1}^{6} W(R_i,G_j)\,h(R_i,G_j)\,A(R_i,G_j)}{\sum_{i=1}^{6}\sum_{j=1}^{6} h(R_i,G_j)\,A(R_i,G_j)}$$

for throws that are weighted with a factor h. Figure 2-2c shows the individual cells functional weight for the loaded dice of Fig. 2-2b and the functional value of Fig. 2-2a. The average height of the graph in Fig. 2-2c would represent the average value of the win–loss function for a large number of first throws of these loaded dice.

We may generalize these statements concerning the throw of dice to continuous functions by noticing that

$$\sum_{i=1}^{6}\sum_{j=1}^{6} (\text{argument})\cdot A(R_i,G_j)$$

may be written in the limit of zero cell size as

$$\int_x\int_y (\text{argument})\,dx\,dy$$

for all allowable values of the variables x and y where $dx\,dy$ represents the base area of each individual cell. In this case the quantity

$$\int_x\int_y h(x,y)\,dx\,dy$$

represent the volume enclosed by the surface $h(x,y)$. In this notation the average value becomes

$$\overline{W} = \frac{\int_x\int_y w(x,y)h(x,y)\,dx\,dy}{\int_x\int_y h(x,y)\,dx\,dy}$$

for independent and weighted processes. This is shown for a particular example in Fig. 2-3. The notation may be further generalized to three or more dimensions by writing

$$\overline{W} = \frac{\int_{x_1}\int_{x_2},\ldots\int_{x_n} w(x_1,x_2,\ldots,x_n)\,h(x_1,x_2,\ldots,x_n)\,dx_1,dx_2,\ldots,dx_n}{\int_{x_1}\int_{x_2},\ldots\int_{x_n} h(x_1,x_2,\ldots,x_n)\,dx_1,dx_2,\ldots dx_n}$$

We will also find cause later in this chapter to be interested in the process of sorting and counting a large number of particles. Let us therefore consider the case of N distinguishable particles to be placed in

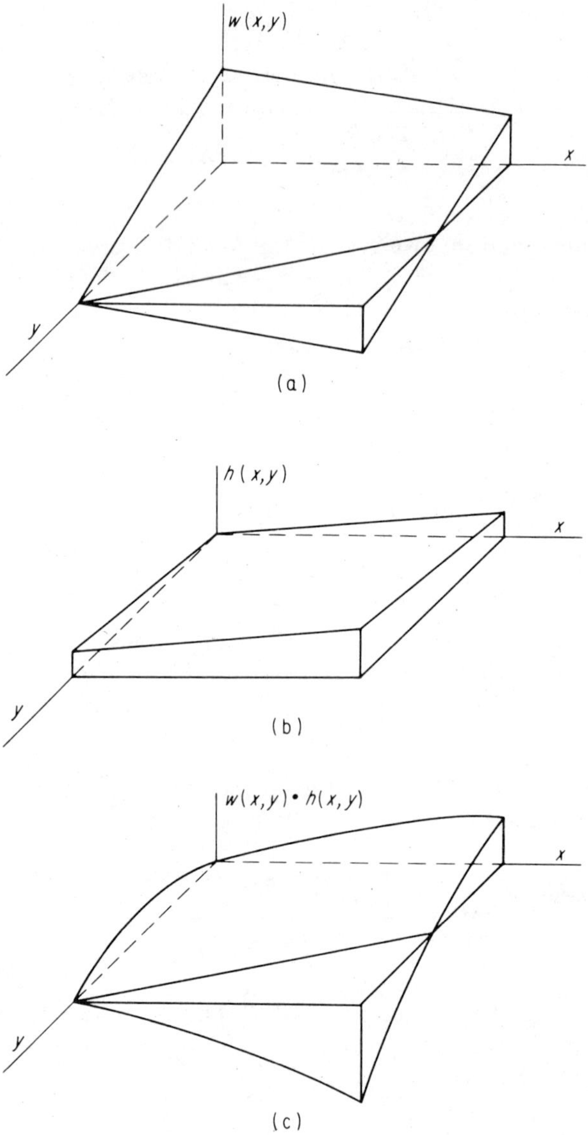

Fig. 2-3. Probability for continuous functions.

N boxes, one to a box. The total number of possible arrangements in this case is simply $N!$ since there are N first choices, $N - 1$ second choices, $N - 2$ third choices, . . . , and finally only one last choice for positioning the N particles. The product of all these choices is $N!$. If, on the other hand, we allow two of these boxes to merge and become one box (containing two particles) the number of possibilities is divided by two, since the interchange of any particles that happen to be placed in the two original boxes does not constitute a new arrangement. Similarly, if three boxes are merged to form one, the number of arrangements is divided by 6 (or 3!), since we can no longer distinguish between what previously had been six distinguishable arrangements of the particles in the three boxes. (This statement is true irrespective of what three particles are in the merged box.) In general, if we wish to arrange N distinguishable objects into p piles containing a specified distribution of $(n_1, n_2, \ldots, n_p)$ particles in each of the p piles, the number of possible arrangements is

$$\Omega(N, n_1, \ldots, n_{p-1}) = \frac{N!}{\prod_{i=1}^{p} (n_i!)} \tag{2-5}$$

where the n_i represent the numbers in the distribution $(n_1, n_2, \ldots, n_p)$. Turning now to a simple example this statement means that there are

$$\frac{52!}{(13!)^4} = 5.2 \times 10^{28}$$

possible deals in a bridge game where each of the possible 24 arrangements of fixed hands between the four players is counted separately. Notice that we can paraphrase this statement by saying that the degeneracy of the (13, 13, 13, 13) distribution of 52 distinguishable cards based on equal *a priori* probability of any card being found in any one hand is 5.2×10^{28}.

2-4. THE PREMISES OF ELEMENTARY KINETIC THEORY FOR A DILUTE GAS

There are three macroscopic observations concerning dilute gases which form the foundation of elementary kinetic theory.

I. A dilute gas has an extremely low density when compared to the solid or liquid phase.

II. A dilute gas at equilibrium exerts an equal pressure on all walls of the container and obeys the equation of state $PV = RT$.

III. The internal energy of a dilute gas is a function of the temperature and independent of the pressure.

These observations when coupled with our concept of indivisible atoms or molecules having fixed properties yield the three basic premises of kinetic theory.

I. The pressure exerted by a dilute gas must be caused by individual molecules colliding with the walls, and therefore the gas molecules must be in constant motion in the container.

II. The molecules, being far apart, must be unaware of the existence of their neighbors during a good portion of the time. In other words, if you were to look at any one molecule of a dilute gas many times, the chances of finding it in collision with another molecule would be extremely small.

III. The internal energy of the gas must be stored in the individual molecules either as internal molecular energy or as translational energy associated with their average speed of motion relative to the container. Significant collisional energy storage is ruled out by the second premise.

If we are to use the laws of probability in deriving the properties of our ideal gas we must, in addition to these three premises, make an assumption concerning the behavior of the individual molecules or the system of molecules in equilibrium. We must assume (without *a priori* justification) that any allowable energy state or position of an individual molecule has an equal *a priori* probability of being observed if we look at a large number of samples of the gas or at one sample for a long enough period of time. This statement implies that the molecule will spend an equal amount of time in every available energy state and every available position over a long period of time. It also implies that we may externally restrict the number of available states or available positions. We restrict positions, for example, by defining the volume and we restrict states, for example, by requiring that a mixture of hydrogen and oxygen at room temperature cannot form any water molecules (even though it is thermodynamically possible for water to be formed). Thus these restrictions are based on conclusions drawn from our ordinary macroscopic observations.

This assumption that a state or position can either exist or not exist—and if it does exist that it has an equal *a priori* probability of being observed—is called the *ergodic hypothesis* of statistical mechanics. It may be applied equally well to an individual molecule in equilibrium with a large number of other molecules or to a macroscopic system of N molecules in equilibrium with a large number of identical macroscopic systems. The only justification for this assumption is that it allows us to use the laws of probability with exactness, and further (as we will see) that it gives results which agree well with experimental observations. Notice that neither of these justifications validates the initial assumption.

2-5. THE DENSITY OF A DILUTE GAS: FLUCTUATIONS

Let us consider a dilute gas of N point molecules in a chamber of volume V which is divided into two equal parts by a thin and completely permeable wall. We now use the assumption of equal *a priori* probability for the position of each molecule and discuss the possible distributions of the N molecules between these two chambers. Let each molecule have an equal *a priori* probability of being found in either half of the chamber, independent of the number of molecules which it already contains. Under these conditions the number of possible arrangements of N molecules in the two chambers is simply

$$\sum_{c=0}^{N} \Omega(N,c)$$

where

$$\Omega(N,c) = \frac{N!}{c! \cdot (N - c)!}$$

and c is the number in the first half. For example, there is only one possible arrangement for which all molecules would be found in one chamber

$$\Omega(N,O) = \frac{N!}{O! \cdot N!} = 1$$

and N possible arrangements for which one molecule is in the first chamber and all the rest are in the second chamber:

$$\Omega(N,1) = \frac{N!}{1!(N - 1)!} = N$$

Similarly, there are $N(N - 1)/2$ possible arrangements for exactly two molecules being found in the first chamber with the rest being found in the second chamber:

$$\Omega(N,2) = \frac{N!}{2! \cdot (N - 2)!} = \frac{N(N - 1)}{2}$$

Therefore, on the basis of equal *a priori* probability, we can say that the probability of observing any one distribution (fixed c) is

$$P_r(N,c) = \frac{\Omega(N,c)}{\sum\limits_{c=0}^{N} \Omega(N,c)} = \frac{\dfrac{N!}{c! \cdot (N - c)!}}{\sum\limits_{c=0}^{N} \dfrac{N!}{c! \cdot (N - c)!}}$$

Notice that the distribution with the largest degeneracy [or largest value of $\Omega(N,c)$] occurs when $c = N/2$, i.e., the most probable distribution is that which corresponds to an equal number of molecules appearing in each of the two cells. In fact for very large N we can easily show that the probability of observing a distribution slightly removed from this most degenerate distribution is extremely low. Consider the distribution where $A = N/2 - c$. The number of possible arrangements is

$$\Omega(N,c) = \frac{N!}{\left(\frac{N}{2} + A\right)! \cdot \left(\frac{N}{2} - A\right)!}$$

and the ratio of the number of distributions of the maximum type ($c = N/2$) to distributions of this type is

$$\frac{\Omega\left(N, \dfrac{N}{2}\right)}{\Omega(N,c)} = \frac{\dfrac{N!}{\dfrac{N}{2}! \cdot \dfrac{N}{2}!}}{\dfrac{N!}{\left(\dfrac{N}{2} + A\right)! \cdot \left(\dfrac{N}{2} - A\right)!}} = \frac{\left(\dfrac{N}{2} + A\right)! \cdot \left(\dfrac{N}{2} - A\right)!}{\dfrac{N}{2}! \cdot \dfrac{N}{2}!}$$

If we work with the natural logarithm of this ratio we may use Stirling's approximation[2] to write

$$\ln \frac{\Omega\left(N, \dfrac{N}{2}\right)}{\Omega(N,c)} = \left(\frac{N}{2} + A\right)\left[\ln\left(\frac{N}{2} + A\right) - 1\right] + \left(\frac{N}{2} - A\right)\left[\ln\left(\frac{N}{2} - A\right) - 1\right] - N\left(\ln \frac{N}{2} - 1\right)$$

However

$$\ln\left(\frac{N}{2} + A\right) = \ln \frac{N}{2}\left(1 + \frac{2A}{N}\right) = \ln \frac{N}{2} + \ln\left(1 + \frac{2A}{N}\right) \cong \ln \frac{N}{2} + \frac{2A}{N}$$

and

$$\ln\left(\frac{N}{2} - A\right) = \ln \frac{\dfrac{N}{2}}{1 + \dfrac{2A}{N}} \cong \ln \frac{N}{2} - \frac{2A}{N}$$

Therefore

$$\ln\left[\frac{\Omega\left(N, \dfrac{N}{2}\right)}{\Omega(N,c)}\right] = \left(\frac{N}{2} + A\right)\frac{2A}{N} + \left(\frac{N}{2} + A\right)\left(\ln \frac{N}{2} - 1\right) - \left(\frac{N}{2} - A\right)\frac{2A}{N} + \left(\frac{N}{2} - A\right)\left(\ln \frac{N}{2} - 1\right) - N\left(\ln \frac{N}{2} - 1\right)$$

$$\cong \left(\frac{N}{2} + A\right)\frac{2A}{N} - \left(\frac{N}{2} - A\right)\frac{2A}{N} = \frac{4A^2}{N}$$

[2]Stirling's approximation for the factorial of a large number is

$$\ln N! = \sum_{n=1}^{N} \ln N \cong \int_1^N \ln n \, dn = N \ln N - N - 1 \cong N \ln N - N$$

Now if $N = 2 \times 10^{20}$ ($V \cong 2.8$ *cc* at STP) and we want to calculate the probability of a density fluctuation $\Delta\rho/\rho = \pm 10^{-6}$ we find that

$$C = 1.000001 \times 10^{20}$$

and

$$A = 10^{14}$$

Therefore

$$\ln \frac{\Omega\left(N, \dfrac{N}{2}\right)}{\Omega(N,c)} = 2 \times 10^{8}$$

or

$$\frac{\Omega\left(N, \dfrac{N}{2}\right)}{\Omega(N,c)} = e^{(2 \times 10^{8})}$$

We infer from this that the probability of observing a density fluctuation even slightly removed from the equilibrium density is very unlikely in an ideal gas at equilibrium. Thus we have shown that a gas should have a uniform macroscopic density if we allow every molecule equal access to all portions of the vessel. This observation also allows us to define a macroscopic system. In the kinetic theory sense a macroscopic system is one that exhibits only extremely small fluctuations when in equilibrium. If the system is small enough to allow sizable statistical fluctuations it is said to be a *microscopic system*. Furthermore, we define a thermodynamic system as one that obeys the ordinary thermodynamic laws, and note that it is always a macroscopic system in the kinetic-theory sense.

2-6. PRESSURE AND THE AVERAGE MOLECULAR VELOCITY

Let us again consider a volume of gas containing N molecules. If we assume N is large so that the system is macroscopic, the equal *a priori* assumption and our definition of macroscopic can be used to yield the principle of detailed balances. The principle of detailed balances states that for a macroscopic system at equilibrium each individual molecular process is balanced by an equal and exactly opposite process taking place at some other point in the system. If this were not true the distribution of the molecules would change with time and for a macroscopic system at equilibrium this type of change is extremely unlikely. For example, the number of molecules leaving one side of the vessel in Sec. 2-5 is always balanced by an influx of molecules coming from the other side, and furthermore this balance has an extremely high probability of being accurate to one part in 10^6. Therefore the principle of detailed balances represents a logical conclusion drawn from the original postulates of kinetic theory and the fact that N is extremely large in an ordinary gas.

Now let us consider the velocity distribution of N molecules contained in a unit volume. We first must assume (for the simple theory) that these molecules cannot store energy internally. In addition, since the gas is rarefied, we assume that each molecule spends most of its time well removed from the other molecules. In the main then, the only way the individual molecules can store the gas's internal energy is as molecular kinetic energy

$$\epsilon = \tfrac{1}{2}mc^2$$

where c is the molecule's speed relative to the main body of the gas. We of course do not know the distribution of molecular velocity $f(u, v, w)\cdot du\,dv\,dw$, where $f(u, v, w) = f(\vec{V})$ is the number of molecules per unit volume in physical space in the velocity range

$$\vec{V} \quad \text{to} \quad (\overrightarrow{V + dV})$$

with components

$$(u, v, w) \quad \text{to} \quad (u + du, v + dv, w + dw)$$

However, we can deduce that the value of $f(u, v, w)$ is independent of time in a quiescent macroscopic system and that it is not a simple constant. It must be independent of time because of the principle of detailed balances (i.e., because the system is macroscopic) and it cannot be a simple constant, since the limits on the integral

$$N = \iiint\limits_{-\infty}^{\infty} f(u, v, w)\,du\,dv\,dw \tag{2-6}$$

must be infinite and the value of N must be finite. Notice that we may define the distribution in terms of one velocity component

$$h(u) = \iint\limits_{-\infty}^{\infty} f(u, v, w)\,dv\,dw$$

Or in terms of two velocity components,

$$g(u, v) = \int_{-\infty}^{\infty} f(u, v, w)\,dw$$

etc. Furthermore, since the gas is isotropic, we know that $h(u) = h(v) = h(w)$, and since the gas is quiescent we can state that the average molecular velocity

$$\overline{\vec{V}} = \frac{\displaystyle\iiint\limits_{-\infty}^{\infty} \vec{V}(u, v, w)\,f(u, v, w)\,du\,dv\,dw}{\displaystyle\iiint\limits_{-\infty}^{\infty} f(u, v, w)\,du\,dv\,dw}$$

is zero. However, the average square speed is not zero, but is related to the average kinetic energy of the molecules

$$\overline{\vec{V}\cdot\vec{V}} = \overline{c^2} = \frac{\displaystyle\iiint_{-\infty}^{\infty} c^2(u,v,w)\, f(u,v,w)\, du\, dv\, dw}{\displaystyle\iiint_{-\infty}^{\infty} f(u,v,w)\, du\, dv\, dw}$$

$$= \frac{3}{N}\iiint_{-\infty}^{\infty} u^2 f(u,v,w)\, du\, dv\, dw$$

since the gas is isotropic and

$$\overline{c^2} = \overline{u^2 + v^2 + w^2} = 3\overline{u^2}$$

Therefore, the energy per unit volume may be written as

$$\mathcal{E} = N\bar{\epsilon} = \frac{1}{2} mN\overline{c^2} = \frac{3}{2} m \iiint_{-\infty}^{\infty} u^2 f(u,v,w)\, du\, dv\, dw$$

for a monatomic gas.

Let us now consider the pressure exerted against one square centimeter of a wall in the yz-plane for the case of a unit volume of a monatomic gas containing N molecules. We assume that the pressure is equal to the time average momentum change of the molecules which strike the wall. If we assume specular reflection, this momentum change is $2mu$ for a molecule with a velocity component u. However, we know that molecules do stick to the walls and in general are emitted in a direction and at a velocity which has little to do with their incident velocity. Thus wall collisions in the microscopic sense are similar to intermolecular interactions, i.e., each causes changes that must be calculated for the individual interaction. However, in the macroscopic sense, the principle of detailed balances allows us to ignore the complicated individual processes. We simply note that if the gas and wall are in equilibrium on the average each molecule striking the wall with a momentum mu is balanced by one emitted with a momentum of $-mu$ and furthermore, since there is no tangential force on a wall enclosing an equilibrium gas, the momentum terms mv and mw are conserved for a large number of collisions.

Now let us determine the integrated momentum change for all molecules which strike 1 cm^2 of the yz-plane in contact with an equilibrium gas. First consider molecules of the type $f(u, v, w)$. The number of these molecules hitting the wall per second are contained in a parallelepiped of base 1 cm^2 and height u(cm/sec). Thus, since $f(u, v, w)\, du\, dv\, dw$ is the number of molecules of this type contained in a unit volume, the number of this type hitting the wall is simply

$$u\,f(u, v, w)\,du\,dv\,dw \qquad \text{molecules of type } (u, v, w) \text{ to hit the wall/cm}^2\text{ sec}$$

(See Fig. 2-4.) Since each of these molecules carries with it an x component of momentum equal to mu, the net momentum change on reflection from the wall is

$$2mu^2 f(u,v,w)\,du\,dv\,dw \qquad \text{change in momentum/sec (gm cm/cm}^2\text{ sec}^2\text{) due to molecules of type } (u,v,w)$$

This expression may be integrated over u, v, and w to include all molecules which initially have positive values of u. The integrated expression gives

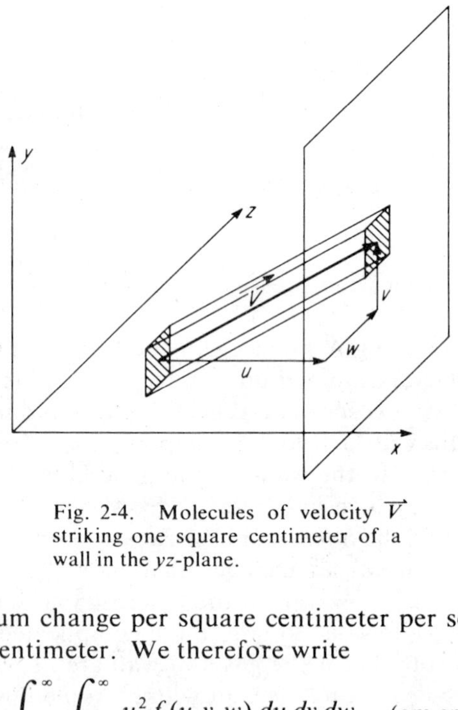

Fig. 2-4. Molecules of velocity $\vec{V}$ striking one square centimeter of a wall in the yz-plane.

the total momentum change per square centimeter per second, which is a force per square centimeter. We therefore write

$$P = 2m \int_{-\infty}^{\infty} \int_{-\infty}^{\infty} \int_{0}^{\infty} u^2 f(u,v,w)\,du\,dv\,dw \qquad \text{(gm cm/cm}^2\text{ sec}^2\text{) due to all molecules}$$

for the pressure on the surface of the chamber. However the integrand in this equation is an even function of u, since

$$(-u)^2 f(-u,v,w) = u^2 f(u,v,w)$$

and the integral may therefore be written

$$P = m \iiint_{-\infty}^{\infty} u^2 f(u,v,w)\,du\,dv\,dw$$

We now note that we have already defined the mean square value of the velocity component as

$$\overline{u^2} = \frac{1}{N} \iiint_{-\infty}^{\infty} u^2 f(u,v,w)\, du\, dv\, dw$$

Therefore

$$P = mN\,\overline{u^2}\ \text{dyne/cm}^2 \tag{2-7}$$

and since the gas is isotropic, we find that

$$P = \frac{mN}{3}\,\overline{c^2} \tag{2-8}$$

However, if V is taken as the molar volume, NV is Avagadros number, mNV is the molecular weight of the gas, M, and we obtain the result that

$$PV = \frac{M}{3}\,\overline{c^2} = RT$$

where R is the molar gas constant. This in turn yields the equation

$$\overline{c^2} = \frac{3RT}{M} \tag{2-9}$$

for the root-mean-square speed of the gas molecules. We may also write the molar internal energy as

$$E = \mathfrak{N}\bar{\epsilon} = \frac{1}{2}\mathfrak{N} m\overline{c^2} = \frac{3}{2} RT \tag{2-10}$$

which yields an expression for the molar heat capacity of this monatomic gas,

$$C_v = \frac{3}{2} R$$

or

$$\frac{C_p}{C_v} = \gamma = 1.6667 \tag{2-11}$$

Therefore, simple kinetic theory as applied to an ideal gas tells us that

I. The molecules are moving at a root-mean-square speed which is greater than the velocity of sound in the gas. Since by definition the velocity of sound in an ideal gas is given by

$$a = \sqrt{\frac{\gamma RT}{M}}$$

$$\frac{\sqrt{\overline{c^2}}}{a} = \sqrt{\frac{3}{\gamma}} > 1$$

II. If the molecules are restricted to not storing internal energy the heat capacity ratio of the gas is $\gamma = 1.6667$.

Notice that these conclusions have been reached without knowing the exact form of the velocity distribution $f(u, v, w)$. Notice also that the first conclusion is eminently logical. A signal cannot be transmitted unless it is being carried by something (in this case the molecules). Furthermore, the second conclusion (that $\gamma = 1.6667$ for a monatomic gas) has been adequately verified by experiment.

Incidentally, the above results also verify Dalton's law of partial pressures for a dilute gas. If we consider two gases at the same pressure in separate chambers which are in thermodynamic equilibrium, the number of molecules per unit volume of each gas is the same and the average molecular kinetic energy must be the same for each gas. If we now remove the separating partition the two types of molecules in the mixture must retain the same average energy because the principle of detailed balancing applies. Therefore, in general, we may write that

$$\sum_i p_i = P_{\text{total}} \tag{2-12}$$

and

$$\bar{\epsilon}_i = \bar{\epsilon}_j; \quad i = 1, \ldots, n; \quad j = 1, \ldots, n$$

for an n component gas.

2-7. THE MAXWELL DISTRIBUTION OF VELOCITY

Let us consider the collision of two molecules in a macroscopic system of identical molecules. For a particular collision, molecules of velocity $\vec{V}_1$ and $\vec{V}_2$ will produce molecules of velocity $\vec{V}_1'$ and $\vec{V}_2'$ after the collision. The number of these particular forward collisions of type $(\vec{V}_1 \cdot \vec{V}_2) \rightarrow (\vec{V}_1' \cdot \vec{V}_2')$ will be proportional to the total number of molecules of type $\vec{V}_1$ and $\vec{V}_2$, i.e., forward rate $= Af(\vec{V}_1) \cdot f(\vec{V}_2)$. One can also write the reverse rate as $Bf(\vec{V}_1') \cdot f(\vec{V}_2')$ where B may be a different proportionality constant. However, the principle of detailed balances for a macroscopic system simply states that these rates must be equal at equilibrium, i.e.,

$$Af(\vec{V}_1) \cdot f(\vec{V}_2) = Bf(\vec{V}_1')f(\vec{V}_2')$$

In addition, we must assume that energy is conserved during the collision process. Therefore, we may write

$$(\vec{V}_1)^2 + (\vec{V}_2)^2 = (\vec{V}_1')^2 + (\vec{V}_2')^2$$

We now consider a particular case: a collision where $V_1 = 0$. For this case

$$(\vec{V}_2)^2 = (\vec{V}_1')^2 + (\vec{V}_2')^2$$

and

$$Af(0) \cdot f\left(\sqrt{(\vec{V}_1')^2 + (\vec{V}_2')^2}\right) = Bf(\vec{V}_1') \cdot f(\vec{V}_2')$$

The only nontrivial form of the function $f(\vec{V})$ which will satisfy this equation is

$$f(\vec{V}) = \alpha e^{-\beta(\vec{V})^2} = \alpha e^{-\beta(u^2+v^2+w^2)}$$

where $A = B$, as may be verified by direct substitution. However,

$$f(\vec{V}) = f(u, v, w) = \alpha e^{-\beta(u^2+v^2+w^2)}$$
$$= \alpha e^{-\beta u^2} \cdot e^{-\beta v^2} \cdot e^{-\beta w^2}$$

Therefore,

$$f(\vec{V}) = h(u) \cdot h(v) \cdot h(w)$$

where from before

$$h(u) = h(v) = h(w)$$

We may now use the fact that

$$N = \iiint_{-\infty}^{\infty} f(u, v, w)\, du\, dv\, dw$$

to eliminate α from this equation. We write

$$N = \iiint_{-\infty}^{\infty} \alpha e^{-\beta(u^2+v^2+w^2)}\, du\, dv\, dw$$
$$= \alpha \left[\int_{-\infty}^{\infty} e^{-\beta u^2}\, du \right]^3 = \alpha \left[\sqrt{\frac{\pi}{\beta}} \right]^3$$

Therefore,

$$\alpha = N \left(\frac{\beta}{\pi} \right)^{3/2}$$

The value of β may be determined by solving the equation for the square of one of the velocity components.

$$\overline{u^2} = \frac{1}{N} \iiint_{-\infty}^{\infty} u^2 f(u, v, w)\, du\, dv\, dw$$
$$= \left(\frac{\beta}{\pi} \right)^{3/2} \int_{-\infty}^{\infty} u^2 e^{-\beta u^2}\, du \left[\int_{-\infty}^{\infty} e^{-\beta v^2}\, dv \right]^2$$
$$= \left(\frac{\beta}{\pi} \right)^{1/2} \int_{-\infty}^{\infty} u^2 e^{-\beta u^2}\, du$$

By differentiating with respect to the parameter β we obtain

$$\int_{-\infty}^{\infty} u^2 e^{-\beta u^2}\, du = -\frac{d}{d\beta} \left[\int_{-\infty}^{\infty} e^{-\beta u^2}\, du \right] = -\frac{d}{d\beta} \left(\sqrt{\frac{\pi}{\beta}} \right)$$
$$= \frac{1}{2} \sqrt{\frac{\pi}{\beta^3}}$$

Therefore

$$\overline{u^2} = \frac{1}{2\beta}$$

However, in Sec. 2-6 we saw that

$$\overline{u^2} = \frac{1}{3}\overline{c^2} = \frac{RT}{M} = \frac{k\mathfrak{N}T}{m\mathfrak{N}} = \frac{kT}{m}$$

where k is the gas constant per molecule (called the *Boltzmann constant*). Therefore,

$$\beta = \frac{m}{2kT}$$

and the Maxwell velocity distribution becomes

$$\begin{aligned} f(\vec{V})\,dV &= f(u,v,w)\,du\,dv\,dw \\ &= N\left(\frac{2\pi kT}{m}\right)^{-3/2} e^{-(u^2+v^2+w^2)m/2kT}\,du\,dv\,dw \end{aligned} \tag{2-13}$$

This represents the number density of molecules of type $(\vec{V})$ with velocity components (u,v,w). That is, if we were to plot the molecular distribution in velocity space [with coordinates (u, v, w)] this would represent the molecular population in a cell of size $du\,dv\,dw$ at position (u,v,w). We would also like to determine the total number of molecules which have a speed c irrespective of their direction in velocity space. This may be represented by a spherical shell of thickness dc and radius c. To do this we make a transformation to spherical coordinates (see Fig. 1-5) and integrate over all ϕ and θ. In spherical coordinates the cell $du\,dv\,dw$ is replaced by a cell of volume $c^2 \sin\theta\,dc\,d\phi\,d\theta$ and Eq. (2-13) becomes

$$f(u,v,w)\,du\,dv\,dw = N\left(\frac{2\pi kT}{m}\right)^{-3/2} c^2 e^{-(c^2m/2kT)} \sin\theta\,dc\,d\phi\,d\theta$$

If we define

$$\zeta(c,\theta) \text{ as } \left(\frac{2\pi kT}{m}\right)^{-3/2} c^2 e^{-(c^2m/2kT)} \sin\theta$$

and integrate $\zeta(c,\theta)$ for all ϕ and θ we obtain

$$\int_0^{2\pi} d\phi \int_0^{\pi} \zeta(c,\theta)\,d\theta = 4\pi\left(\frac{2\pi kT}{m}\right)^{-3/2} c^2 e^{-(c^2m/2kT)}$$

Therefore

$$\eta(c)\,dc = 4\pi\left(\frac{2\pi kT}{m}\right)^{-3/2} c^2 e^{-(c^2m/2kT)}\,dc \tag{2-14}$$

Integrating to determine the mean square speed yields

$$\overline{c^2} = \frac{N\int_0^\infty c^2 \eta(c)\,dc}{N} = \frac{3kT}{m} = 3\mathcal{R}T$$

which is the result obtained earlier in Sec. 2-6. In a similar manner one may obtain the mean speed (which is not zero) by evaluating

$$\bar{c} = \frac{N\int_0^\infty c\,\eta(c)\,dc}{N} = \sqrt{\frac{8\mathcal{R}T}{\pi}}$$

by integrating by parts.

The Maxwell distribution has been verified by experiment. Zartman and Ko in 1930–1934 measured $\eta(c)$ using a modification of a technique developed by Stern in 1920. An oven at a temperature T with a slit opening is used as a source of molecules. These molecules are allowed to fall on a rotating drum which also has a narrow slit. Those molecules entering the drum travel to the opposite side at their respective velocities and plate out on the interior wall, leaving a deposit whose density can be measured. (See Fig. 2-5.) Thus one can directly measure the molecular

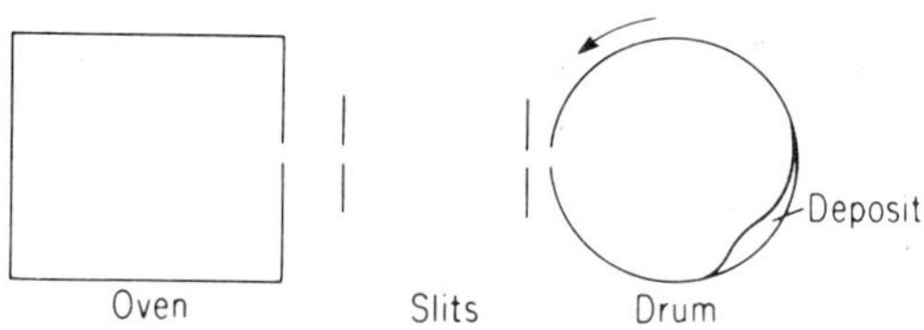

Fig. 2-5. The experiment which confirmed the Maxwell distribution in an ideal gas. (Adapted with permission from J. F. Lee, F. W. Sears, and D. L. Turcotte, *Statistical Thermodynamics*, Addison Wesley, Reading, Mass., 1963.)

velocity distribution at the temperature of the oven. The experimental results agree very well with the theoretical distribution. The confirmation is all the more striking because there are no adjustable constants involved in plotting the experimental or theoretical curves.[3]

[3]A very complete discussion of experimental methods may be found in J. Jean's *An Introduction to the Kinetic Theory of Gases*, The University Press, Cambridge, pp. 124–130, (1940).

2-8. MOLECULAR PROPERTIES IN A DILUTE GAS

We have determined the average speed and the root-mean-square speed for the molecules of an equilibrium dilute gas in the previous section. We may also determine some other properties using simple kinetic theory. Let us assume that all the molecules are moving at their average speed $\overline{c}$ and that $1/6$ of them are moving toward each wall of the container. If this were true the number of collisions with the wall per square centimeter per second would simply be

$$\frac{1}{6} N\overline{c}$$

where N as before is the number of molecules per unit volume. If we remove the restriction of constant velocity in the six normal directions and use the Maxwell velocity distribution the multiplying constant is modified to yield

$$Z_{\text{wall}} = \frac{N\sqrt{\overline{c}^2}}{\sqrt{6\pi}} \text{ molecules/cm}^2\text{sec} \qquad (2\text{-}15a)$$

or

$$Z_{\text{wall}} = N\overline{c}/4 \text{ molecules/cm}^2\text{sec} \qquad (2\text{-}15b)$$

We may also calculate the number of molecular collisions occurring in one cm^3 of gas by using a similar argument. However, in order for collisions to occur the molecules must have a finite size. Let us assume that each molecule is a hard sphere of diameter σ_0 and that one molecule is moving at the velocity $\overline{c}$. This molecule will hit another molecule when its centerline passes within σ_0 cm of the other molecule's centerline. It therefore sweeps out a volume $\pi\sigma_0^2\overline{c}$ between collisions. However, there are N molecules per cubic centimeter so it will strike $N\pi\sigma_0^2\overline{c}$ other molecules per second. At the same time though each other molecule is sweeping its own collision path. Therefore, there are $\frac{N^2\pi\sigma_0^2\overline{c}}{2}$ collisions per cubic centimeter per second. The factor of $1/2$ enters because we have counted each collision twice. Thus

$$Z_{\text{molecules}} = \frac{\pi}{2} N^2\sigma_0^2\overline{c} \text{ molecules colliding per cm}^3\text{/sec}$$

for our simple model. If one uses the Maxwell distribution in the analysis the constant changes and the number of molecular collisions becomes

$$Z_{\text{molecules}} = \sqrt{2}\pi N^2\sigma_0^2\overline{c} \text{ collisions per cm}^3\text{/sec}$$

for hard-sphere molecules.

The average transit distance between collisions is called the *mean free path*. Since we showed that a molecule moving at a velocity $\overline{c}$ suffers $N\pi\sigma_0^2\overline{c}$ collisions/sec, the mean free path is

$$l = \frac{\overline{c}}{\pi\sigma_0^2 N\overline{c}} = \frac{1}{\pi\sigma_0^2 N} \text{ cm/collision}$$

If one assumes hard spheres with a Maxwell distribution this becomes

$$l = \frac{1}{\sqrt{2}\,\pi\,\sigma_0^2 N}$$

Representative values of σ_0 for various molecules are tabulated in Appendix E. These are not truly effective diameters for the hard-sphere model but are the zero-potential approach distance for the Lennard-Jones interaction potential. The properties of this potential will be discussed in the next chapter.

2-9. STATISTICAL THERMODYNAMICS OF AN IDEAL GAS

We have seen that elementary kinetic theory, as applied to a dilute gas, yields results which agree well with our experience. We further saw that when the molecules are not allowed to store energy internally, the heat capacity and gamma of the gas is accurately predicted by simple kinetic theory. However, the accurate thermodynamic functions needed to handle combustion problems adequately cannot be obtained using only this simple approach. The sources of these thermodynamic functions are experimental heats of formation and dissociation energies coupled with spectroscopic measurements interpreted by means of a statistical thermodynamic approach as it applies to the calculation of the thermodynamic properties of a dilute gas.

We found in our discussion of the Maxwell velocity distribution that this distribution is of the form

$$f(\vec{V}) = \alpha e^{-\beta(\vec{V})^2}$$

where α and β are constant. When one considers the statistical mechanics of a large number of "independent" particles which are in equilibrium it may be shown that this statement can be generalized to yield the distribution law in terms of the energy stored by each particle:

$$\Xi\,(\epsilon) = \text{const.} \cdot e^{-\epsilon/kT} \tag{2-16}$$

where for a classical atomic model ϵ is assumed to be a continuous function while for a quantum model ϵ is assumed to be limited to a number of discrete values. For a dilute or ideal gas we have already discussed the observation that molecules behave as though they are "independent" of each other even though they are in statistical equilibrium. In short, this behavior occurs because we only seldom observe a molecule undergoing a collision process even though there are sufficient collisions to cause equilibrium in the thermodynamic sense.

Since the molecules are assumed to be independent and since we have assumed that vibration and rotation are decoupled we find that the energy of an individual molecule may be written as:

$$\epsilon = \epsilon_t + \epsilon_v + \epsilon_r$$

where ϵ_t is the translational energy considered earlier, ϵ_v is the energy stored in the internal vibrations and ϵ_r is the energy stored in rotation (the energy stored in internal electronic excitation being neglected in this treatment). Notice that this type of decoupling allows us to discuss the distribution function for each type of energy storage separately. We have already discussed energy storage in the translational degrees of freedom and we summarize these results by saying that the treatment was classical and that we found that the kinetic energy stored in each degree of translational freedom contributed to a value of $R/2$ to the heat capacity of the gas.

We now turn our attention to molecular vibration. We will discuss only the case of the simple harmonic oscillator model of a diatomic molecule and use this example to compare the classical and quantum approach as it is used in the calculation of heat capacities. The classical harmonic oscillator can store energy as both kinetic and potential energy.

$$\epsilon = \tfrac{1}{2} m\dot{x}^2 + m\omega_0 x^2$$

where $\omega_0 = 2\pi\nu_0$ and ν_0 is defined in Chapter 1. We note that the energy which is stored by this degree of freedom is a quadratic function of both the particle's velocity (momentum) and the particle's position (displacement from its equilibrium position). The average energy storage for a weighted equilibrium distribution Ξ of a number of such oscillators is simply

$$\bar{\epsilon} = \frac{\iint_{-\infty}^{\infty} \epsilon\,\Xi(\epsilon)\,d\dot{x}\,dx}{\iint_{-\infty}^{\infty} \Xi\,d\dot{x}\,dx}$$

If the expression for the energy of a harmonic oscillator is substituted into this equation it reduces to the form

$$\bar{\epsilon} = \frac{\int_{-\infty}^{\infty} \tfrac{1}{2} m\dot{x}^2 e^{-(m\dot{x}^2/2kT)}\,d\dot{x}}{\int_{-\infty}^{\infty} e^{-(m\dot{x}^2/2kT)}\,d\dot{x}} + \frac{\int_{-\infty}^{\infty} \tfrac{1}{2} m\omega_0^2 x^2 e^{-(m\omega_0^2 x^2/2kT)}\,dx}{\int_{-\infty}^{\infty} e^{-(m\omega_0^2 x^2/2kT)}\,dx}$$

Notice that these integrals are of the same type that occurred in Sec. 2-7 and therefore may be easily evaluated to obtain the result

$$\bar{\epsilon} = \tfrac{1}{2} kT + \tfrac{1}{2} kT = kT/\text{molecule} \qquad \text{(2-17)}$$

It may be shown that this result can be generalized to yield the law of equipartition of energy for classical mechanics: Any energy term which is quadratic in either displacement or velocity (momentum) contributes

$kT/2$ to the average energy stored in each of a large number of independent particles at equilibrium.

Let us now look at the quantum model. In this case the energy states are discrete and their energies are given by the relation

$$\epsilon_n = (n + 1/2)h\nu_0$$

Since there are now an infinite number of discrete states we obtain the average energy by writing the equation

$$\bar{\epsilon} = \frac{\sum_{n=1}^{\infty} (\epsilon_n - \epsilon_0)\Xi_n}{\sum_{n=1}^{\infty} \Xi_n}$$

We are now counting all the allowable states of the system and the average energy $\bar{\epsilon}$ is measured relative to the zero-point energy of vibration. Thus we obtain the expression.

$$\bar{\epsilon} = \frac{\sum_{n=1}^{\infty} nh\nu_0 e^{-(nh\nu_0/kT)}}{\sum_{n=1}^{\infty} e^{-(nh\nu_0/kT)}}$$

However, the denominator of this expression is a geometric series which may be expressed analytically while the numerator is the negative derivative of the denominator with respect to $1/kT$ as the argument. Therefore

$$\bar{\epsilon} = \frac{h\nu_0}{e^{(h\nu_0/kT)} - 1}$$

We now evaluate the heat capacity by noting that

$$(C_v)_{\text{vib}} = \mathfrak{N} \frac{d\bar{\epsilon}}{dT}$$

which yields the relationship

$$\frac{(C_v)_{\text{vib}}}{R} = \left(\frac{\theta_v/2T}{\sinh \theta_v/2T}\right)^2 \tag{2-18}$$

where $\theta_v = h\nu_0/k\,°\text{K}$, and is called the *characteristic vibrational temperature* of the harmonic oscillator. We note that Eq. 2-18 predicts that the vibrational contribution to the heat capacity in a system of quantum harmonic oscillators will range from 0 to R as T ranges from 0 to ∞. This disagrees rather strikingly with the classical equipartition law which states that vibration simply contributes R to the heat capacity irrespective of the system's temperature.

Typical vibrational temperatures for diatomic molecules can be calculated from the values of ω_e tabulated in Appendix C_1. We note that

$\theta_v = 2{,}274°K$ for oxygen, $\theta_v = 3{,}393°K$ for nitrogen and $\theta_v = 807°K$ for chlorine.

In general, if a large number of levels of a particular degree of freedom of a mechanical system are populated according to the distribution law, Eq. 2-16, this degree of freedom is said to behave in a classical manner. Translation is a good example. It may be also shown that at room temperature and above rotation is classically excited in most molecules. That is, each degree of rotational freedom contributes $R/2$ to the heat capacity. We may summarize these statements for a diatomic molecule by noting that since it has two degrees of rotational freedom its heat capacity as predicted by quantum-statistical mechanics may be written as:

$$C_p = C_v + R = \underbrace{\frac{3}{2}R}_{\text{translation}} + \underbrace{R}_{\text{rotation}} + \underbrace{R\left(\frac{\theta_V/2T}{\sinh(\theta_v/2T)}\right)^2}_{\text{vibration}} + R$$

This contrasts with the heat capacity predicted by classical mechanics, which is

$$C_p = C_v + R = \underbrace{\frac{3}{2}R}_{\text{translation}} + \underbrace{R}_{\text{rotation}} + \underbrace{R}_{\text{vibration}} + R = \frac{9}{2}R$$

Fig. 2-6 is a plot of the dimensionless heat capacity C_p/R of three diatomic gases as a function of a dimensionless temperature based on θ_v for each gas. This plot shows that the introduction of a vibrational temperature for each diatomic molecule allows one to handle the heat capacity of diatomic gases quite adequately by means of a corresponding states theory.

We have seen that quantum-statistical thermodynamics predicts reasonable heat capacities for diatomic molecules and that the energy stored in vibration in oxygen and nitrogen molecules is essentially zero at room temperature. Thus the quantum theory predicts that the heat-capacity ratio γ of air should be in the neighborhood of 1.4. In the case of classical statistical mechanics, we obtain the prediction that heat-capacity ratio of air should be equal to $9/7 = 1.29$. The fact that the value of gamma for air is very close to 1.4 at room temperature further supports the quantum-mechanical approach.

The above discussion, at best, represents only a brief introduction to the use of statistical thermodynamics for the calculation of thermodynamic properties. As we indicated in Chapter 1, the harmonic oscillator-rigid rotor description of molecular structure is only a first approximation to the true structure of the molecule. Nevertheless, the above exercise using the harmonic oscillator shows the utility of the quantum-mechanical approach in the calculation of thermodynamic properties. This approach

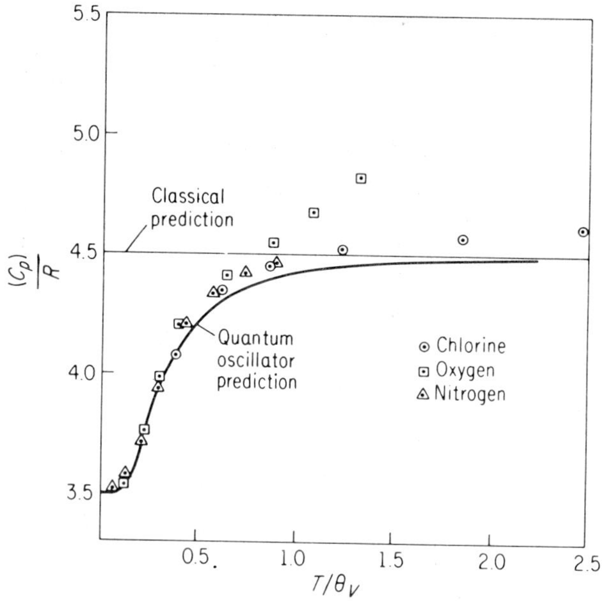

Fig. 2-6. Comparison of heat capacities of three diatomic molecules with the quantum mechanical and classical mechanical predictions. The high temperature deviations for oxygen are due to low-level electronic states in this molecule.

has, in fact, reached such a degree of sophistication that it is now usually agreed[4] that experimentally measured heats of combustion and dissociation energies coupled with quantum-statistical mechanical calculations based on spectroscopic data yield calculated thermodynamic properties which are more accurate than directly measured thermodynamic properties.

At present the most extensive and up-to-date compilation of thermodynamic properties is to be found in the JANAF (Joint Army-Navy-Air Force) tables compiled and published by the Dow Chemical Corporation, Midland, Michigan. Appendix F contains JANAF tables for some selected compounds of interest in the combustion field.

2-10. SOME THERMODYNAMIC CONSIDERATIONS

In elementary treatments of thermodynamics one is usually introduced to eight fundamental thermodynamic variables. These are pressure, P, temperature, T, volume, V, internal energy, E, enthalpy, $H = E + PV$,

[4]See, for example, J. S. Rowlinson, *The Perfect Gas*, Pergamon, New York, 1963, p. 57.

entropy, S, Helmholtz free energy or work function, $A = E - TS$, and the Gibbs free energy, $F = H - TS = E + PV - TS$. Five of these variables are fundamental to the subject and three are defined for operational convenience.

In this section we will review and possibly extend the elementary treatment in two important directions. We will discuss (1) a rather simple technique which may be used to evaluate thermodynamic first derivatives and (2) the general thermodynamic definition of equilibrium and the utility of the defined functions A and F.

The eight fundamental thermodynamic variables mentioned above are all state variables and may each be expressed uniquely as a function of any other two thermodynamic state variables, which are called the *independent variables* of the system in that representation. This is equivalent to the mathematical statement that the value of the line integral connecting two points in the plane of the independent variables is independent of the path used in its evaluation. However, this last statement is the definition of an exact differential and therefore the eight primary thermodynamic variables may all be expressed as exact differentials in terms of any two other state variables.

The above properties of thermodynamic state variables may be used to allow the easy determination of the value of the first partial derivative of any property with respect to any other property with a third held constant in terms of any two variables for any system. This is not really a trivial problem, since there are $8 \cdot 7 \cdot 6 = 336$ possible first derivatives to express in terms of 28 possible combinations of any two variables. A very general method involving the use of Jacobians has been discussed by Shaw[5] but we will present Bridgman's method in which we are restricted to expressing the derivatives as a junction of P and T as the independent variables.

We first discuss the properties of Jacobians. The Jacobian

$$J(x,y) = \frac{\partial(x,y)}{\partial(\alpha,\beta)} = (\partial x)_y$$

is defined as the following determinant

$$J(x,y) = \frac{\partial(x,y)}{\partial(\alpha,\beta)} = \begin{vmatrix} \left(\frac{\partial x}{\partial \alpha}\right)_\beta & \left(\frac{\partial x}{\partial \beta}\right)_\alpha \\ \left(\frac{\partial y}{\partial \alpha}\right)_\beta & \left(\frac{\partial y}{\partial \beta}\right)_\alpha \end{vmatrix} = \left(\frac{\partial x}{\partial \alpha}\right)_\beta \left(\frac{\partial y}{\partial \beta}\right)_\alpha - \left(\frac{\partial x}{\partial \beta}\right)_\alpha \left(\frac{\partial y}{\partial \alpha}\right)_\beta$$

As the reader may verify, a Jacobian defined in this manner has the following properties:

(1)

$$J(x,y) = -J(y,x)$$

[5]A. N. Shaw, *Phil. Trans. Royal. Soc. A*, *334* 299 (1935).

(2)

(3)

$$J(x,x) = 0$$

$$\left(\frac{\partial y}{\partial x}\right)_z = \frac{J(y,z)}{J(x,z)} = \frac{J(z,y)}{J(z,x)} = -\frac{J(y,z)}{J(z,x)} = -\frac{J(z,y)}{J(x,z)}$$

and (4)

$$J(\alpha,\beta) = -J(\beta,\alpha) = 1$$

where $J(x,y)$ is taken to be relative to the independent variables (α,β) in all the above relations.

Bridgman's tables for Jacobians in terms of the independent variables P and T are given in Appendix A. The right-hand sides of these expressions are all given in terms of the properties T, P, V, and S and the three first derivatives $(\partial V/\partial T)_P$, $(\partial V/\partial P)_T$ and $(\partial H/\partial T)_P = C_P$. Thus there is a redundancy on the right-hand side of the expressions which seems unnecessary at first glance. This redundancy arises because of convenience. While it is theoretically true that one equation of state, for example $E(V,T)$, is sufficient to allow the derivation of all state properties of a substance one usually specifies both a thermal equation of state, $P(V,T)$ (which is not sufficient by itself) and a caloric equation of state, $E(V,T)$. This is because it is usually quite difficult to determine detailed $P(V,T)$ behavior from an energy relationship. Also, this double specification is convenient for gases because the internal energy E is independent of V for an ideal gas. Therefore for slightly nonideal gases it is very convenient to correct the caloric equation of state $E(T)_{\text{ideal gas}}$ by using the thermal equation of state $P(V,T)_{\text{real gas}}$ to describe the system. The redundancy on the righthand side of the equations in Appendix A is of the same form; since the seven quantities used are easily determined from experiment they have simply not been eliminated from the expressions.

As an example of the use of Bridgman's table, consider the thermodynamic relation for the velocity of sound for a gas with a general equation of state and a general caloric equation of state. Recall that

$$a^2 = \left(\frac{\partial P}{\partial \rho}\right)_S = -V^2\left(\frac{\partial P}{\partial V}\right)_S = -V^2\frac{J(P,S)}{J(V,S)}$$

when $V = 1/\rho$ and is the volume per unit mass. Thus, from the table, we get the general relationship for the velocity of sound

$$a^2 = \frac{-V^2}{\left(\dfrac{\partial V}{\partial P}\right)_T + \dfrac{T}{\mathfrak{C}_p}\left(\dfrac{\partial V}{\partial T}\right)_P^2} \tag{2-19}$$

which contains terms that may be evaluated for any specific equation of state. Notice that, for a polytropic ideal gas,

$$\left(\frac{\partial V}{\partial P}\right)_T = -\frac{V}{P};\ \left(\frac{\partial V}{\partial T}\right)_P = \frac{\mathcal{R}}{P} \quad \text{and} \quad \mathcal{C}_p = \text{Const.}$$

and the general expression reduces to

$$a^2 = \gamma \mathcal{R} T$$

where $\mathcal{R}$ is the gas constant per unit mass.

The question of equilibrium in a homogeneous thermodynamic system will be discussed next. From introductory thermodynamics recall that the entropy change associated with a transition from state A to B is defined as

$$S_B - S_A = \int_A^B \frac{đq_{\text{rev}}}{T}$$

where $đq_{\text{rev}}$ is an imperfect (i.e., path-dependent) differential which represents the infinitesimal and reversible heat absorption of the system along any specified reversible path. Also recall that for an irreversible process, such as the irreversible expansion of an ideal gas into a vacuum the value of $S_B - S_A$ calculated using the above formula is always positive. This may be easily shown with the help of Fig. 2-7 in which we con-

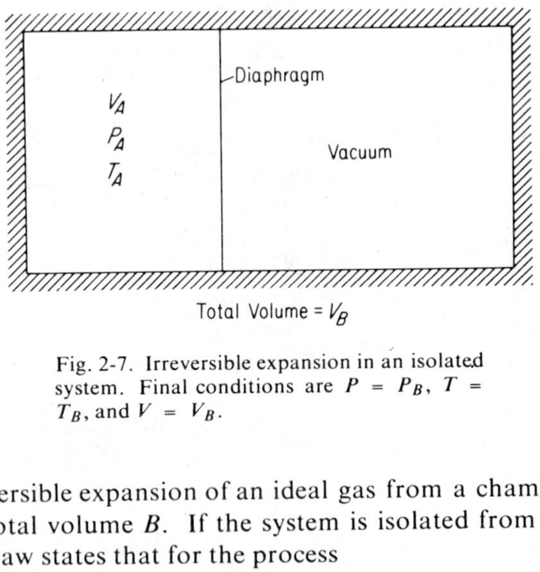

Fig. 2-7. Irreversible expansion in an isolated system. Final conditions are $P = P_B$, $T = T_B$, and $V = V_B$.

sider the irreversible expansion of an ideal gas from a chamber of volume A to fill the total volume B. If the system is isolated from the surroundings, the first law states that for the process

$$E_B - E_A = 0$$

Since the gas is perfect, the receprocity relation of thermodynamics

$$\left(\frac{\partial E}{\partial V}\right)_T = T\left(\frac{\partial P}{\partial T}\right)_V - P$$

yields the result that $E = E(T)$. Therefore the energy equation reduces to the statement

$$T_B = T_A$$

This means that $P_B = P_A \cdot \frac{V_A}{V_B}$ for this irreversible process. However, the entropy change in an ideal gas may be calculated directly from the equation

$$S_B - S_A = \int_{T_A}^{T_B} C_P \frac{dT}{T} - R \ln \frac{P_B}{P_A}$$

which yields the result

$$S_B - S_A = R \ln \frac{V_B}{V_A}$$

and the entropy of the isolated system is seen to increase during this irreversible change. In general, we find that if we were to calculate the entropy of the system and its surroundings for a variety of configurations, the state with the maximum entropy is the state observed when the system is in equilibrium, within the framework of the constraints imposed. For example, if we specify that there be no internal barriers in the vessel of Fig. 2-7, our experience tells us that the pressure will be constant throughout the vessel. Thus the state-variable entropy is found to be a convenient mathematical tool for quantitatively discussing the condition of equilibrium in an isolated system within the constraints that are present. In statistical mechanics we would make the equivalent statement that the system has the highest possible degeneracy consistent with the imposed constraints. (See, for example, Sec. 2-5.) Note that the use of the modifying term "imposed constraints" is very important to the definition of equilibrium. For example, consider two separate mixtures at room temperature, one of hydrogen and oxygen and another of argon and helium. Both of these appear to be in equilibrium in the ordinary thermodynamic sense (i.e., both have a definable uniform pressure and temperature). However, the hydrogen–oxygen mixture is known to be in a state of restricted equilibrium, the restriction being that water may not be formed. Thus statements concerning thermodynamic equilibrium must always be phrased so that the constraints are known.

The fact that S is a maximum for an isolated system in a state of thermodynamic equilibrium also leads to its utility in discussing spontaneous processes. A process is said to be spontaneous if it tends to drive the system and surroundings toward a state of equilibrium. We may also use the equivalent statement that a spontaneous process leads to an increase of the entropy of the universe. For example, in the vessel of Fig. 2-7 our experience tells us that the state A, in which all the gas is located in only one portion of the vessel, is a very improbable state unless the diaphragm separating the two portions is leaktight. However, we saw above that the state of having all the gas in the portion A does not violate the first law but simply has a higher entropy than the equilibrium state. Thermodynamically we say that the process $A \rightarrow B$ is spontaneous, whereas the process $B \rightarrow A$ is impossible (or at least extremely unlikely). Thus a straightforward calculation of the entropy of two end states for a process which occurs in an isolated system will tell us the direction in which the process occurs.

Unfortunately, experimental measurements are usually made in systems which are held at constant temperature and therefore are not isolated in the thermodynamic sense. Thus the entropy argument, even though useful, may not be conveniently applied in a direct manner. This was first pointed out by Massieu in 1869 approximately 35 years after the concept of entropy was first introduced by Clausius. Massieu defined two new functions for discussing equilibrium in constant (P, T) and (V, T) systems. However, it was Gibbs who first demonstrated the utility of the newly defined functions and extensively discussed their possible applications. The two defined functions are, of course (1) the Helmholtz free energy or work function

$$A = E - TS$$

and (2) the Gibbs free energy

$$F = H - TS$$

The Helmholtz free energy is useful for discussing equilibrium in a system held at constant volume and temperature, since we may write

$$\begin{aligned} dA &= dE - T\,dS - S\,dT \\ &= đq - P\,dV - T\,dS - S\,dT \end{aligned}$$

where the only work is PV work. For a process occurring at constant volume and temperature this equation reduces to the equation

$$dA = đq - T\,dS$$

However, since

$$dS = \frac{đq_{\text{rev}}}{T} \quad \text{and} \quad dS > \frac{đq_{\text{irr}}}{T}$$

we obtain the statements that $dA = 0$ for a reversible process and that A decreases for a spontaneous process occurring at constant volume and temperature.

In a similar manner we may write

$$\begin{aligned} dF &= dH - T\,dS - S\,dT \\ &= dE + V\,dP + P\,dV - T\,dS - S\,dT \\ &= đq + V\,dP - T\,dS - S\,dT \\ &= đq - T\,dS \text{ at constant pressure} \\ &\qquad \text{and temperature} \end{aligned}$$

This leads to the statements that $dF = 0$ for a reversible process and that F decreases for a spontaneous process occurring at constant pressure and temperature. Thus if we are able to express either the free-energy or work function for all possible states of a homogeneous system mathematically in terms of any two other state variables, the state of minimum A (for V, T held constant) or minimum F (for P, T held constant) will be the thermodynamic equilibrium state of that system within the imposed constraints. This general principle will be applied to chemical equilibrium in Chapter 4.

BIBLIOGRAPHY

1. T. G. Cowling, *Molecules in Motion.* (New York: Harper and Row, 1960.)

2. J. Jeans, *Introduction to the Kinetic Theory of Gases.* (New York: The University Press, Cambridge, 1959.)

3. E. H. Kennard, *Kinetic Theory of Gases.* (New York: McGraw-Hill, 1938.)

4. S. Glasstone, *Theoretical Chemistry.* (New York: Van Nostrand, 1944.)

5. J. E. Mayer and M. G. Mayer, *Statistical Mechanics.* (New York: Wiley, 1940.)

6. T. L. Hill, *Introduction to Statistical Thermodynamics.* (Reading, Mass.: Addison-Wesley, 1960.)

7. J. S. Rowlinson, *The Perfect Gas.* (New York: Pergamon, 1963.)

8. J. F. Lee, F. W. Sears, and D. L. Turcotte, *Statistical Thermodynamics.* (Reading, Mass.: Addison-Wesley, 1963.)

PROBLEMS

2-1. Calculate the relative probability of being dealt 2, 5, 7, or 12 spades in a bridge hand.

2-2. Calculate the vertical distribution of air above the earth, assuming constant temperature and constant gravity.

2-3. Show that the earth looses hydrogen from its atmosphere rather rapidly if the average temperature of the atmosphere is 270°K.

2-4. Use the Maxwell distribution law to calculate the drag of a flat plate traveling in a direction normal to its surface if the mean free path is much larger than the plate diameter. Assume specular reflection and assume that the reflected molecules do not change the velocity distribution of the incoming molecules. Calculate the drag for velocities that range from 0 to $10c$ when c is the average molecular speed.

2-5. Calculate the number of molecular collisions per square centimeter of wall area, the number of molecular collisions per cubic centimeter of gas, the molecular velocity and the mean free path for hydrogen, oxygen, and carbon dioxide at standard conditions ($T = 273.16°\text{K}$, $P = 1$ atm).

2-6. Calculate the fraction of molecules with a velocity of at least $3c$ and $10c$ where c is the average speed. Use the Maxwell distribution.

2-7. Verify Eq. 2-15a and 2-15b by integrating the Maxwell distribution given in Eq. 2-14.

2-8. Calculate the quantum number of a single particle in a cubic box one cm on a side at a temperature $T = 300°K$. Assume that the energy is divided into three equal parts for the three normal directions.

2-9. Show that nitrogen exists almost entirely in its lowest vibrational state at room temperature. Calculate the probability of observing a nitrogen molecule in its second or third vibrational state at 300°K.

2-10. Calculate the heat capacity of oxygen and nitrogen at 300°, 2500°, and 5000° using the methods of Sec. 2-9. Compare your answers to the exact values tabulated in Appendix F.

2-11. Calculate the heat capacity of water vapor (ideal gas) and carbon dioxide at 300°, 1000°, and 3000°K using the methods of Sec. 2-9. Compare your answers to the exact values tabulated in Appendix F.

2-12. Use the energy relations of Chapter 1 (Eq. 1-4), the tabulated rotational constants of Appendix C1 and the distribution law (Eq. 2-16) to show that rotation is classically excited at room temperature for all molecules (except possibly hydrogen).

3

The Physical Properties of Real Gases

3-1. INTRODUCTION

We saw in the previous chapter that the ideal gas law and ideal-gas thermodynamic properties can be obtained without considering the details of collision processes. In this chapter we will consider the more important properties of real gases which are collision determined. These include static or equilibrium nonideality (i.e., $PV \neq RT$), the primary transport properties (diffusion, viscosity, and thermal conductivity), and the lag in equilibriation of internal degrees of freedom of the gas molecule when the translational temperature is changed rapidly (relaxation processes). We are in essence restricting ourselves to those properties that are determined by the details of collisions which do not cause chemical change. We defer until the next chapter the discussion of collisions which produce changes in the chemical composition of the gas.

3-2. NONIDEAL GASES

The ideal gas law is well supported by experimental evidence under conditions where the mean free path is very much larger than the diameter of a molecule. However, deviations do occur at high densities and at low temperatures, and in fact all real gases are observed to condense to the liquid state if the temperature is low enough and the density sufficiently high. Experimentally this condensation behavior is observed to be similar for all pure substances (see Fig. 3-1). In this figure the double cross-hatched area represents the region where only pure liquid is present while the single cross-hatched area under curve ACB represents the coexistence region where liquid and gas are in equilibrium. It has been observed that there is a critical temperature T_c (31.0°C in Fig. 3-1) for each gas above which condensation (i.e., the formation of a liquid phase in equilibrium with the gas phase) does not occur, even though the application of sufficient pressure will produce densities well above ordinary liquid densities.

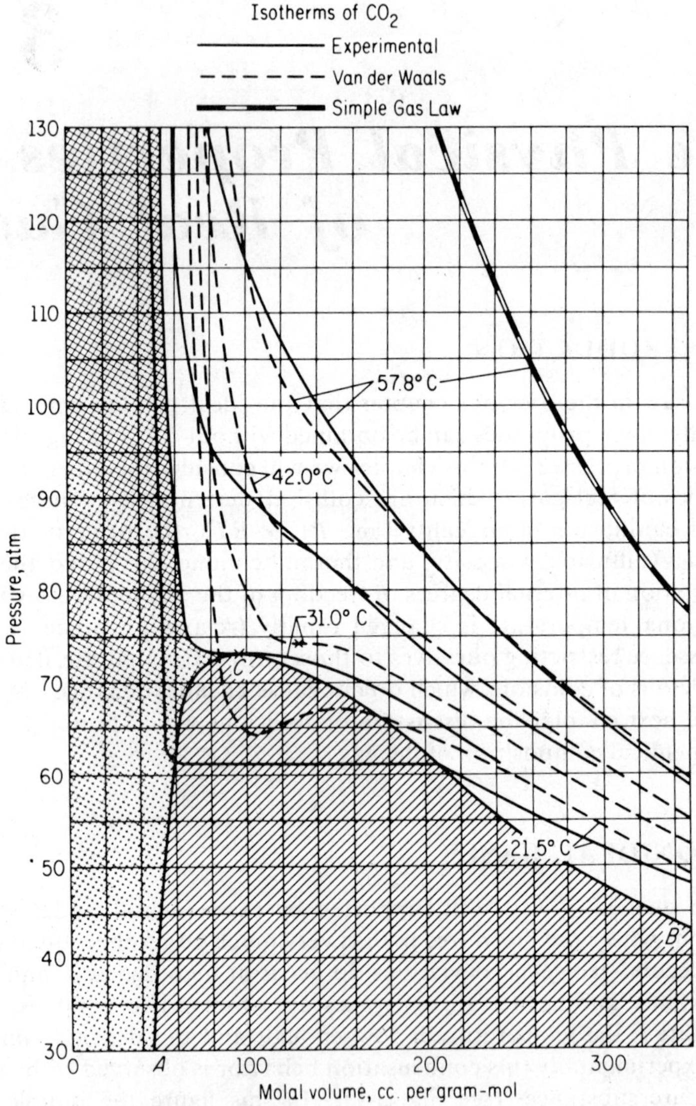

Fig. 3-1. Isotherms of carbon dioxide. (With permission from O. A. Hougen et al., *Chemical Process Principles*; Part II, *Thermodynamics*, Wiley, New York, 1959.)

Furthermore, on the critical isotherm ($T = T_c$) one observes an inflection point at which $(\partial P/\partial V)_T = 0$ and $(\partial^2 P/\partial V^2)_T = 0$. This inflection point is called the *critical point* for the substance and the volume and pressure at this point may be measured and are unique constants for each pure substance. At temperatures below the critical temperature the gas is observed to condense when the pressure reaches its vapor pressure, and to continue to condense at this pressure as the volume is decreased until only the pure liquid phase is present. Further decreases in the volume then cause extremely large increases in the pressure of the pure liquid phase.

Many modifications of the ideal gas law have been proposed to account for these experimentally observable effects. For example, the most useful equation of state for exact work is a simple geometric series expansion of the ideal gas law in terms of either the pressure or the volume:

$$\frac{PV}{RT} = 1 + b(T)\,P + c(T)\,P^2 + \cdots \tag{3-1}$$

or

$$\frac{PV}{RT} = 1 + \frac{B(T)}{V} + \frac{C(T)}{V^2} + \cdots \tag{3-2}$$

where the coefficients are ordinarily taken to be temperature-dependent. Experimental PVT data may be fitted quite accurately to either of these equations and the first and second coefficients for these equations may be calculated with reasonable accuracy from a theoretical consideration of the collision process for an equilibrium gas. This approach will be discussed in more detail in Sec. 3-4.

Among the other empirical equations of state there is one, the Van der Waals equation, which has a rather sound theoretical basis, is a closed form (analytical) equation, and quite nicely predicts the observed behavior of gases through the condensation region. This equation was first derived in the latter part of the nineteenth century. Clausius first proposed that the volume term in an equation of state should be the net volume available to the molecules and therefore modified the ideal gas law to read:

$$P(V - b) = RT$$

where b is an effective volume of one gram-mole of molecules. In 1873 Johannes Van der Waals further modified this equation to allow for the slight attractive force between neighboring molecules. This attractive force is very weak and for ordinary molecules it decreases rapidly with the separation distance (for nonpolar molecules the attractive force varies as $1/r^6$). Thus only the molecules nearest neighbors affect its motion appreciably, and those molecules near the surface of the vessel are in an unbalanced potential field. The pressure at any surface is proportional to the number of molecules per unit volume in the gas near the surface, and the net attractive force (on the molecules away from the surface) is also proportional to the number density of the molecules in the layer near the surface. Thus a weak attractive force will tend to make the measured pressure less than the actual pressure by an amount proportional to $(\mathfrak{N}/V)^2$ where $\mathfrak{N}/V$ equals the number of molecules/unit volume in the gas ($\mathfrak{N}$ is Avogadro's number). Therefore, the true gas pressure (i.e., on an imaginary internal surface) will be higher than the measured pressure and the Van der Waals equation is written as

$$\left(P + \frac{a}{V^2}\right)(V - b) = RT \tag{3-3}$$

where P is the externally measured pressure. The units on a and b are (pressure volume²/gram-mole) and (volume/gram-mole). Appendix D contains a tabulation of critical constants and Van der Waals constants for various gases.

The Van der Waals equation is a cubic in the volume and predicts the phenomena of condensation and the existence of a critical point where the gaseous and liquid state are indistinguishable. To demonstrate this, let us define the critical properties as $P = P_c$, $V = V_c$ and $T = T_c$. At the critical point $(\partial P/\partial V)_T = 0$ and $(\partial^2 P/\partial V^2)_T = 0$ and we can write the Van der Waals equation as

$$V_c^3 - V_c^2\left(b + \frac{RT_c}{P_c}\right) + V_c\frac{a}{P_c} - \frac{ab}{P_c} = 0$$

But since $V = V_c$ at the critical point we can also write

$$V^3 - 3V_cV^2 + 3V_c^2V - V_c^3 = 0$$

which yields by equating coefficients

$$a = 3V_c^2P_c; \quad b = V_c/3$$

and

$$R = \frac{8}{3}\frac{P_cV_c}{T_c} \text{ or } \frac{P_cV_c}{RT_c} = 0.375$$

Values of P_cV_c/RT_c for typical gases are tabulated in Table 3-1. Thus the Van der Waals constants are simply related to the critical constants of the gas and the behavior of all Van der Waals gases is similar when related to critical properties. This may be shown explicitly by defining the dimensionless variables $\alpha = P/P_c$, $\beta = V/V_c$, and $\delta = T/T_c$ which, when substituted into Eq. 3-3, yield the dimensionless Van der Waal equation of state:

$$\left(\alpha + \frac{3}{\beta^2}\right)(3\beta - 1) = 8\delta \tag{3-4}$$

In practice the Van der Waals equation is most useful in applications where the deviations from ideality are small since it does not conform quantitatively to the behavior of real gases. It is a useful equation of state,

TABLE 3-1
CRITICAL CONSTANTS RATIOS FOR GASES
(Theoretical value = 0.375)

Gas	$P_c V_c / R T_c$
Ar	0.291
N_2	0.292
O_2	0.292
H_2	0.304
Cl_2	0.286
NO	0.256
CO	0.287
HCl	0.286
CO_2	0.287
H_2O	0.224
CH_4	0.290
C_2H_2	0.275
C_2H_4	0.291
NH_3	0.238

however, because it indicates the source of the primary corrections to the ideal gas law and because it neatly illustrates the principle of corresponding states or similitude for the PVT behavior of gases. Equation 3-4 is a dimensionless equation of state in terms of critical properties of the gas and is generally applicable to all gases with reasonable accuracy. Thus, in the same manner that the ideal gas law shows that equal volumes of different gases under the same conditions contain equal number of molecules, the Van der Waals equation shows that the first-order correction terms to the ideal gas law have the same origin in all gases.

The principle of corresponding states has also led to the development of graphical methods of determining the properties of real gases. In this approach one uses the equation $PV = ZRT$ where Z is called the *compressibility factor* and is determined graphically as a function of reduced pressure α and reduced temperature δ. The accuracy of such an empirical all-inclusive fit for pure gases is estimated to be about $\pm 5\%$ over most of the range from condensation to $\delta = 15$ and $\alpha = 30$.

A discussion of the nonideal behavior of gas mixtures is outside the scope of this text.

3-3. TRANSPORT PHENOMENA—ELEMENTARY THEORY

There are three fundamental transport properties, one for each of the properties conserved in the three conservation equations of fluid dynamics. The three conservation equations may be derived with the assumption that transport is zero or small enough to be neglected. In fact, transport phenomena are only important when the solution to the conservation equations predicts that the fluid must support a large gradient in either concentration, velocity, or temperature. Under these conditions, mass transport, momentum transport, or energy transport (respectively) will occur and the proportionality constants which relate the quantity transported to the gradient (or the driving force which causes transport) are called the *diffusion coefficient*, the *viscosity coefficient*, and the *coefficient of thermal conductivity*.

Let us consider viscosity. If two large plates are placed a short distance y apart and are moved relative to each other in the x direction (keeping y constant) one finds that a force is required to keep their relative velocity in the x direction, V_x, at a constant value. Furthermore, experimental evidence has shown that when gases or simple liquids are trapped between the plates, the steady velocity that is maintained is proportional to the force per unit area applied in the direction of slippage times the distance between the plates. i.e.,

$$y\frac{F}{A} \propto V_x$$

or

$$\frac{F}{A} = -\mu\frac{dV_x}{dy} \tag{3-5}$$

This is called *Newton's law of viscosity*. Fluids with a constant proportionality constant (independent of dV_x/dy) are called Newtonian fluids. In cgs units,

$$\mu = \left|\frac{F}{A}\frac{dy}{dV_x}\right| = \text{gm/cm sec}$$

where the cgs unit of 1 gm/cm sec is called the *poise*. Most viscosity data is reported in units of centipoises, where one centipoise equals 0.01 poise. Kinematic viscosity is defined as

$$\nu = \frac{\mu}{\rho}$$

where ρ is the fluid density. Therefore kinematic viscosity has units of cm^2/sec.

Using the kinetic theory of gases from Chapter 2 we may relate the viscosity coefficient to the molecular properties of an ideal gas consisting of hard spheres. Recall that the average speed of a molecule is

$$\bar{c} = \sqrt{\frac{8}{\pi}\frac{RT}{M}}$$

and its mean free path is

$$l = \frac{1}{\sqrt{2}\,\pi\sigma_0^2 N}$$

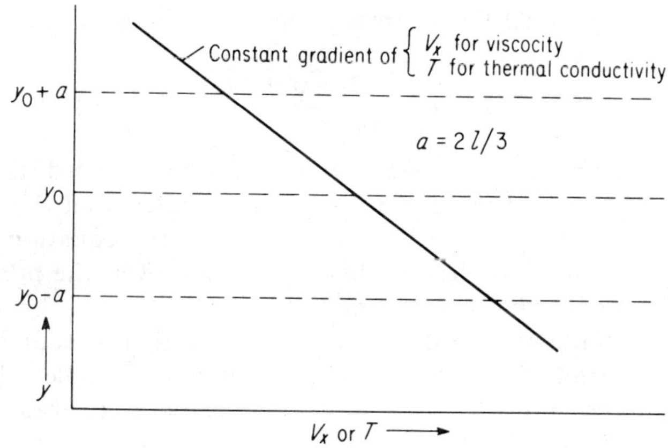

Fig. 3-2. Constant gradient assumed in the kinetic theory derivation of the transport properties viscosity and thermal conductivity.

Now consider an imaginary plane in a gas at a position y normal to a velocity gradient (see Fig. 3-2). Molecules are continually crossing this plane from above and below. One may calculate that on the average those that cross the plane had their last collision at a distance $a = {}^2\!/_3 l$ away from the plane. As we have seen (page 44), the number of molecules crossing the plane per second from one side is $Z = N\bar{c}/4$. We assume that each molecule is carrying an x momentum characteristic of the x momentum in the plane of its last collision. The momentum flux across the plane is therefore

$$\frac{F}{A} = Z(mV_x)\,|_{y-a} - Z(mV_x)\,|_{y+a}$$

But since we assume a constant gradient,

$$V_x\,|_{y-a} = V_x\,|_y - \frac{2}{3}l\frac{dV_x}{dy}$$

and

$$V_x\,|_{y+a} = V_x\,|_y + \frac{2}{3}l\frac{dV_x}{dy}$$

therefore

$$\frac{F}{A} = -Zm\frac{4}{3}l\frac{dV_x}{dy} \cong -\frac{1}{3}Nm\bar{c}l\frac{dV_x}{dy}$$

or

$$\mu = \frac{1}{3}Nm\bar{c}l = \frac{1}{3}\rho\bar{c}l \qquad (3\text{-}6)$$

Maxwell first obtained this expression in 1860. Substituting for l and $\bar{c}$, we find that

$$\mu = \frac{2}{3\pi^{3/2}} \frac{\sqrt{mkT}}{\sigma_0^2} \tag{3-7}$$

Viscosity measurements in gases can therefore be used to determine molecular diameters. Equation 3-7 correctly predicts that the viscosity of dilute gases is independent of the pressure and the equation also yields approximately correct molecular diameters. However, the predicted temperature dependence has been found to be too weak.

Thermal conductivity without convection is quite similar to viscosity, since the heat conduction per unit area through any simple substance has been found to be simply proportional to the temperature gradient.

$$\frac{Q}{A} = -\kappa \frac{dT}{dy} \tag{3-8}$$

This is called *Fourier's law of heat conduction,* where κ is the thermal conductivity. In cgs units,

$$\kappa = \left| \frac{Q}{A} \frac{dy}{dT} \right| = \text{cal/cm sec °K}$$

Thermal diffusivity is defined as

$$\alpha = \frac{\kappa}{\rho \mathcal{C}_p} = \text{cm}^2/\text{sec}$$

where $\mathcal{C}_p$ = cal/gm °K.

Kinetic theory can also be used to relate the thermal conductivity k to the molecular parameters of the gas. Using the same approach as for viscosity, we discuss the transport of kinetic energy across the plane normal to the temperature gradient. For this elementary discussion we limit ourselves to monatomic molecules which cannot store energy in internal degrees of freedom. In this case the translational temperature is simply proportional to the translational kinetic energy.

$$\frac{1}{2} m\overline{c^2} = \frac{3}{2} kT$$

Using the same assumptions as for viscosity we allow the heat flux to equal the net energy of the molecules arriving from above and below the plane:

$$\frac{Q}{A} = Z\left(\frac{1}{2} m\overline{c^2}\right)\bigg|_{y-a} - Z\left(\frac{1}{2} m\overline{c^2}\right)\bigg|_{y+a}$$

or

$$\frac{Q}{A} = \frac{3}{2} kZ(T|_{y-a} - T|_{y+a})$$

where, with a linear temperature variation,

$$T|_{y-a} = T|_y - \frac{2}{3} l \frac{dT}{dy}$$

$$T|_{y+a} = T|_y + \frac{2}{3} l \frac{dT}{dy}$$

and

$$\frac{Q}{A} = -\frac{1}{2} Nk\bar{c}l \frac{dT}{dy}$$

This yields

$$\kappa_{\text{mono}} = \frac{1}{2} Nk\bar{c}l = \frac{1}{3} \rho \mathcal{C}_v \bar{c} l$$

for the thermal conductivity of a monatomic gas. This reduces to the expression

$$\kappa_{\text{mono}} = \frac{1}{\sigma_0^2} \sqrt{\frac{k^3 T}{\pi^3 m}} \tag{3-9}$$

Equation 3-9 predicts the correct pressure sensitivity for thermal conductivity (none) and the correct relationship with molecular diameter and molecular weight. However, once again the predicted temperature sensitivity has been found to be too weak.

The diffusion coefficient is defined as a proportionality constant which relates the net transport of a species to the concentration gradient under conditions where the total net transport of molecules is zero. Because of the requirement that there be no net transport in the mixture as a whole, binary-diffusion coefficients are dependent on the properties of both gases. There are, therefore, self-diffusion coefficients, i.e., for the gas into itself; binary-diffusion coefficients, for the gas into another gas (which must be specified), or multicomponent-diffusion coefficients which determine the average motion of one type of gas molecules in a mixture containing two or more other gases.

Let us consider only self- and binary diffusion. Assume the mixture has a constant concentration gradient as shown in Fig. 3-3. *Fick's law of*

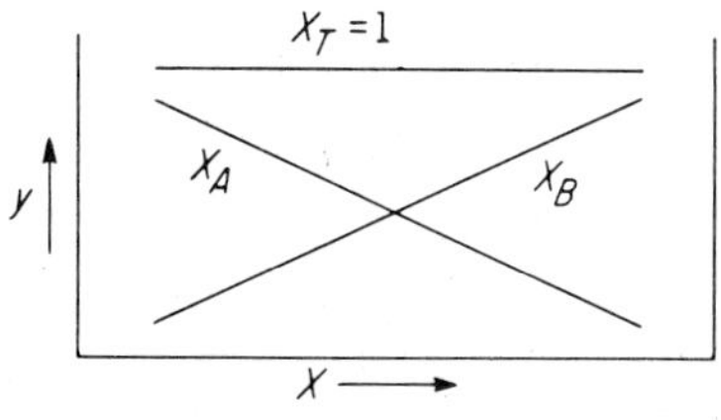

Fig. 3-3. Mole fractions in a binary gas mixture containing a concentration gradient.

diffusion states that the current density of flow (or net molecular flow of one species) past a plane normal to the gradient is proportional to the gradient, i.e.,

$$\Gamma_A = -D\frac{dN_A}{dy} \quad \text{and} \quad \Gamma_B = -D\frac{dN_B}{dy} \tag{3-10}$$

Notice that both D's are the same. Here Γ has units of molecules/cm^{2}sec, and dN/dy has units of molecules/cc cm. Therefore D must have units of cm^2/sec. These units are identical with the units for kinematic viscosity and thermal diffusivity as defined above. We now can use our earlier kinetic theory approach to calculate a value of the diffusion coefficient. Let us define the mole fraction of molecules A as $X_A = N_A/N_T$. The net current density of A is equal to the net number of molecules of A crossing the reference plane from below minus those crossing from above, i.e.,

$$\Gamma_A = \Gamma_B = ZX_A\,|_{y-a} - ZX_A\,|_{y+a}$$

For simplicity we assume that the two different molecules have the same molecular properties, i.e., $\sigma_A = \sigma_B$, $m_A = m_B$, etc. Therefore:

$$X_A\,|_{y-a} = X_A\,|_y - \frac{2}{3}l\frac{dX_A}{dy}$$

and

$$X_A\,|_{y+a} = X_A\,|_y + \frac{2}{3}l\frac{dX_A}{dy}$$

or

$$\Gamma_A = -Z\frac{4}{3}l\frac{dX_A}{dy}$$

where, as before, Z equals total number of molecules crossing the boundary, $N_T\bar{c}/4$. This yields

$$\Gamma_A = -\frac{1}{3}N_T\bar{c}l\frac{dX_A}{dy}$$

$$\frac{dX_A}{dy} = \frac{1}{N_T}\frac{dN_A}{dy}$$

but since

$$\Gamma_A = -\frac{1}{3}\bar{c}l\frac{dN_A}{dy}$$

or

$$D_{AA} = \frac{1}{3}\bar{c}l$$

where the subscript AA refers to the self-diffusion coefficient. Substituting in the more ordinary state variables yields the equation

$$D_{AA} = \frac{2}{3}\left(\frac{k^3}{\pi^3 m_A}\right)^{1/2} \frac{T^{3/2}}{P\sigma_A^2}$$

This simple theory may be extended, by using the proper assumptions, to derive an expression for binary diffusion coefficients. The derivation (which is rather lengthy) yields the formula

$$D_{AB} = \frac{2}{3}\left(\frac{k^3}{\pi^3}\right)^{1/2}\left(\frac{1}{2m_A} + \frac{1}{2m_B}\right)^{1/2} \frac{T^{3/2}}{P\left(\dfrac{\sigma_A + \sigma_B}{2}\right)^2}$$

Once again the simple theory yields a result which is adequate except that the predicted temperature dependence is found to be too weak.

The dimensional similarity of these three transport properties suggests that their ratios may be used to define a set of convenient dimensionless parameters for discussing the properties of gases. This is indeed the case, and the three resulting parameters have had sufficient utility to each be identified with a scientist's name. They are (1) the ratio of the kinematic viscosity and the thermal diffusivity, or Prandtl number,

$$\Pr = \frac{\mathcal{C}_p \mu}{\kappa} = 1.667 \tag{3-12}$$

(2) the ratio of the kinematic viscosity and the self-diffusion coefficient, or Schmidt number,

$$\mathrm{Sc} = \frac{\mu}{\rho D} = 1.0 \tag{3-13}$$

and (3) the ratio of the self-diffusion coefficient and the thermal diffusivity, or Lewis number

$$\mathrm{Le} = \frac{\mathcal{C}_p D \rho}{\kappa} = 1.667 \tag{3-14}$$

Some texts define Le as the reciprocal of Eq. 3-14. The numerical values are for the interaction of hard sphere monatomic molecules using the simple kinetic theory approach. These values do not agree with the experimentally measured ratios for dilute gases.

Thus, in summary, we find that elementary kinetic theory yields the correct pressure dependence for the dilute-gas transport properties but does not yield either a correct temperature dependence or a correct set of values for the dimensionless ratios of these properties. The reason for these discrepancies will be discussed in the next section.

3-4. TRANSPORT PHENOMENA AND THE COLLISION INTERACTION POTENTIAL

The elementary treatment of transport, developed in Sec. 3-3, adequately explains the gross behavior of the transport coefficients of dilute gases and has been found to yield order-of-magnitude agreement with experimentally determined values. However, the results are not really useful for estimating the transport coefficients of real gases with reasonable accuracy. The more rigorous Chapman-Enskog theory of transport does, however, yield reasonably accurate transport coefficients. It is based on the same hypothesis as the elementary theory (i.e., dilute gas/only binary collisions contribute) but it includes the effect of an assumed interaction potential during the molecular collision process, and in addition it assumes that each interaction is weighted in relation to all other possible elastic collisions using a Maxwell distribution of molecular velocities. Thus the more rigorous kinetic theory requires that an interaction potential be specified for each molecular encounter, and this theory is therefore only as realistic as the assumed potential. Using this theory, the ultimate result of a rather lengthy calculation is a set of correction factors to the elementary formulas which are dependent on the chosen interaction potential and the gas temperature. Because of the complexity of the theory only the interaction of spherically symmetric molecules has been calculated at this time.

The true molecular interaction potential (even between truly spherical molecules such as pure argon) cannot be represented exactly by a simple mathematical expression. However, many simple equations come quite close to the correct potential. This is usually assumed to consist of a long-distance attractive force (for a simple spherical molecule with no permanent dipole moment this is caused by an induced-dipole–induced-dipole interaction and is proportional to r^{-6}) and a close-approach repulsive force which is thought to be exponential. The net result is an interaction potential similar to that shown in Fig. 3-4. Notice that the potential-energy minimum (i.e., the "well") formed by this type of interaction will markedly change the character of low-velocity collision processes. In general one can describe this "well" by determining its depth ϵ and a zero-energy collision diameter σ_0 within the framework of a particular mathematical model.

The most successful and best known two-parameter relationship for the potential energy for which the Chapman-Enskog theory has been solved is the Lennard-Jones potential:

$$\phi(\sigma) = 4\epsilon\left[\left(\frac{\sigma_0}{\sigma}\right)^{12} - \left(\frac{\sigma_0}{\sigma}\right)^{6}\right] \tag{3-15}$$

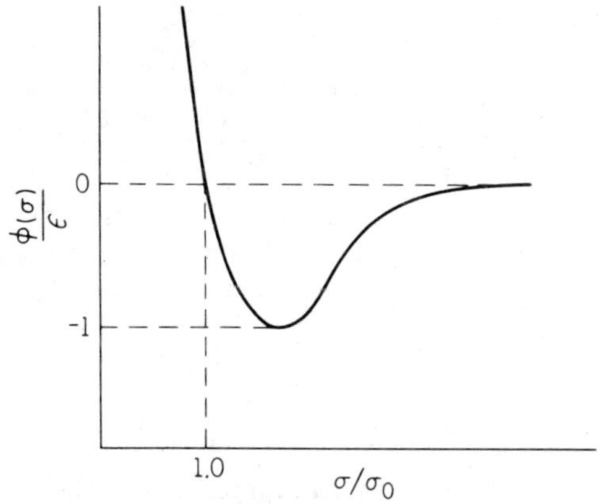

Fig. 3-4. Typical interaction potential for a molecular collision. The Leunard-Jones potential (Eq. 3-15) is plotted in this figure.

The constants are illustrated in Fig. 3-4. Notice in Fig. 3-4 that the depth of the well is important to a collision only as it relates to the kinetic energy of the collision. Since in a Maxwell gas the average kinetic energy is proportional to the temperature, ϵ is made dimensionless by writing it as kT/ϵ where k is the Boltzmann constant. The theory is, therefore, a corresponding states theory, since it assumes a universal potential energy function for all molecules. Correction factors calculated from the Chapman-Enskog theory using the Lennard-Jones potential are tabulated in Appendix E. Since in the theory the viscosity and thermal conductivity corrections are determined by the same averaging process, these correction factors (called the *collision integrals* Ω_μ and Ω_κ) are identical. The collision integral for diffusion Ω_D is obtained by a slightly different averaging process and therefore is tabulated separately. Values of ϵ/k in degrees Kelvin and σ_0 in Å (10^{-8} cm) for various molecules are tabulated in Appendix E. The transport coefficients may be calculated from the values of ϵ/k, °K and σ_0, Å found in these tables using the following formulas:

Viscosity

$$\mu = 2.6693 \times 10^{-5} \frac{\sqrt{MT}}{\sigma_0^2 \Omega_\mu} \text{ poises} \tag{3-16}$$

where M = gm/mole and T = °K.

Thermal conductivity for monatomic gases

$$\kappa_{\text{mono}} = 1.9891 \times 10^{-4} \frac{\sqrt{\dfrac{T}{M}}}{\sigma_0^2 \Omega_k} \text{ cal/cm sec °K} \tag{3-17}$$

and diffusion coefficients (binary)

$$D_{AB} = 0.0018583 \frac{\sqrt{T^3\left(\frac{1}{M_A} + \frac{1}{M_B}\right)}}{P(\sigma_0)^2_{AB}\Omega_{AB_D}} \text{ cm}^2/\text{sec} \tag{3-18}$$

where P = atmospheres and Ω_{AB_D} is a dimensionless collision integral calculated from a value of $(kT/\epsilon)_{AB}$. Here $(\epsilon/k)_{AB}$ and $(\sigma_0)_{AB}$ are calculated from the individual parameters using the approximate equations

$$(\sigma_0)_{AB} = \tfrac{1}{2}(\sigma_{0_A} + \sigma_{0_B})$$

and

$$(\epsilon/k)_{AB} = \sqrt{(\epsilon/k)_A \cdot (\epsilon/k)_B}$$

Self-diffusion coefficients may be calculated using the above formulas by assuming that A equals B and M_A equals M_B, etc.

In the case of thermal conductivity, the Chapman-Enskog theory allows the correct calculation of a coefficient only for the case of a monatomic gas. This is true because even the rigorous theory ignores the transport of energy stored in the internal degrees of freedom. Eucken first proposed an empirical formula to calculate the contribution of this internal-energy transport. His formula may be written

$$\kappa = \kappa_{\text{mono}}\left[\frac{4}{15}\frac{C_v}{R} + \frac{3}{5}\right] \tag{3-19}$$

where C_v equals the constant volume heat capacity of the real gas. Hirschfelder[1] has recently derived a similar expression on a sound theoretical basis. The Hirschfelder formula for thermal conductivity is

$$\kappa = \kappa_{\text{mono}}\left[1 + \delta_f\left(\frac{\mathcal{C}_p}{\mathcal{C}_{p_f}} - 1\right)\right]$$

where

$$\delta_f = \frac{\rho D \mathcal{C}_{p_f}}{\kappa_{\text{mono}}} = \text{Le}_{\text{mono}}$$

and

$$\mathcal{C}_{p_f} = \tfrac{5}{2}\mathcal{R}$$

The Hirschfelder formula seems to agree well with the limited experimental data.

The more rigorous kinetic-theory approach yields quite reasonable values of the dimensionless ratios of the transport properties. A few typical values for the Prandtl number are listed in Table 3-2. In the case of self-diffusion the data are much more meager, so that extensive com-

[1] J. O. Hirschfelder, *J. Chem. Phys*, **26**, 282(1957).

TABLE 3-2

PREDICTED AND OBSERVED VALUES OF $\left(\frac{\mathcal{C}_p \mu}{\kappa}\right)$ FOR GASES AT ATMOSPHERIC PRESSURE†

Gas	T, °K	$\left(\frac{\mathcal{C}_p \mu}{\kappa}\right)$ From Eqs. 3-16, 3-17, and 3-19	$\left(\frac{\mathcal{C}_p \mu}{\kappa}\right)$ from observed values of C_p, μ, and κ
Ne	273.2	0.667	0.66
Ar	273.2	0.667	0.67
H_2	90.6	0.68	0.68
	273.2	0.73	0.70
	673.2	0.74	0.65
N_2	273.2	0.74	0.73
O_2	273.2	0.74	0.74
Air	273.2	0.74	0.73
CO	273.2	0.74	0.76
NO	273.2	0.74	0.77
Cl_2	273.2	0.76	0.76
H_2O	373.2	0.77	0.94
	673.2	0.78	0.90
CO_2	273.2	0.78	0.78
SO_2	273.2	0.79	0.86
NH_3	273.2	0.77	0.85
C_2H_4	273.2	0.80	0.80
C_2H_6	273.2	0.83	0.77
$CHCl_3$	273.2	0.86	0.78
CCl_4	273.2	0.89	0.81

†With permission from R. B. Bird, W. E. Stewart, and E. N. Lightfoot, *Transport Phenomena*, Wiley, New York, 1960.

parisons cannot be made. However, the rigorous theory is quite successful in fitting experimental diffusion data and therefore the same type of fit for the dimensionless Schmidt and Lewis numbers should be obtained using the rigorous theory.

The Chapman-Enskog theory may also be used to determine the second and third virial coefficients in the equation of state, Eq. 3-2:

$$\frac{PV}{RT} = 1 + \frac{B(T)}{V} + \frac{C(T)}{V^2} + \cdots$$

The formulas for this estimation are

$$B(T) = B^*\left(\frac{kT}{\epsilon}\right)\cdot b_0$$

$$C(T) = C^*\left(\frac{kT}{\epsilon}\right)\cdot b_0^2$$

where

$$b_0 = \tfrac{2}{3}\pi \mathfrak{N} \sigma_0^3 \times 10^{-24}\ \text{cm}^3/\text{mole}$$

and $\mathfrak{N}$ = Avagadro's number.
Notice that V must have the same units as b_0 has in Eq. 3-2. Values of B^*, C^*, and b_0 are tabulated in Appendix E.

3-5. HIGH-TEMPERATURE TRANSPORT COEFFICIENTS

The Chapman-Enskog theory of transport, while based on a sound interaction model, suffers some loss of generality in its application. This is caused by the lack of flexibility that the theory allows in the choice of potential models. Once a model is fixed—say a two-parameter Lennard-Jones potential—there is a great deal of calculation needed to evaluate the correction factors. After this effort has been made, the predicted temperature dependence is compared to experimental data to determine the model constants for this particular substance, model, and transport property. Thus the theory, because of its inherent complexity, does not allow one to determine the best molecular model but only the best parameters in an *a priori* model. This is why the Lennard-Jones parameters as evaluated from experimental viscosity and equation of state data are different for most substances (see Appendix E). Furthermore, it has been found that the evaluation of "best fit" parameters for other "realistic" potential models yield interaction potentials which are markedly different for the same substance, even though the different models yield transport coefficients which fit the experimental data with similar standard deviations. This is why only Lennard-Jones parameters are presented in this text. For ordinary work they should be adequate to estimate transport coefficients and equation of state nonideality.

There is one area where caution is needed in the application of this model. In general, the parameters have been estimated from viscosity data taken at or slightly above room temperature. Thus the "best fit" to experiment is determined in this temperature range and high-temperature transport properties cannot, in general, be estimated accurately with these "best fit" constants. Two classes of molecules exhibit a particularly bad extrapolation behavior. These are polar molecules and oblong molecules, both of which have an interaction potential which is markedly different from the Lennard-Jones potential. Experimental evidence on polar molecules points to the fact that they exhibit a stronger temperature-dependence than the theory predicts. High-temperature data on long molecules are virtually nonexistent. The reader is referred to references in the bibliography and to the current literature on transport properties for a more detailed discussion of this problem.

3-6. MULTICOMPONENT MIXTURES TRANSPORT PROPERTIES

It has been observed experimentally that the transport coefficient of a binary mixture of dilute gases does not vary linearly with the composition. This is particularly true for those cases where the transport coefficients of the pure substances are markedly different. In the case of viscosity and thermal conductivity the properties of a mixture may be estimated with reasonable accuracy by using the semiempirical formula of Wilke[2]

$$\lambda_{\text{mix}} = \sum_{i=1}^{n} \frac{x_i \lambda_i}{\sum_{j=1}^{n} x_j \phi_{ij}} \tag{3-20}$$

where

$$\phi_{ij} = \frac{1}{\sqrt{8}} \left(1 + \frac{M_i}{M_j}\right)^{-1/2} \left[1 + \left(\frac{\mu_i}{\mu_j}\right)^{1/2} \left(\frac{M_j}{M_i}\right)^{1/4}\right]^2$$

and where λ is either the viscosity coefficient μ or the thermal conductivity coefficient κ. In this equation x is the mole fraction and M is the molecular weight of the species involved. The i and j refer to the specific species in the mixture.

In the case of diffusion the problem is more complicated. The reader will recall that the diffusion coefficient of a binary mixture is dependent on the properties of both species because the coefficient is defined for no net transport of molecules in the mixture as a whole. In the same manner the diffusion coefficient of the ith species in a multicomponent mixture is dependent on the local composition of the mixture. The problem is compounded, however, by the fact that each individual component in a multicomponent mixture diffuses at its own rate relatively independent of the other species. Thus, observation of the diffusion of carbon dioxide into a helium–nitrogen mixture[3] shows clearly that the helium concentration gradient disappears very rapidly whereas the nitrogen–carbon dioxide gradient remains. In other words, the rapid diffusion of helium into the pure carbon dioxide will initially cause a negative net mass flux of the nitrogen in the direction of the predominate diffusion gradient. The reader is referred to texts listed in the bibliography for further information on this subject.

[2]C. R. Wilke, *J. Chem. Phys.*, **18,** 517 (1950). Also see E. A. Mason and C. S. Saxena, *Phys. Fluids*, pp. 1, 361 (1958).

[3]See, for example, R. E. Walker, N. de Haas, and A. A. Westenberg, *J. Chem. Phys.*, *32*, 1314 (1960).

3-8. RELAXATION PHENOMENA

In Chapter 2 we discussed the technique for calculating the energy stored in the internal degrees of freedom of an equilibrium gas sample. In addition we found that the quanitization of the energy states of a rigid rotor or harmonic oscillator in some cases led to the internal degrees of freedom storing less than their classically allowable energy. Most important to the discussion in this section, however, is the fact that the treatment in Chapter 2 leads us to the conclusion that energy storage in the internal degrees of freedom is intrinsically different from energy storage in translation. Thus we should expect that if we were to change the temperature of the gas rapidly the internal degrees of freedom might come to equilibrium at a different rate than the translational degrees.

Experimentally it is rather easy to change the translational temperature of a gas rapidly because rapid adiabatic compression leads to relatively high heating rates in the gas sample. In the limit, the fastest way to heat each gas element is to compress it with a step shock wave in a shock tube. Experiments on the thickness of shock waves in gases with no internal degrees of freedom (i.e., neon, argon, or krypton) have shown that translational equilibration is extremely fast. Shock-wave thicknesses in these gases correspond to complete equilibration in less than 5 mean free paths. In addition, we note that during the transit from cold to hot gas conditions, we cannot define a temperature in the usual sense. All we can really say is that the Maxwell distribution for a cold gas yields a Maxwell distribution for a hot gas after passage through a region which is less than 5 mean free paths thick.

For the rotational degrees of freedom in diatomic and polyatomic molecules relaxation is almost as rapid as translational equilibration. It is rapid enough in fact that one usually discusses how the shock wave is "thickened" by the rotational relaxation phenomena. Indeed, it is difficult to discuss this relaxation phenomena independent of translational relaxation because translational relaxation times are always a good fraction of the rotational relaxation time.

The vibrational degrees of freedom show relaxation times which in general are very much longer than the characteristic shock transient time for the gas. Theoretically this result is explained by noticing that a single quanitized vibrational level usually stores considerably more energy than a single translational degree of freedom. This means that there is a very low probability that any single collision can result in internal energy storage by exciting the next vibrational level of the molecule, because the kinetic energy of an average collision is usually much less than the energy needed for a single vibrational quantium jump.

Vibrational relaxation is usually slow enough so that one can measure the gas properties in a narrow region behind the shock where the vibrational "temperature" is still equal to the cold-gas translational temperature, and one can then follow the energy loss of the translational degrees of freedom to determine the characteristic vibrational relaxation time for the gas. This process is externally adiabatic and has been shown to be an exponential decay process within experimental error. The characteristic time for this decay process is defined as the *relaxation time*. It can be measured using step shock waves; however, we wish to defer a discussion of the structure of shock waves until Chapter 5. It is sufficient here to know that vibrational relaxation times exist and can be measured.

There have been many theoretical models proposed for estimating relaxation times in pure gases and in gas mixtures. The best correlation obtained so far is a semiempirical correlation obtained by Millikan and White.[4] They find that the relaxation of diatomic molecules can be estimated rather well with the formula.

$$\log_{10}(P\tau_v) = 5.0 \times 10^{-4}\mu^{1/2}\Theta_v^{4/3}(T^{-1/3} - 0.015\mu^{1/4}) - 8.0 \qquad (3\text{-}21)$$

where P = atmospheres, μ = reduced mass of the responsible collision pair $M_1M_2/(M_1 + M_2)$ with units of gm/mole, Θ_v = the characteristic oscillator frequency of the vibrational degree of freedom with units of °K and τ_v = the relaxation time for vibration in seconds. This formula has been found to fit experimental data for 27 diatomic gas pairs over a large temperature range.

BIBLIOGRAPHY

1. J. O. Hirshfelder, C. F. Curtiss, and R. B. Bird, "Molecular Theory of Gases and Liquids." (New York: Wiley, 1954.)

2. R. B. Bird, W. E. Stewart, and E. N. Lightfoot, *Transport Phenomena*. (New York: Wiley, 1960.)

3. O. A. Hougen, K. M. Watson, and R. A. Ragatz, *Chemical Process Principles*, 2d ed., Part II, *Thermodynamics*. (New York: Wiley, 1959.)

4. A. A. Westenberg, "Present Status of Information on Transport Properties Applicable to Combustion Research," *Combustion and Flame*, **1**, 346 (1957).

5. R. M. Fristrom and A. A. Westenberg, "Molecular Transport Properties for Flame Studies," *Fire Research Abstracts and Reviews*, **8**, 3 pp. 155–183 (1966).

PROBLEMS

3-1. Derive an equation for (a) $\gamma(P, T)$ – the heat capacity ratio and (b) $a(P, T)$—velocity of sound, for a Van der Waals gas with no internal degrees of freedom using Bridgeman's tables (Appendix A).

3-2. Calculate the binary-diffusion coefficient for methane-oxygen (CH_4 –

[4]Millikan and White, *J. Chem. Phys.*, **39**, 3209 (1963).

O_2) at 300 °K, 800 °K, and 1200 °K using the Lennard-Jones interaction potential and compare with experimental values reported by Walker and A. A. Westenberg, *J. Chem. Phys.*, **32**, 436 (1960).

3-3. Calculate the thermal conductivity coefficient for pure hydrogen gas at 1,500 °K using (1) Monatomic assumption, (2) Euken's formula, and (3) Hirschfelder's formula. Compare your answers with the results of N. C. Blais and J. B. Mann, *J. Chem. Phys.*, **32,** 1459 (1960).

3-4. Calculate the pressure at which the density of air at ambient temperature deviates from an ideal gas by (a) 1%, (b) 10%.

3-5. Calculate vibrational relaxation times in pure oxygen and nitrogen using the formula Sec. 3-8 and compare with V. Blackman, *J. Fluid Mech*, **1**, 61 (1956).

4

Chemical Reactions in Gases

4-1. INTRODUCTION

In this chapter we wish to discuss the consequences of reactive collisions occurring in an ideal gas mixture. A reactive collision may be defined as one which directly causes a local change in the molecular composition of the gas. If a reactive collision occurs, the colliding molecules do not retain their identity throughout the course of the collision and new species are formed at the microscopic level in the gas. Thus the possibility of this type of process occurring in a gas sample introduces a new complex of possible processes which must be accounted for in our macroscopic view of the gases' behavior.

There are three limit types of macroscopic behavior relative to reactive gases which can be rather sharply defined. These are (1) inert gas—all collisions are nonreactive; (2) nonequilibrium reactive gas—all collisions are not inert, and local composition changes occur which are not balanced macroscopically by reverse processes, causing the composition of the gas to change with time; and (3) equilibrium gas—reactive collisions occur but are balanced on the macroscopic scale by equal and opposite reverse processes so that, for constant external conditions, the composition of the gas is observed to remain unchanged with time. In case (3) there are two types of equilibrium systems which will be defined and discussed later, either restrictive or complete equilibrium systems.

Because of the microscopic and dynamic nature of the processes occurring in the last two categories, they admit to the existence of reactive species which can only be observed *in situ*. For example, one can imagine the possible reactive collision

$$M + H_2O \rightarrow M + H + OH \tag{4-1}$$

where M is any inert collision partner. As written this microscopic process is producing a neutral hydrogen atom and a neutral hydroxyl radical. These species are both highly reactive and therefore cannot be extracted from the mixture and stored in any ordinary sense. However, their occurrence in the reacting mixture contributes markedly to the systems macroscopic behavior and their presence cannot be ignored if we wish to obtain an accurate description of reactive processes.

Notice that Eq. 4-1 does not describe the only reactive process available for an $M + H_2O$ collision where M is completely inert. If we limit ourselves to products which are neutral species, we can also imagine the reactions

$$M + H_2O \rightarrow M + H_2 + O \tag{4-1a}$$

and

$$M + H_2O \rightarrow M + H + H + O \tag{4-1b}$$

However, there is no reason why a collision of this type should produce only neutral species, and we may consider the possibility that the products are charged particles, i.e.,

$$\begin{aligned} M + H_2O &\rightarrow M + H_2^+ + O^- \\ M + H_2O &\rightarrow M + H^+ + OH^- \\ M + H_2O &\rightarrow M + H^- + OH^+ \\ M + H_2O &\rightarrow M + H^+ + OH + e, \text{ etc.} \end{aligned} \tag{4-2}$$

where e is a free electron. In general, we observe that these ionization processes (Eq. 4-2) are much less likely to occur because the formation of ionized species requires more energy than processes involving only the separation of neutral fragments. Thus in most cases the first departure from inert collision behavior is a "chemical reaction" occurring through the formation and interaction of neutral fragments. Ionization processes are important at extremely high temperatures and will be discussed briefly in Sec. 7-8.

4-2. STOICHIOMETRY

The term *stoichiometry* comes from the Greek "stoicheion," meaning a first principle or element. The chemical use of the word is therefore quite apt because the word describes the requisite mass-balance criteria for reaction processes. For our discussion we restrict ourselves to a sample which does not lose or gain material by diffusion. Since we allow chemical reactions to occur, the mass conservation equations must require only that the total quantity of each atom in the mixture remain constant. Therefore let us consider one gram of a mixture of molecules containing c distinct elements as components with the ith element having the symbol ${}^iE$ and an atomic weight A_i: $i = 1, 2, \ldots, c$. Furthermore, let the mixture contain s species, the general mole formula of the jth specie being

$$^{1}E_{\xi_{1j}}\,^{2}E_{\xi_{2j}}\cdots\,^{c}E_{\xi_{cj}}$$

where the ξ_{ij} are all simple integers which may be zero. The molecular weight of the jth specie is therefore

$$M_j = \sum_{i=1}^{c} \xi_{ij} A_i$$

Thus in a system where chemical change is allowed there are c mass balance criteria which may be written

$$Y_{i_E} = \text{const.} = \sum_{j=1}^{s} \xi_{ij} y_j \frac{A_i}{M_j} \tag{4-3a}$$

where Y_{i_E} is the mass of the ith element and y_j is the mass of the jth specie present in one gram of the mixture. Notice that since each element has a distinct atomic weight, this weight may be divided out of each mass-balance equation (Eq. 4-3a) to yield the i equations

$$\frac{Y_{i_E}}{A_i} = N_{i_E} = \text{const.} = \sum_{j=1}^{s} \xi_{ij} \frac{y_j}{M_j} = \sum_{j=1}^{s} \xi_{ij} n_j \tag{4-3b}$$

where N_{i_E} is the number of gram-atoms of the ith element present in one gram of the mixture and the n_j's are the number of gram-moles of each species present in one gram of the mixture. Equations 4-3b have mixed units (moles/gm), however, these units are usually found to be most convenient for representing mass balance in reactive flow systems.

For numerical simplicity it is quite often advantageous to write the mass balance equation for a quantity of material different than one gram. For example, it is common in combustion problems to consider an air-fuel mixture as containing one mole of oxygen plus 3.76 moles of nitrogen plus x moles of fuel. The total number of grams for this case is then $32 + 3.76 \cdot 28 + x \cdot MW_{\text{fuel}} = \mathcal{G}$ where $\mathcal{G}$ is a constant throughout the calculation. If we define $\mathcal{H}_i$ as the number of gram-atoms of the ith element present in $\mathcal{G}$ grams of the mixture and η_j as the number of the gram-moles of each specie in $\mathcal{G}$ grams of the mixture, we obtain a third form for the mass-balance requirement.

$$\mathcal{G} N_{i_E} = \mathcal{H}_{i_E} = \text{const.} = \sum_{j=1}^{s} \xi_{ij}\, \mathcal{G} \frac{y_j}{M_j} = \sum_{j=1}^{s} \xi_{ij} \eta_j \tag{4-3c}$$

A set of general-process equations may be obtained by differentiating Eqs. 4-3. For the moment we write only the equations obtained by differentiating the set 4-3b:

$$0 = \sum_{j=1}^{s} \xi_{ij} dn_j; \quad i = 1, 2, \ldots, c \tag{4-4}$$

Equation 4-4 represents a constraint on the species during a chemical process. During changes in the specie composition of the mixture (i.e., during the course of any chemical reaction) the principle of conservation of mass simply requires that these equations be satisfied. Since the Eqs. 4-4 are a set of c equations in s unknowns, many linear parametric equations may be written to satisfy them and there are, in fact, many sets of $s - c$ linearly independent parametric equations which may be used to completely describe the reactive system. We will find that each of these parametric equations is equivalent to a balanced·stoichiometric equation as usually written by the chemist.

This may best be illustrated by means of an example. Consider a system containing the elements, carbon, C, oxygen, O, and hydrogen, H ($c = 3$), and the species carbon dioxide, CO_2, carbon monoxide, CO, hydrogen, H_2, hydrogen atoms, H, oxygen, O_2, oxygen atoms, O, hydroxyl radicals OH, and water vapor, H_2O ($s = 8$). The three independent mass-balance equations written on a mole per gram basis are:

$$\begin{aligned} N_C &= n_{CO} + n_{CO_2} \\ N_O &= 2n_{O_2} + n_{H_2O} + n_{CO} + 2n_{CO_2} + n_{OH} + n_O \\ N_H &= 2n_{H_2} + 2n_{H_2O} + n_H + n_{OH} \end{aligned} \tag{4-5}$$

These may be differentiated to yield

$$\begin{aligned} 0 &= dn_{CO} + dn_{CO_2} \\ 0 &= dn_{CO} + 2dn_{CO_2} + dn_{H_2O} + dn_{OH} + dn_O + 2dn_{O_2} \\ 0 &= 2dn_{H_2} + 2dn_{H_2O} + dn_H + dn_{OH} \end{aligned} \tag{4-6}$$

Equations 4-6 are satisfied by many different parametric coordinates (or reaction coordinates, λ_k). Consider, for example, the set

$$\begin{aligned} dn_{H_2O} &= -d\lambda_1 \\ dn_{H_2} &= d\lambda_1 \\ dn_{CO} &= -d\lambda_1 \\ dn_{CO_2} &= d\lambda_1 \end{aligned}$$

with the $d\lambda_1$ coefficients for the other species equal to zero. This set satisfies Eq. 4-6 as may be verified by substitution and it represents the balanced stoichiometric equation

$$H_2O + CO \rightarrow H_2 + CO_2$$

Notice that a reaction coordinate has the units of moles of forward reaction (as expressed by an accompanying stoichiometric equation) per mass unit of the system considered. Since, in this example the system contains one gram, the units on λ are moles of reaction per gram of mixture.

We may also choose the set

$$\begin{aligned} dn_{CO_2} &= -2d\lambda_2 \\ dn_{CO} &= 2d\lambda_2 \\ dn_{O_2} &= d\lambda_2 \end{aligned}$$

with all other $d\lambda_2$ coefficients equal to zero. This represents the balanced stoichiometric equation

$$2CO_2 + 2CO + O_2$$

Notice that we may make three other independent choices of balanced stoichiometric equations to yield five parameters which would completely specify the system.[1] Also notice that another choice, the stoichiometrically balanced equation

$$2H_2O \rightarrow 2H_2 + O_2$$

is not independent of the above two equations because, while we may write

$$\begin{aligned} dn_{H_2O} &= -2d\lambda_3 \\ dn_{H_2} &= 2d\lambda_3 \\ dn_{O_2} &= d\lambda_3 \end{aligned}$$

we find that $d\lambda_3$ may also be written as a linear combination of the first two reaction coordinates. That is,

$$\begin{aligned} dn_{H_2O} &= 2(-d\lambda_1) + (0d\lambda_2) = -2d\lambda_2 \\ dn_{H_2} &= 2(d\lambda_1) + (0d\lambda_2) = 2d\lambda_3 \\ dn_{O_2} &= 2(0d\lambda_1) + (d\lambda_2) = d\lambda_3 \\ dn_{CO} &= 2(-d\lambda_1) + (2d\lambda_2) = 0d\lambda_3 \\ dn_{CO_2} &= 2(d\lambda_1) + (-2d\lambda_2) = 0d\lambda_3 \end{aligned}$$

which yields

$$2d\lambda_1 + d\lambda_2 = d\lambda_3$$

The proof of dependence may also be obtained by adding the stoichiometric equations to obtain

$$\begin{array}{lll} 2[H_2O + CO & \rightarrow & H_2 + CO_2] \\ +[2CO_2 & \rightarrow & 2CO + O_2] \\ \hline 2H_2O & \rightarrow & 2H_2 + O_2 \end{array}$$

We have thus shown by example that balanced chemical equations written either in the usual chemist's notation or in the reaction co-

[1]This would represent an interesting and instructive exercise.

ordinate notation satisfy the general mass-balance requirements (Eqs. 4-3 or 4-4) for the system under consideration. Unfortunately, using these rather simple stoichiometric equations to discuss the composition of reactive systems presupposes a knowledge of (1) the species present in the system, and (2) the processes which produce the chemical change. For example, in case 2, Sec. 4-1, where the composition is allowed to change with time, all the individual reactive collision processes must be known and considered before stoichiometry can be used to discuss how the composition changes with time.

The application of a stoichiometric restriction, once the species are known and the process is defined, may best be illustrated with two simple examples. In the first example, consider one gram of a mixture initially containing $y_{H_{2_i}}$ gm of hydrogen, $y_{O_{2_i}}$ gm of oxygen, and $y_{H_2O_i}$ gm of water capable of reacting to either produce or destroy water with the simultaneous disappearance or appearance of hydrogen and oxygen. In the chemist notation, the reaction is

$$2H_2O \rightleftharpoons 2H_2 + O_2 \quad \text{(on a mole basis)}^2 \tag{4-7a}$$

or

$$36H_2O \rightleftharpoons 4H_2 + 32O_2 \quad \text{(on a gram basis)}^3 \tag{4-7b}$$

Since for this system the number of species s is equal to 3, all allowable composition changes may be represented by plotting the mass of the water molecules, the mass of the hydrogen molecules, and the mass of the oxygen molecules present in the fixed weight of reactive mixture as the three coordinate directions in a cartesian coordinate system. This is illustrated in Fig. 4-1. The plane in Fig. 4-1a, represents all allowable compositions in one gram of mixture and a point on the plane, or inside the triangular region of Fig. 4-1b, represents a fixed species composition. A line which represents the chemical reaction process expressed in Eq. 4-7b may be drawn through any point in this triangular region. The same representation is given on a mole/gram of mixture basis in Fig. 4-2. In some cases the mole/gram or mole/ $\mathcal{G}$ gram basis can be a more useful representation.

The term "stoichiometric mixture" as used in these figures is common in the combustion field. In carbon-hydrogen-oxygen-nitrogen combustion systems it is ordinarily defined as the mixture of fuel and oxidizer which has the correct initial composition to theoretically produce (only on the basis of stoichiometric considerations) a product mixture which contains only the species N_2, CO_2, and H_2O. Lean and rich mixtures are then defined as mixtures which contain either a deficiency or excess of fuel on the basis of this definition.

[2]It should be pointed out that this direct process is not physically realizable and is used here only as a simple example to illustrate the meaning of a stoichiometric restriction.

[3]Molecular weight has been rounded to the nearest gram for convenience.

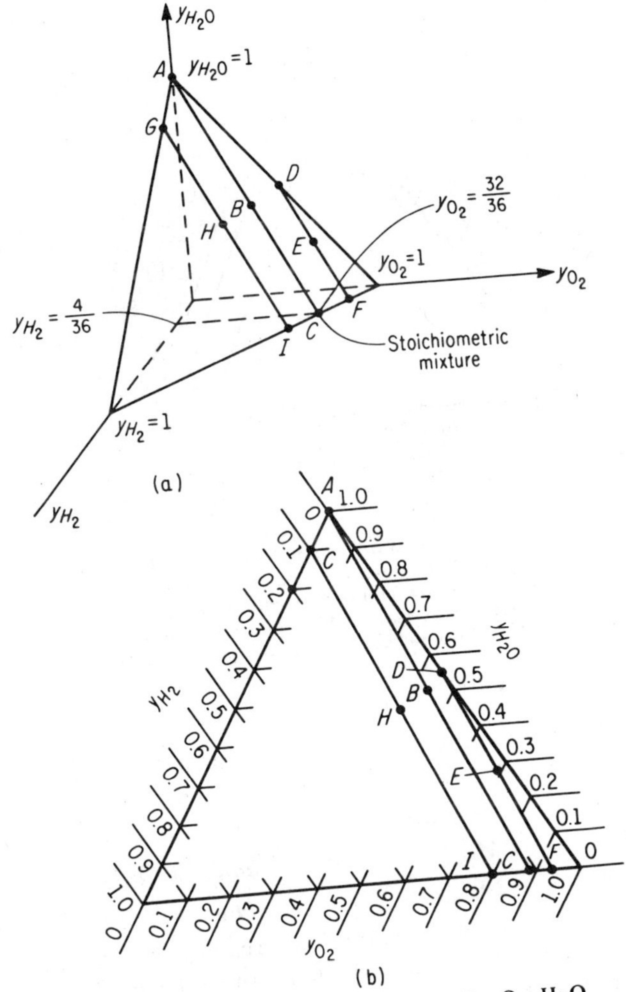

Fig. 4-1. Compositions of the H_2–O_2–H_2O system on a gram basis. (a) The plane in three dimensions. (b) The triangular coordinate representation. The line *ABC* represents the hydrogen-oxygen mixture which is stoichiometrically correct to produce water. All the straight lines are reaction coordinates (see Table 4-1).

Since for a single fuel the carbon-hydrogen ratio is fixed, we can ordinarily define a single parameter Φ for each mixture of any one fuel. We will define Φ = [(mass of fuel)/(mass of oxidizer)]/[(mass of fuel)/(mass of oxidizer)]$_s$, where the subscript *s* refers to a stoichiometric mixture. We call this the equivalence ratio of that combustible mixture. On the basis $\Phi < 1$ for lean mixtures and $\Phi > 1$ for rich mixtures.

*Symbols refer to points labeled on Fig. 4-2.
†Symbols refer to points labeled on Fig. 4-1.

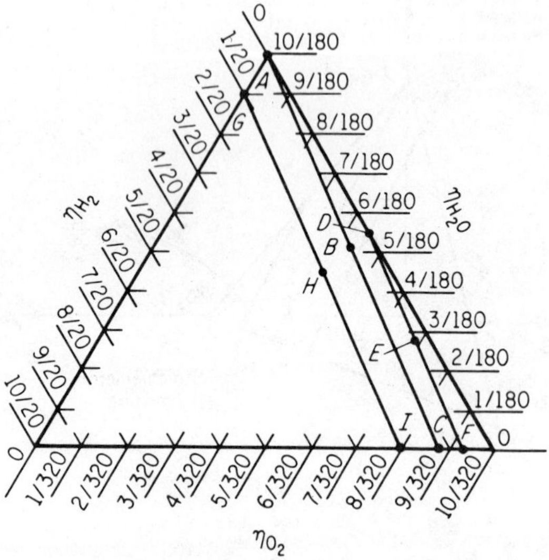

Fig. 4-2. Allowable compositions of H_2 + O_2 + H_2O mixtures on the mole/gm basis. Notice that the number of moles/gm is not constant in the plane. Also notice the similarity of the reaction coordinate lines in this figure and Fig. 4-1. This figure is keyed for comparison with Table 4-1.

These stoichiometric equations (4-7) may also be written more explicitly as progress equations using the reaction coordinate notation.

$$
\begin{aligned}
dy_{H_2O} &= -2M_{H_2O}\,d\lambda & \qquad & dn_{H_2O} = -2d\lambda \\
dy_{O_2} &= +M_{O_2}\,d\lambda & \text{or} \qquad & dn_{O_2} = +d\lambda \\
dy_{H_2} &= +2M_{H_2}\,d\lambda & \qquad & dn_{H_2} = +2d\lambda
\end{aligned}
\tag{4-8}
$$

where λ is the common progress variable or reaction coordinate for this one reaction and where y and n are the mass or moles of the subscripted species in one gram of the mixture. The reaction coordinate $d\lambda$ has the same units (moles of reaction/gram of mixture) in both representations.

We may also write the equation set in the η notation. For example, let $\mathcal{G}$ = mass of 3 moles of initial mixture irrespective of stoichiometry and define a new reaction coordinate $\alpha = \mathcal{G}\,\lambda$. With these definitions we obtain the equations

$$
\begin{aligned}
d\eta_{H_2O} &= -2\mathcal{G}\,d\lambda = -2d\alpha \\
d\eta_{O_2} &= +\mathcal{G}\,d\lambda = +d\alpha \\
d\eta_{H_2} &= +2\mathcal{G}\,d\lambda = +2d\alpha
\end{aligned}
$$

Note that the (η,α) representation is functionally identical to the (n,λ) representation. The (η,α) representation is quite often preferred for set-

and

$$Y_H = [y_{H_2O_i} + 2M_{H_2O}\lambda]\frac{M_{H_2}}{M_{H_2O}} + [y_{H_{2_i}} - 2M_{H_2}\lambda]$$

$$= y_{H_2O_i}\frac{M_{H_2}}{M_{H_2O}} + y_{H_{2_i}}$$

on the mass basis and

$$N_O = n_{H_2O} + 2n_{O_2}$$
$$= [n_{H_2O_i} + 2\lambda] + 2[n_{O_{2_i}} - \lambda]$$
$$= n_{H_2O_i} + 2n_{O_{2_i}}$$
$$N_H = 2n_{H_2O} + 2n_{H_2}$$
$$= 2[n_{H_2O_i} + 2\lambda] + 2[n_{H_{2_i}} - 2\lambda]$$
$$= 2n_{H_2O_i} + 2n_{H_2}$$

on the mole/gram basis. Detailed equations for the $\mathcal{G}$ basis are not presented because in this case the base is only chosen for convenience. However, recall the similarity of the (n,λ) and (η,α) representations. Notice that while stoichiometry requires that the total mass of the system be fixed, it does not require that the total number of moles remain fixed during the course of a reaction. A numerical example corresponding to Figs. 4-1 and 4-2 is presented in Table 4-1.

In the second example we will consider 48 gm of mixture which initially contains $\eta_{O_{3_i}}$ moles of ozone, $\eta_{O_{2_i}}$ moles of oxygen molecules and η_{O_i} moles of oxygen atoms, all capable of reacting. This system again contains three species but now contains only one component. Therefore, there is only one mass-balance equation which may be written

$$3 = 3\eta_{O_3} + 2\eta_{O_2} + \eta_O$$

Since we now have only one equation in three unknowns, two reaction coordinates are necessary to completely specify the system, each corresponding to a balanced stoichiometric equation for the system. However three stoichiometric equations are obviously available.[4]

$$2O \rightarrow O_2$$
$$3O \rightarrow O_3$$
$$3O_2 \rightarrow 2O_3$$

Since these three equations are linearly dependent, we may choose any two. We choose the first two equations which may be written as follows using the reaction coordinates α and β respectively.

[4]As in the previous example, these do not necessarily correspond to processes which occur in nature but are used here only to present a simple example of stoichiometric restrictions.

$2O \rightarrow O_2$	$3O \rightarrow O_3$
$d\eta_O = -2d\alpha$	$d\eta_O = -3d\beta$
$d\eta_{O_2} = d\alpha$	$d\eta_{O_2} = 0$
$d\eta_{O_3} = 0$	$d\eta_{O_3} = d\beta$

The above relationships are constructed on the assumption that each reaction is occurring independent of the others. However, the net change of composition is due to the net effect of both processes occurring and one therefore obtains the following stoichiometric relations for the reactions written in terms of the reaction coordinates.

$$\begin{aligned} d\eta_O &= 2d\alpha - 3d\beta \\ d\eta_{O_2} &= d\alpha \\ d\eta_{O_3} &= d\beta \end{aligned} \tag{4-9}$$

We now have an arbitrary choice for initial composition at our disposal. We choose the set $\eta_{O_i} = 3$, $\eta_{O_{2_i}} = 0$, $\eta_{O_{3_i}} = 0$ and assume that $\alpha = \beta = 0$ for this initial composition. Integration of the equations with these limits therefore yields the set

$$\begin{aligned} \eta_O &= 3 - 2\alpha - 3\beta \\ \eta_{O_2} &= \alpha \\ \eta_{O_3} &= \beta \end{aligned}$$

We note that the requirement that $\eta_j \geq 0$ means that in our notation

$$0 < \alpha < \frac{3}{2}(1 - \beta)$$

$$0 < \beta < 1 - \frac{2}{3}\alpha$$

A plot of the reaction coordinate behavior for this system is shown in Fig. 4-3. Notice that (1) since there is only one mass-balance equation, the entire triangular area is accessible to the system, (2) the limits on α and β are coupled, and (3) the simplicity of the equations $\eta_j(\alpha,\beta)$ is very dependent on the assumptions used in their derivation. However, every correctly constructed set will yield correct compositions, irrespective of its internal complexity.

These results may now be generalized. In the case of an inert system the composition is fixed and inviolate (if we disallow second-order transport processes such as thermal diffusion). Thus an inert system may be treated as a simple gas having an average molecular weight and a set of composite thermodynamic properties. However, in the case of a nonequilibrium reactive gas we must know the number of components c, the number of species s, and the stoichiometry of the individual reaction processes which determine the overall molecular composition changes—i.e., we must know the stoichiometric coefficients in the equations

$$dy_i = \sum_{j=1}^{k} \nu_{ij} M_i d\lambda_j; \qquad i = 1, 2, \ldots, s \tag{4-10a}$$

$$dn_i = \sum_{j=1}^{k} \nu_{ij} d\lambda_j; \qquad i = 1, 2, \ldots, s \tag{4-10b}$$

or

$$d\eta_i = \sum_{j=1}^{k} \nu_{ij} d\alpha_j; \qquad i = 1, 2, \ldots, s \tag{4-10c}$$

where there are k distinct reaction processes being considered, each with its own reaction coordinate λ_j or α_j. Note that the ν_{ij}'s are the coefficients that ordinarily appear in the balanced stoichiometric equations as written by the chemist, and that the Eqs. 4-10 are a generalization of the equation set 4-9 as it was written for the ozone–oxygen system. Here, if all the ν_{ij}'s for an ith specie are identically zero the specie is assumed to be inert in the system under consideration and if one or more of the ν_{ij}'s are not zero the equation represents the net production of that specie in terms of the net changes of the individual reaction coordinates λ_j or α_j. In addition, we should point out that a nonequilibrium reactive gas always

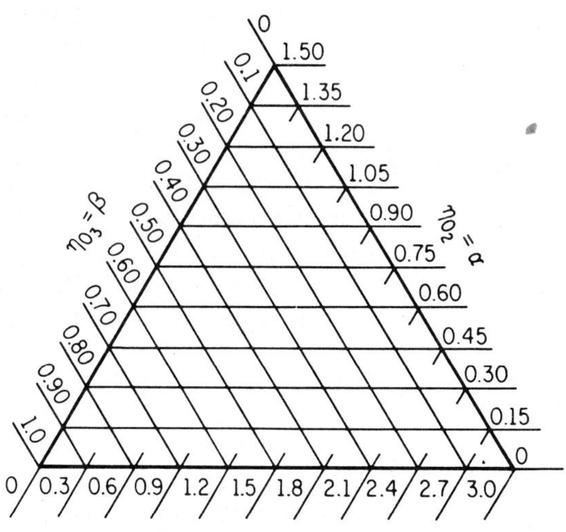

Fig. 4-3. Reaction coordinate diagram for the O_3–O_2–O system. Lines of constant α and β are shown. Note that all regions are accessible if α and β are both variables.

exhibits a net change in the molecular composition with time and we may therefore write Eq. 4-10b (for example) as

$$\frac{dn_i}{dt} = \sum_{j=1}^{k} \nu_{ij} \frac{d\lambda_j}{dt} \tag{4-11}$$

where the $d\lambda_j/dt$ are related to the reaction rates. Under the most advantageous conditions these may be directly evaluated for the individual reactive collision processes that are occurring in the gas sample. If this is possible, direct integration of the i equations will allow one to determine the composition of the mixture as a function of time. Also note that in a system containing two or more independent reaction coordinates it is generally true that their limit values are coupled in a complex manner.

In the case of an equilibrium gas mixture at any specified temperature or pressure, all the dn_i/dt's are zero. We are thus not directly concerned with the individual processes which occur in the gas because we may make an *a priori* specification that the sum effect of all the rate processes must lead to no net composition change. In fact, for equilibrium one may make this *a priori* assumption without proof or one may examine the known reaction processes to determine whether or not they may be reduced to $s-c$ linearly independent reaction coordinates. If this is either found or assumed to be possible, *any* set of $s-c$ linearly independent reaction coordinates may be chosen to discuss the equilibrium process. This is true because all stoichiometrically allowable compositions may be expressed in terms of any set of $s-c$ linearly independent reaction coordinates. The more general case of a reactive gas could also be reduced mathematically to a set of $s-c$ linearly independent rate equations of the form 4-11. However, the mathematical complication that this involves is usually avoided by writing each reaction as it occurs in nature.

In certain systems a detailed investigation of the reactive processes leads one to the conclusion that complete equilibrium is unattainable because the possible reactions do not reduce to $s-c$ linearly independent equations. In this case the system will be observed to relax to a restricted equilibrium composition. The stoichiometry of restricted equilibria is discussed in Sec. 4-10.

4-3. REACTION KINETICS

The quantitative study of chemical reactions, i.e., the evaluation of the $d\lambda_i/dt$ terms in Eq. 4-11, is called chemical kinetics or reaction kinetics. It involves the experimental determination and theoretical explanation of the rates of production or disappearance of chemical species during a chemical reaction and the effects of pressure, temperature, impurities, etc., on these rates. Reaction kinetics was first placed on its modern quantitative basis in 1884 when Van't Hoff stated the simple laws of uni-, bi-, and trimolecular reactions and predicted the exponential temperature dependence of reaction rate which was verified by Arrhenius in 1889. Van't Hoff's statements concerning the rate of formation or disappearance of a species during a chemical reaction are still valid (if slightly modified) today. Today we apply his definition of uni-, bi-, or trimolecular reactions, not to the overall reaction, but to the individual kinetic steps of a reaction process. In the following sections we will relate the occurrence of these steps to individual collision processes through the use of kinetic theory. However, we will first relate the chemists notation for reaction processes to the reaction coordinate notation and define a few terms.

The general form of the rate law preferred by the chemist is

$$\frac{d[A]}{dt} = k_n(T)\frac{\nu_A}{|\nu_A|}\prod_{i=1}^{n}[B_i] \tag{4-12}$$

where n is the reaction order, k_n the rate coefficient, and the symbol [ ] represents a concentration expressed in moles per unit volume (we choose cm^3). Strictly speaking this expression is valid only when applied to a rate measured in a constant-volume, constant-temperature environment. This is because the term on the left-hand side of this equation is expressed in a restrictive form in terms of the modern interpretation of reaction kinetic processes. The early chemists could only observe overall processes and their rate equations were usually empirical equations determined by fitting experimental data obtained in a specific reaction environment. Thus, the convention exists that the reaction rate is expressed in terms of the rate of change of concentration of a specific specie. However, as we will see the kinetic theory of gases yields a rather fundamental relation between the rate of change of a *reaction coordinate* and the *concentrations* of the reacting species.

In general (i.e., $V, T \neq$ constant) the rate of change of concentration of a single specie may be written as

$$\frac{d[A]}{dt} = n_A\frac{d\rho}{dt} + \rho\frac{dn_A}{dt}$$

However, substituting Eq. 4-11 we obtain

$$\frac{d[A]}{dt} = n_A\frac{d\rho}{dt} + \rho\sum_{j=1}^{k}\nu_{A_j}\frac{d\lambda_j}{dt} \tag{4-13}$$

where ν_{A_j} is the stoichiometric coefficient of the specie A in the jth chemical reaction and λ_j is the reaction coordinate of the jth reaction. The first term on the right-hand side of this equation is a contribution to concentration change due to density changes in the system. The reaction coordinates that occur in the second term are related to the concentration changes caused by the individual reactions as ordinarily reported by the chemist.

$$\frac{d\lambda_j}{dt} = \frac{1}{\rho\nu_{A_j}}\left[\frac{d[A]}{dt}\right]_{V,T,j} = \frac{1}{\rho|\nu_{A_j}|}\left[k_{n_j}(T)\prod_{i=1}^{n}[B_{ij}]\right]_{V,T,j} \qquad (4\text{-}14)$$

The subscript V,T,j in Eq. 4-14 implies (or at least should imply) that for the jth reaction the rate coefficient k_{n_j} was measured at constant volume and temperature. We will use the chemist's notation in our discussion primarily because it is the conventional notation. We now turn our attention to the definition of the elementary rate laws.

A unimolecular or first-order reaction is one where the rate of disappearance of a specie is proportional to the instantaneous concentration of that specie. The reaction is written as

$$A \longrightarrow \text{products}$$

and the rate is given by

$$\frac{d[A]}{dt} = -k_1[A]$$

At constant temperature and volume we may integrate this expression to obtain

$$k_1 = \frac{1}{t}\ln\frac{[A]_0}{[A]} \text{ sec}^{-1}$$

or

$$[A] = [A]_0 e^{-k_1 t}$$

Thus, the decay of a radioactive nucleus is a special case of a unimolecular reaction. The half-life (the time when half of the original material is gone) is independent of the initial concentration for a unimolecular reaction.

$$t_{1/2} = \frac{1}{k_1}\ln 2$$

In a similar fashion a bimolecular or second-order reaction is written as

$$A + B \longrightarrow \text{products}$$

and has a rate

$$\frac{d[A]}{dt} = \frac{d[B]}{dt} = -k_2[A]\cdot[B]$$

Since the concentration of A and B are not necessarily equal, this equation may be integrated for a constant-temperature, constant-volume process after writing $[A] = a - x$ and $[B] = b - x$ to obtain

$$\frac{dx}{dt} = k_2(a - x)(b - x)$$

This yields

$$k_2 t = \frac{1}{a - b} \ln \frac{b(a - x)}{a(b - x)} \quad \text{cm}^3/\text{mole}$$

or if $a = b$

$$k_2 = \frac{x}{ta(a - x)} \quad \text{cm}^3/\text{mole sec}$$

The half-life of this reaction is given by the expression

$$t_{1/2} = \frac{1}{ak_2}$$

A *termolecular,* or third-order, reaction is written as

$$A + B + C \rightarrow \text{products}$$

and the rate is given by

$$\frac{d[A]}{dt} = \frac{d[B]}{dt} = \frac{d[C]}{dt} = -k_3[A]\cdot[B]\cdot[C]$$

Once again, in general, $[A] \neq [B] \neq [C]$. For the case where they are equal, we can write $A = B = C = (a - x)$ and the equation becomes

$$\frac{dx}{dt} = k_3(a - x)^3$$

This may be integrated to yield

$$k_3 = \frac{1}{2t}\left[\frac{1}{(a - x)^2} - \frac{1}{a^2}\right] \text{cm}^6/\text{mole}^2\text{sec}$$

and the equation for half-life becomes

$$t_{1/2} = \frac{3}{2k_3 a^2}$$

Therefore, in general, the half-life of a reaction occurring at constant temperature and volume is

$$t_{1/2} \propto \frac{1}{a^{(n-1)}}$$

where n is the order of the reaction.

The effect of temperature on the rate constant k of simple first- and second-order reactions is given quite accurately by the Arrhenius equation

$$k_i = A_i e^{-E_i/RT} \tag{4-15}$$

where E_i is called the activation energy and A_i is the frequency factor and has the same units as k_i. For most third order reactions the temperature dependence is found to be very small. It may usually be expressed by the relation

$$k_3 = BT^{\zeta}$$

where $$\zeta \cong -1.$$

In the next sections we will relate these elementary macroscopic rate equations to the collision processes which must be responsible for the macroscopic behavior of the system.

4-4. REACTIVE COLLISION POTENTIALS—ACTIVATION ENERGY

Our kinetic theory concept of a gas as an aggregate of colliding molecules separated on the average by large distances leads us immediately to the observation that any measurable reaction rate must be the result of relatively infrequently occurring "reactive collisions." For example, consider a mixture at room temperature and pressure containing an equal number of molecules of gas A and B which are reactive ($A + B \rightarrow$ products). The kinetic theory which we developed in Chapter 2 states that one molecule of A will collide with a molecule of B approximately 10^9 times per second. Thus, if each collision led to reaction, the process would be completed in an extremely short time. Therefore we can assume that even rapid reactions (half-lives of about 10^{-4} sec) must be occurring only on extremely infrequent collisions between the reactive species. This fact allows us to consider a reacting gas mixture as being slightly perturbed from thermodynamic (but not chemical) equilibrium at all times during the course of the reaction. There are some limitations to this approach but they will not concern us in the present section.

In order to discuss the reasons for the apparent inertness of most collisions between inherently reactive molecules, we must look at some of the details of the collision process for an encounter of reactive species. We do this by considering the collision potential for an encounter. A typical collision potential for unreactive spherical molecules is given by the Lennard-Jones potential discussed in Chapter 3. In the case of reactive collisions we are interested mainly in the close approach portion of the potential because the slowness of reactions indicates that highly energetic collisions are necessary before chemical changes will occur. However, reactive collision potentials cannot be constructed easily for a general collision of complicated molecules. This is true because orientation and molecular distortion are important in describing a reactive collision. Nevertheless, reactive collision potentials have been constructed for a few simple collision processes, and the behavior of these potential curves leads to a very useful generalization concerning reaction processes.

If we consider the elementary reaction

$$H + H_2 \rightarrow H_2 + H$$

as a three-body problem (3H atoms) we can calculate the potential energy of this system for any three separation distances r_1, r_2, r_3 (see Fig. 4-4.)

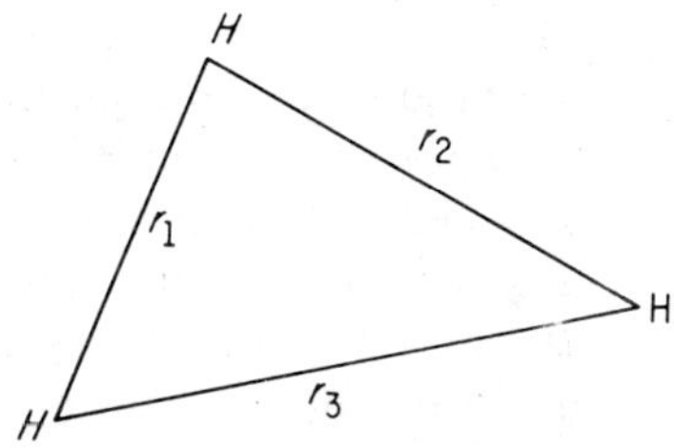

Fig. 4-4. Typical configuration for calculating an interaction potential.

Unfortunately, this extremely simple process is not sufficiently simple to allow a convenient graphical representation of the potential energy as a function of the configuration. However, the potential energy for the two-body problem consisting of a fixed hydrogen molecule (i.e., r_1 fixed) and a neighboring hydrogen atom has been calculated and is shown in Fig. 4-5. The most interesting feature of this potential energy diagram is the potential well which is observed at the end of the fixed hydrogen molecule. This peculiarity in the potential field leads one to the conclusion that a chemical reaction is most likely to occur during an in-line collision of H_2 and H. Fortunately, the potential energy for such an in-line collision can be represented graphically because it is a two-parameter problem. A plot

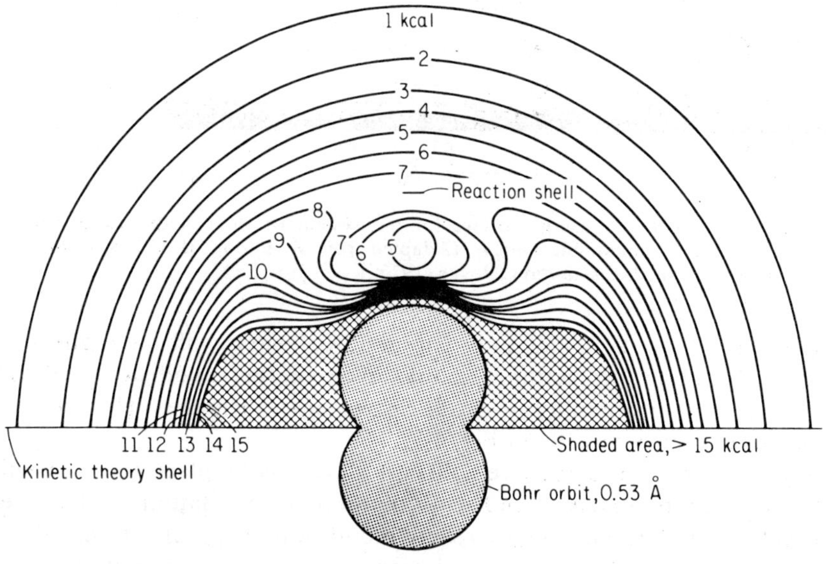

Fig. 4-5. Potential-energy contours for approach of a hydrogen atom to a rigid hydrogen molecule. (Adapted from J. O. Hirschfelder, H. Eyring, and B. Topley, *J. Chem. Phys.*, *4*, 170 (1936).)

of the potential energy for the case of an in-line collision is shown in Fig. 4-6. Notice the existence of a saddle point on this plot at $r_1 = r_2$. The restricted (in-line) collision behavior of a hydrogen molecule and hydrogen atom may be discussed using the potential energy surface (Fig. 4-6) by following the motion, on this surface, of a mass point in a gravitational field with an initial potential energy and arbitrary directed velocity. Approaching the saddle point from either r_1 or $r_2 \simeq 0.5$ Å represents a $H_2 + H$ collision process and leaving the potential well in the r_2 or r_1 direction respectively represents the occurrence of a reactive collision.

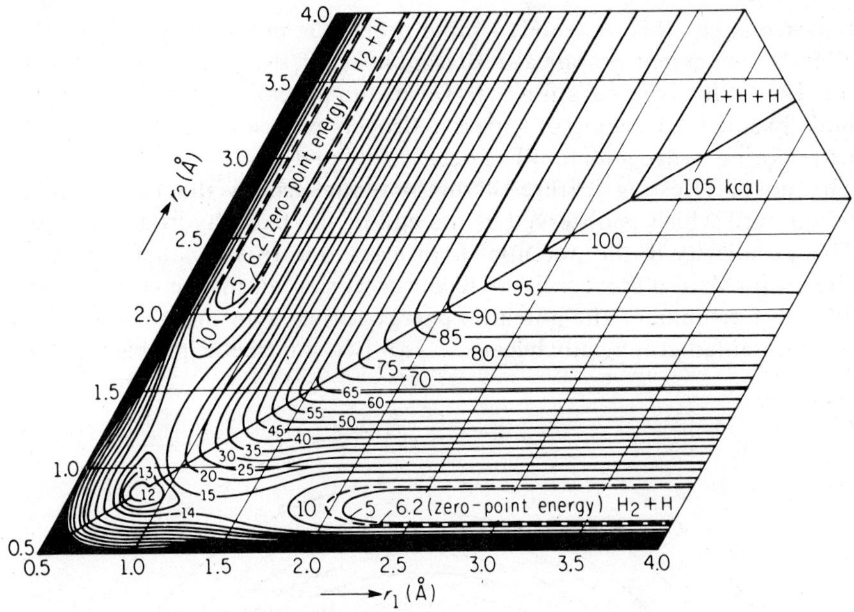

Fig. 4-6. Potential-energy surface for the system of three hydrogen atoms based on 14% coulombic energy. (Adapted from H. Eyring, H. Gershinowitz, and C. E. Sun, *J. Chem. Phys.*, *3*, 786 (1935).)

Thus Fig. 4-6 predicts that only energetic collisions will have any probability of leading to reaction and that even energetic collisions will only exhibit a finite probability of producing chemical change. This behavior is illustrated schematically in Fig. 4-7.

This concept of a reactive collision allows us to place the Arrhenius equation on a relatively sound semiquantitative foundation. All we need require is that the number of reactive collisions be proportional to the total number of collisions with an intermolecular kinetic energy above the critical collision energy for a particular reactive process. We also must

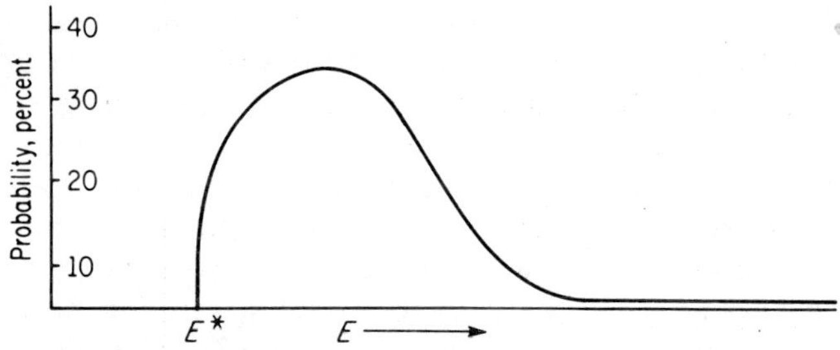

Fig. 4-7. Probability of reactive collision vs. collision energy (schematic).

assume that this critical energy for reaction is high when compared to the average kinetic energy of the molecules so that only a few collisions will yield products. If these assumptions are true, we note that we can replace the relative collisional kinetic energy by the kinetic energy of one of the molecules relative to the gas sample as a whole without greatly altering the results. Let us therefore calculate the fraction of the total number of molecules in an equilibrium gas at temperature T having a translational kinetic energy above a certain value E^*, where $E^* = \frac{1}{2} mc^{*2}$. This fraction may be determined by integrating Eq. 2-14 from E^* to ∞. Replacing $\eta(c)dc$ by dn/n_T where n = number of molecules with velocity c to $c + dc$ and n_T = total number of molecules, we obtain

$$\frac{dn}{n_T} = 4\pi\left(\frac{M}{2\pi RT}\right)^{3/2} \exp\left(\frac{Mc^2}{2RT}\right)c^2\,dc$$

or

$$\frac{n^*}{n_T} = \int_{E^*}^{\infty} \frac{2}{\sqrt{\pi}} \frac{1}{RT^{3/2}} e^{-E/RT} E^{1/2}\,dE$$

$$= \frac{2}{\sqrt{\pi}}\left(\frac{E^*}{RT}\right)^{1/2} e^{-E^*/RT}\left[1 + \left(\frac{RT}{2E^*}\right) - \left(\frac{RT}{2E^*}\right)^2 + \cdots\right]$$

which yields

$$\frac{n^*}{n_T} \equiv \frac{2}{\sqrt{\pi}}\left(\frac{E^*}{RT}\right)^{1/2} e^{-E^*/RT}$$

However, since the exponential factor varies extremely rapidly with temperature and the $T^{-1/2}$ factor varies only slowly, as a first approximation the number of molecules above an energy E^* may be written as

$$\frac{n^*}{n_T} \propto e^{-E^*/RT}$$

Therefore this simple concept yields a functional temperature dependence which is in good agreement with the experimentally observable Arrhenius equation. It also presents the concept of a unique activation energy for each collision pair and the concept of an activated complex as a loose molecular aggregate which has some identity during the reactive-collision process. Unfortunately, the experimental study of individual reactive collisions has only recently become practical and the theoretical explanation of macroscopic reaction kinetics in terms of a detailed discussion of reactive-collision processes has yet to yield meaningful activation energies or rate constants for all but the simplest reaction processes. In other words, reaction kinetics remains an experimental science.

A further consideration of the energetics of a reactive collision allows us to discuss limits on the magnitude of the activation energy in some detail. Consider the reactive-collision process illustrated in Fig. 4-8:

$$XY + Z \rightarrow X + YZ \tag{4-16}$$

which destroys the molecule XY and produces in its stead the molecule YZ. Each of these molecules has a dissociation energy or heat of formation and in general the process represented by Eq. 4-16 is not thermally neutral. Referring again to Fig. 4-8 we will assume, for discussion, that the lowest possible potential energy of the system $XY + Z$ is 5 arbitrary units above the lowest possible potential energy of the system $X + YZ$. In other words, if the dissociation energy of molecule $XY = 65$ units, the dissociation energy of molecule $YZ = 70$ units and the plateau at large r_1 and r_2 is at a height of 70 units on the diagram. This means that occurrence of this reaction will cause the liberation or absorption of heat energy from the system in which the reaction is occurring. On the microscopic scale, each reactive collision will either produce or absorb some kinetic energy from the collision partners, since the collision is conservative overall and the product molecules will have a new potential energy associated with their bond structure.

We can also discuss this process on a molar basis or on a gram basis. On a molar basis, the energy absorbed from the system at constant temperature and presssure by the process (Eq. 4-16) is called the *enthalpy of reaction*, ΔH. Since we are interested in the relative energy or enthalpy storage in different compounds, we need only tabulate the energy or enthalpy of each compound relative to a set of reference compounds—as long as we use a self-consistent and linearly independent set of reference compounds. The most universally accepted set of standard reference compounds are the elements in their most stable form at one atmosphere pressure and the temperature of interest. If their most stable form is gaseous, the state is assumed to be the ideal gaseous state because of the convenience of calculating ideal gaseous properties using statistical thermodynamics. Thus each molecule or atom has a molar enthalpy of formation which is a function of temperature (but not of pressure because

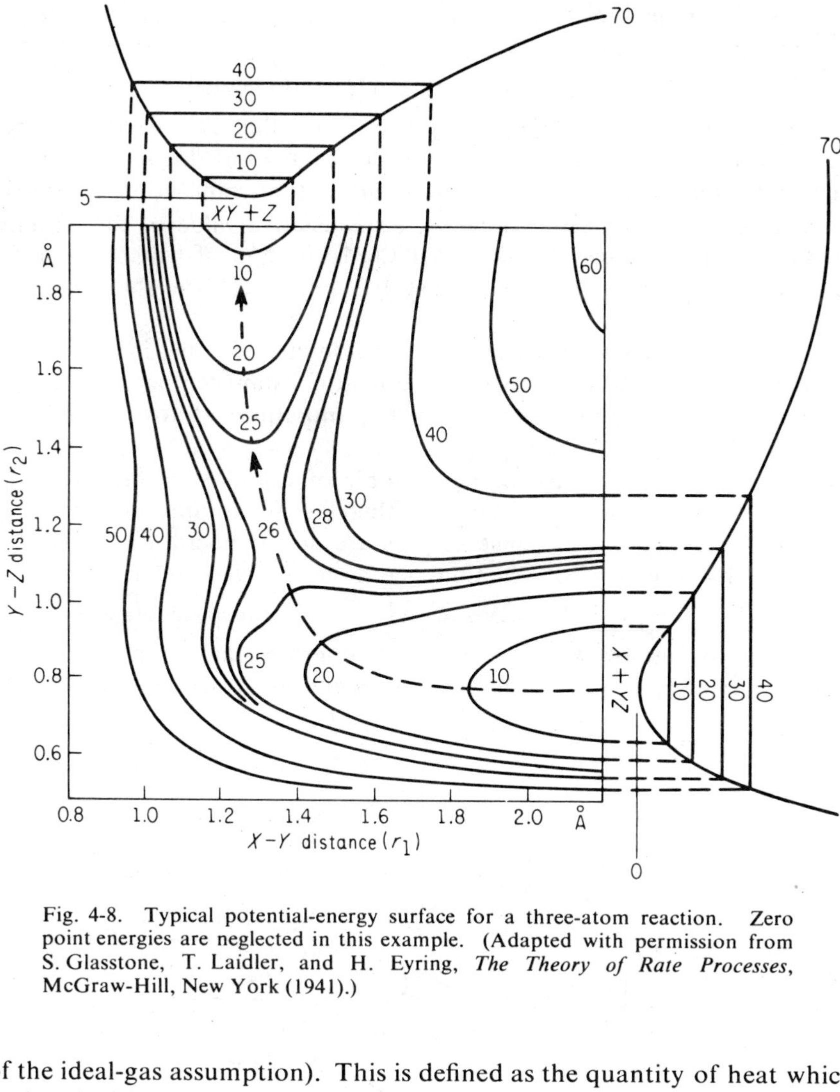

Fig. 4-8. Typical potential-energy surface for a three-atom reaction. Zero point energies are neglected in this example. (Adapted with permission from S. Glasstone, T. Laidler, and H. Eyring, *The Theory of Rate Processes*, McGraw-Hill, New York (1941).)

of the ideal-gas assumption). This is defined as the quantity of heat which is absorbed from the system at constant temperature and pressure during the production of one gram-molecular weight of that substance in the desired state at that temperature. For example, the enthalpy of formation of water vapor $H_2O(g)$ at 25° is written as

$$H_2(g) + \tfrac{1}{2}\,O_2(g) \rightarrow H_2O(g); \quad \Delta H_f^{25^\circ C} = -57.85 \text{ kcal}$$

whereas the enthalpy of formation of liquid water $H_2O(1)$ at 25°C and one atmosphere pressure is written as

$$H_2(g) + \tfrac{1}{2}O_2(g) \rightarrow H_2O(1); \quad \Delta H_f^{25^\circ C} = -68.40 \text{ kcal}$$

and the difference between these two equations yields the enthalpy of vaporization of water at 25°C

$$H_2O(l) \rightarrow H_2O(g); \quad \Delta H = 10.55 \text{ kcal}$$

The positive sign for ΔH in this case indicates that the reaction (vaporization) is an endothermic process (i.e., heat must be added to the system from an external source in order to maintain the temperature during this constant-pressure process) while the large negative value of the enthalpy of formation of water indicates that the formation of water is a highly exothermic process (i.e., heat must be removed from the system to maintain the temperature).

Using these simple rules and the thermodynamic tables in Appendix F we may calculate the enthalpy of reaction at constant temperature for any process by simply adding or subtracting enthalpies of formation of all substances involved in the process.

Returning now to our simple example of Fig. 4-8 we first note that $\Delta H = \Delta E$ for the reaction because there is no volume change if the reaction is carried out at constant pressure and temperature. Therefore we find that the heat of reaction per mole is given by the relation

$$XY + Z \rightarrow X + YZ \quad \Delta H = \Delta E = -5 \text{ arbitrary energy units}$$

Referring to Fig. 4-8, we see that this exothermic reaction has an activation energy of $E_{act} = 21$, while the reverse endothermic reaction will have

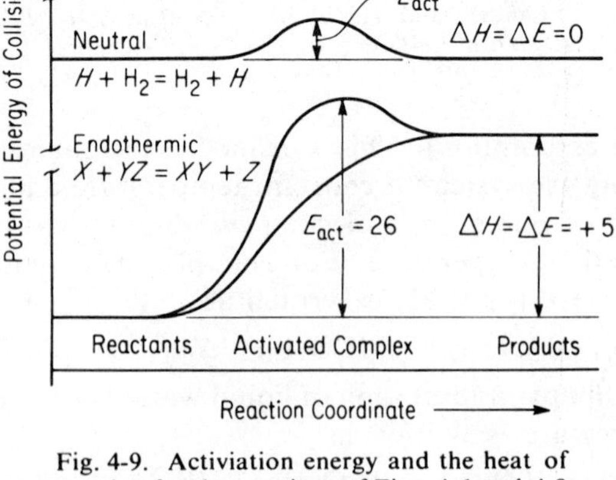

Fig. 4-9. Activiation energy and the heat of reaction for the reactions of Figs. 4-6 and 4-8 (not to scale).

an activation energy of $E_{act} = 26$ arbitrary units. This is summarized in Fig. 4-9.

There are many reactions which are either highly exothermic or highly endothermic and in these cases $\Delta E \cong \Delta H \cong \mathfrak{N}\Delta\phi$ where $\Delta\phi$ is the potential energy change in a single reactive encounter. Thus we note from Fig. 4-9 that every endothermic reaction must have an activation energy at least as great as its endothermicity, whereas highly exothermic reactions can have extremely low activation energies and their behavior is not predictable from the ΔH of the reaction.

4-5. THE COLLISION CONCEPT AND KINETIC ORDER

It is quite easy to see how collision-determined reactions can yield a second-order rate dependence at constant temperature and volume. Let us consider molecules A and B which are capable of reacting to produce products

$$A + B \rightarrow \text{products}$$

The reactive collision concept yields the conclusion that the rate of this reaction is proportional (at any one temperature) to the total number of collisions of molecules A with molecules B. However, the total number of available collisions is simply proportional to the product of the number density of molecules A multiplied by the number density of molecules B, and these number densities are proportional to concentrations. Thus, second-order rate constants arise as a direct consequence of the collision determined nature of a reaction process.

True first-order rate constants are observed experimentally only for a very specific type of decomposition reaction—one where the molecule is capable of storing a considerable quantity of energy internally as vibrational energy and then decomposing when this vibrational energy resides in a weak bond. In essence, this reaction becomes first-order because sufficient energy for dissociation can be stored for an appreciable length of time in the molecule before spontaneously causing decomposition. The stoichiometric rate equation is written as

$$A + A \underset{k_2}{\overset{k_1}{\rightleftharpoons}} A + A^* \xrightarrow{k_3} \text{products}$$

where three separate and distinct processes are occurring. Two of these are collision determined. These are (1) activation by collision

$$\left(\frac{d[A^*]}{dt}\right)_1 = k_1 [A]^2$$

and (2) deactivation by collision

$$\left(\frac{d[A^*]}{dt}\right)_2 = -k_2[A]\cdot[A^*]$$

The third process is the spontaneous decomposition of the activated species and is therefore first-order:

$$\left(\frac{d[A^*]}{dt}\right)_3 = -k_3[A^*]$$

In these simultaneous processes we cannot, of course, measure either $[A^*]$ or $d[A^*]/dt$. However, we see that

$$\frac{d[\text{Prod}]}{dt} = k_3[A^*]$$

and, therefore knowledge of $[A^*]$ will allow us to calculate the rate of production of products. We note that three reactions are competing for either the production or destruction of the activated species A^* and we may therefore write that

$$\frac{d[A^*]}{dt} = k_1[A]^2 - k_2[A]\cdot[A^*] - k_3[A^*]$$

As a first approximation this equation may be solved by assuming that the composite rate of production of A^* during the course of the reaction is small compared to the rate of the three competing steps. This is called the *steady-state approximation* of chemical kinetics and is reasonably valid for most simple decomposition reactions. With this assumption we obtain

$$[A^*] = \frac{k_1[A]^2}{k_2[A] + k_3}$$

or

$$\frac{d[\text{Prod}]}{dt} = \frac{k_1[A]^2}{\frac{k_2}{k_3}[A] + 1}$$

Let us examine the denominator of this equation more closely. Since the $[A]$ term depends on the concentration of A, this term should be pressure dependent. Thus at high pressures $\frac{k_2}{k_3}[A] > 1$ and

$$\frac{d[\text{Prod}]}{dt} = \frac{k_3k_1}{k_2}[A]$$

which is a first-order rate equation. However, at low pressure $\frac{k_2}{k_3}[A] < 1$ and our kinetic equation becomes second-order in A

$$\frac{d[\text{Prod}]}{dt} = k_1[A]^2$$

Thus the collision theory of first-order processes predicts that the reaction should behave as a second-order process at low pressures where the collision deactivation process is slower than the spontaneous decomposition process and should be first-order at high pressure where the reverse is true. This is observed experimentally. The gas-phase decomposition of a complicated molecule is second-order at low pressures and slowly becomes first-order as the pressure increases to the point where collisional deactivation is important.

Third-order rates are observed for recombination reactions where the species which is formed cannot conveniently store the heat of formation internally. Consider the collision of two hydrogen atoms. The potential curve for this process is given by Fig. 1-7 where the energy of a stable molecule is represented by the energy level of one of the vibrational levels. However, on this plot the energy of the collision pair can be represented by a horizontal line well above the dissociation energy of the molecule. Thus, for the collision of two isolated hydrogen atoms there is simply no way to remove energy[5] and allow a stable configuration to form. However, if a third body is present at the moment of collision we may write the equation

$$\mathrm{H} + \mathrm{H} + M \rightarrow M + \mathrm{H_2}$$

where the heat of reaction appears as recoil velocity of the M and H_2 species. Thus recombinations occur mainly through three-body collision processes where the third body is any available species in the gas mixture. Characteristically, these processes are slow because three-body collisions are infrequent. Notice that our collision concept shows that the possibility of forward reaction for this process should not be greatly dependent on the relative collision energy because it is a highly exothermic reaction. Thus we expect to find only a weak temperature dependence for atom recombination, which is confirmed by experiment. In fact, experimental evidence shows that most elementary recombination processes exhibit a rate which decreases slightly as temperature increases.

In addition to the gas-phase collisions that yield chemical reaction, there can be wall collisions which lead to composition changes in the gas. This is particularly true because wall collisions are usually "sticky" in the sense that the molecule resides in the neighborhood of the wall for a relatively long time and therefore has an increased probability of contacting another reactive molecule. The occurrence of wall reactions greatly complicated the work of the earlier kineticists, since they were restricted to work on relatively slow reactions in small chambers. Thus almost every molecule in the system had an opportunity to view the wall during the course of the reaction and the investigator was required to study wall reactions as an important part of the overall process. Wall

[5]Except by the emission of radiation.

reactions, if relatively rapid, can lead to a diffusion controlled rate, but if sufficiently slow simply yield a reaction rate which is proportional to the internal surface area of the container. Recombination processes, which are by nature three-body processes in the gas, can conveniently occur on the wall of the vessel because the wall can act as a requisite heat sink for the reaction. Since the number of wall collisions (per square centimeter) increases linearly with the gas pressure but the number of tertiary collisions increases as the square of the pressure, wall collisions become more important for recombination reactions as the pressure is lowered.

4-6. CHAIN REACTIONS

After Van't Hoff introduced his postulates concerning order and activation energy, an attempt was made to force or at least to catalogue all reaction schemes in accordance with these elementary mechanisms. However, only a few reactions were observed whose overall rate fitted the postulate of integral order, and detailed studies of these reactions always showed them to be somewhat complex. Between 1928 and 1935 the results of experimental rate determinations were found to fit the first-, second-, and third-order postulates of Van't Hoff if one assumed that the laws applied only to individual steps and that chain-reaction mechanisms were important. Briefly, a chain reaction consists of a series of independent, elementary (usually bimolecular) reactions which cause relatively rapid conversion of reactants to products through the use of regenerated intermediates. A typical example of a reaction which may occur through a chain mechanism is the reaction of a halogen with hydrogen to form the appropriate halogen acid. The overall process stated in terms of stable species is

$$H_2 + X_2 \rightarrow 2HX$$

where X represents either F, Cl, Br, or I. If we allow the unstable atoms H and X to exist in the reacting mixture we can obtain a chain process. In this case we have a system which contains five species. Let us discuss the prognosis of each possible binary-collision process in this system to determine how chain processes originate. Since there are five species, there are fifteen possible binary-collision processes which we must consider. Three of these are atom-atom collisions and therefore must be third-order processes if they are to yield products.

1. $H + H + M \rightarrow H_2 + M$
2. $H + X + M \rightarrow HX + M$
3. $X + X + M \rightarrow X_2 + M$

These three-body recombinations occur only slowly and tend to remove reactive species from the mixture. Six of the other possible binary-collision processes occur between an atom and a diatomic molecule. These may be listed with their possible results

4. $H + H_2 \rightarrow H + H + H$ (energetically difficult)
   $\searrow H_2 + H$ (no change)
5. $H + X_2 \rightarrow HX + X$ (possible process)
   $\searrow H + X + X$ (formation of halogen atoms)
6. $H + HX \rightarrow H_2 + X$ (possible process)
   $\searrow H + H + X$ (energetically difficult)
7. $X + H_2 \rightarrow HX + H$ (possible process)
   $\searrow X + H + H$ (energetically difficult)
8. $X + X_2 \rightarrow X + X + X$ (formation of halogen atoms)
   $\searrow X_2 + X$ (no change)
9. $X + HX \rightarrow H + X_2$ (energetically more difficult than 5a, 6a, and 7a).
   $\searrow X + X + H$ (energetically difficult)

These six distinct collision processes therefore yield three possible reaction paths and two dissociation paths on the basis of energetics. Notice that the only dissociation process that is being considered is the collisional dissociation of the halogen molecule. Appendix C shows that the halogen molecule has by far the lowest dissociation energy of the three molecules involved in this system, and therefore its collisional dissociation is more likely than dissociation of either a hydrogen molecule or a halogen acid molecule. Process 9a is more unlikely than either process 5a, 6a, or 7a because, in this case, we are decomposing a molecule with a relatively strong bond and forming a molecule with a realtively weak bond. Therefore, since this reaction must be highly endothermic, it must have a high activation energy and a low probability of occurrence.

The last six possible collision pairs are molecule–molecule collisions and therefore offer more possible paths.

10. $H_2 + H_2 \rightarrow H + H + H_2$ (energetically difficult)
    $\searrow H + H + H + H$ (energetically extremely difficult)
11. $H_2 + X_2 \rightarrow HX + HX$ (direct conversion)
    $\searrow H_2 + X + X$ (formation of halogen atoms)
    $\searrow H + H + X_2$ (energetically difficult)
    $\searrow HX + H + X$ (energetically difficult)
    $\searrow H + H + X + X$ (energetically extremely difficult)
12. $H_2 + HX \rightarrow HX + H_2$ (no change)
    $\searrow H + H + HX$ (energetically difficult)
    $\searrow H + X + H_2$ (energetically difficult)
    $\searrow H + H + H + X$ (energetically extremely difficult)

13. $X_2 + X_2 \rightarrow X + X + X_2$ (formation of halogen atoms)
$\searrow X + X + X + X$ (energetically difficult)

14. $X_2 + HX \rightarrow X + X + HX$ (formation of halogen atoms)
$\searrow X_2 + H + X$ (energetically difficult)
$\searrow X + X + H + X$ (energetically extremely difficult)
$\searrow HX + X_2$ (no effect)

15. $HX + HX \rightarrow H_2 + X_2$ (direct dissociation)
$\searrow H_2 + X + X$ (formation of halogen atoms)
$\searrow HX + H + X$ (energetically difficult)
$\searrow H + X + H + X$ (energetically extremely difficult)

Notice that, of this large number of possible processes arising from molecule–molecule collisions, we are left with only halogen atom formation (processes 11b, 13a, 14a, 15b), direct conversion (process 11a), or direct dissociation (process 15a) as possible reactive processes. Thus the fifteen conceivable collisions result in the following set of possible kinetic processes for the reaction

$$X_2 + M \xrightarrow{k_1} 2X + M \text{ (chain initiation)}$$

$$\left.\begin{aligned} X + H_2 &\xrightarrow{k_2} HX + H \text{ (reaction 7a)} \\ H + X_2 &\xrightarrow{k_3} HX + X \text{ (reaction 5a)} \\ H + HX &\xrightarrow{k_4} X + H_2 \text{ (reaction 6a)} \end{aligned}\right\} \text{(chain reaction)} \qquad (4\text{-}17a)$$

$$M + 2X \xrightarrow{k_5} X_2 + M \text{ (three-body termination or deactivation)}$$

or the alternate route represented by the equation

$$H_2 + X_2 \xrightarrow{k_6} 2HX$$

The chain process (Eq. 4-17a) may be analyzed using the steady-state assumption by setting the overall rates $d[X]/dt$ and $d[H]/dt = 0$ relative to their rate of production and disappearance in the individual reaction steps. This produces a kinetic rate equation for the formation of halogen acid which has the form

$$\frac{d[HX]}{dt} = \frac{2k_2\left(\dfrac{k_1}{k_5}\right)^{1/2}[H_2][X_2]^{1/2}}{1 + \dfrac{k_4}{k_3}\dfrac{[HX]}{[X_2]}} \qquad (4\text{-}17b)$$

The simple direct combination reaction yields a rate expression of the form

$$\frac{d[HX]}{dt} = k_6[H_2]\cdot[X_2] \qquad (4\text{-}18)$$

The actual experimental data for this reaction for the halogens F_2, Cl_2, Br_2, and I_2 shows typically how the energetics of the individual collision processes can cause the character of the reaction mechanism to change. Fluorine reacts spontaneously and rapidly with hydrogen to form hydrogen fluoride at all temperatures in the presence of traces of moisture. Since the reaction is excessively rapid it has not been studied in detail. A chlorine-hydrogen mixture is stable at room temperature *in the dark*. However, a ray of light causes an extremely rapid reaction to occur and to continue for a short time after the light is removed. This process may be explained if we assume that chlorine molecules can be photodissociated and that the chain length of the reaction (i.e., the average number of chain cycles per chlorine atom before it is collisionally deactivated) is large. Chain lengths of 10^6 have been measured for the $H_2 + Cl_2$ reaction. At high temperatures the $H_2 + Cl_2$ reaction is rapid and cannot be studied. The hydrogen-bromine reaction proceeds relatively slowly at temperatures slightly above room temperature and exhibits an experimental rate behavior which agrees with Eq. 4-17b. This reaction has been studied extensively and it exhibits a chain length of about 3,000. The inhibition of the rate of formation of products by the presence of the product is a typical kinetic behavior for chain reactions. In the case of the hydrogen–iodine reaction, the formation of hydrogen iodide appears to be primarily bimolecular in $H_2 + I_2$ but recent experimental measurements[6] have shown that it may follow the steps

$$2I + H_2 \rightarrow 2HI$$

or

$$M + I + H_2 \rightarrow IH_2 + M$$

followed by

$$IH_2 + I \rightarrow 2HI$$

these steps are kinetically undistinguishable from the direct process in thermal systems since iodine dissociation is rapid.

It is interesting to comment on the change in nature of the hydrogen–halogen reactions as we go from halogen to halogen. In the case of bromine, the reaction 7a is known to be the slow reaction in the chain, since it is a relatively endothermic reaction and therefore must have a rather high activation energy. In the chlorine system this reaction is rather less endothermic and the chain length becomes quite long because of the rapidity of the cycle. However, in the case of iodine, step 7a becomes prohibitively slow and the majority of the reaction occurs by the direct processes described above. Thus we have a set of rather semiquantitative reasons for the experimentally observed behavior of the hydrogen–

[6] J. H. Sullivan, *J. Chem. Phys.*, **46**, 73 (1967). Notice that the specie IH_2 was not considered in our discussion.

halogen reactions based on both the existence of reactive species in the mixture during the reaction and on the qualitative energetics of the possible reactive collision processes.

There is another type of chain reaction which is extremely important in combustion processes. This is the branching-chain reaction. There are certain systems capable of reacting with a chain mechanism but which have the added feature that each cycle of the chain produces a net increase in the number of reactive chain carriers (either radicals or atoms). These "branching-chain" reactions can be completely uncontrollable because radical concentrations tend to rise exponentially with time. A typical example is the hydrogen–oxygen branching-chain explosion

$$H_2 + M \rightarrow 2H + M \qquad \text{(initiating)}$$

$$\left.\begin{aligned} H + O_2 &\rightarrow OH + O \\ O + H_2 &\rightarrow OH + H \\ OH + H_2 &\rightarrow H_2O + H \end{aligned}\right\} \text{branched chain}$$

$$H + H + M \rightarrow H_2 + M \qquad \text{(terminating)}$$

One cycle of this reaction produces a net increase of two hydrogen atoms in the reactive system, and if terminating processes are not sufficiently rapid to use up two hydrogen atoms in the same length of time, this reaction will increase its rate exponentially with time and behave as a homogeneous explosion.

The general conditions for a branching-chain explosion may be discussed by considering a general reaction of the branching-chain type. Let C represent a chain carrier; then the reaction set

$$nA \xrightarrow{k_1} C \qquad \text{(initiation)}$$

$$A + C \xrightarrow{k_2} \text{products} + \alpha C \qquad \text{(chain cycle)}$$

$$C \xrightarrow{k_w} \text{products or stable molecules (termination at wall)}$$

$$C + M + M \xrightarrow{k_g} \text{products or stable molecules (termination in the gas).}$$

represents a typical branching-chain reaction. In the steady state,

$$\frac{d[C]}{dt} = O = k_1[A]^n - k_2[C]\cdot[A] + \alpha k_2[C]\cdot[A] - k_g[C][M]^2 - k_w[C]$$

Therefore at steady state

$$[C] = \frac{k_1[A]^n}{k_2[A](1 - \alpha) + k_g[M]^2 + k_w}$$

In order for $[C]$ to be finite and positive it is obvious that

$$\alpha < 1 + \frac{k_g\cdot[M]^2 + k_w}{k_2[A]}$$

Therefore if

$$\alpha \geq 1 + \frac{k_g \cdot [M]^2 + k_w}{k_2[A]}$$

the reaction will become a branching-chain explosion.

This behavior is to be sharply contrasted to the behavior of a typical homogeneous thermal explosion. In the case of a thermal explosion, heat release causes a rise in the temperature of an isolated system which in turn causes the reaction rate to increase, and this self-accelerating process is observed as the explosion. In the case of a branching-chain explosion the branching-chain process can be thermally neutral and therefore not produce any sizeable temperature changes during the initial stages. Thus it is possible to have a branching-chain explosion which does not produce any initial temperature rise in an isolated system. These processes will be discussed more extensively in Chapter 6.

4-7. FAST REACTIONS

One of the major difficulties in reaction kinetic studies is that of making a direct extrapolation of rates to a temperature outside the experimental range. There are three main reasons for this difficulty. First, most reactions are quite complex and the overall rate is usually determined by a relatively complicated combination of the reaction constants for a series of individual reaction steps, and these individual reaction steps all have different activation energies. Thus all the individual steps must be understood to justify extrapolation. Second, it is possible that the entire character of the reaction will change because elementary steps which were not contributing to the rate in one part of the temperature range become important in another. Third, the reaction may become extremely rapid. Here we consider reactions with half-lives of less than 10 μsec to be very rapid. For these reactions an understanding of the classical theory is, in all probability, not sufficient to determine the course of the reaction because thermodynamic equilibria no longer exists during the reaction process. In the classical (Arrhenius) sense, a reaction occurs rapidly for one of two reasons. These are (1) an extremely high temperature exists or (2) the activation energy is small. In both cases the quantity E/RT becomes small, thereby causing a large fraction of the collisions between reactive species to be effective in producing reaction. This will, of course, negate our original derivation of the Arrhenius equation since the assumptions can no longer be considered valid. There is another difficulty which appears when an extremely rapid reaction occurs. This concerns the equipartition of the energy in the reaction product molecules. It is quite obvious that for a very rapid reaction our assumption of classical equipartition is invalid. For example, rather recently rapid reactions of the type

$$A + XYZ \rightarrow AX^* + YZ$$

have been investigated with flash photolysis. In this technique the species A is formed at large concentrations by flashing a high intensity light for a few microseconds. Typical of this technique is the study of the reaction

$$O + O_3 \rightarrow O_2^* + O_2$$

by photodecomposing some of the O_3 with a high-intensity light source.

$$O_3 + h\nu \rightarrow O_2 + O$$

Observations (at room temperature) have shown that the O_2^* that is formed is almost entirely in the 14th, 15th, and 16th vibrational levels and only slowly decays to the ground state, which is its thermal equilibrium state. In a similar manner the hydroxyl radical produced by the rapid reaction

$$H + O_3 \rightarrow OH^* + O_2$$

has been found to have an effective rotational temperature of 3200°K in a system where the translational gas temperature is 300°K.

In addition, recent work on the high-temperature hydrogen–oxygen reaction and separate work on the high-temperature dissociation of oxygen has shown that the processes are altered by the lag in vibrational equilibriation of the molecules. In these systems the gas is heated rapidly from room temperature and a slow vibrational relaxation process causes the system to react from a different state than the full equilibrium state.

4-8. CHEMICAL EQUILIBRIA

We indicated earlier that the occurrence of chemical equilibrium could be considered as a possible restrictive behavior for a reactive system (case 3, Sec. 4-1). In this case many detailed composition changes are occurring on the local or molecular level but the overall composition of the gas remains invariant with time. We infer that the system contains competing reactions which do not allow any one reaction to go to completion. If this is true, the principle of detailed balancing, discussed in Chapter 2, allows us to define equilibrium concentrations in terms of the rate constants for the reversible reactions responsible for equilibrium. Consider as an example the forward and reverse reactions

$$A + B \xrightarrow{k_f} C + D$$

and

$$C + D \xrightarrow{k_r} A + B$$

Local concentration changes due to these reactions may be written as

$$-\frac{d[A]}{dt} = -\frac{d[B]}{dt} = \frac{d[C]}{dt} = \frac{d[D]}{dt} = k_f[A][B]$$

$$\frac{d[A]}{dt} = \frac{d[B]}{dt} = -\frac{d[C]}{dt} = -\frac{d[D]}{dt} = k_r[C][D]$$

Therefore at equilibrium Eq. 4-11 predicts that

$$\left(\frac{d[J]}{dt}\right)_{\text{net}} = 0 = k_f[A][B] - k_r[C][D]$$

where

$$J = A, B, C \text{ or } D$$

This equation may be arranged to yield

$$\frac{k_f}{k_r} = \frac{[C][D]}{[A][B]} = K_c \text{ for the reaction}$$

$$A + B \rightleftharpoons C + D$$

This definition of K_c involves a convention in that the concentration of the products in the forward stoichiometric equation are always placed in the numerator of the defining equation while the reactant concentrations are placed in the denominator.

Even though this is a useful definition of equilibria it does not ordinarily allow one to determine an equilibrium constant experimentally (our macroscopic measuring instruments do not allow a measurement of the individual rates at equilibrium). However, from stoichiometry, we know that in general in an equilibrium system we can define $s - c$ reaction coordinates whose possible range of numerical values defines all the mixture compositions available to the system. We will now see that thermodynamics may be used to determine the composition that the system actually reaches within these restraints. This is possible because chemical equilibrium may be considered as a specific case of thermodynamic equilibrium.

Before we discuss chemical equilibria we introduce the thermodynamics of open systems. Consider an ideal gas mixture consisting of s species and containing n_i moles of the ith species where the n_i are not constants but are s independent extensive variables. We may write the equations for the differential change of each of the four extensive thermodynamic energies as

$$dE = TdS - pdV + \sum_{i=1}^{s} \mu_i dn_i \tag{4.19a}$$

where

$$T = \left(\frac{\partial E}{\partial S}\right)_{V,n_i}$$

$$-p = \left(\frac{\partial E}{\partial V}\right)_{S,n_i}$$

and

$$\mu_i = \left(\frac{\partial E}{\partial n_i}\right)_{S,V,n_j;j\neq i}$$

$$dH = T\,dS + V\,dp + \sum_{i=1}^{s} \mu_i dn_i \tag{4-19b}$$

where

$$T = \left(\frac{\partial H}{\partial S}\right)_{P,n_i}$$

$$V = \left(\frac{\partial H}{\partial p}\right)_{S,n_i}$$

and

$$\mu_i = \left(\frac{\partial H}{\partial n_i}\right)_{S,p,n_j;j\neq i}$$

$$dA = -S\,dT - p\,dV + \sum_{i=1}^{s} \mu_i dn_i \tag{4-19c}$$

where

$$-S = \left(\frac{\partial A}{\partial T}\right)_{V,n_i}$$

$$-p = \left(\frac{\partial A}{\partial V}\right)_{T,n_i}$$

and

$$\mu_i = \left(\frac{\partial A}{\partial n_i}\right)_{T,V,n_j;j\neq i}$$

and

$$dF = -S\,dT + V\,dp + \sum_{i=1}^{s} \mu_i dn_i \tag{4-19d}$$

where

$$-S = \left(\frac{\partial F}{\partial T}\right)_{p,n_i}$$

$$V = \left(\frac{\partial F}{\partial p}\right)_{T,n_i}$$

and

$$\mu_i = \left(\frac{\partial F}{\partial n_i}\right)_{T,p,n_j;j \neq i.}$$

The fact that

$$\mu_i = \left(\frac{\partial E}{\partial n_i}\right)_{S,V,n_j;i\neq j} = \left(\frac{\partial H}{\partial n_i}\right)_{S,P,n_j;i\neq j} = \left(\frac{\partial A}{\partial n_i}\right)_{T,V,n_j;i\neq j} = \left(\frac{\partial F}{\partial n_i}\right)_{T,P,n_j;i\neq j}$$

may be proved by substituting the definitions of H, A, and F into the energy equation. The quantity μ_i is called the chemical potential of the ith specie and is the energy change of the system if one mole of the ith specie is added to an infinite amount of the mixture under the various restrictions described in Eqs. 4-19. As we will see later, for ideal gas mixture the μ_i are simply the Gibbs free energy of the ith specie calculated at the partial pressure of the ith specie and the temperature of the system.

We now return to a closed but reactive system. The conditions for equilibrium in such a system at constant temperature are either that the Gibbs free energy F be a minimum at constant pressure or that the Helmholtz free energy A be a minimum at constant volume. We will use the Gibbs free energy and therefore discuss a closed system at constant pressure and temperature. Here a closed reactive system is obtained from the open system defined with Eqs. 4-19 by restricting the changes in the n_i's through the definition of $s - c$ reaction coordinates. For simplicity, let us consider a single general reaction whose stoichiometry is given by the equation

$$\nu_1 \text{I} + \nu_2 \text{II} + \cdots \rightleftharpoons \nu_{n+1}(N + 1) + \nu_{n+2}(N + 2) + \cdots \qquad (4\text{-}20)$$

which may also be written as the equation set

$$dn_i = \nu_i d\lambda \qquad (4\text{-}21)$$

Here, by convention, if any ν_i is negative it means that the specie appears on the left in Eq. 4-20. Since the reactive mixture is an ideal mixture of gases which obeys Dalton's law, we can write simple expressions for the thermodynamic properties of this mixture. The enthalpy is given by the expression

$$H = \sum_{i=1}^{s} n_i H_i$$

where H_i is a molar enthalpy of the ith species. Similarly, the entropy and free energy may be written as

$$S = \sum_{i=1}^{s} n_i S_i$$

$$F = \sum_{i=1}^{s} n_i F_i = \sum_{i=1}^{s} n_i(H_i - TS_i)$$

However, we must allow for the appearance or disappearance of materials in determining the thermodynamic properties. For example, the heat release accompanying chemical reaction cannot be neglected in determining the enthalpy. Thus, the enthalpy of any specie must be defined as

$$H_i(T) = \int_{T_0}^{T} C_{p_i} dt + [\Delta H_f(T_0)]_i$$

where $[\Delta H_f(T)]_i$ is the standard heat of formation of the ith species at temperature T_0. Similarly, the entropy for the specie S_i at a pressure p_i and a temperature T is given by the expression

$$S_i(T, p_i) = \int_{T_0}^{T} \frac{C_{p_i} dT}{T} - R \ln p_i + [\Delta S_f(T_0, P_0)]_i + R \ln P_0$$

where $[\Delta S_f(T_0, P_0)]_i$ is evaluated at some standard temperature T_0 and a pressure of one atmosphere. By convention the specie pressure p_i is expressed in atmospheres in this equation and the last term therefore becomes zero. The free energy of the species F_i is therefore

$$F_i = H_i - TS_i = \int_{T_0}^{T} C_{p_i} dT + [\Delta H_f(T_0)]_i - T\int_{T_0}^{T} C_{p_i} \frac{dT}{T} + RT \ln p - T[\Delta S_f(T_0, P_0)]_i$$

or

$$F_i(T, p_i) = \omega_i(T) + RT \ln p_i = \mu_i(T, p_i) \tag{4-22}$$

as may be seen by examining Eq. 4-19d. We are now ready to discuss equilibrium conditions for the general reaction expressed in Eq. 4-20 or 4-21. For Eq. 4-21 we may write

$$F = F(T, P, \lambda)$$

and since we are working at constant pressure and temperature the equilibrium condition $dF = 0$ implies that

$$\frac{dF}{d\lambda} = 0$$

or that

$$\sum_{i=1}^{s} \frac{dn_i}{d\lambda} [\omega_i(T) + RT \ln p_i] + \sum_{i=1}^{s} n_i RT \frac{d}{d\lambda} (\ln p_i) = 0 \tag{4-23}$$

However, the last term vanishes because the total pressure is assumed constant.

$$\sum_{i=1}^{s} n_i RT \frac{d}{d\lambda} (\ln p_i) = \sum_{i=1}^{s} \frac{n_i RT}{p_i} \frac{dp_i}{d\lambda} = V \sum_{i=1}^{s} \frac{dp_i}{d\lambda} = 0$$

Therefore, Eq. 4-23 becomes

$$\sum_{i=1}^{s} \frac{dn_i}{d\lambda} [\omega_i(T) + RT \ln p_i] = 0$$

However, since

$$\frac{dn_i}{d\lambda} = \nu_i$$

we obtain

$$\sum_{i=1}^{s} \nu_i [\omega_i(T) + RT \ln p_i] = 0 \tag{4-24}$$

Equation 4-24 may also be derived directly from Eq. 4-19 by writing the equilibrium condition using that equation,

$$\left(\frac{\partial F}{\partial \lambda}\right)_{P,T} = 0 = \sum_{i=1}^{s} \mu_i \left(\frac{\partial n_i}{\partial \lambda}\right)_{P,T} = \sum_{i=1}^{s} \nu_i \mu_i$$

for s species. However, for an ideal system Eq. 4-24 is valid and, therefore, both these derivations yield the equation

$$-\frac{1}{RT} \sum_{i=1}^{s} \nu_i \omega_i(T) = \sum_{i=1}^{s} \ln p_i^{\ \nu_i} = \ln K_p \tag{4-25}$$

Equation 4-25 relates a thermodynamic quantity (on the left) to a ratio of the partial pressures of the reacting species at equilibrium (in the middle) and this in turn is used to define an equilibrium constant (on the right). Thus, through the use of Eq. 4-25, a knowledge of the thermodynamic properties of the individual species in the mixture allows an evaluation of the equilibrium composition of the mixture. Equation 4-25 has been verified experimentally for many different systems over a wide range of conditions and its use in making theoretical performance calculations is quite well justified.

The use of Eq. 4-25 for the calculation of chemical equilibrium requires that one employ a self-consistent set of rules in each calculation. These formalities have been standardized to a certain extent by the chemists and, at present, the most common set of standardized definitions are those used in preparing the JANAF Tables, a few of which are reproduced in Appendix F. Unfortunately, if other sources of thermodynamic data are used they may not be consistent with these tables. Because of this, one should always be careful to check the standards when data from two different sources are used in an equilibrium calculation. Notice that the $\omega_i(T)$ term in Eq. 4-25 has a numerical value which is dependent on the units used to define pressure and the quantity RT in Eq. 4-22. In the JANAF Tables p_i always has units of atmospheres and R has units of cal/mole °K. Thus if $p_i = 1$ atmosphere the pressure dependence in Eq.

4-22 disappears and we may write

$$\omega_i(T) = \mu_i^\circ(T) = F_i^\circ(T) = [\Delta F_f^\circ(T)]_i$$

where those symbols with a superscript "°" represent a standard state value of the thermodynamic function for the compound relative to the thermodynamic properties of the constituent elements in their standard states. Thus, Eq. 4-25 may be written

$$\ln K_p = \frac{-1}{RT}\sum_{i=1}^{s} \nu_i(\Delta F_f^\circ)_i \tag{4-26}$$

where the $(\Delta F_f^\circ)_i$'s are obtained from the appropriate table in Appendix F and the ν_i's are obtained from the equation defining the reaction path (Eq. 4-21). To simplify the use of the tables in Appendix F even further, values of $\log_{10} K_p$, the equilibrium constants of formation, are tabulated in the last column of these tables. These may be used to calculate the equilibrium constant for a specific reaction by simply writing

$$(\log_{10} K_p)_{\text{reaction}} = \sum_{i=1}^{s} \nu_i(\log_{10} K_p)_i \tag{4-27}$$

We make three more observations pertinent to the use of Eq. 4-26. First, the numerical value of K_p is dependent on the assumed stoichiometric relationship, Eqs. 4-20 or 4-21, and this relationship must be specified before a stated value of K_p will have meaning. For example, the statement, "The equilibrium constant for the formation of water at 2500°K is 167.5," is meaningless. One should say instead, that for the reaction

$$H_2 + \frac{1}{2} O_2 \rightleftharpoons H_2O$$

$$K_p = 167.5 \text{ at } 2500°\text{K}$$

This means that at the indicated temperature

$$167.5 = \frac{p_{H_2O}}{p_{H_2} \cdot p_{O_2}^{1/2}}$$

Notice that this is equivalent to the statement that

$$K_p = (167.5)^2 = \frac{p^2_{H_2O}}{p^2_{H_2} \cdot p_{O_2}}$$

for the reaction

$$2H_2 + O_2 \rightleftharpoons 2H_2O$$

Furthermore, in these two relations the p_i's must be expressed in units of atmospheres, at least for the convention adopted in the JANAF tables and this text. Second, we observe that our derivation of Eq. 4-26 places no restriction on the occurrence of simultaneous equilibria. For complex systems the original thermodynamic equations may be written as a set of s–c linearly independent equations each with a definite reaction coordinate. Therefore, once a set of linearly independent reaction coordinates are discovered, $s{-}c$ independent equilibrium relations may be deduced, each governing the extent of reaction along one reaction coordinate. Unfortunately, as we will see later, the resulting s equations in s unknowns are a coupled set of c linear and $s{-}c$ nonlinear equations and this set is usually very difficult to solve for cases where $s{-}c > 2$.

Thirdly, we have seen that for each of $s{-}c = j$ independent reactions occurring in a system containing s species we may write

$$dn_i = \nu_{ij} d\lambda_j \qquad j = 1, 2, \ldots, (s{-}c)$$

and we may define j equilibrium constants for these reactions as

$$(K_p)_j = \prod_{i=1}^{s} (p_i)^{\nu_{ij}} \qquad j = 1, 2, \ldots, (s{-}c) \tag{4-28}$$

However, for an ideal gas Dalton's law states that

$$P_T = \sum_{i=1}^{s} p_i$$

and since

$$n_T = \sum_{i=1}^{s} n_i$$

the partial pressure of any specie may be expressed as

$$p_i = n_i \frac{P_T}{n_T} = [I]\,RT$$

where $[I]$ represents the concentration of the ith specie.

Therefore we may rewrite Eq. 4-28 in terms of molar quantities to yield

$$(K_p)_j = \left(\frac{P_T}{n_T}\right)^{\sum_{i=1}^{s} \nu_{ij}} \cdot \prod_{i=1}^{s} (n_i)^{\nu_{ij}} = (RT)^{\sum_{i=1}^{s} \nu_{ij}} \prod_{i=1}^{s} [I]^{\nu_{ij}} \tag{4-29}$$

where

$$\prod_{i=1}^{s} (n_i)^{\nu_{ij}} = (K_n)_j$$

and

$$\prod_{i=1}^{s} [I]^{\nu_{ij}} = (K_c)_j$$

The quantity K_c was defined at the beginning of this section. The first portion of Eq. 4-29 may also be written

$$(K_p)_j = \left(\frac{P_T}{\eta_T}\right)^{\sum_{i=1}^{s} \nu_{ij}} \prod_{i=1}^{s} (\eta_i)^{\nu_{ij}} \tag{4-30}$$

where the quantity η is written on the basis of $\mathcal{G}$ gm of mixture.

Equations 4-29 and 4-30 explicitly illustrate the effect of pressure changes on the composition of an equilibrium mixture. They state that if the sum of the ν_{ij}'s is zero for a specific reaction, pressure changes will not affect the composition and that, if the sum is not zero, the equilibrium composition will be shifted by a change in pressure.

4-9. THE EFFECT OF INERTS OR CONDENSATION ON EQUILIBRIA

The value of an equilibrium constant is not affected by the presence of an inert gas. Equation 4-25 is still valid. However, in this case the partial pressures of the reactive components do not add to give the total pressure of the system because the inert gas exerts a partial pressure in the mixture. Therefore, n_T in Eq. 4-29 now becomes

$$n_T = n_1 + n_2 + \cdots + n_{s-1} + n_s + n_I$$

where n_I represents the number of moles of inert gas present in one gram of mixture. Obviously, if $\Sigma \nu_{ij}$ in Eqs. 4-29 and 4-30 is zero the addition of an inert will not shift the equilibrium composition. However, for a constant total pressure, if this exponent is not zero the equilibrium composition will shift with the addition of an inert.

The presence of a solid or liquid at equilibrium greatly affects the composition of the equilibrium mixture. If the solid or liquid is *pure*, one can assume that it exerts its equilibrium vapor pressure in the gaseous portion of the mixture independent of the quantity of solid or liquid present. To solve an equilibrium problem in this case one must first determine if a solid or liquid can be present at equilibrium. An equilibrium calculation is made assuming no condensed phases are present. The equilibrium partial pressures of all the species are then calculated and if these are all below the vapor pressures of the respective liquid or solid phases the calculation is valid. However, if any one of these partial pressures is above the vapor pressure of that component, it will condense and a new equilibrium calculation must be made.

Chemists have established a convenient convention in defining the standard-state free energies of formation of solids or liquids. Since a solid or liquid is in equilibrium with its vapor at its vapor pressure, the free energy must be the same in the two phases. Furthermore, this free energy is independent of the total pressure in an ideal gaseous system. Therefore, by definition.

$$[\Delta F_f^\circ(T)]_{j\text{th solid or liquid}} = \omega_j(T) + RT \ln p_{j_V}$$

where p_{j_V} is the equilibrium vapor pressure of the solid or liquid. Now consider a general system of s species, g of which are ideal gases and s–g of which are solids or liquids, each exerting its own vapor pressure in the system. Substitution into Eq. 4-25 yields the relation

$$RT\sum_{i=1}^{g} \ln (p_i)^{\nu_i} + RT \sum_{i=g+1}^{s} \ln (p_{i_V})^{\nu_i}$$
$$= - \sum_{i=1}^{s} \nu_i[\Delta F_f^\circ(T)]_i + RT \sum_{i=g+1}^{s} \ln (p_{i_V})^{\nu_i}$$

where the second terms on each side cancel and we may therefore write

$$RT\sum_{i=1}^{g} \ln (p_i)^{\nu_i} = -\sum_{i=1}^{s} \nu_i[\Delta F_f^\circ(T)]_i$$

or

$$RT \ln K_p' = -\sum_{i=1}^{s} \nu_i[\Delta F_f^\circ(T)]_i \tag{4-31}$$

where the $\Delta F_f^\circ(T)$ are now for solids, liquids, or gases as appropriate. The new equilibrium constant K_p' contains only terms for the partial pressure of the gaseous components. When these partial pressures satisfy this value of K_p' the solid or liquid formed will be in exact equilibrium with its vapor and complete chemical equilibrium will exist. The convention, therefore, replaces the extra step of calculating the vapor pressures of solid or liquid components and their substitution into Eq. 4-25. Notice that the numerical value of K_p in Eq. 4-25 is not equal to the value of K_p' in Eq. 4-31. In Eq. 4-25, K_p is equal to the sum of the logarithms of all the partial pressures and they are all variable. In Eq. 4-31 the K_p' is equal to the sum of the logarithms of the partial pressures of only the completely gaseous components. The remaining partial pressures are fixed at the equilibrium vapor pressure of that component and divided out of the equation. The above discussion is valid only for the formation of pure solids or liquids in equilibrium with the gas phase. If an intimate mixture (a solution) of solids and/or liquids is formed, another difficulty arises because the vapor pressure of a component in a solution is not equal to the vapor pressure of the pure component. In this case allowances can be made by defining an activity a_i of the component where

$$a_i = p_i\,(\text{mixture})/p_i\,(\text{pure solid or liquid})$$

and Eq. 4-31 must contain activities a_i raised to the proper stoichiometric exponents ($a_i^{\nu_i}$) before equilibrium concentrations may be calculated.

4-10. ACCURACY AND RESTRICTIONS IN EQUILIBRIUM CALCULATIONS

In the previous sections we discussed chemical kinetics and chemical equilibria as limit phenomena with the implication that rate processes in some way yield equilibrium systems after the passage of sufficient time. Furthermore, we found that reaction kinetics is a rather experimental science, in which a directly measured elementary rate is generally more useful than a theoretically calculated rate. We also observed that this contrasts rather sharply with the sophistication of our theoretical understanding of chemical equilibria, which in many cases appears to allow a more accurate calculation of the equilibrium state than can be measured experimentally. There are five major limitations to our ability to calculate the equilibrium behavior of chemical systems:

1. Thermodynamic functions must be known with extremely high accuracy to obtain realistic values for equilibrium constants. There are two main sources for this data. Spectroscopic information allows one to calculate the ordinary thermodynamic functions of a compound, and calorimetry is required to determine heats and entropies of formation. However, at high temperatures the molecules cannot be described as containing a series of "normal" independent energy storage modes, and cross-term contributions become important. This constitutes a challenge to the spectroscopist; high temperature thermodynamic functions are only as good as the interpretation of complex spectra. In addition, the heat of reaction and the entropy of reaction cannot always be determined unambiguously using spectroscopic evidence. A great deal of careful calorimetric work must be performed to establish heats and entropies of formation. The importance of obtaining correct thermodynamic values can be shown using Eq. 4-26. At $T = 300°$K, an uncertainty of 0.057 in the value of $\Sigma\nu_i(\Delta F_f^\circ)_i$ will cause K_p to have an uncertainty of 10%, while at 3000°K an uncertainty of 0.57 will cause the same uncertainty in K_p. Here the units on the Σ are kilocalories. Since most standard heats of formation have values which lie between 25 and 150 kcal they must be measured with extreme accuracy to estimate equilibrium constants reasonably well. This problem is continually under attack and the thermodynamic values tabulated in Appendix F represent only the best current values of the thermodynamic functions.

2. All species must be included in the calculation. We mentioned earlier that the presence of a specie is in no way determined by the writing of a stoichiometric equation. These equations can only be written after the participating species are chosen. Furthermore, one can determine those species which are present at equilibrium only from experimental experience. This problem is particularly difficult if we are attempting to determine the thermodynamic behavior of a system which contains an atom whose properties and whose high-temperature compounds have not been studied extensively. This problem is not trivial because the formation of even a few percent of a compound with a high heat of formation can drop an equilibrium explosion temperature hundreds of degrees and thus greatly modify performance calculations.

3. Nonideal effects can be large. This problem becomes important at high pressures and it can be treated with fair accuracy by introducing fugacities and fugacity coefficients. The fugacity of the ith species f_i is a fictitious pressure (in atmospheres) which the species must exert to satisfy Eq. 4-25. It is related to the mole fraction by the fugacity coefficients ξ_i

$$f_i = \xi_i \frac{n_i}{n_T} P_T$$

Therefore in an ideal gas mixture all $\xi_i = 1$ and $f_i = p_i$. In a nonideal system the ideal value of K_p (from Eq. 4-25) is given in terms of fugacities

$$K_p = \prod_{i=1}^{s} (f_i)^{\nu_i} = \prod_{i=1}^{s} (n_i)^{\nu_i} \cdot \prod_{i=1}^{s} (\xi_i)^{\nu_i} \cdot \left(\frac{P_T}{n_T}\right)^{\sum_{i=1}^{s} \nu_i}$$

where

$$K_\xi = \prod_{i=1}^{s} (\xi_i)^{\nu_i}$$

The mole fraction at equilibrium may, therefore, be determined once a value for K_p and K_ξ are known. K_ξ may be determined with reasonable accuracy in most cases by using nonideal thermodynamic functions derived from an assumed equation of state.[7] However, K_ξ is a function of both temperature and pressure and the calculations become considerably more laborious.

4. Equilibrium might not be reached. This will occur when the number of available independent reactions is less than the $s-c$ required reaction paths. In certain simple systems it is relatively easy to construct a set of restricted equilibrium equations to account for this behavior. However, in a general case, the specification of appropriate fictitious components is not obvious and a generalized technique is required.

Let us first consider a case of full equilibrium in a system containing s species, c components and therefore $s-c = r$ independent reactions. Assume that there are $r_i > r$ possible reversible reaction paths and construct an $r_i \times s$ matrix of the r_i available stoichiometric coefficients.

	s_1	s_2	s_3		s_j
r_1	ν_{11}	ν_{12}	ν_{13}	$\cdots$	ν_{1j}
r_2	ν_{21}	ν_{22}	ν_{23}	$\cdots$	ν_{2j}
r_3	ν_{31}	ν_{32}	ν_{33}	$\cdots$	ν_{3j}
	$\cdot$	$\cdot$	$\cdot$		$\cdot$
	$\cdot$	$\cdot$	$\cdot$		$\cdot$
	$\cdot$	$\cdot$	$\cdot$		$\cdot$
r_i	ν_{i1}	ν_{i2}	ν_{i3}		ν_{ij}

[7]See, for example, O. A. Hougen et al., *Chemical Process Principles*, Part II, *Thermodynamics*, 2nd ed., Wiley, New York, 1959.

Since this matrix represents a set of linear equations containing only r variables, it must have a rank no greater than r. If inspection shows that this matrix has a rank r, we are dealing with a case of nonrestricted or full equilibrium. If this matrix has a rank $r^* < r$ we have a case of restricted equilibrium.

When the rank is r we can add or subtract rows to reduce this matrix to an $r \times s$ matrix. We may then diagonalize any r independent columns, writing the matrix so that a (-1) appears at each position on the main diagonal of the r columns. When this is done, we will have, in effect, written the stoichiometric equations for the formation of the r species using the remaining $s-r = c$ species as components.

$$
\begin{array}{c|cccccccc|}
 & c_1 & c_2 & \cdots & c_n & s_1 & s_2 & \cdots & s_i \\
\hline
r_1 & \nu'_{11} & \nu'_{12} & \cdots & \nu'_{1n} & -1 & 0 & \cdots & 0 \\
r_2 & \nu'_{21} & \nu'_{22} & \cdots & \nu'_{2n} & 0 & -1 & \cdots & 0 \\
\vdots & \vdots & \vdots & & \vdots & \vdots & \vdots & & \\
r_r & \nu'_{r1} & \nu'_{r2} & \cdots & \nu'_{rn} & 0 & 0 & & -1
\end{array}
$$

This matrix may be used to construct the mass-balance equations for these components in terms of the species present at equilibrium by constructing an $s \times c$ matrix consisting of a $c \times c$ unit matrix written over the component coefficient matrix.

$$
\begin{array}{c|cccc|}
 & c_1 & c_2 & & c_n \\
\hline
c_1 & 1 & 0 & \cdots & 0 \\
c_2 & 0 & 1 & \cdots & 0 \\
\vdots & \vdots & \vdots & & \vdots \\
c_n & 0 & 0 & \cdots & 1 \\
r_1 & \nu'_{11} & \nu'_{12} & \cdots & \nu'_{1n} \\
r_2 & \nu'_{21} & \nu'_{22} & \cdots & \nu'_{2n} \\
\vdots & \vdots & \vdots & & \vdots \\
r_r & \nu'_{r1} & \nu'_{r2} & & \nu'_{rn}
\end{array}
$$

An example will be useful. Consider the system containing H_2, O_2, H_2O, H,O and OH. Here $s = 6$, $c = 2$, and $r = 4$. Consider the equilibrium reactions

$$H_2 + O_2 \rightleftharpoons 2OH$$
$$H_2 + \tfrac{1}{2}O_2 \rightleftharpoons H_2O$$
$$H_2 \rightleftharpoons 2H$$
$$O_2 \rightleftharpoons 2O$$
$$H + OH \rightleftharpoons H_2 + O$$
$$O + H_2 \rightleftharpoons OH + H$$

These reactions yield the matrix

H_2	O_2	H_2O	H	O	OH
−1	−1	0	0	0	+2
−1	−½	+1	0	0	0
−1	0	0	+2	0	0
0	−1	0	0	+2	0
+1	0	0	−1	+1	−1
−1	0	0	+1	−1	+1

which is of rank 4 and may therefore be altered by adding and subtracting rows to yield the matrix

O	H	OH	H_2	O_2	H_2O
+1	+1	−1	0	0	0
0	+2	0	−1	0	0
+2	0	0	0	−1	0
+1	+2	0	0	0	−1

This matrix is equivalent to the reactions

$$OH \rightleftharpoons H + O$$
$$H_2 \rightleftharpoons 2H$$
$$O_2 \rightleftharpoons 2O$$
$$H_2O \rightleftharpoons 2H + O$$

which represent the independent formation of the species OH, H_2, O_2, and H_2O from the components H and O. Rewriting to express the mass-balance equations, we obtain

	O	H
O	1	0
H	0	1
OH	1	1
H_2	0	2
O_2	2	0
H_2O	1	2

which represents the equations

$$N_{\mathrm{O}} = n_{\mathrm{O}} + n_{\mathrm{OH}} + 2n_{\mathrm{O_2}} + n_{\mathrm{H_2O}}$$
$$N_{\mathrm{H}} = n_{\mathrm{H}} + 2n_{\mathrm{H_2}} + n_{\mathrm{OH}} + 2n_{\mathrm{H_2O}}$$

where the N's represent component moles and are therefore constants of the system. Notice that while this choice of components is not unique, we are restricted to choose independent species for the diagonalization procedure and therefore do not have a completely unrestricted choice of components. For example, the determinant represented by the columns O_2, H, O and OH has a value of zero and therefore H and H_2 cannot be components in this system. However, we may rewrite the matrix to yield the new matrix,

OH	O	H_2O	H	O_2	H_2
+2	−1	−1	0	0	0
+1	−1	0	−1	0	0
0	+2	0	0	−1	0
+2	−2	0	0	0	−1

The equilibrium equations for this case are

$$\mathrm{O} + \mathrm{H_2O} \rightleftharpoons 2\mathrm{OH}$$
$$\mathrm{O} + \mathrm{H} \rightleftharpoons \mathrm{OH}$$
$$\mathrm{O_2} \rightleftharpoons 2\mathrm{O}$$
$$2\mathrm{O} + \mathrm{H_2} \rightleftharpoons 2\mathrm{OH}$$

and the mass-balance matrix becomes

	OH	O
OH	1	0
O	0	1
H_2O	+2	−1
H	+1	−1
O_2	0	+2
H_2	+2	−2

This matrix represents the equations

$$N_{\mathrm{OH}} = n_{\mathrm{OH}} + 2n_{\mathrm{H_2O}} + n_{\mathrm{H}} + 2n_{\mathrm{H_2}}$$
$$N_{\mathrm{O}} = n_{\mathrm{O}} - n_{\mathrm{H_2O}} - n_{\mathrm{H}} + 2n_{\mathrm{O_2}} - 2n_{\mathrm{H_2}}$$

This is a valid set of mass-balance equations for the six specie system in terms of the components OH and O. Notice that for $\Phi > 0.5$, N_{O} will be a negative number in this representation.

Restricted equilibrium occurs where the $r' \times s$ matrix has a rank $r^* < r$. In a hydrogen–oxygen explosion the early reactions yield a state of restricted equilibrium because complete equilibrium requires a net reduction in the number of moles of gas present in the mixture. Since, in this system, mole reduction can only occur via three-body collisions, the rapid reactions result in the momentary appearance of a restricted equilibrium state. Let us examine a representative set of available binary exchange reactions:

$$\begin{aligned} H + O_2 &\rightleftharpoons OH + O \\ O + H_2 &\rightleftharpoons OH + H \\ OH + H_2 &\rightleftharpoons H_2O + H \\ H_2 + O_2 &\rightleftharpoons 2OH \\ &\text{etc.} \end{aligned}$$

These may be reduced to the 3 × 6 matrix

$$\begin{array}{cccccc} O_2 & H_2 & H_2O & H & O & OH \end{array}$$
$$\left|\begin{array}{cccccc} +\frac{1}{2} & +1\frac{1}{2} & -1 & -1 & 0 & 0 \\ +1 & +1 & -1 & 0 & -1 & 0 \\ +\frac{1}{2} & +\frac{1}{2} & 0 & 0 & 0 & -1 \end{array}\right|$$

representing the reactions

$$\begin{aligned} H + H_2O &\rightleftharpoons \tfrac{1}{2}O_2 + 1\tfrac{1}{2}H_2 \\ O + H_2O &\rightleftharpoons H_2 + O_2 \\ OH &\rightleftharpoons \tfrac{1}{2}O_2 + \tfrac{1}{2}H_2 \end{aligned}$$

The mass-balance matrix becomes

	O_2	H_2	H_2O
O_2	1	0	0
H_2	0	1	0
H_2O	0	0	1
H	½	1½	−1
O	1	1	−1
OH	½	½	0

which yields the equations

$$\begin{aligned} N_{O_2} &= n_{O_2} + \tfrac{1}{2}n_H + n_O + \tfrac{1}{2}n_{OH} \\ N_{H_2} &= n_{H_2} + 1\tfrac{1}{2}n_H + n_O + \tfrac{1}{2}n_{OH} \\ N_{H_2O} &= n_{H_2O} - n_H - n_O \end{aligned}$$

In this case one must specify three components to discuss the restricted equilibrium behavior. The *initial* specie composition of this mixture is therefore important in the determination of a restricted equilibrium composition. Once again, negative component mole numbers are possible.

5. Kinetic rate equations are integrated to determine the equilibrium composition. In this case one must be careful to adjust the rate constants

for the forward and for the separate reverse reactions so that the system will relax to the correct equilibrium composition. This is not a trivial problem because the inherent accuracy of experimentally determined rate constants is not high. Furthermore, there is no *a priori* reason why a measured forward rate in the early stages of reaction should yield a rate constant which is valid for that reaction in the equilibrium mixture. The reader is referred to Sec. 4-13 for further discussion of this problem.

4-11. DISSOCIATIVE EQUILIBRIA

One of the more important types of simple equilibrium is the dissociation equilibrium which occurs when a simple homonuclear diatomic molecule decomposes at high temperature. Consider, for example, the dissociation of pure hydrogen:

$$H_2 \rightleftharpoons 2H$$

This system contains two species and one component. We can analyze this system on either the mole/gram basis, the gram/gram basis or the mole/$\mathcal{G}$ gram basis. We choose to use the mole/$\mathcal{G}$ gram basis. If we consider a system which initially contains one mole of H_2 and no H atoms and set the reaction coordinate $\alpha = 0$ at $\eta_{H_2} = 1$, $\eta_H = 0$ we may integrate the two progress equations to obtain the moles/$\mathcal{G}$ of the species in terms of α (called the *degree of dissociation* in this case).

$$\eta_{H_2} = \int_1^\alpha - d\alpha = 1 - \alpha$$

$$\eta_H = \int_0^\alpha + 2\, d\alpha = 2\alpha$$

Thus we find that one mole of hydrogen dissociates to produce 2α moles of hydrogen atoms and $1 - \alpha$ moles of hydrogen molecules at equilibrium. The system, therefore, must contain $1 + \alpha$ moles of gas at equilibrium. Dalton's law allows us to write

$$p_H = \frac{2\alpha}{1 + \alpha} P \quad \text{and} \quad p_{H_2} = \left(\frac{1 - \alpha}{1 + \alpha}\right) P$$

or

$$K_p = \frac{p_H^2}{p_{H_2}} = \frac{4\alpha^2 P}{(1 + \alpha)(1 - \alpha)} = \frac{4\alpha^2}{1 - \alpha^2} P$$

which is the equilibrium equation in terms of the degree of dissociation α and the total pressure of the system. Thus at a specified temperature we can find a value of K_p from the tables in Appendix F and, with the total pressure, we can directly determine the value of α and thereby all the thermodynamic properties of the system. This dissociative-equilibrium problem is simple enough, however, that we may discuss it in terms of dimensionless parameters. Let us write the reciprocity relation from thermodynamics for the equilibrium system:

$$\left(\frac{\partial H}{\partial P}\right)_T = V - T\left(\frac{\partial V}{\partial T}\right)_P$$

However, the enthalpy of this system is governed by the relation

$$\begin{aligned} H &= 2\,\alpha H_A + (1 - \alpha)H_m \\ &= H_m + \alpha(2H_A - H_m) = H_m + \alpha H_D \end{aligned}$$

Here the subscripts A and m refer to the atomic and molecular species respectively. H_D is the dissociation enthalpy of the molecule and is, in general, a function of the temperature. We also note that the equations

$$P = \frac{RT}{V}(1 + \alpha) \tag{4-32a}$$

and

$$K_p = \frac{4\,\alpha^2}{1 - \alpha^2}P \tag{4-32b}$$

are governing equations for the system. Substituting these into the reciprocity relation yields

$$H_D\left(\frac{\partial \alpha}{\partial P}\right)_T = \frac{RT}{P}(1 + \alpha) - T\left[\frac{R}{P}(1 + \alpha) - \frac{RT}{P}\left(\frac{\partial \alpha}{\partial T}\right)_P\right]$$

or

$$\frac{H_D}{RT^2} = -\frac{\left(\frac{\partial \alpha}{\partial T}\right)_P}{\left(\frac{\partial \alpha}{\partial P}\right)_T} \cdot \left(\frac{1}{P}\right)$$

The two partial derivatives may be evaluated using the equilibrium equation to yield

$$\frac{\partial(\ln K_p)}{\partial T} = \frac{H_D}{RT^2}$$

This is called Van't Hoff's equation and may be integrated if H_D is independent of temperature to yield

$$K_p = P_D \exp\left(-\theta_D/T\right) \tag{4-33}$$

When P_D is a characteristic dissociation pressure with units of atmospheres and $\theta_D = (H_D/R)$ °K is a characteristic dissociation temperature.

Substituting into Eq. 4-32b gives the degree of dissociation of a homonuclear molecule in terms of the dimensionless parameters P/P_D and θ_D/T.

$$\alpha = \sqrt{\frac{1}{1 + \frac{4P}{P_D}\exp\left(\frac{\theta_D}{T}\right)}} \tag{4-34}$$

The enthalpy of a dissociating gas may be written as

$$H = H_m + \alpha H_D$$

where

$$H_m = \frac{7}{2}RT + RT\left[\frac{\theta_v/2T}{\sinh \theta_v/2T}\right]^2 \simeq 4RT$$

Thus

$$\frac{H}{H_D} = \frac{(4+\alpha)T}{\theta_D} + \alpha \simeq \frac{4T}{\theta_D} + \alpha \tag{4-35}$$

This approximation was first introduced by Lighthill[8] and is satisfactory in light of the earlier assumption that H_D is constant.

Table 4-2 lists dissociation temperatures and pressures for a number of homonuclear diatomic species. The multiple values listed for oxygen

TABLE 4-2.
DISSOCIATION PARAMETERS FOR DIATOMIC MOLECULES

	Dissociation Energy*	Best fit constants to Eq. 4-33 using a least-squares fit to K_p's taken from Appendix F		
Molecule	θ_D, °K	θ_D, °K	$\log P_D$ P_D-Atm	Temperature Range, °K
H_2	52,000	54,500	2.09	1000–6000
O_2	59,400	61,400	10.00	1000–6000
O_2	59,400	60,700	6.87	500–3000†
O_2	59,400	61,700	10.95	2000–5000†
N_2	113,300	116,300	13.78	2000–6000
F_2	18,500	19,700	4.15	500–3000
Cl_2	28,700	29,900	1.92	500–4500
Br_2	22,900	23,700	0.739	500–3500
I_2	17,900	18,500	0.372	500–3000

*Taken from either Appendix C, $\theta_D(°K) = 1.439 \times D_0(\text{cm}^{-1})$ or from Appendix F, $\theta_D(°K) = \Delta H_f(0°K)_{\text{Atoms}} \cdot 2/R$ (Br_2 and I_2 values have been corrected for the heat of sublimation at 0°K).

†For comparison to indicate the accuracy of the fit constants assuming P_D and θ_D are constant.

give an indication of the validity of this approach. Figures 4-10 and 4-11 are plots of the degree of dissociation and the dimensionless enthalpy of an ideal dissociating gas in terms of the dimensionless pressure P/P_D and temperature (T/θ_D).

[8] M. J. Lighthill, *J. Fluid Mech.*, *2*, 1(1957).

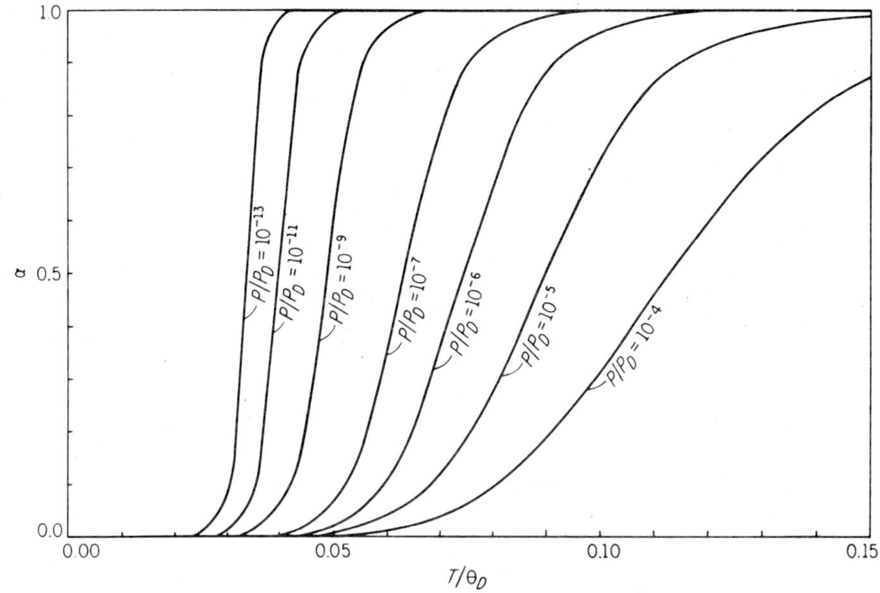

Fig. 4-10. Degree of dissociation for an ideal dissociating diatomic gas in terms of the reduced pressure and temperature (Eq. 4-34).

4-12. A CALCULATION OF EXPLOSION TEMPERATURE

In this section we present two explosion-temperature calculations based on the assumptions that the products of the explosion process are at equilibrium and that the explosion process is either (1) a constant-pressure process or (2) a constant-volume process. We choose as the system 1 mole of nitric oxide at $T_i = 298.16$ °K and $p_i = 1$ atm. To simplify the calculation we limit the equilibrium mixture to one containing only nitrogen molecules, N_2, oxygen molecules, O_2, and nitric oxide, NO.[9]

We use the stoichiometric equation

$$\tfrac{1}{2}N_2 + \tfrac{1}{2}O_2 \rightleftharpoons NO$$

and call α the number of moles of NO remaining at equilibrium. Thus

$$K_p = \frac{2\alpha}{1 - \alpha}$$

[9]The final temperature calculated in these examples will be too high because oxygen atoms, O, should be included to obtain a realistic final state. However, if oxygen atoms are included the enthalpy and internal energy of the system become pressure-dependent and this complicates the calculation.

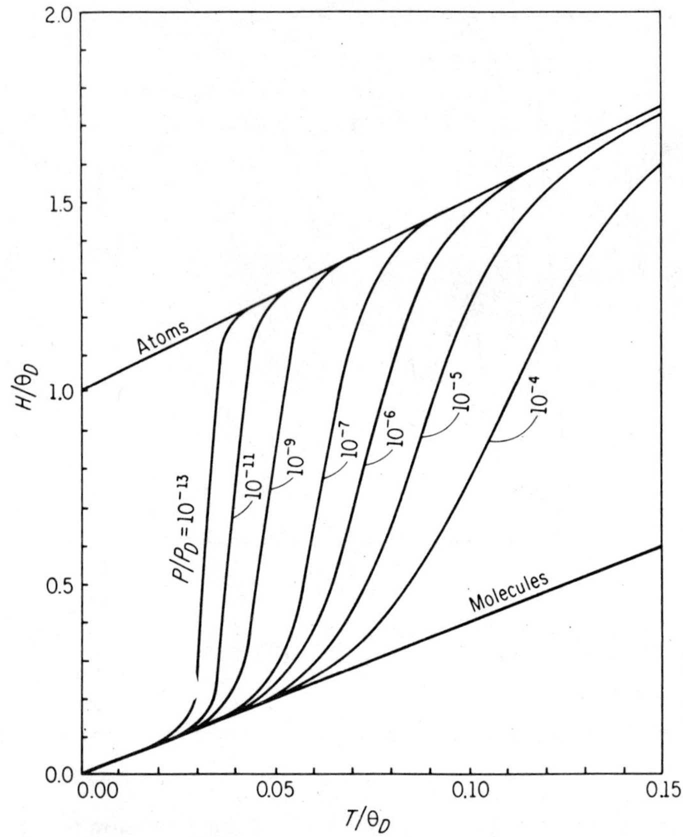

Fig. 4-11. The dimensionless enthalpy of an ideal dissociating gas (Eq. 4-35).

and α is a function of temperature only:

$$\alpha = \frac{K_p}{2 + K_p}$$

In this notation, at any temperature, the extent of reaction in the forward direction is

$$(\alpha - 1) \text{ moles of reaction (note that this is negative)}$$

and the composition is given by

$$\eta_{NO} = \alpha$$

and

$$\eta_{O_2} = \eta_{N_2} = {}^1\!/_2(1 - \alpha)$$

Notice that because of our specific choice of a stoichiometric equation the K_p which we are using is the K_p tabulated as $\log_{10} K_p$ in the nitric oxide table in Appendix F. The quantity $\alpha(T)$ is tabulated in Table 4-3 and graphed in Fig. 4-12.

For a constant-pressure process the enthalpy of the system is conserved and the equilibrium explosion temperature may be determined by requiring that $\Delta H_{\text{process}} = 0$. There are a number of possible ways of

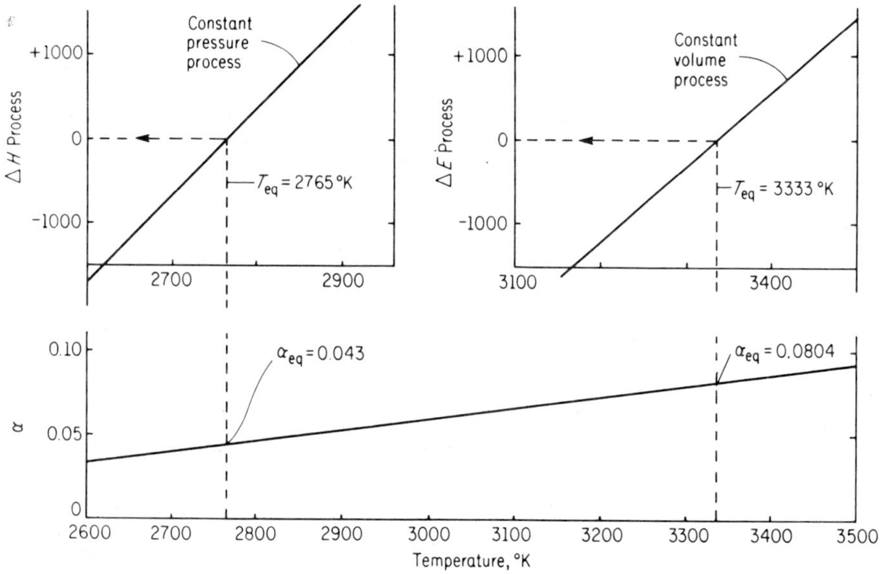

Fig. 4-12. Degree of reaction α, ΔH process, ΔE process, plotted for the calculation of Sec. 4-12.

performing this calculation. First, for purposes of calculating an enthalpy or energy balance for the system we write an equation to summarize the chemical changes that occur.

$$\underbrace{(1)\mathrm{NO}}_{\substack{\text{initial}\\\text{system}\\\text{(mole basis)}}} \rightarrow \underbrace{\alpha(T_f)\mathrm{NO} + \tfrac{1}{2}(1 - \alpha(T_f))\mathrm{O}_2 + \tfrac{1}{2}(1 - \alpha(T_f))\mathrm{N}_2}_{\substack{\text{equilibrium mixture composition at temperature } T_f\\\text{final system}\\\text{(mole basis)}}} \tag{4-36}$$

These coefficients are not stoichiometric coefficients in the usual sense but simply represent the net quantity of each specie present in the initial mixture and at an assumed final temperature T_f. In general, a final pressure would also be needed to fully evaluate the right-hand side of this equation.

We now note that in terms of tabulated values available to evaluate the condition $\Delta H = 0$, there are two contributions to the enthalpy change in this system. These are the chemical enthalpy change caused by com-

TABLE 4-3
Constant-Pressure Explosion-Temperature Calculation*

T °K	α	Method a ($T_r = T_f$)		$\Delta H_{process}$	Method b ($T_r = T_i$)	
		$(\alpha - 1)\left[(\Delta H_f^\circ)_{T_f}\right]_{NO}$	$(H_{T_f} - H_{298.16})_{NO}$		$(\alpha - 1)\left[(\Delta H_f)_{T_i}\right]_{NO}$	Σ†
2600	0.03388	−20827.6	19129.0	−1698.6	−20848.9	19150.3
2700	0.03932	−20693.0	20020.0	−673.0	−20731.5	20058.5
2800	0.04508	−20549.9	20911.0	361.1	−20607.2	20968.3
2900	0.05119	−20397.5	21805.0	1407.5	−20475.3	21882.8
3000	0.05758	−20237.5	22700.0	2462.5	−20337.5	22800.0

*Interpolation yields $T_{eq} = 2765$ °K; $\alpha_{eq} = 0.0430$ (see Fig. 4-12).
† $\Sigma = \frac{1}{2}(1 - \alpha)(H_{T_f} - H_{298})_{N_2} + \frac{1}{2}(1 - \alpha)(H_{T_f} - H_{298})_{O_2} + \alpha(H_{T_f} - H_{298})_{NO}$.

TABLE 4-4
Constant-Volume Explosion Calculation‡
Method A Only

T °K	α	$(\alpha - 1)\left[(\Delta E_f^\circ)_{T_f}\right]_{NO}$	$(E_{T_f} - E_{298.16})_{NO}$	$\Delta E_{process}$
3100	0.06415	−20073.0	18028.8	−2044.2
3200	0.07111	−19897.8	18727.1	−1170.7
3300	0.07815	−19120.2	19427.3	−292.9
3400	0.08534	−19538.1	20127.7	589.6
3500	0.09264	−19353.1	20830.0	1476.9

‡Interpolation yields $T_{eq} = 3333$ °K; $\alpha_{eq} = 0.0804$ (See Fig. 4-12).

position changes at constant temperature and the sensible enthalpy change caused by temperature changes at constant composition. Therefore, for calculation purposes one usually evaluates the enthalpy change in the following manner.

$$(H_{T_r} - H_{T_i})_{\text{reactants}} + (\Delta H_{\text{chem}})_{T_r} + (H_{T_f} - H_{T_r})_{\text{products}} = \Delta H_{\text{process}}(T_f) \quad (4\text{-}37)$$

Where the choice of the temperature T_r is either arbitrary (if sufficient thermodynamic data is available) or is limited to a short range (by the lack of thermodynamic data for either the sensible enthalpy or the enthalpy of formation of the reactants). As written, $\Delta H_{\text{process}}$ is a function of the assumed final temperature; in general it is also a function of pressure. The equilibrium temperature T_{eq} may now be determined by using this equation and adjusting T_f until $\Delta H_{\text{process}} = 0$.

In our example we evaluate Eq. 4-37 for two assumed values of T_r: (a) $T_r = T_f$ and (b) $T_r = T_i$. These results are tabulated in Table 4-3 and summarized in Fig. 4-12. Note that both techniques of evaluation yield the same value of T_{eq} and in fact yield identical values of $\Delta H_{\text{process}}$. In general, if thermodynamic data are available, the evaluation of T_{eq} by process *a* is simpler.

If we want to calculate a constant volume explosion temperature internal energy is conserved and we write

$$(E_{T_r} - E_{T_i})_{\text{reactants}} + (\Delta E_{\text{chem}})_{T_r} + (E_{T_f} - E_{T_r})_{\text{products}} = \Delta E_{\text{process}}(T_f) \quad (4\text{-}38)$$

Values of E for Eq. 4-38 may be determined from values of H tabulated in Appendix F in the following manner. In general the individual reaction processes will be pressure dependent and $\Delta E_{\text{chem}} \neq \Delta H_{\text{chem}}$. From the definition of enthalpy we obtain (for any specific stoichiometric relation)

$$(\Delta E_{\text{chem}})_{T_r} = (\Delta H_{\text{chem}})_{T_r} - RT_r \sum_{i=1}^{s} \nu_i$$

therefore, in our case

$$(\Delta E_{\text{chem}})_{T_r} = (\Delta H_{\text{chem}})_{T_r}$$

Sensible internal energies may also be evaluated on a molar basis by using the definition of enthalpy

$$E_T = H_T - RT$$

Table 4-4 contains a tabulation of the internal energy terms from Eq. 4-38 as used to determine the constant-volume explosion temperature of this system. The results are graphed in Fig. 4-12.

The value of

$$(T_{\text{eq}})_{V=V_i}$$

is higher than

$$(T_{\text{eq}})_{P=P_i}$$

because external work was performed during the process in the later case. A calculation of the entropy change of the system for these two processes will show that entropy increases, indicating that the processes are thermodynamically spontaneous.

4-13. A REACTION-RATE CALCULATION

In this section we will consider the course of a complex reaction to indicate the technique of calculating the concentration changes that occur. In order to simplify the calculation as much as possible yet still retain the essence of the problem, we will consider a gas-phase reaction occurring in a constant-volume–constant-temperature environment. Therefore let us assume that the following four reactions are occurring simultaneously in a mixture that initially contains only the specie A.

$$A \xrightarrow{k_1} B + C$$

$$B + C \xrightarrow{k_2} A$$

$$A + B \xrightarrow{k_3} D$$

and

$$D + C \xrightarrow{k_4} E$$

where the first two reactions represent a single reversible reaction and the last two are each irreversible. Notice that at the end of all reaction in this system the overall stoichiometry of the above equations requires that only E remain in the system. We will assign the reaction coordinates α, $-\alpha$, β and γ respectively to these four reactions. Here α appears twice because the first two reactions are stoichiometrically equivalent.

The rate of these reactions as written in both the chemist's notation and the reaction coordinate notation are:

Chemist's Notation | Reaction Coordinate Notation

$$(1)\quad \left(\frac{d[A]}{dt}\right)_1 = -\left(\frac{d[B]}{dt}\right)_1 = -\left(\frac{d[C]}{dt}\right)_1 = -k_1[A] = \frac{-1}{\rho}\left(\frac{d\alpha}{dt}\right)_1$$

$$(2)\quad \left(\frac{d[A]}{dt}\right)_2 = -\left(\frac{d[B]}{dt}\right)_2 = -\left(\frac{d[C]}{dt}\right)_2 = k_2[B][C] = \frac{-1}{\rho}\left(\frac{d\alpha}{dt}\right)_2$$

$$(3)\quad \left(\frac{d[A]}{dt}\right)_3 = \left(\frac{d[B]}{dt}\right)_3 = -\left(\frac{d[D]}{dt}\right)_3 = -k_3[A][B] = \frac{1}{\rho}\left(\frac{d\beta}{dt}\right)_3$$

$$(4)\quad \left(\frac{d[D]}{dt}\right)_4 = \left(\frac{d[C]}{dt}\right)_4 = -\left(\frac{d[E]}{dt}\right)_4 = -k_4[D][C] = \frac{1}{\rho}\left(\frac{d\gamma}{dt}\right)_4$$

where k_1 has units of $\sec^{-1}$, k_2, k_3, and k_4 have units of $\text{cm}^3/\text{mole sec}$ and the chemist's notation on the left-hand side of these equations implies that the rate of concentration change measured at constant volume and temperature was used to determine the numerical values of the k's.

We now use the notation of Eq. 4-11 to obtain relations for the *net* change of specie concentration at constant volume and temperature at time t_i. Recall that $\rho n_A = [A]$.

$$\frac{d[A]}{dt} = -k_1[A] + k_2[B][C] - k_3[A][B] = \zeta_{1i}$$

$$\frac{d[B]}{dt} = k_1[A] - k_2[B][C] - k_3[A][B] = \zeta_{2i}$$

$$\frac{d[C]}{dt} = k_1[A] - k_2[B][C] - k_4[C][D] = \zeta_{3i}$$

$$\frac{d[D]}{dt} = k_3[A]B - k_4[C][D] = \zeta_{4i}$$

$$\frac{d[E]}{dt} = -k_4[C][D] = \zeta_{5i}$$

These equations may now be integrated numerically by assuming a small time increment and writing them as finite-difference equations:

$$\Delta[A]_i = \zeta_{1i}\Delta t$$
$$\Delta[B]_i = \zeta_{2i}\Delta t$$
$$\vdots$$
$$\Delta[E]_i = \zeta_{ni}\Delta t$$

and

$$t_{i+1} = t_i + \Delta t$$

where i refers to the ith time step. For simple forward integration we may write

$$[A]_{i+1} = [A]_i + \Delta[A]_i$$
$$[B]_{i+1} = [B]_i + \Delta[B]_i$$
$$\vdots$$
$$[E]_{i+1} = [E]_i + \Delta[E]_i$$

and

$$t_{i+1} = t_i + \Delta t$$

Notice that the ζ_{ni} in these equations are variables and that Δt may be varied from step to step in the calculation. The reader is referred to books on numerical techniques for a discussion of more sophisticated numerical integration methods. We simply note here that the error in each step of a forward integration technique, while small, is in general cumulative and therefore represents a limitation to this simple method.

In a constant-volume system, not constrained to constant temperature, an internal energy balance must be used to determine the proper temperature at each successive point in the calculation. In the case where V and T are not constrained, the concentration of each specie can change because the density changes. The proper conservative equations (for the specific assumptions of the problem, i.e., constant-pressure or constant-area steady-duct flow or nonsteady flow etc.) must be applied at each step of the calculation (sometimes in an iterative manner) to determine a set of new state properties and concentrations. In this more general case it is usually easier to work with moles/gram as the independent variable. For this case, the finite-difference equations become

$$(\Delta n_j)_i = \rho \zeta_{ji} \Delta t; \quad j = 1, 2, \ldots, s$$

where n_j refers to the jth specie and i refers to the ith integration step.

For illustration, the rate equations introduced at the start of this section have been numerically integrated for $[A]_{t=0} = 1 \times 10^{-5}$ moles/cm^3, $[B]_{t=0} = [C]_{t=0} = [D]_{t=0} = [E]_{t=0} = 0$ for a constant-volume, constant-temperature reaction for the three cases in Table 4.5.

TABLE 4-5.
RATE CONSTANTS FOR EXAMPLE PROBLEM

Case	$\sec^{-1}$	cm^3/mole sec			K_c moles/cm^3	
	k_1	k_2	k_3	k_4	$\frac{[B][C]}{[A]}$	k_1/k_2
I	0.1	10^5	10^4	10^3	0.9979×10^{-6}	1×10^{-6}
II	0.1	10^4	10^3	10^4	1.0076×10^{-5}	1×10^{-5}
III	0.1	10^3	10^4	10^4	1.1256×10^{-4}	1×10^{-4}

The results are illustrated in Figs. 4-13 through 4-15. Notice that the concentrations of the intermediate species increase at first to a maximum value and then decrease as the concentration of E continues to increase. The relative behavior of the intermediate specie concentrations in the three cases is caused by changes in the relative magnitude of the rate constants.

This example may be used to discuss briefly two commonly used concepts in reaction kinetic–equilibrium studies. The first of them pertains to the relationship between the reaction rates and the equilibrium constant introduced at the beginning of Sec. 4-8. We check this concept at late time by calculating the ratio $[B] \cdot [C]/[A]$ at 1,700 sec for comparison with the ratio k_1/k_2. This comparison is shown in Table 4.5. In Case III,

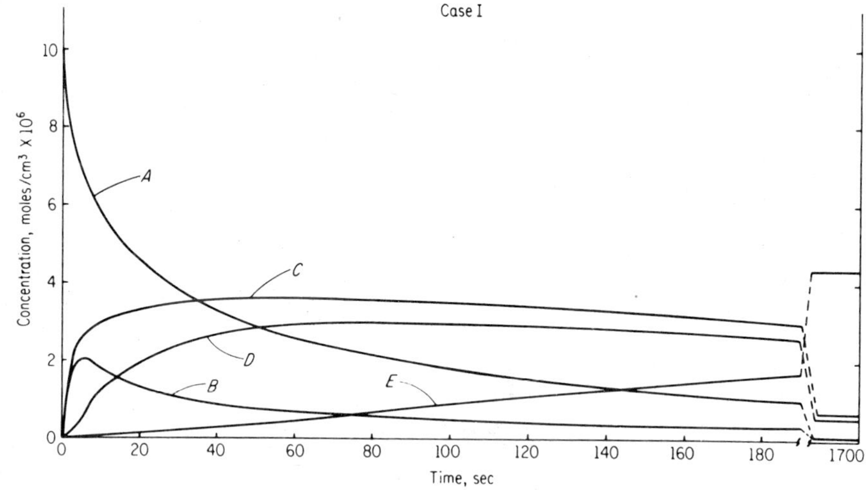

Fig. 4-13. Concentration of the species as a function of time for the rate constants of Table 4-5.

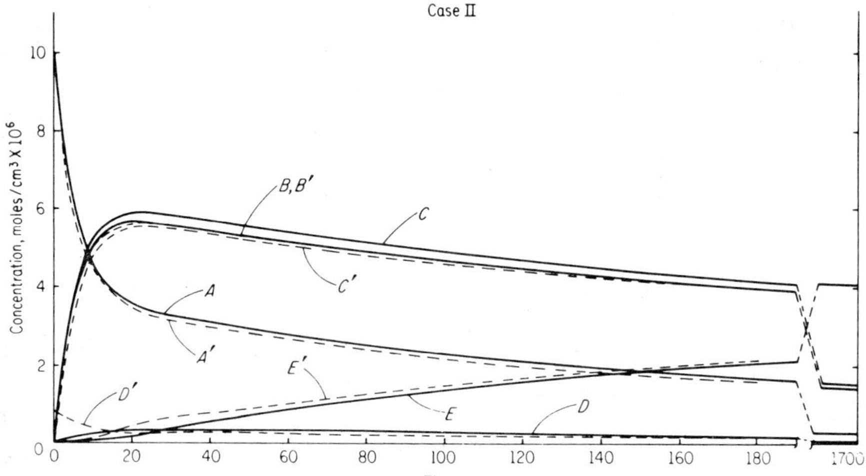

Fig. 4-14. Concentration of the species as a function of time for the rate constants of Table 4-5. (Solid lines represent full forward numerical integration, dashed lines the result when the steady-state approximation is applied to D throughout the course of the reaction.)

equilibrium is not approached because the side reactions are constantly perturbing the concentrations. Note that the relative rate of these side reactions is highest in this case.

The utility of the steady-state approximation introduced in Sec. 4-6 will be investigated next. We will apply this approximation to the con-

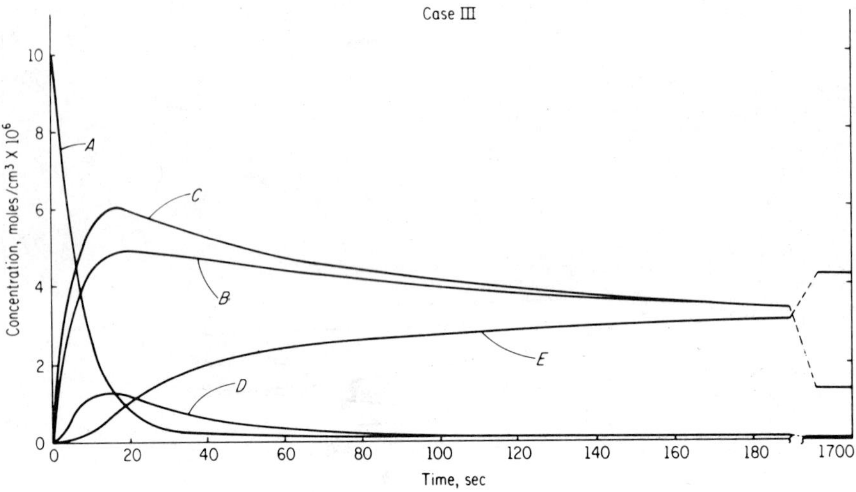

Fig. 4-15. Concentration of the species as a function of time for the rate constants of Table 4-5.

centration of specie D in Case II. If the steady-state approximation is valid we may write

$$\frac{d[D]}{dt} = k_3[A][B] - k_4[C][D] \cong 0 \tag{4-40}$$

which yields

$$[D] \cong \frac{k_3}{k_4}\frac{[A][B]}{[C]} \tag{4-41}$$

Note that Eq. 4-40 violates mass conservation for the "steady-state" specie.

For purposes of this example we perform a new calculation of Case II in which we apply this requirement throughout the course of the reaction. The dashed lines in Fig. 4-14 were calculated by using Eq. 4-41 to first determine the concentration of D at any time increment and then using the remainder of the rate equations to determine concentration changes for the other species. We note from this figure that only the concentration of D and E deviate markedly from the concentrations calculated with the full equations. We also note from Fig. 4-14 that if we were to defer application of the steady-state approximation for 40 sec in Case II we could expect to duplicate the full calculation rather accurately. Numerical programs have been developed in which the steady-state approximation (Eq. 4-40 or its equivalent) is examined at every point in the calculation. When this criterion is satisfied to within a desired accuracy the program automatically uses the equivalent of Eq. 4-41 to calculate the concentration of the "steady-state" species and the remaining rate equations to calculate other composition changes.[10]

4-14. MULTICOMPONENT EQUILIBRIA

In a multicomponent system the calculation of an equilibrium composition becomes quite complex. As an example, let us consider the carbon–hydrogen–oxygen–nitrogen system (with no solid carbon formation). This system is particularly interesting as all hydrocarbon–air mixtures contain only these four components. Let us illustrate how to calculate an equilibrium composition at a temperature in the range 1500–3500°K. As a first step we assume that all species stable in the 1500–3500°K temperature range are present at equilibrium. These are carbon dioxide, CO_2; carbon monoxide, CO; nitrogen, N_2; nitrogen atoms, N; water, H_2O; hydrogen, H_2; hydrogen atoms, H; oxygen, O_2; oxygen atoms, O; hydroxyl radicals, OH; and nitric oxide, NO. Next, using stoichiometry, the mass-balance constants of the problem are defined by the equations

$$N_C = n_{CO_2} + n_{CO}$$

$$N_O = 2n_{CO_2} + n_{CO} + n_{H_2O} + 2n_{O_2} + n_O + n_{OH} + n_{NO}$$

$$N_H = 2n_{H_2} + 2n_{H_2O} + n_H + n_{OH}$$

$$N_N = 2n_{N_2} + n_N + n_{NO}$$

At this point, we have eleven unknows (the n_i) and only four equations. Therefore, we need seven independent equilibrium equations to solve the problem. The choice of the particular seven equations is arbitrary. Let us choose these seven independent equations:

(1) The water-gas equilibrium:

$$CO_2 + H_2 \rightleftharpoons CO + H_2O$$

$$K_1 = \frac{p_{CO}p_{H_2O}}{p_{CO_2}p_{H_2}} = \frac{n_{CO}n_{H_2O}}{n_{CO_2}n_{H_2}}$$

(2) The dissociation of water:

$$2H_2O \rightleftharpoons 2H_2 + O_2$$

$$K_2 = \frac{p_{H_2}^2 p_{O_2}}{p_{H_2O}^2} = \frac{n_{H_2}^2 n_{O_2}}{n_{H_2O}^2}\left(\frac{P}{n_T}\right)$$

(3) The dissociation of nitrogen:

$$N_2 \rightleftharpoons 2N$$

$$K_3 = \frac{p_N^2}{p_{N_2}} = \frac{n_N^2}{n_{N_2}}\left(\frac{P}{n_T}\right)$$

[10] For a rather complete discussion, see R. H. Snow, *J. Phys. Chem.*, *70*, 2780 (1966).

(4) The dissociation of hydrogen:

$$H_2 \rightleftharpoons 2H$$

$$K_4 = \frac{p_H^2}{p_{H_2}} = \frac{n_H^2}{n_{H_2}}\left(\frac{P}{n_T}\right)$$

(5) The dissociation of oxygen:

$$O_2 \rightleftharpoons 2O$$

$$K_5 = \frac{p_O^2}{p_{O_2}} = \frac{n_O^2}{n_{O_2}}\left(\frac{P}{n_T}\right)$$

(6) The formation of hydroxyl radical:

$$H_2O \rightleftharpoons H + OH$$

$$K_6 = \frac{p_H p_{OH}}{p_{H_2O}} = \frac{n_H n_{OH}}{n_{H_2O}}\left(\frac{P}{n_T}\right)$$

(7) The formation of nitric oxide:

$$N_2 + O_2 \rightleftharpoons 2NO$$

$$K_7 = \frac{p_{NO}^2}{p_{N_2}p_{O_2}} = \frac{n_{NO}^2}{n_{N_2}n_{O_2}}$$

These equations are independent since none can be simply expressed as a combination of the others. Using the above equations and the free-energy or equilibrium constants of formation of the eleven species, one may calculate equilibrium compositions at any temperature and pressure. In general, the computation is rather tedious.[11] (Notice that there are eleven equations and eleven unknowns and that seven of these equations are nonlinear algebraic equations.)

Once the composition is determined at an assumed temperature and pressure, the thermodynamic properties and the volume of the equilibrium mixture may be calculated using the equations discussed earlier. To subsequently determine a homogeneous explosion temperature at constant pressure or constant volume we follow the procedure outlined in Sec. 4-12. Notice that this calculation requires a double iteration procedure; the inner calculation is an iteration on the composition at an assumed temperature, the outer an iteration on the temperature to yield $\Delta H = 0$ or $\Delta E = 0$.

As an example of a simplified approach to the calculation of equilibrium in the C, O, H, N system, consider a system containing a hydro-

[11] Accurate solutions can be calculated quickly with the use of modern high-speed digital computers. Programs and techniques are available. See, for example, R. G. Anthony and D. M. Himmelblau, *J. Phys. Chem.*, *67*, 1080 (1963); F. J. Zeleznik and S. Gordon, "A General IBM 704 or 7090 Code for Equilibrium," NASA Tech Note D 1454 (1962); P. F. Bird et al., "Hug a Fortran Code," Report LA-2980, Los Alamos Scientific Laboratory, Los Alamos, New Mexico (February 1964); D. M. Himmelblau, Kirk-Othmer Encyclopedia of Chemical Technology, 2nd ed., vol. 10, pp. 191–220 (New York: Wiley Interscience, 1966); W. B. White, *J. Chem. Phys.*, *46*, 4171 (1967).

carbon plus excess air. The eleven species needed in the general treatment can then be reduced to five because of the low adiabatic explosion temperatures in the presence of excess air. The five species are carbon dioxide, carbon monoxide, hydrogen, oxygen, and water vapor. A sixth species, nitrogen, is taken as inert. The reaction equation for any general hydrocarbon may then be written

$$C_uH_v + (u + v/4 + e)(O_2 + 3.76\,N_2) \rightarrow$$

$$ux\,CO + u(1 - x)CO_2 + \frac{v}{2}\,yH_2 + \frac{v}{2}(1 - y)H_2O$$

$$+ \left(\frac{ux}{2} + \frac{vy}{4} + e\right)O_2 + 3.76\left(u + \frac{v}{4} + e\right)N_2$$

where x and y represent the fraction of carbon remaining as carbon monoxide and the fraction of hydrogen remaining as hydrogen gas and may be considered as reaction coordinates for this system. Notice that stoichiometry requires that $0 < x < 1$ and $0 < y < 1$. Two equilibrium constants are needed to specify this system. We will use

$$K_{p_{CO_2}} = \frac{p^2_{CO_2}}{p^2_{CO}p_{O_2}} = \frac{n^2_{CO_2}}{n^2_{CO}n_{O_2}} \cdot \left(\frac{n_T}{P}\right)$$

and

$$K_{p_{wg}} = \frac{p_{CO_2} \cdot p_{H_2}}{p_{H_2O} \cdot p_{CO}} = \frac{n_{CO_2} \cdot n_{H_2}}{n_{H_2O} \cdot n_{CO}}$$

Using the stoichiometric coefficients these yield

$$K_{p_{CO_2}} = \frac{(1 - x)^2[u(1 + x/2) + v(1 + y/2)/2 + e + 3.76(u + v/4 + e)]}{x^2(ux/2 + vy/4 + e)P}$$

and

$$K_{p_{wg}} = \frac{(1 - x)y}{(1 - y)x}$$

We, therefore, have reduced the problem to two nonlinear equations in two unknowns. Notice that the quantities

$$\alpha = u(1 + x/2) + v(1 + y/2)/2 + e + 3.76(u + v/4 + e)$$

$$\beta = ux/2 + vy/4 + e$$

vary only slightly with x and y. We may, therefore solve these equations by first assuming that α and β are truly constant and then iterating on the result to determine the correct value of x and y. Let α_0 and β_0 be the values of α and β for $x = y = 0$. Then

$$K_{p_{CO_2}} = \frac{(1 - x)^2\,\alpha_0}{x^2\beta_0 P}$$

or

$$x = -\lambda_0 + \sqrt{\lambda_0^2 + \lambda_0}$$

where

$$\lambda_0 = \frac{\alpha_0}{K_{p\mathrm{CO}_2}\beta_0 P - \alpha_0}$$

and therefore

$$y = \frac{K_{p_{wg}} x}{1 - x(1 - K_{p_{wg}})}$$

Using these values of x and y we can now calculate the second approximation to α and β and then calculate a second estimate of x and y, using these values.

As soon as the equilibrium composition is known at the assumed temperature, the enthalpy balance is checked as discussed in Sec. 4-12. Interpolation on the temperature until $\Delta H = 0$ will finally yield the explosion temperature.

Notice that the nature of this particular iteration procedure is such that it will work well for mixtures containing a large excess of air. It converges rather slowly for nearly stoichiometric mixtures. If one were to consider rich mixtures, there would be very little oxygen at equilibrium and one would have to set up a new calculation scheme. This scheme would be simpler, however, because the carbon monoxide–carbon dioxide equilibrium would not have to be considered at first and only the water–gas equilibrium would be needed in the initial calculation.

Once the equilibrium composition is known at any temperature and pressure, it is easy to check and determine if an appreciable error was made by assuming that the other species were absent at equilibrium. In the case treated above we assumed that nitric oxide, hydroxyl radical, nitrogen atoms, oxygen atoms, and hydrogen atoms are absent. Consider nitric oxide as an example. To estimate its formation under the conditions of the above calculation, find the equilibrium constant $K_{p\mathrm{NO}}$ for the reaction

$$N_2 + O_2 \rightleftharpoons 2NO$$

$$K_{p\mathrm{NO}} = \frac{p_{\mathrm{NO}}^2}{p_{\mathrm{N}_2} \cdot p_{\mathrm{O}_2}}$$

and solve to find

$$p_{\mathrm{NO}} = \sqrt{K_{p\mathrm{NO}} p_{\mathrm{N}_2} p_{\mathrm{O}_2}}$$

This approach yields a first estimate of the partial pressure of a specie not considered in the initial calculation and thereby determines how critical is its omission. If the calculation has already yielded an explosion temperature, the maximum effect of the presence of this species on the equilibrium temperature may be estimated by using a simple enthalpy balance. This may be done by calculating the enthalpy change associated with the formation of the estimated amount of the trace specie, keeping all other equilibria unchanged.

BIBLIOGRAPHY

1. K. Denbigh, *The Principles of Chemical Equilibria.* (London: Cambridge, 1961.)

2. S. Glasstone, K. J. Laidler, and H. Eyring, *The Theory of Rate Processes.* (New York: McGraw-Hill, 1941.)

3. S. Glasstone, *Thermodynamics for Chemists.* (New York: Van Nostrand, 1947.)

4. N. N. Semenov, *Some Problems of Chemical Kinetics and Reactivity*, Vol. 2. (New York: Pergamon, 1959.)

5. E. L. King, *How Chemical Reactions Occur.* (New York: Benjamin, 1964.)

6. D. D. Fitts, *Nonequilibrium Thermodynamics.* (New York: McGraw-Hill, 1962.)

PROBLEMS

4-1. Calculate the equilibrium composition of air (composition, 21% O_2 + 79% N_2 by volume) at 2500°K and 5000°K and P = 0.1 atm. Assume that O, N, N_2, O_2, and NO are present at equilibrium.

4-2. In problem 1 above calculate

A. Volume ratio $V_{\text{equilibrium}}/V_{\text{unreacted}}$.

B. Enthalpy of the equilibrium mixture in cal/mole of unreacted mixture relative to the unreacted mixture at 298.16°K.

C. Heat-capacity ratio (frozen composition) $\gamma_{\text{equilibrium}}/\gamma_{\text{unreacted}}$.

4-3. Consider pure oxygen at equilibrium at 3000°K, P = 0.01 atm.

A. Calculate the degree of dissociation α.

B. Calculate the frozen composition heat-capacity ratio of this mixture.

C. Calculate the shifting equilibrium heat-capacity ratio of this mixture.

4-4. Calculate a value for the degree of dissociation α for Br_2 at 1000°K, for O_2 at 5000°K and for F_2 at 5000°K and 1 atm. pressure, using Table 4-2 and Fig. 4-10. Compare your values to values calculated from the thermodynamic functions given in Appendix F.

4-5. Using the system presented in Sec. 4-12, show that isentropic expansion of the constant-volume explosion products to $P = P_i$ does not yield the constant-pressure explosion temperature for either an equilibrium composition or frozen composition expansion. Discuss the reason for this behavior.

4-6. Perform the calculation of Sec. 4-12 at $P = P_i = 1$ atm assuming that oxygen atoms are present at equilibrium. Should one also include nitrogen atoms?

5

One-Dimensional Reactive-Gas Dynamics

5-1. INTRODUCTION

In the previous chapters we have discussed in some detail the properties of ideal and real gases, including the behavior of reactive systems. In those chapters the emphasis was on homogeneous processes and the molecular basis for our understanding of these processes. However, a number of combustion phenomena occur as relatively distinct wave processes, and thus we must consider some aspects of reactive-gas dynamics to complete our coverage of fundamentals. In this chapter we will therefore consider a few simple one-dimensional flow situations for a gas capable of reacting chemically. For simplicity, transport and internal relaxation phenomena will be ignored and only inserted as needed for the discussion of specific cases in later chapters. With these restrictions the fundamental conservation equations may be written as

$$\frac{D\rho}{Dt} + \rho\nabla \cdot u = 0 \quad \text{(mass)} \tag{5-1}$$

$$\rho\frac{Du}{Dt} = -\nabla P \quad \text{(momentum)} \tag{5-2}$$

$$\rho\frac{De}{Dt} = -P\nabla \cdot u \quad \text{(energy)} \tag{5-3}$$

$$\rho\frac{Dy_i}{Dt} = \mathcal{W}_i \quad i = 1, 2, \ldots, N \tag{5-4}$$

(mass balance on each specie; in this chapter we use N for the total number of species to avoid confusion with s, the specific entropy)

where

$$\sum_{i=1}^{N} \frac{Dy_i}{Dt} = 0 \text{ everywhere} \tag{5-5}$$

since by definition $\sum_{i=1}^{N} y_i = 1$. The $\mathcal{W}_i$ term in Eq. 5-4 is a composite rate term for each specie with the units, gm/cc sec. In these equations the total or Eulerian derivative is defined as

$$\frac{D}{Dt} = \frac{\partial}{\partial t} + u \cdot \nabla$$

Since these equations represent inviscid flow we know that they also represent equilibrium flow for an unreactive or fixed-composition system. Thus Eqs. 5-1 and 5-3 yield the first law of thermodynamics for a conservative system written as rate equation:

$$\frac{De}{Dt} = \frac{P}{\rho^2}\frac{D\rho}{Dt} \tag{5-6}$$

This verifies the statement that

$$\frac{Ds}{Dt} = 0$$

in the unreactive flow case. Implicit in this statement is the assumption that we can define all the thermodynamic properties at every point in the flow. This concept will now be extended to include the reactive case as one which is everywhere definable as a thermodynamic system, but which may or may not be in full chemical equilibrium. Let us write Eq. 4-19a on a mass basis:

$$de = T\,ds - Pd\left(\frac{1}{\rho}\right) + \sum_{i=1}^{N}\left(\frac{\mu_i}{M_i}\right) dy_i \tag{5-7}$$

for a differential and reversible change in the systems properties. Equation 5-7 is correct either for an open system (as defined in Chapter 4) or for the particular closed system where the dy_i's can vary only along the reaction coordinates given by Eq. 4-10a. Combining Eqs. 5-6 and 5-7 we obtain

$$\frac{Ds}{Dt} = -\frac{1}{T}\sum_{i=1}^{N}\left(\frac{\mu_i}{M_i}\right)\frac{Dy_i}{Dt} \tag{5-8}$$

which states that entropy may be produced in this thermodynamic system by the occurrence of irreversible chemical reactions. Note that Eq. 5-8 reduces to an isentropic law for either fixed composition (all $dy_i = 0$) or for the occurrence of chemical equilibria throughout the flow (i.e., $\Sigma \mu_i dn_i = 0$).

5-2. THE SPECIFYING EQUATION

The specifying equation for reactive flow may be derived if we assume that every point in the flow has a thermodynamically definable state and that the flow is isentropic excepting the contribution of the irreversible chemical reaction. We can then write

$$\rho = \rho(P, s, y_i) \qquad i = 1, 2, \ldots, N$$

which allows us to write the equation

$$\frac{D\rho}{Dt} = \left(\frac{\partial \rho}{\partial P}\right)_{s,y_i} \frac{DP}{Dt} + \left(\frac{\partial \rho}{\partial s}\right)_{P,y_i} \frac{Ds}{Dt} + \sum_{i=1}^{N} \left(\frac{\partial \rho}{\partial y_i}\right)_{P,s,y_j} \frac{Dy_i}{Dt}$$

where $(\partial/\partial y_i)_{y_j}$ means that all the y_j's are held constant for $j \neq i$. Therefore

$$\frac{D\rho}{Dt} = \frac{1}{a_f^2} \frac{DP}{Dt} + \left(\frac{\partial \rho}{\partial s}\right)_{P,y_i} \left[\frac{Ds}{Dt} - \sum_{i=1}^{N} \left(\frac{\partial s}{\partial y_i}\right)_{P,\rho,y_j} \frac{Dy_i}{Dt}\right]$$

Where a_f is the frozen-composition velocity of sound. This may be written

$$\frac{D\rho}{Dt} = \frac{1}{a_f^2} \frac{DP}{Dt} - T\rho \frac{\beta_f}{\mathfrak{C}_{Pf}} \left[\frac{Ds}{Dt} - \sum_{i=1}^{N} \left(\frac{\partial s}{\partial y_i}\right)_{P,\rho,y_j} \frac{Dy_i}{Dt}\right]$$

where

$$\beta_f = \frac{1}{V} \left(\frac{\partial V}{\partial T}\right)_{P,y_i}$$

the frozen-composition coefficient of thermal expansion and $\mathfrak{C}_{Pf}$ is the frozen-composition heat capacity. Each term of the sum may be expanded by noting that

$$s = s(P, T, y_i)$$

or

$$ds = \left(\frac{\partial s}{\partial P}\right)_{T,y_i} dP + \left(\frac{\partial s}{\partial T}\right)_{P,y_i} dT + \sum_{i=1}^{N} \left(\frac{\partial s}{\partial y_i}\right)_{P,T,y_j} dy_i$$

which yields

$$\left(\frac{\partial s}{\partial y_i}\right)_{P,\rho,y_j} = \left(\frac{\partial s}{\partial y_i}\right)_{T,P,y_j} + \left(\frac{\partial s}{\partial T}\right)_{P,y_i} \left(\frac{\partial T}{\partial y_i}\right)_{P,\rho,y_j} + \left(\frac{\partial s}{\partial P}\right)_{T,y_i} \left(\frac{\partial P}{\partial y_i}\right)_{P,\rho,y_j}$$

where the last term is zero and the second term becomes

$$\frac{\mathfrak{C}_{Pf}}{\beta_f T\rho} \left(\frac{\partial \rho}{\partial y_i}\right)_{P,T,y_j}$$

Therefore

$$\frac{D\rho}{Dt} = \frac{1}{a_f^2} \frac{DP}{Dt} - \frac{T\rho\beta_f}{\mathfrak{C}_{Pf}} \left[\frac{Ds}{Dt} - \sum_{i=1}^{N} \left\{\left(\frac{\partial s}{\partial y_i}\right)_{T,P,y_j} \frac{Dy_i}{Dt} + \frac{\mathfrak{C}_{Pf}}{\beta_f T\rho} \left(\frac{\partial \rho}{\partial y_i}\right)_{P,T,y_j} \frac{Dy_i}{Dt}\right\}\right]$$

and, substituting Eq. 5-8 for Ds/Dt we obtain

$$\frac{D\rho}{Dt} = \frac{1}{a_f^2} \frac{DP}{Dt} + \frac{T\rho\beta_f}{\mathfrak{C}_{Pf}} \sum_{i=1}^{N} \left[\frac{\mathfrak{C}_{Pf}}{\beta_f T\rho} \left(\frac{\partial \rho}{\partial y_i}\right)_{P,T,y_j} + \frac{1}{T} \frac{\mu_i}{M_i} + \left(\frac{\partial s}{\partial y_i}\right)_{P,T,y_j}\right] \cdot \frac{Dy_i}{Dt}$$

However,

$$\left(\frac{\mu_i}{M_i}\right) = \left(\frac{\partial f}{\partial y_i}\right)_{P,T,y_j} = \left(\frac{\partial h}{\partial y_i}\right)_{P,T,y_j} - T\left(\frac{\partial s}{\partial y_i}\right)_{P,T,y_j}$$

Therefore, in its final form the specifying equation becomes

$$\frac{D\rho}{Dt} = \frac{1}{a_f^2}\frac{DP}{Dt} + \rho\sum_{i=1}^{N}\left[\frac{1}{\rho}\left(\frac{\partial \rho}{\partial y_i}\right)_{P,T,y_j} + \frac{\beta_f}{\mathcal{C}_{Pf}}\left(\frac{\partial h}{\partial y_i}\right)_{P,T,y_j}\right]\cdot\frac{Dy_i}{Dt} \tag{5-9}$$

For an ideal gas mixture $\beta_f = 1/T$ and

$$\frac{D\rho}{Dt} = \frac{1}{a_f^2}\frac{DP}{Dt} + \sum_{i=1}^{N}\left[\left(\frac{\partial \rho}{\partial y_i}\right)_{P,T,y_j} + \frac{\rho}{\mathcal{C}_{Pf}T}\left(\frac{\partial h}{\partial y_i}\right)_{P,T,y_i}\right]\frac{Dy_i}{Dt} \tag{5-10}$$

For the case of no reaction Eq. 5-10 reduces to the well-known isentropic relation for conservative flow:

$$\frac{D\rho}{Dt} = \frac{1}{a^2}\frac{DP}{Dt}$$

The extra terms represent respectively a contribution arising from a change in the density of the gas at constant pressure and temperature due to the net production or destruction of species during reaction and a contribution representing the sensible enthalpy release during chemical change. If we assume that the gas mixture is polytropic and that there is no molecular weight change during chemical reaction, we can represent the chemical reaction by a programmed heat addition along a fluid particle path:

$$\frac{Dq}{Dt} = -\sum_{i=1}^{N}\left(\frac{\partial h}{\partial y_i}\right)_{P,T,y_j}\frac{Dy_i}{Dt}$$

where, by convention, Dq/Dt is positive for an exothermic reaction. By replacing $1/T$ by $\gamma\mathcal{R}/a^2$ we obtain the usual specifying equation for the heat-addition–working-fluid model

$$\frac{D\rho}{Dt} = \frac{1}{a^2}\frac{DP}{Dt} - \rho\frac{(\gamma - 1)}{a^2}\frac{Dq}{Dt} \tag{5-11}$$

where the units on q and a^2 must be the same. Equations 5-11, 5-1, 5-2, and 5-3 constitute an equation set which has been used extensively for studying combustion systems. Using these equations, thermal energy or heat is assumed to appear spontaneously in the flow from some external source, thus representing the effect of the chemical reaction. Accordingly, the apparent entropy of the flow increases markedly when a heat-addition model is used. This artifice of the heat-addition model has no physical significance because the entropy change in the universe must be discussed to determine reversibility or spontaneity and the model by its simplification, allows no such discussion. However, the heat-addition model can still be considered as a useful approximation for discussing many combustion processes.

5-3. NONSTEADY ONE-DIMENSIONAL FLOW— THE METHOD OF CHARACTERISTICS

Equations 5-2, 5-4, 5-8 and Eqs. 5-1 and 5-4 substituted into Eq. 5-10 reduce to the following set for the case of one-dimensional nonsteady flow.

$$\rho \frac{\partial u}{\partial t} + \rho u \frac{\partial u}{\partial x} + \frac{\partial P}{\partial x} = 0 \tag{5-12}$$

$$\frac{\partial y_i}{\partial t} + u \frac{\partial y_i}{\partial x} = \frac{\mathcal{W}_i}{\rho} \qquad i = 1, 2, \ldots, N \tag{5-13}$$

$$\frac{\partial s}{\partial t} + u \frac{\partial s}{\partial x} = -\frac{1}{T} \sum_{i=1}^{N} \left(\frac{\mu_i}{M_i}\right) \frac{Dy_i}{Dt} \tag{5-14}$$

and

$$\rho \frac{\partial u}{\partial x} + \frac{1}{a_f^2} \frac{\partial P}{\partial t} + \frac{u}{a_f^2} \frac{\partial P}{\partial x} = + \sum_{i=1}^{N} \sigma_i \mathcal{W}_i \tag{5-15}$$

where

$$\sigma_i = \frac{1}{\rho} \left[\left(\frac{\partial \rho}{\partial y_i}\right)_{P,T,y_j} - \frac{\rho}{C_{Pf} T} \left(\frac{\partial h}{\partial y_i}\right)_{P,T,y_j} \right]$$

Multiplying Eq. 5-12 by a_f and Eq. 5-15 by a_f^2 and adding and subtracting yields the equation:

$$\frac{\partial P}{\partial t} + (u \pm a_f) \frac{\partial P}{\partial x} \pm \rho a_f \left[\frac{\partial u}{\partial t} + (u \pm a_f) \frac{\partial u}{\partial x} \right] = a_f^2 \sum_{i=1}^{N} \sigma_i \mathcal{W}_i \tag{5-16}$$

The set of Eqs. 5-13, 5-14, and 5-16 constitute the characteristic equations for the analysis of one-dimensional nonsteady flow situations with reaction. Equations 5-13 and 5-14 are $(N + 1)$-fold degenerate equations along a particle path, while Eqs. 5-16 are single characteristic equations along the two directions of sound propagation. The important result of this derivation is the observation that the characteristic directions at any point in the fluid are given by the expression $(u \pm a_f)$ and therefore are determined by the frozen-sound velocity for conservative flow. These equations may be simplified to yield the characteristic set for the heat-addition model, using the approximations described in Sec. 5-2 above. A discussion of the detailed numerical techniques used in solving these equations is beyond the scope of this text but has been described adequately in references listed in the bibliography.

5-4. STEADY ONE-DIMENSIONAL FLOW

For the steady flow of an unreactive gas in a constant area stream tube the equations of motion are known to yield a nontrivial solution: the jump conditions for a steady shock wave. Furthermore, for this adiabatic flow a calculation of the entropy difference between stations located on opposite sides of the shock wave adequately justifies our experimental observation of compression shocks, representing transitions from supersonic to subsonic flow, and our inability to observe steady expansion discontinuities.

For the more general case of reactive flow the jump equations may be obtained by considering one-dimensional flow through a control volume with the additional stipulation that $h = e + P/\rho$. This operation yields the following equations:

$$\rho_1 u_1 = \rho_2 u_2 \quad \text{(mass)} \tag{5-17}$$

$$P_1 + \rho_1 u_1^2 = P_2 + \rho_2 u_2^2 \quad \text{(momentum)} \tag{5-18}$$

$$h_1 + {}^1\!/_2 u_1^2 = h_2 + {}^1\!/_2 u_2^2 \quad \text{(energy)} \tag{5-19}$$

where h is the enthalpy per gram and includes chemical enthalpy.

In our analysis of this system of equations let us first determine the final states (2) allowable from any initial state (1) in the pressure specific volume (P, V) plane. Flow velocity may be written in terms of the initial and final pressures and volumes by rearranging the mass and momentum equations

$$P_1 + \rho_1 u_1^2 = P_2 + \frac{\rho_1^2}{\rho_2} u_1^2$$

to yield

$$u_1^2 = V_1^2 \left(\frac{P_2 - P_1}{V_1 - V_2} \right) \tag{5-20}$$

and similarly

$$u_2^2 = V_2^2 \left(\frac{P_2 - P_1}{V_1 - V_2} \right) \tag{5-21}$$

Notice that these equations state that the mass-flow velocity (ρu) is equal to the square root of the negative slope of a line in the (P, V) plane connecting the final state and the initial state for the particular process under consideration. Thus a truly steady transformation from state 1 to 2 with no transport is restricted to occur through states which lie on a straight line, usually called the *Rayleigh line* in the English literature and the *Mikhel'son line* in the Russian literature.

The energy and momentum equations may be solved to yield the expressions

$$h_2 - h_1 = \tfrac{1}{2}(u_1^2 - u_2^2)$$

and

$$(P_2 - P_1)\left(\frac{1}{\rho_2} + \frac{1}{\rho_1}\right) = u_1^2 - u_2^2 + \frac{\rho_1}{\rho_2}u_1^2 - \frac{\rho_2}{\rho_1}u_2^2$$

Substituting, the mass-balance equation yields

$$(P_2 - P_1)\left(\frac{1}{\rho_2} + \frac{1}{\rho_1}\right) = u_1^2 - u_2^2$$

or

$$h_2 - h_1 = \tfrac{1}{2}(P_2 - P_1)(V_1 + V_2) \qquad (5\text{-}22)$$

This is called the Rankine-Hugoniot equation or simply the Hugoniot equation for steady flow. Notice that for steady, constant area flow the Hugonoit and Rayleigh line equations are completely independent of the choice of an equation of state or of the process occurring between stations 1 and 2. Without imposing any further limiting assumptions we can easily see one limitation to steady-flow processes. In the P, V plane a steady-flow process cannot simultaneously lower or raise both the pressure and volume; thus there are two restricted areas in this plane which cannot contain end states for any steady-wave process, see Fig. 5-1.

To discuss the behavior of these equations in more detail we will now limit ourselves to a constant molecular weight polytropic gas as the working fluid and replace the chemical change by a simple heat addition to the

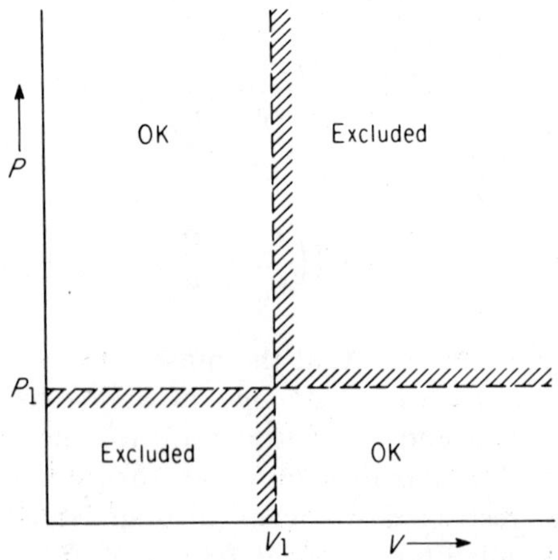

Fig. 5-1. Excluded areas for steady flow.

working fluid. (See Fig. 5-2.) With these assumptions,

$$p_1/\rho_1 T_1 = p_2/\rho_2 T_2 \quad \text{(state)}$$

$$h_1 = h_1' = \mathcal{C}_P T_1$$

and

$$h_2 = h_2' - q = \mathcal{C}_P T_2 - q$$

where a positive value of q represents heat addition to the flow. Therefore

$$\frac{\gamma}{\gamma - 1}(P_2 V_2 - P_1 V_1) - q = \frac{1}{2}(P_2 - P_1)(V_2 + V_1) \tag{5-23}$$

Equation 5-23 represents the locus of final states (P_2, V_2) for any initial state (P_1, V_1) and heat addition q.

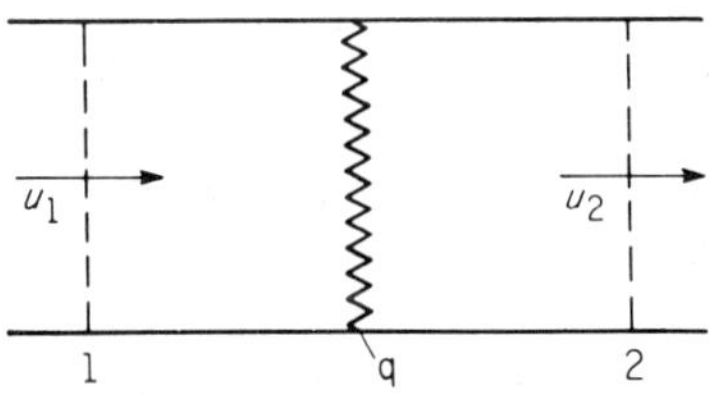

Fig. 5-2. Steady flow with heat addition.

Asymptotic values of both P_2 and V_2 exist for this equation for any value of P_1, V_1, and q. Assume that P_2 approaches infinity. Then

$$\frac{\gamma}{\gamma - 1} P_2 V_2 - q = \frac{1}{2}(P_2 V_2 + P_2 V_1)$$

or

$$\frac{2q}{P_2 V_2} = \frac{\gamma + 1}{\gamma - 1} \cdot \frac{V_2}{V_1} - 1$$

Therefore

$$\frac{V_2}{V_1} \rightarrow \frac{\gamma - 1}{\gamma + 1} \quad \text{as} \quad P_2 \rightarrow \infty$$

Similarly, if V_2 approaches infinity we see that

$$\frac{\gamma}{\gamma - 1} P_2 V_2 - q = \frac{1}{2}(P_2 V_2 - P_1 V_2)$$

or

$$\frac{2q}{p_1 V_2} = \frac{\gamma + 1}{\gamma - 1} \cdot \frac{P_2}{P_1} + 1$$

Therefore

$$\frac{P_2}{P_1} \rightarrow -\frac{\gamma - 1}{\gamma + 1} \quad \text{as} \quad V_2 \rightarrow \infty$$

Since the asymptotic value of P_2/P_1 as V_2 approaches infinity is negative, the solution curve must cross the $P = 0$ line at some point. The value of V_2/V_1 at this intersection may be easily determined by substituting $P_2 = 0$ in the original equations:

$$\frac{\gamma}{\gamma - 1} P_1 V_1 + q = \frac{1}{2} (P_1 V_2 + P_1 V_1)$$

or

$$\frac{2q}{P_1 V_1} = -\frac{\gamma + 1}{\gamma - 1} + \frac{V_2}{V_1}$$

Therefore

$$\frac{V_2}{V_1} = \frac{\gamma + 1}{\gamma - 1} + \frac{2q}{P_1 V_1} \quad \text{at} \quad P_2 = 0$$

The values of P_2 at $V_2 = V_1$ and of V_2 at $P_2 = P_1$ may also be determined quite easily. At $V_2 = V_1$,

$$\frac{\gamma}{\gamma - 1} V_1 (P_2 - P_1) - q = \frac{1}{2} (P_2 - P_1) 2 V_1$$

Therefore

$$\frac{P_2}{P_1} = 1 + \frac{(\gamma - 1) q}{P_1 V_1} \quad \text{at} \quad V_2 = V_1$$

At $P_2 = P_1$,

$$\frac{\gamma}{\gamma - 1} P_1 (V_2 - V_1) - q = 0$$

Therefore

$$\frac{V_2}{V_1} = 1 + \frac{\gamma - 1}{\gamma} \frac{q}{P_1 V_1} \quad \text{at} \quad P_2 = P_1$$

We have now determined the value of P_2 and V_2 on the Rankine-Hugoniot wave at five points in the pressure-volume plane for the working fluid, heat addition model. To show the functional form of the Rankine-Hugoniot equation for the case of a constant heat capacity gas with heat addition we substitute the variables

$$\eta = \frac{V_2}{V_1} - \frac{\gamma - 1}{\gamma + 1}$$

$$\xi = \frac{P_2}{P_1} + \frac{\gamma - 1}{\gamma + 1}$$

which transposes the coordinates such that the equation will asymptote the axes in the new coordinate system. After some algebraic manipulation the Rankine-Hugoniot relation reduces to the form of set of simple hyperbolas in the (η, ξ) plane.

$$\eta\xi = \frac{4\gamma}{(\gamma+1)^2} + \frac{2q}{P_1 V_1}\left(\frac{\gamma-1}{\gamma+1}\right) \tag{5-24}$$

The results of this analysis are illustrated in Fig. 5-3. An inspection of the velocity Eq. 5-20, and the Hugoniot Eq. 5-24, shows, as before,

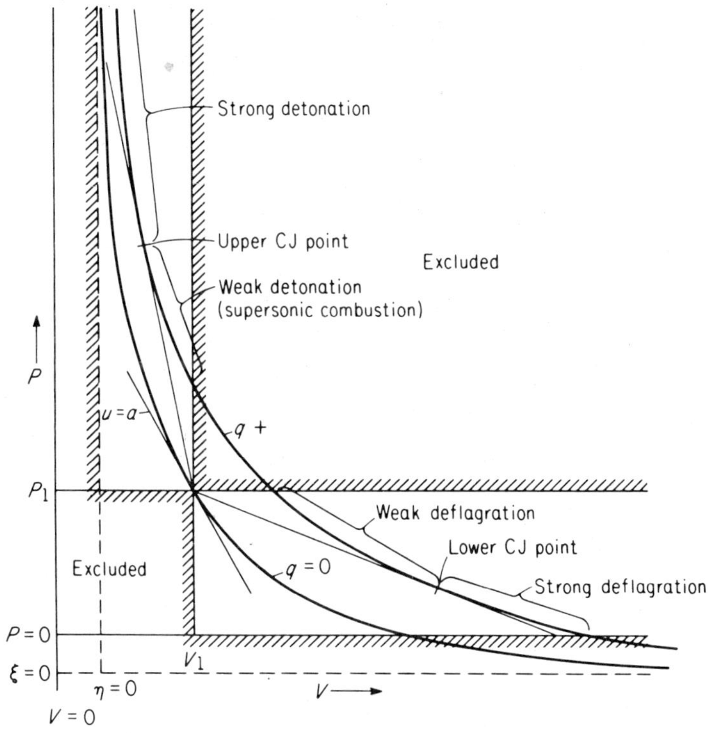

Fig. 5-3. Pressure-volume plot of end states for a one-dimensional steady process with heat addition indicating excluded regions and upper and lower Chapman-Jouget states.

that there are two excluded quadrants in the (P, V) plane corresponding to imaginary wave propagation velocities and, of course, since negative pressures are impossible there is a lower solution limit imposed by the $P_2 = 0$ intersection. Since the isotherms for this gas are represented by hyperbolas which are asymptotic to the axes of the P, V diagram, the Hugoniot curves always cross the isotherms such that increasing P_2 along a Hugoniot increases temperature. The $P_2 = 0$, $V_2 =$ constant intersection therefore represents a value of $T_2 = 0$. This particular solution to the equation represents the hypothetical and impossible case where all the random kinetic energy of the molecules has been converted to an ordered velocity in the flow direction.

Notice that if no heat is added the equations reduce to the simple steady-shock equations and there is exactly one solution (i.e., (P_2, V_2) value) for each supersonic and subsonic incident-flow velocity. However, if heat is added the situation changes markedly. In the velocity region above a limiting subsonic velocity and below a limiting supersonic velocity, there are no possible solutions to the steady equations, while below and above these two critical velocities there are two possible end states for each velocity of propagation. Furthermore, on the subsonic branch, due to the intersection of the Hugoniot with a $P = 0$ curve, very low subsonic-flow velocities yield only one solution. From the earlier relation for the velocity of the flow we see that we can write the expression for the Mach number of the Rayleigh process at stations 1 and 2 as

$$M_1^2 = \frac{V_1}{\gamma P_1}\left(\frac{P_2 - P_1}{V_1 - V_2}\right) = \frac{1}{\gamma}\frac{(y - 1)}{(1 - \epsilon)} \tag{5-25a}$$

and

$$M_2^2 = \frac{V_2}{\gamma P_2}\left(\frac{P_2 - P_1}{V_1 - V_2}\right) = \frac{\epsilon}{\gamma y}\frac{(y - 1)}{(1 - \epsilon)} \tag{5-25b}$$

where

$$\epsilon = V_2/V_1 \quad \text{and} \quad y = P_2/P_1$$

To show that the points of tangency of the Hugoniot and Rayleigh line in Fig. 5-3 correspond to exactly sonic flow at station 2 we differentiate the Hugoniot, Eq. 5-23, to determine its slope. This yields

$$\left.\frac{dy}{d\epsilon}\right)_{\text{Hugoniot}} = -\left(\frac{\dfrac{\gamma + 1}{\gamma - 1}y + 1}{\dfrac{\gamma + 1}{\gamma - 1}\epsilon - 1}\right)$$

Since the slope of the Rayleigh line is

$$\left.\frac{dy}{d\epsilon}\right)_{\text{Rayleigh}} = \frac{y - 1}{\epsilon - 1}$$

we obtain the equation

$$y = \frac{\epsilon}{(\gamma + 1)\epsilon - \gamma}$$

as the condition for tangency. However, this may be rearranged to yield

$$1 = \frac{\epsilon}{\gamma y}\frac{(y - 1)}{(1 - \epsilon)}$$

which is identical to Eq. 5-25b and therefore proves that $M_2 = 1$ at the upper and lower tangency points. The existence of these unique points for steady flow was discovered around the turn of the century, and this type of flow is called *Chapman-Jouguet* flow, abbreviated *CJ*.

For any other possible steady flows Fig. 5-3 shows that

$$\left.\frac{dy}{d\epsilon}\right)_{\text{Hugoniot}} > \left.\frac{dy}{d\epsilon}\right)_{\text{Rayleigh}}$$

for a strong detonation or weak deflagration, while

$$\left.\frac{dy}{d\epsilon}\right)_{\text{Hugoniot}} < \left.\frac{dy}{d\epsilon}\right)_{\text{Rayleigh}}$$

for a weak detonation or strong deflagration.

Thus the equal sign is replaced by an inequality in Eq. 5-25b and we have proved the general statement that

$$M_2 < 1$$

for a strong detonation or weak deflagration, and

$$M_2 > 1$$

for a weak detonation or strong deflagration.

Also note that for a weak shock wave the slope of the Hugoniot is $-\gamma$ and the equation for M_1, Eq. 5-25a reduces to $M_1 = 1$ at $y = \epsilon = 1$. Thus the slope of the Rankine-Hugoniot curve for an infinitesimal adiabatic shock wave is the same as that of an acoustic wave.

We have thus developed a set of equations which show multiple solutions for steady one-dimensional flow with heat addition. Only certain of these steady-flow solutions are observable experimentally and represent truly physical processes. While it is true that any steady one-dimensional flow phenomena which is observed must satisfy these equations, the justification for the physical existence of these particular solutions cannot be made wholly on the basis of an examination of these equations. Indeed, the stability of any of the above solutions can be answered only by investigating their true transient behavior in the neighborhood of these steady solutions. This we will attempt to do when we individually discuss the theory of any specific heat release wave.

5-5. WAVES WITH ENDOTHERMIC REACTION

Individual wave-propagation problems in exothermic systems will be discussed in later chapters in considerable detail. There remains one problem, however—the propagation of compression waves in endothermic systems—which does not classify as a combustion process but should be discussed because of its importance. The most common endothermic reactive system is an equilibrium gas mixture which is capable of reacting with a finite rate, or a nonreactive gas which has a finite relaxation time. In either of these cases the system admits to the definition of two distinct and limiting velocities of sound—the frozen and equilibrium velocities of sound:

$$a_f^2 = \left(\frac{\partial P}{\partial \rho}\right)_{s,\lambda_1,\lambda_2,\lambda_3,\ldots,\lambda_n}$$

$$a_{eq}^2 = \left(\frac{\partial P}{\partial \rho}\right)_{s\text{(along an equilibrium isentrope)}}$$

For a system with a thermodynamically stable equilibrium state, $a_{eq} < a_f$ because the reaction is always endothermic as the temperature is raised by adiabatic compression. Both of these equations represent isentropic wave-propagation processes because the adiabatic compression process in the wave is assumed to follow the isentropic conditions discussed in Sec. 5-1. The equilibrium velocity of sound may be calculated using Eq. 2-19 once the state equations for the reacting or relaxing system are known. For the case of a simple dissociating gas obeying the Eqs. 4-32 and 4-33 we obtain

$$a_{eq} = \left[\frac{C_{P_{eq}} P/\rho}{C_{P_{eq}}\left(1 + \dfrac{\alpha(1-\alpha)}{2}\right) - (1+\alpha) R \left[1 + \dfrac{\alpha(1-\alpha)}{2(T/\theta_D)}\right]^2}\right]^{1/2}$$

where

$$C_{P_{eq}} = (4 + \alpha) + (\theta_D/T)^2 \frac{\alpha(1-\alpha)}{2}$$

Therefore we may evaluate the velocity decrement $(a_f - a_{eq})/a_f$ for a constant-gamma gas using the parameters T/θ_D and P/P_D as dimensionless parameters.

$$\text{Percent decrement} = 100 \times \frac{[P(4+\alpha)/3\rho]^{1/2} - a_{eq}}{[P(4+\alpha)/3\rho]^{1/2}}$$

Figure 5-4 is a plot of the velocity decrement $(a_f - a_{eq})/a_f$ and shows that for a dissociating gas there can be quite a difference between the equilibrium and frozen-sound velocities. If we now modify our steady-flow concepts (Sec. 5-3) to discuss the case of an equilibrium gas we find that there are two Rankine-Hugoniot curves through the point (P_1, V_1) on a P, V plot; the equilibrium Hugoniot always lying to the left and below the frozen-composition Hugoniot (see Fig. 5-5). Furthermore, we find that a Rayleigh line crosses both Hugoniots at high flow velocities but crosses only the equilibrium Hugoniot for flow velocities which lie be-

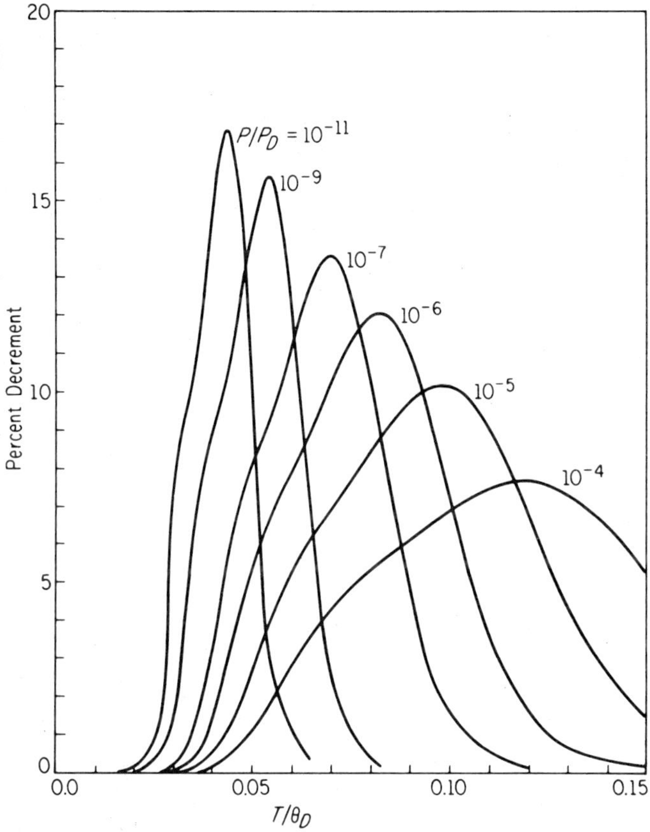

Fig. 5-4. The percent velocity of sound decrement $(a_f - a_{eq}) \cdot 100/a_f$ for an ideal dissociating gas.

tween a_f and a_{eq}. Experimental behavior representing both these possibilities has been observed (see Fig. 5-6).[1] In the strong shock case vibrational relaxation or finite rate endothermic reactions yield shock profiles which may be adequately represented as a discontinuous shock jump from O to E, followed by a continuous relaxation along the Rayleigh line ED to the end state D on the equilibrium Hugoniot. Steady diffuse shock waves are represented by Rayleigh lines which lie between slopes CO and AO in Fig. 5-5. These also represent an observable process. The transition from O to B, for example, cannot be abrupt because of the finite rate of the relaxation or reaction process, nor can it be preceded by a high-temperature frozen-composition shock wave because signals can

[1]A discussion of these experimental results may be found in W. C. Griffith and A. Kenny, *J. Fluid Mech.*, *3*, 286 (1957).

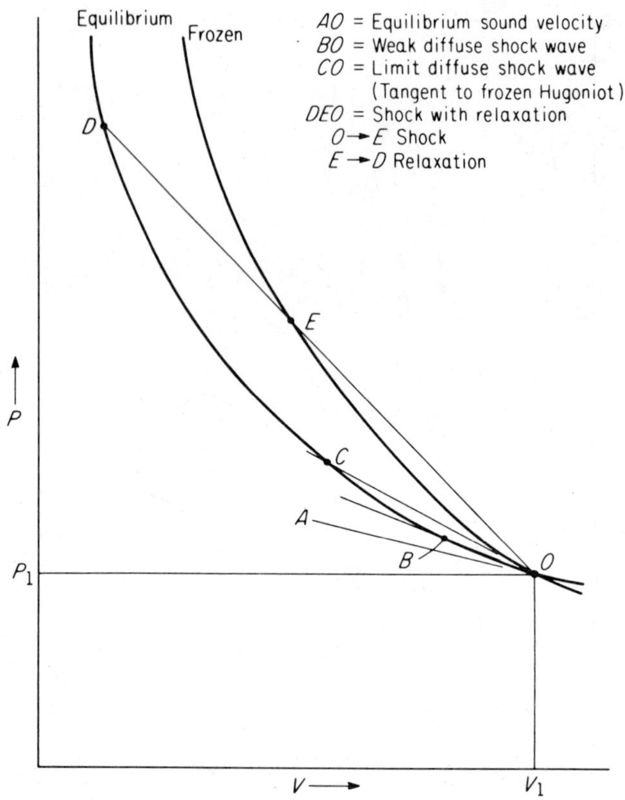

Fig. 5-5. Steady one-dimensional flow in a reacting or relaxing gas.

propagate locally at a higher velocity than the overall compression process. However, these signals (traveling at the frozen-sound velocity) attentuate rapidly in a reactive system and the net effect is the eventual appearance of a diffuse compression wave traveling as a steady phenomenon at an intermediate velocity.

Thus, steady wave propagation though an endothermic reactive system yields two distinct compression processes—a shock followed by a distinguishable reaction zone at high Mach numbers, and at lower Mach numbers a diffuse shock wave with no preceding discontinuity.

5-6. REACTIVE NOZZLE FLOW

There is one steady inviscid reactive flow which has a considerable interest to the combustion field. This is the rocket-nozzle flow which occurs when a high-temperature equilibrium gas is exhausted through a

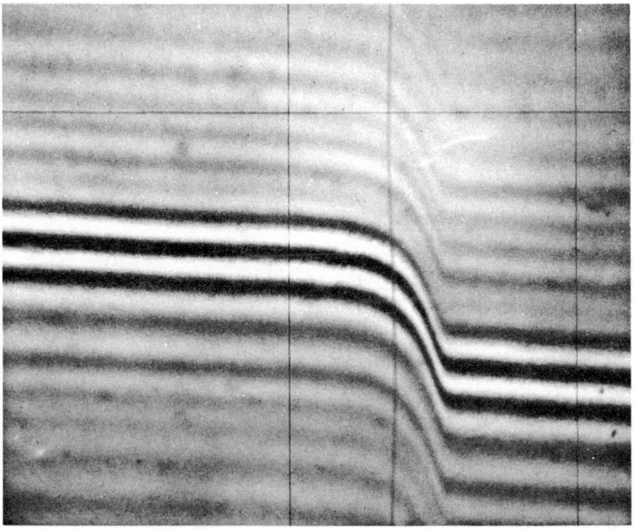

(a) Diffuse shock wave in carbon dioxide, photographed with white-light fringes. The vertical lines are 1 in. apart.

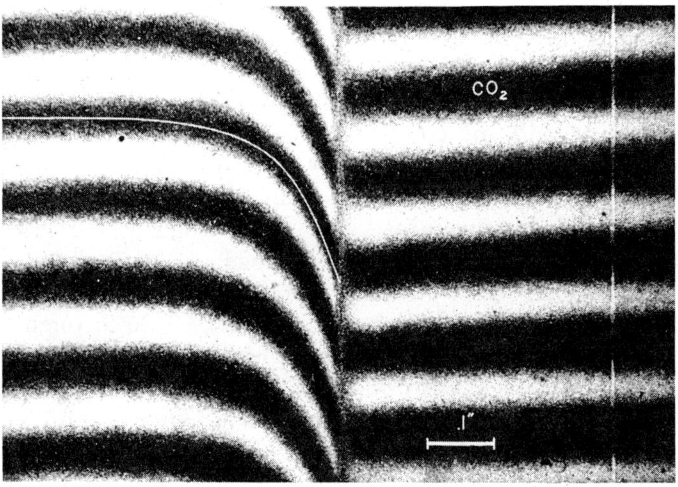

(b) Strong shock wave in carbon dioxide showing vibrational relaxation after the shock transition. The fringe shift at the shock is 7 fringes.

Fig. 5-6. Optical interferogram of diffuse and strong shock waves, showing the density profile through the waves. The shock wave is propagating to the right. The white curved line in (b) is the theoretical fringe shift using vibrational relaxation theory. (Courtesy of Professor Walker Bleakney, Princeton University.)

sonic nozzle and subsequently expanded supersonically to a high velocity and low pressure. For purposes of this section we will treat this flow as a quasi–one-dimensional (variable area) steady flow of a reactive gas containing a single recombination reaction, the recombination of a dissociated Lighthill gas (see Sec. 4-11). For this case Eqs. 5-1 through 5-4 reduce to the following equations:

$$\frac{d\rho}{\rho} + \frac{du}{u} + \frac{dA}{A} = 0 \tag{5-26}$$

$$\rho u\, du + dP = 0 \tag{5-27}$$

$$u\, du + dh = 0 \tag{5-28}$$

and

$$u\frac{d\beta}{dx} = f(P,T,\beta) \tag{5-29}$$

Equations 5-27 and 5-28 are directly obtainable from Eqs. 5-2 and 5-3. However, the mass-balance equation for quasi-one-dimensional flow must be obtained from the integral form of Eq. 5-1. Equation 5-29 is a general expression for the rate of change of the single reaction coordinate with distance. The integration of these equations requires a set of throat conditions, the rate function $f(P, T, \beta)$ and an equation of state $h(P, T, \beta)$. Here β is a reaction coordinate based on mass fraction of atoms present in a homonuclear dissociating diatomic gas. Numerically β has the same value as the α of Sec. 4-11.

These equations reduce to a set of algebraic equations for the two limiting cases of inert and equilibrium flow. In these cases the flow may be represented by a constant-entropy line on an appropriate Mollier diagram, and the properties are uniquely determined as a function of initial conditions and the local value of any one of the state or flow variables. In the general case (that is, when the flow is fully reactive) the chemical lags will preclude the occurrence of either full equilibrium or inert flow at any point in the nozzle, and Eqs. 5-26 through 5-29 must be numerically integrated to determine the behavior for any particular nozzle shape. The flow at any point now depends in a complex manner on the previous history, and general conclusions are not easily drawn.

In 1959 Bray[2] pointed out that for most nozzle-flow problems a very simple approximation yields results which agree quite adequately with the result obtained from a full numerical integration of the equations. We summarize his model and conclusions here.

Bray assumed that downstream of the throat the nozzle has an area given by the equation

$$A = A_{\text{throat}} + \pi(\tan^2\theta)x^2 \tag{5-30}$$

[2] K. N. C. Bray, *J. Fluid Mech.*, *6*, 1 (1959).

and that the rate law for the recombination of a Lighthill gas could be written in a simplified form (due to Freeman)[3]

$$\frac{d\beta}{dt} = C\rho\left[(1 - \beta)\exp(-\theta_E/T) - \frac{\rho}{\rho_D}\beta^2\right] \tag{5-31}$$

where $C\rho$ has units of $\sec^{-1}$, θ_E is the effective characteristic dissociation temperature based on energy and related to θ_D by the equation

$$\theta_E = \theta_D - T(1 + \alpha)$$

and ρ_D is a characteristic dissociation density which is equivalent to the dissociation pressure introduced in Sec. 4-11. It may be shown in fact that Lighthill's ρ_D and the P_D of Sec. 4-11 are related by the equation

$$\rho_D = P_D\left[\frac{M}{328.2T\cdot\exp(1 + \alpha)}\right] \tag{5-32}$$

where M is the molecular weight of the diatomic specie and T is the temperature in °K. We have already seen that P_D as determined by a best fit to the equilibrium constant is not strictly a constant. Lighthill's ρ_D has also been found to be a function to temperature. Thus the temperature-dependence noted in Eq. 5-32 is not too surprising.

Notice that the quantity in the brackets of Eq. 5-31 reduces to the equation

$$\frac{\beta^2}{1 - \beta} = \frac{\rho_D}{\rho}\exp(-\theta_E/T) \tag{5-33}$$

for $d\beta/dt = 0$ (i.e., for equilibrium). Eq. 5-33 may be obtained from Eq 4-32 by substituting the equation of state and Eq. 5-32. For this fixed-nozzle geometry and rate equation Bray found that all nozzle behaviors could be expressed in terms of a single dimensionless factor which contains both nozzle-size and rate-constant terms:

$$\Phi = \frac{C\rho_D}{2\sqrt{\pi}\tan\theta}\sqrt{\frac{A_{\text{throat}}M}{\theta_E}} \tag{5-34}$$

The results of numerical integration through the nozzle for a series of fixed values of Φ are shown in Fig. 5-7. Notice that for values of Φ which lie between $\Phi = 0$ (frozen flow) and $\phi = \infty$ (equilibrium flow) the flow behaves as though the chemical reactions were in equilibrium near the throat and then suddenly changes to a flow which acts as though the chemical reactions are frozen. Also note that the transition point moves away from the nozzle as the effective reaction rate Φ becomes larger. Bray noted this behavior and postulated that to a first approximation the nozzle expansion of a reactive flow with a single reaction coordinate could be handled as equilibrium flow to a point and thereafter as frozen

[3]N. C. Freeman, *J. Fluid Mech.*, *4*, 407 (1958).

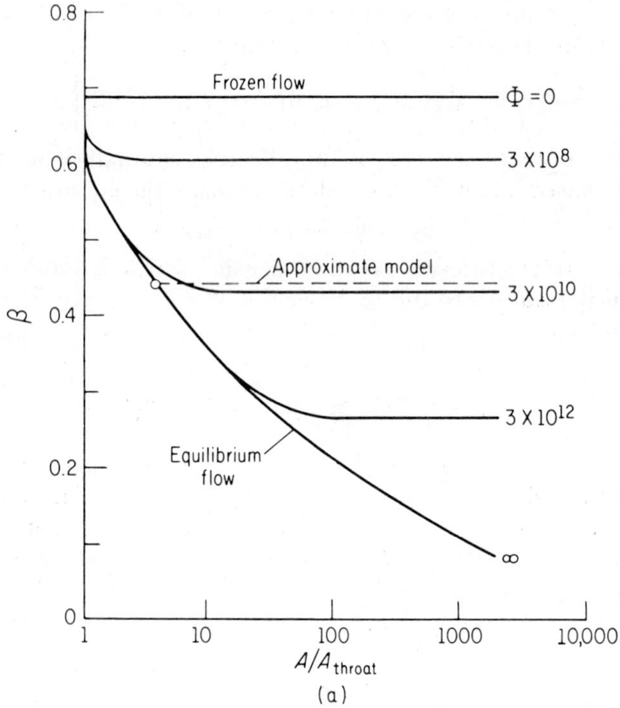

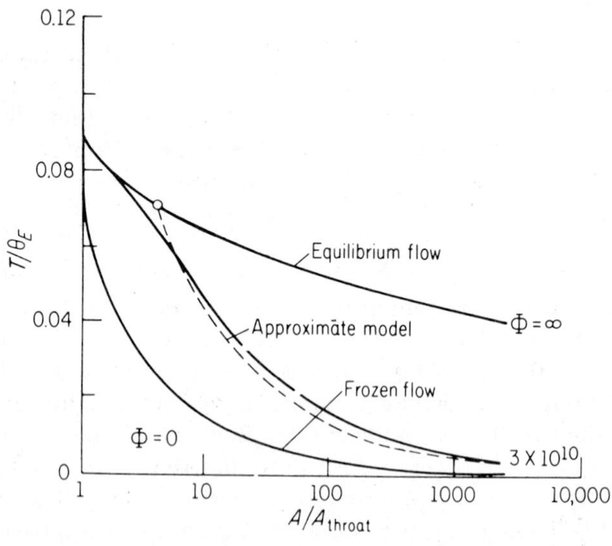

Fig. 5-7. Flow variables in a hyperbolic nozzle. $T_o/\theta_E = 0.1$, $P_o M/R_{\rho D}\theta_E = 5.0 \times 10^{-6}$. (a) Degree of dissociation (b) Temperature. (Adapted with permission from W. G. Vincenti and C. H. Kruger, Jr. *Introduction to Physical Gas Dynamics*, John Wiley and Sons, New York: 1965.)

flow. He postulated that the reaction "froze" primarily because the recombination rate dropped very rapidly during the expansion process. Bray postulated that the position of a "freezing line" in the nozzle could be estimated by performing an equilibrium calculation until the condition

$$-\left(u\,\frac{d\beta}{dx}\right)_{\text{equil}} = KC[\rho(1 - \beta)\exp(-\theta_E/T)]_{\text{equil}}$$

is satisfied. Here K is an adjustable fitting constant whose value has been found to be near unity. The flow that results from the application of this criterion (with $K = 1$) is shown in Fig. 5-7 as the dotted line for comparison to the case when $\Phi = 3 \times 10^{10}$.

This criterion allows one to construct a composite Mollier diagram which will allow the prediction of the behavior of the flow of a specific reactive gas through a specific nozzle. In this diagram one constructs a Mollier diagram for equilibrium flow in the high enthalpy region. This region is then bounded by a line representing Bray's criterion for that specific nozzle and chemical system. The portions of the diagram below this line are constructed for the frozen compositions found at the freezing line. This composition varies with the entropy, but the construction is performed along lines of constant entropy for each specific composition.

BIBLIOGRAPHY

1. F. A. Williams, *Combustion Theory*. (Reading, Mass.: Addison-Wesley, 1965).

2. G. Rudinger, *Wave Diagrams for Nonsteady Flow in Ducts*. (Princeton, N. J.: Van Nostrand, 1955.)

3. H. W. Liepman and A. Roshko, *Elements of Gas Dynamics*. (New York: Wiley, 1957.)

4. W. W. Wood and J. G. Kirkwood, "Hydrodynamics of a Reacting and Relaxing Fluid," *J. Appl. Phys.*, *28*, pp. 395–398 (1957).

5. B. T. Chu, "Wave Propagation and the Method of Characteristics in Reacting Gas Mixtures with Applications to Hypersonic Flow," WADC TN 57-213 ASTIA Doc. No. Ad 118350, May, 1957.

6. V. Blackman, "Vibrational Relaxation in Oxygen and Nitrogen," *J. Fluid Mech.*, *1*, pp. 61–85 (1956).

7. W. G. Vincenti and C. H. Kruger, Jr., *Introduction to Physical Gas Dynamics*. (New York, Wiley and Sons, 1965.)

PROBLEMS

5-1. Show that at constant-mass flow the heat-addition model for steady one-dimensional flow exhibits a local minimum in the dimensionless entropy production $(S_2 - S_1)/R$ at the upper CJ point and a local maximum at the lower CJ point. Show that $(S_2 - S_1)/R$ is always positive for positive q at both the lower and upper CJ points.

5-2. Show that Eq. 5-16 reduces to an equation containing Reimann invariants in the unreactive case. Show that the same Reimann functions become variables in the heat-addition model. Why is it impossible to make such a simplification in the general case?

5-3. Derive a relationship for the velocity decrement $(a_f - a_{eq})/a_f$ of an ideal dissociating gas containing an inert diluent. Show that at constant pressure and temperature, $(a_f - a_{eq})/a_f$ increases with dilution for small additions of diluent.

5-4. For air, as the temperature is raised, we may assume that oxygen dissociation is the first important reactive process to occur. Using the dimensionless plots 4-9, 4-10, and 5-4, estimate the highest temperature (as a function of pressure) where nitrogen dissociation may be neglected (to 1% in composition and to 1% in its effect on temperature and equilibrium velocity of sound).

5-5. The Hugoniot equation (Eq. 5-22) also represents an integrated equation for a thermodynamically reversible (i.e., static) transformation from station 1 to station 2. Discover the path in the (P, V) plan for this thermodynamically reversible transformation.

6

General Combustion Phenomena

6-1. INTRODUCTION

The term "combustion" applies to a large variety of natural and artificially controlled processes. These processes are as varied as the occurrence of a forest fire, the burning of a candle or the detonation of a composite solid high-explosive charge. It is therefore quite difficult to catalogue all the observed modes of combustion in a completely self-consistent and adequate manner. However, we first observe that most combustion reactions are ultimately exothermic gas-phase reactions. If solids or liquids are present at the start of combustion, the process virtually always consists of the evaporation of the solid or liquid, followed by an exothermic vapor-phase reaction between the fuel components and the oxidizer. The combustion of coke is an important exception; here the carbon reacts from the solid phase.

Even though this first observation represents an important step in defining the nature of combustion processes, it does not by itself lead to a logical breakdown for further study. For this purpose we will define six subclasses of combustion which will be discussed in some detail as limit or "pure" phenomena in this text. The six subclasses are (1) homogeneous or quasi-homogeneous processes such as vessel explosions or reactions occurring in a well-stirred reactor, (2) diffusion flames or diffusion controlled processes, (3) premixed gas flames, (4) the aerodynamics of flames, (5) detonation and (6) rocket-engine combustion. We note that this breakdown is by necessity somewhat arbitrary. However, it is hoped that the subclasses have been chosen so that an understanding of each of these "pure" phenomena will help the reader in his approach to the study of more complex combustion phenomena. We will, of course, discuss a few of the more important complex phenomena which cannot be distinctly categorized and point out, as appropriate, the relationship between these and the elementary categories defined above. These more complex phenomena will always be described in their most appropriate category.

The first two of the six subclasses are distinct from the last four in that they do not contain easily definable wave processes. They will therefore be discussed in this chapter on general combustion phenomena. The last four subclasses are each allowed an individual chapter in the text. We note that the fourth subclass is really not a proper category in the spirit of the definition. However, the aerodynamics of the subsonic flame is such an exceedingly important subject that it has been given status as a separate category in this text. In addition the sixth subclass, rocket-engine combustion, is discussed as a distinct subject because of its technological importance.

6-2. THE THEORY OF VESSEL EXPLOSIONS

In early combustion studies many investigators turned to "vessel" experiments, partially in an attempt to measure an "ignition" temperature for application in the early flame-propagation theories and partially to study the details of reaction kinetics in exothermic systems. These experimentalists truly opened a Pandora's box of experimental results, the consequences of which have not been fully resolved as yet. In this section we wish to develop some of the simpler conclusions available from these experiments.

Three distinct types of vessel experiments have been used for the study of homogeneous or quasi-homogeneous explosions and the reaction kinetics leading to these explosions. These are (1) introduction into a preheated vessel, (2) adiabatic compression by a piston device, and (3) adiabatic but not isentropic compression by a traveling shock wave. These three experimental techniques have been used to study the explosion behavior of virtually all available exothermic systems with varying degrees of experimental sophistication. Four types of observations are usually made in studies of this type:

1. Does an explosion occur for a certain set of experimental conditions?

2. If an explosion occurs, how much time has elapsed between the onset of heating and the time of its occurrence; i.e., what is the explosion delay?

3. Unusual events, such as the appearance of cool flames during the preexplosion period, may be recorded.

4. The rate of change of concentration of the reactants and the pressure and temperature history of the mixture during the preexplosion and explosion period may be studied.

The first step that must be taken to relate these experimental observations to the elementary processes that must be responsible for their occurrence is the construction of an appropriate model, with approximations. At the end of Sec. 4-6, we discussed two distinct processes important to the occurrence of an explosion in a system capable of exothermic reaction; these are thermal and chain-branching explosion processes. Now let us first examine the case of a truly adiabatic thermal explosion.

Assume that for ordinary explosive mixtures the heat release on explosion is so large relative to the heat capacity of the system that the temperature rise will "run away" before the initial concentration of the reactants is changed appreciably. This is usually a good assumption. For a thermal explosion it is also usual to assume that chain processes are not occurring and that the reaction rate can therefore be expressed simply as having a general order in both the fuel and oxidizer concentration and an Arrhenius temperature dependence. For a constant-volume adiabatic system containing fuel, oxidizer, and an inert gas, we may write the initial concentration of the fuel as:

$$[F] = \rho n_F = \frac{(p_F)_0}{RT_0} = x_F \frac{P_0}{RT_0} \tag{6-1}$$

Here x_F is the mole fraction of fuel. For the oxidizer

$$[O] = \rho n_O = \frac{(p_O)_0}{RT_0} = x_O \frac{P_0}{RT_0} = (1 - x_F - x_I)\frac{P_0}{RT_0} \tag{6-2}$$

Here x_I refers to the mole fraction of inert gases in the mixture. The rate of reaction is therefore given by the general expression

$$\frac{d[F]}{dt} = -A\left(\frac{P_0}{RT_0}\right)^{n+m} x_F^n (1 - x_F - x_I)^m \exp(-E/RT) \tag{6-3}$$

where n and m are the order of the reaction for the fuel and oxidizer and E is the effective activation energy for the overall reaction. If Q is the heat of combustion of the fuel (cal/mole), and c_v is the heat capacity of the mixture (cal/cm^3 °K), we find that

$$c_v \frac{dT}{dt} = -Q\frac{d[F]}{dt} \tag{6-4}$$

and therefore that

$$\frac{dT}{dt} = \lambda \exp(-E/RT)$$

where λ may be assumed to be a positive constant. Integrating from $t = 0$ we obtain

$$\lambda t = \int_{T_0}^{T} \exp(E/RT)\, dT$$

We define $\theta = T/T_0$ and $\alpha = E/RT_0$ and change the integration variable to obtain the expression

$$-\frac{\lambda t}{T_0 \alpha} = \int_{\alpha}^{\alpha/\theta} \frac{e^{\alpha/\theta}}{(\alpha/\theta)^2} d\left(\frac{\alpha}{\theta}\right)$$

We note that α and α/θ are usually large (i.e., > 15) and we integrate by parts to obtain a series in α/θ

$$-\frac{\lambda t}{T_0\alpha} = \left[\frac{e^{(\alpha/\theta)}}{(\alpha/\theta)^2}\left(1 + \frac{2!}{(\alpha/\theta)} + \frac{3!}{(\alpha/\theta)^2} + \cdots + \frac{(n+1)!}{(\alpha/\theta)^n} + \cdots\right)\right] \Big/_{\alpha}^{\alpha/\theta}$$

which when evaluated yields

$$\frac{\lambda t}{T_0\alpha} = \frac{e^{\alpha}}{\alpha^2}\left[1 + \frac{2}{\alpha} + \frac{6}{\alpha^2} + \cdots\right] - \frac{e^{(\alpha/\theta)}}{(\alpha/\theta)^2}\left[1 + \frac{2}{(\alpha/\theta)} + \frac{6}{(\alpha/\theta)^2} + \cdots\right]$$

This may be written in a different form by dividing both sides by the constant term that appears on the right-hand side.

$$\frac{\lambda t\alpha}{T_0\left[1 + \frac{2}{\alpha} + \frac{6}{\alpha^2} + \cdots\right]e^{\alpha}} = 1 - \theta^2 e^{(\alpha/\theta-\alpha)} \cdot \left[1 + \frac{2}{\alpha}(\theta - 1) + \cdots\right]$$

Substituting the original variables we obtain an expression for the temperature-time history of an isolated sample

$$\frac{t}{\beta} = \frac{t}{\frac{RT_0^2}{E\lambda}\left[1 + \frac{2RT_0}{E} + \frac{6R^2T_0^2}{E^2}\right]e^{E/RT_0}}$$

$$= 1 - \left(\frac{T}{T_0}\right)^2\left[1 + \frac{2RT}{E}\cdot\left(\frac{T}{T_0} - 1\right) + \cdots\right]\cdot \exp\left[\frac{E}{RT_0}\left(\frac{T_0}{T} - 1\right)\right]$$

where β is a constant whose value is dependent upon the values of T_0, λ, and E. The dimensionless time t/β is plotted versus the dimensionless temperature T/T_0 in Fig. 6-1 for two values of E/RT_0. Notice from this plot that T/T_0 increases only slowly at first but increases extremely rapidly at t/β approaches 1. At $t/\beta = 1$ the system is said to explode and the time to explosion is found to be unique. Since, for most systems, $E/RT_0 > 15$ the ignition delay is given by the expression

$$t_{\text{ign}} = \beta \simeq \frac{RT_0^2}{E\lambda}\exp(E/RT_0) \tag{6-5}$$

We have, therefore, shown that any isolated system containing a finite-rate exothermic reaction with a positive temperature sensitivity will always "explode" in a homogeneous manner. We further note that the initial assumptions are probably quite valid for this type of explosion. We base this on the observation that complete reaction in an exothermic system usually increases the temperature by a factor of 2 or more. Thus a 10% temperature rise represents at most only a 10% depletion of the reactants in the mixture.

Equation 6-5 may be used to investigate the overall order and activation energy of thermal explosions only in systems which do not lose heat to the surroundings. If the system is not isolated, but is allowed to trans-

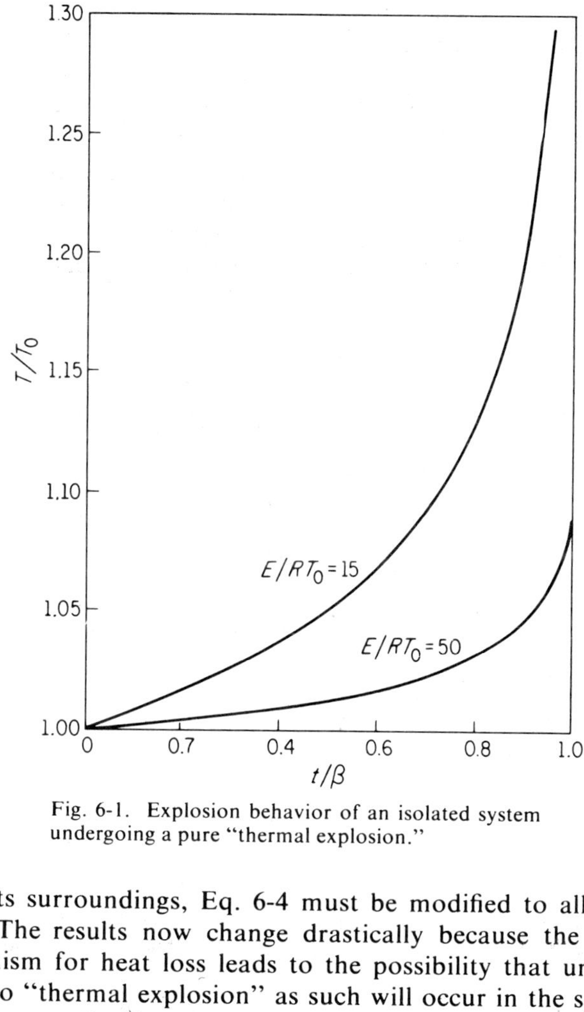

Fig. 6-1. Explosion behavior of an isolated system undergoing a pure "thermal explosion."

fer heat to its surroundings, Eq. 6-4 must be modified to allow for this heat flow. The results now change drastically because the availability of a mechanism for heat loss leads to the possibility that under certain conditions no "thermal explosion" as such will occur in the system, even though it meets all the other requirements which lead to Eq. 6-5.

As a very first approximation, let us assume that the system's temperature is uniform throughout, but that it loses heat to its surroundings (i.e., to the vessel walls at a temperature T_0) by simple conduction or by convection with a constant coefficient. Eq. 6-4 thus becomes

$$c_v \frac{dT}{dt} = -Q \frac{d[F]}{dt} - \frac{s}{V} h(T - T_0) \tag{6-6}$$

for a vessel of volume V where s is the surface area of the vessel and h is an effective heat-transfer coefficient which is assumed to be independent of the pressure (units: cal/cm^2 sec °K). We immediately note that Eq. 6-6 predicts the attainment of a steady system temperature under conditions where the coefficient of the second term is large. However, since the first term on the right-hand side is always finite and increases with increasing system temperature, there is also the possibility that dT/dt will never approach zero or go negative. In this case, an explosion will occur. Equation 6-6 thus predicts that, in a finite vessel, one should see "thermal explosion limits" which are a function of the pressure, temperature, and the fuel and oxidizer concentrations. Let us examine one facet of the theoretical dependence of these limits. Figure 6-2 is a plot of the two terms on the right in Eq. 6-6 and illustrates that for a particular T_0 value,

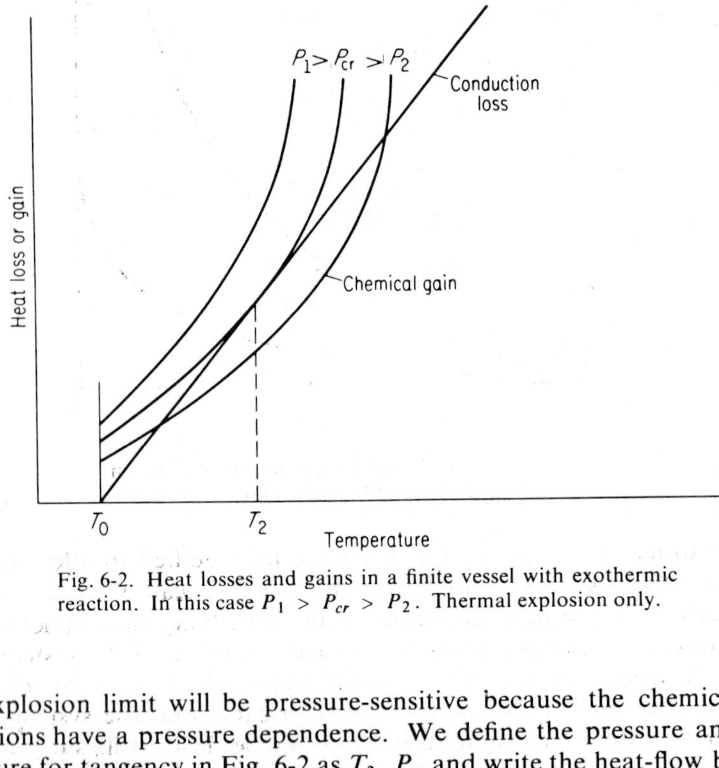

Fig. 6-2. Heat losses and gains in a finite vessel with exothermic reaction. In this case $P_1 > P_{cr} > P_2$. Thermal explosion only.

the explosion limit will be pressure-sensitive because the chemical rate equations have a pressure dependence. We define the pressure and temperature for tangency in Fig. 6-2 as T_2, P_{cr} and write the heat-flow balance equation at tangency as

$$-Q\frac{d[F]}{dt} = \frac{s}{V}h(T_2 - T_0)$$

and the condition for tangency as

$$\frac{d}{dt}\left(-Q\frac{d[F]}{dt}\right) = \frac{d}{dt}\left[\frac{s}{V}h(T_2 - T_0)\right]$$

Substituting the kinetic rate equation and solving these equations simultaneously yields

$$\frac{RT_2^2}{E} = T_2 - T_0$$

or

$$T_2 = \frac{1 \pm \left(1 - 4\,\frac{RT_0}{E}\right)^{1/2}}{2R/E}$$

We choose the minus sign to obtain the lower value of T_2, since our analysis of a pure thermal explosion showed that the reaction rate for an explosion in an isolated system became very rapid after only a slight temperature rise. Once again we assume that $E/RT_2 \gg 1$ and therefore that the expression for T_2 may be expanded to yield

$$T_2 = T_0\left(1 + \frac{RT_0}{E}\right)$$

Substituting into Eq. 6-6 for $T = T_2$ therefore yields the explosion limit equation for a thermal explosion:

$$\ln\left[\frac{P_{cr}}{T_0\left(1 + \dfrac{2}{n+m}\right)}\right] = \frac{E}{(n+m)RT_0} + \text{const.} \qquad (6\text{-}7)$$

Equation 6-7 is represented schematically in Fig. 6-3. We see from Fig. 6-3 that this simple theory predicts that the slope of the explosion limit

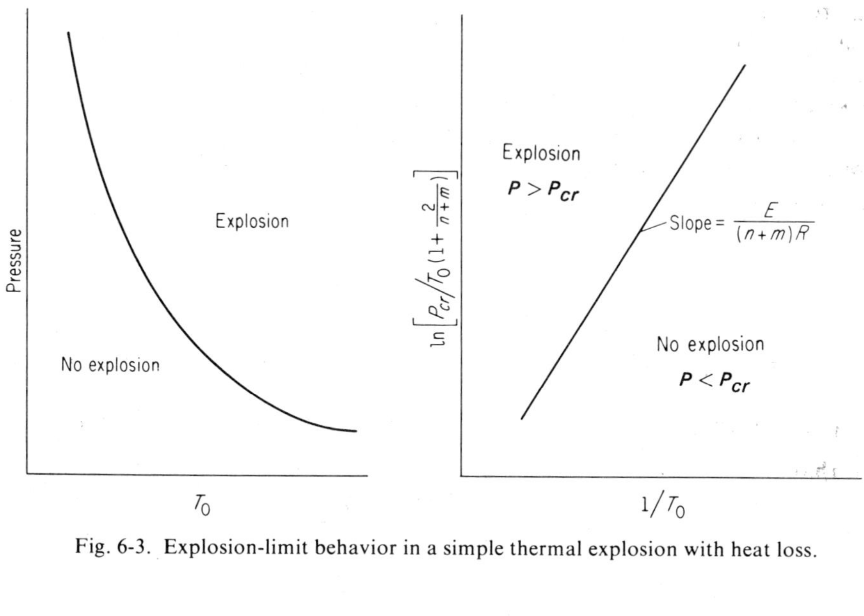

Fig. 6-3. Explosion-limit behavior in a simple thermal explosion with heat loss.

line in the proper ln vs. $1/T$ plot will directly yield the effective activation energy of the preexplosion chemical reaction if the order of the reaction is known. Other manipulations of the equations show that one may determine the effective order of the reaction by plotting P_{cr} vs. concentration at constant vessel temperature.

We note for completeness that if an explosion occurs, the integration of Eq. 6-6 will always yield an explosion delay time. However, even for very simple systems showing heat loss the solution of Eq. 6-6 is difficult and the ignition delay time, t_{ign}, is found to be a complicated function of the external variables.

Equation 6-5 may also be modified by using a less restrictive assumption for heat loss to the vessel walls. In this case we assume that a steady state is reached in the vessel if no explosion occurs, and that the steady-state conduction to the walls just balances the local rate of heat production at every point in the gas sample. We may therefore write the steady-state heat-conduction equation for the particular vessel shape:

$$\nabla^2 T = -\frac{Q}{k}\left(\frac{d[F]}{dt}\right)_{\text{local}}$$

with the boundary condition $T = T_0$ at the walls. In this model explosion is assumed to occur when the solution to this steady-state equation yields an infinite temperature at any point in the vessel. This model may be used to evaluate the effect of vessel shape on explosion limits. The significance of the approximations used in the above two analyses will be discussed later in relation to experimental technique.

In addition to the obvious assumptions used in the above analyses, there was the explicit assumption that the explosion is indeed thermal in nature. However, as we pointed out in Sec. 4-6, chemical systems also admit to the occurrence of chain-branching reactions. We would now like to discuss the effect of a chain-branching reaction on the explosion limit and explosion delay behavior of exothermic systems. We saw in Sec. 4-6 that the condition for explosion of the general chain-branching reaction with

$$nA \xrightarrow{k_1} C \text{ (initiation)}$$

$$A + C \xrightarrow{k_2} \text{Products} + \alpha C \text{ (chain cycle)}$$

$$C \xrightarrow{k_w} \text{Products} + \text{stable molecules (wall termination)}$$

$$C + M + M \xrightarrow{k_g} \text{Products} + \text{stable molecules (gas termination)}$$

is that

$$\alpha \geq 1 + \frac{k_g[M]^2 + k_w}{k_2[A]} = \delta \tag{6-8}$$

and we now wish to examine the consequences of this equation under conditions found in a typical vessel experiment.

We first note that Eq. 6-8 does not contain a term involving a rate constant for the initiation reaction. This is because Eq. 6–8 is the denominator of the steady state equation that was derived on pg. 108, and it does not contain the $k_1[A]$ term. Figure 6-4 is a schematic plot of δ versus pressure at a series of fixed temperatures. We note that at fixed temperature the value

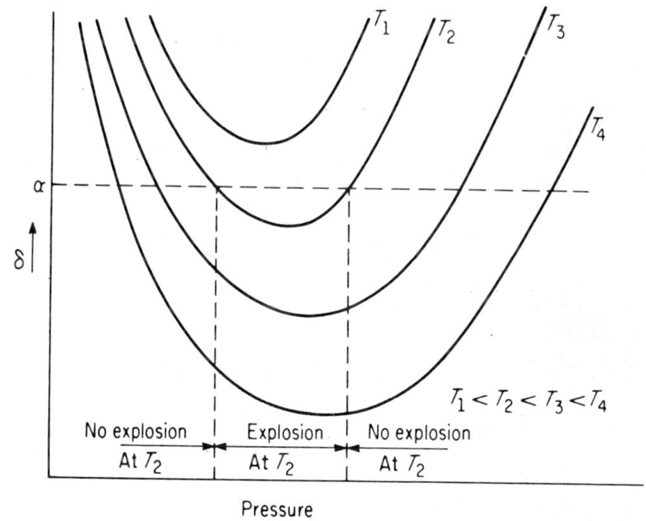

Fig. 6-4. The value of δ in Eq. 6-8 plotted against pressure for various temperatures; points outside the range of each curve represent no explosion for a particular value of α.

of δ approaches infinity as P approaches either zero or infinity. Furthermore, δ exhibits a minimum which is related to the relative rates of the three competing reactions. We also note that both the wall and gas-phase recombination reactions have a low-temperature dependence, while the rate-controlling reaction in the chain cycle must be an endothermic reaction with a relatively large activation energy, and therefore this reaction will be quite temperature-dependent. Accordingly, as the temperature is increased the quantity k_w/k_g will be relatively constant in Eq. 6-8, but the quantity k_g/k_2 will decrease rapidly. This leads to the temperature dependence of δ shown schematically in Fig. 6-4. We now recall that $a \geq \delta$ is the criterion for the occurrence of a spontaneous chain-branching explosion in the vessel, and that $\alpha < \delta$ is the condition for a slow reaction with a relatively constant concentration of the chain carrier C. Figure 6-4 may therefore be used to construct a typical explosion-limit diagram for a

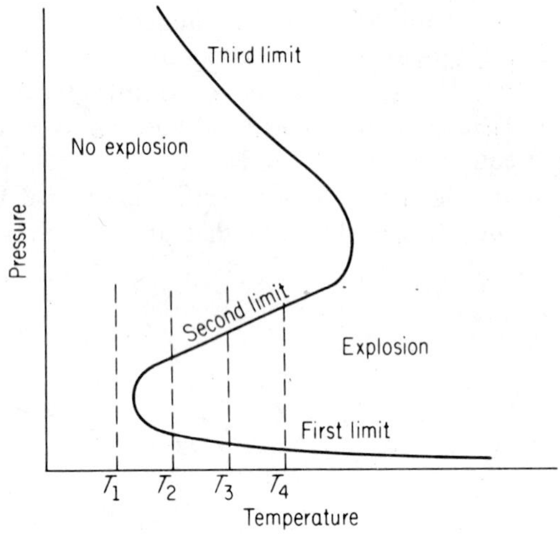

Fig. 6-5. The explosion peninsula behavior of a chain-branching explosion as constructed from Eq. 6-8 and Fig. 6-4.

chain-branching reaction. This is shown schematically in Fig. 6-5. Note the occurrence of an explosion "peninsula" and the existence of three distinct pressure limits over a short temperature range. These limits are usually called the *first limit* (due to wall recombination) the *second limit* (due to gas-phase recombination) and the *third* (or high-pressure) *limit.* The chain-branching theory does not predict the occurrence of this third limit; it is simply the ordinary thermal explosion limit governed by Eq. 6-7. The occurrence of an explosion peninsula has been observed in many vessel experiments for a large variety of chemical systems. Its occurrence is usually considered to be unambiguous evidence that chain branching mechanisms are occuring during the preexplosion period.

If an explosion occurs in a system capable of chain branching, the delay may be measured as a function of temperature, pressure, and the reactant concentrations, and it is proper to ask how these measured delays are related to the fundamental processes leading to the explosion. One observes, however, that the construction of an appropriate model is necessary before ignition delays may be related to the reaction parameters, and that for the general chain-branching reaction the analysis might be expected to be even more complex than for a simple thermal explosion. Because of this complexity, we will develop an expression for the explosion delay of a chain-branching explosion for only one limiting case, the low-pressure, high-temperature case where the terminating reactions become very slow relative to the chain-branching step. If this is the case, we find that after the initiation reactions cease to be important the increase in concentration of the chain carrier is governed by the equation

$$\frac{d[C]}{dt} = (\alpha - 1)k_2[A]\,[C]$$

If we assume that $[A]$ is constant this equation may be integrated to yield the equation

$$\frac{[C]}{[C_0]} = \exp\{(\alpha - 1)k_2[A]t\}$$

The assumption that $[A]$ is constant is usually found to be quite good. We now assume that for an explosion to occur, the ratio $[C]/[C_0]$ must become extremely large. This is equivalent to assuming that the reaction goes through many cycles before $[C]$ increases appreciably. This yields the result that

$$\log([A]\cdot t_{\text{ign}}) = \frac{E}{RT_0} + \text{const.} \tag{6-9}$$

where E is the activation energy of the rate controlling step in the branching-chain cycle. Equation 6-9 has been verified experimentally for many high-temperature, low-pressure homogeneous explosions involving a chain-branching mechanism.

Lastly, we note that the assumption that $[C]/[C_0]$ becomes large is not equivalent to the assumption of the thermal theory that the temperature increase without bound, because it is quite possible for the chain cycle to be nearly thermally neutral. If the cycle is indeed thermally neutral, we obtain the simplest type of chain-branching explosion in which the chain-carrier concentration becomes very large before recombination reactions liberate heat energy to the system. In this case, the state variables of the system show an extremely slow variation with time during the induction period. After this time, the exothermic recombination reactions cause a relatively rapid temperature rise and the formation of products.

6-3. VESSEL EXPLOSIONS—EXPERIMENTAL OBSERVATIONS AND DIFFICULTIES

In the previous section we listed the experimental techniques for observing vessel explosions, and the types of data usually obtainable from explosion studies. We also discussed the theory of the ignition delay time and of explosion limits for certain simple situations. Unfortunately, the cases we discussed represent limit behaviors which have been observed in a clear-cut manner in only a few specific systems using very careful experimental techniques. The complex vessel explosion behavior usually documented by the experimentalist is caused either by the chemical complexity of the system he is studying or by the lack of an adequate model for the experimental technique he is using. We now discuss the three basic experimental techniques in an effort to document some of these difficulties.

The introduction method is the oldest technique for studying vessel explosions and has yielded by far the most complex results. The earlier work using this technique led to the discovery of explosion peninsulas in many explosive mixtures. Thus the prevalence of chain-branching explosions was established before their mechanism was fully understood. There are in fact only a few systems, mainly pure organic azides or nitrate compounds, which show simple thermal-explosion behavior in introduction experiments. A great deal of careful work on many systems finally led to the general conclusion that the occurrence of the explosion peninsula was evidence for a chain-branching reaction during the preexplosion period. Unfortunately, the simple chain-branching mechanism used in developing this concept does not always adequately explain the experimental observations. For example, in hydrocarbon–oxygen systems at high pressures degenerate branching occurs during the preexplosion period and leads to the formation of high concentrations of very reactive organic peroxides. Experimentally, it is observed that when the concentration of these species become sufficiently high, a visible "cool flame" flashes through the mixture and reduces their concentration without markedly altering either the pressure or temperature. After the propagation of this cool flame the primary chain cycle continues even though its rate is possibly modified. The "cool flame" always appears at pressures and temperatures which are above the chain-branching peninsula, along the thermal-explosion boundary in a (P,T) plot. Multiple cool flames may occur before the slow reaction goes to completion. Figure 6-6 illustrates the behavior of a typical system and the location of the cool-flame region relative to the explosion peninsula. Cool flames have also been observed and studied in piston-compression devices.

The introduction method has been extensively used to study explosion limits. However, in studies of this type both the first and third limits are intimately related to the vessel geometry and the first limit is also very sensitive to the effectiveness of the vessel surface in capturing radicals. Many kineticists have found, for example, that it is necessary to "condition" a new explosion vessel by running many explosions to obtain reproducibility before taking quantitative data. Also, in many cases, the walls must be coated with a substance which has a rather low radical-capture efficiency to sufficiently retard the wall reactions so that the nature of the homogeneous second limit may be studied. Because of these effects, second-limit studies are by far the most fruitful and they have been used to determine the mechanism of homogeneous radical capture in some of the simpler systems.

In general, the study of explosion limits in an introduction vessel must

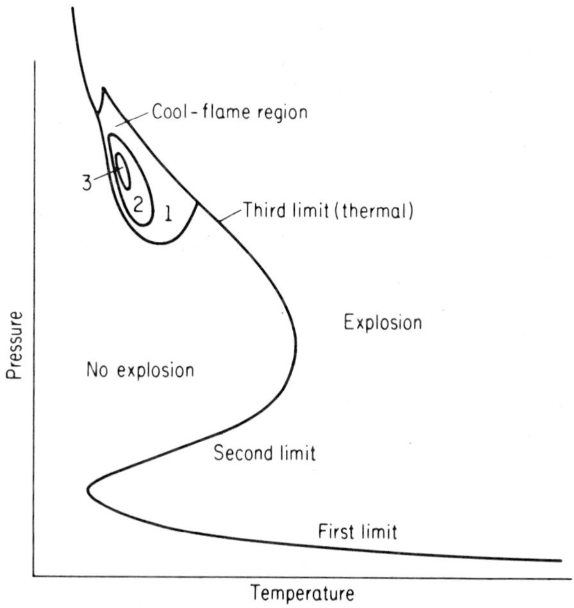

Fig. 6-6. A schematic diagram showing the relative location of explosion, no explosion, and cool flame regions for a hydrocarbon–oxygen–nitrogen mixture. Numbers in the cool-flame regions indicate the number of cool flames observed during the vessel experiment.

be accompanied by the results of other experimental techniques, before they may be used to determine unambiguous rate equations for the elementary steps of the chain cycle.

The introduction technique is generally not suited to the measurement of ignition-delay times. This is because when the delay times are long, wall effects play an important role and for short delay times, the time for heating the gas (mainly by conduction) becomes an appreciable fraction of the measured delay. In this connection we note that the simple theories discussed in the previous section completely ignore the initial transients in the system, and therefore represent only a first approximation to the real experimental behavior. This condition can lead to particularly serious errors when the introduction method is used for explosion studies. For example, the rapid introduction of a room-temperature gas sample into a heated and evacuated vessel is usually accomplished by connecting the preheated vessel to a reservoir at the proper pressure to yield the desired final pressure. This type of introduction technique produces a rather large initial temperature gradient in the vessel due to adiabatic compression in addition to the gradient which results from heat-transfer effects. If one were to measure short delay times in an introduction vessel these effects would have to be included in any theory that is developed to relate the measured delays to the kinetic rates.

There are three types of piston-compression devices that have been used to study homogeneous explosions. It is quite common to study explosion behavior in an internal combustion or diesel engine in which one cylinder has been modified to allow the introduction and compression of an experimental sample. There is also the specialized inertial-piston device in which a piston is driven into a closed tube by compressed air or helium and is allowed to bounce. The third type is a device that mechanically accelerates a piston toward the back wall of a tube and then locks the piston in the neighborhood of its maximum displacement.

The use of the first two types of devices to study either explosion limits or delay times produces the heating cycle shown schematically in Fig. 6-7

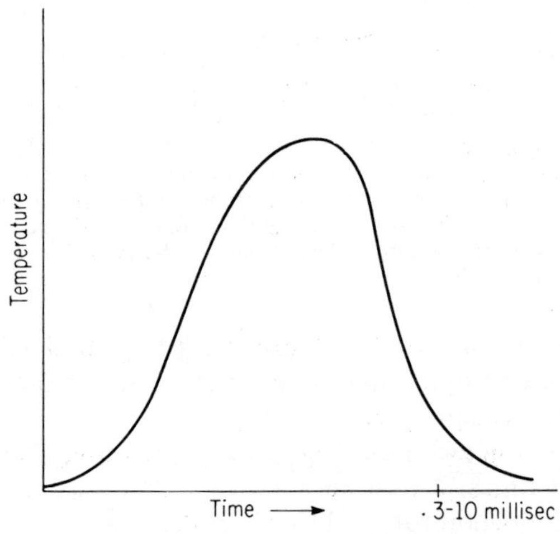

Fig. 6-7. Heating cycle in an adiabatic compression device without a locking device.

only if the piston motion is slow enough to preclude the formation of strong shock waves during the compression portion of the cycle. The occurrence of explosion during this cycle is usually recorded either by following the luminosity of the mixture through a window or by recording the pressure rise associated with the explosion by using fast-response pressure gauges. Schlieren observations have also been made.

Experiments of this type are fundamentally different from ordinary introduction experiments. In the first place, the time scale is markedly different. In ordinary vessel explosions the delay to explosion may be as much as 10 sec, while in adiabatic compression devices the total heating time is hardly ever greater than ten millisec. Thus the no-explosion area of the usual (P,T) explosion limit plot will be increased because of the extremely short times available for explosion. However, one observes all the features ordinarily seen when using introduction techniques, such as the occurrence of cool flames and explosion peninsulas normally associated with degenerate and simple chain-branching reactions. One also observes a new phenomenon, the occurrence of detonation. However, multiple cool flames are not usually observed in adiabatic compression devices.

In these devices the short adiabatic heating time allows a determination of delay times because the heating is due entirely to a relatively homogeneous compression of the gas sample. Therefore the measured delay time need not be corrected for a conductive heat lag. However, if one wants to understand the significance of delay times, the second major difficulty becomes important. In this type of experiment the heating cycle always represents a significant portion of the total possible delay time, and the measured delays must therefore be related to the explosion kinetics with a model that includes the finite heating rate. Because of this difficulty these types of piston devices have been used mostly to obtain engineering type data on fuels in connection with the actual design and operation of internal combustion or diesel engines. There have been few fundamental rate studies in explosive systems using these devices. The reader is referred to the bibliography for further information on this subject.

In contrast to the single-cycle devices, the piston-catching devices have the added capability of measuring a delay time which is significantly longer than the heating-cycle time. While this type of device should be quite useful for determining explosion limits and delay times, its use in studies of the rate of elementary reactions has been almost completely superseded by the advent of the chemical shock tube. We now wish to discuss this third experimental technique.

A conventional constant-area shock tube is a long pipe that produces incident and reflected shock waves in a gas or gas mixture by bursting a thin diaphragm separating a high- and low-pressure region. Figure 6-8 contains an x,t diagram illustrating the operation of an ideal one-dimensional shock tube, and temperature-time curves representing the behavior of gas samples at two initial positions in the expansion chamber after the diaphragm burst. The usefulness of the shock tube in kinetic studies of either endothermic or exothermic reactions is based primarily on the temperature-time history illustrated in Fig. 6-8. It has been shown that in a step shock wave the translational temperature of an element of the gas is raised rapidly (in less than 10^{-7} sec) to a high value and then may be held at this temperature for relatively long times (up to 4 millisec). Thus with

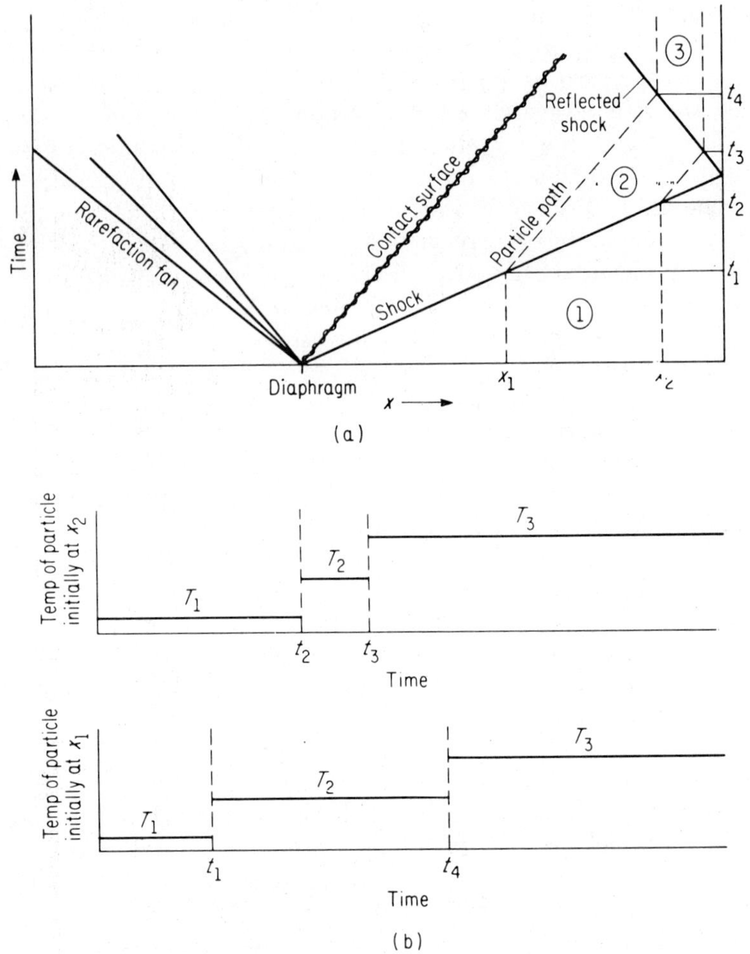

Fig. 6-8. (a) An (x, t) diagram of an ideal shock-tube flow. (b) The time-temperature history of gas samples initially at positions x_1 and x_2 for the shock-tube flow shown in Fig. 6-8a.

the use of rapid-response instrumentation a relatively simple shock tube can become a high-temperature vessel for studying kinetic rates. Furthermore, because of the nature of the apparatus it is relatively easy to obtain temperatures as high as 5000 °K for these studies.

During the past 15 years there has been a tremendous amount of research on chemical systems using shock-tube techniques which we cannot adequately discuss in this text. The reader is referred to the review books listed in the bibliography for further information on specific kinetic studies using shock tubes. It is appropriate however, to make a few general observations concerning the technique and its relation to the study of explosion phenomena.

1. Heating is obtained by a traveling shock wave and therefore is not a strictly homogeneous process. Furthermore, since the flow behind the shock wave is always subsonic relatively to the shock, the occurrence of an exothermic or endothermic reaction behind the front will affect the behavior of the front. In highly exothermic systems rapid reaction leads to a rapid acceleration of the front and the shock wave eventually becomes a detonation wave. Thus the shock tube is not easily adaptable to the study of explosion delays in highly exothermic mixtures. There is one exception to this rule: at the back wall the initial conditions for the heating cycle are well defined, and homogeneous explosions have been observed in this region under certain circumstances. However, away from the back wall this explosion quickly becomes a detonation. Detonations may also be triggered locally in shock-tube experiments on highly exothermic mixtures by the diaphragm burst or by the presence of crevasses or wall discontinuities at joints in the tube. Detonation behavior is discussed in Chapter 9.

2. The shock wave leaving the back wall is not nearly as ideal as the incident shock wave. The primary cause of this effect is the reflected shocks interaction with the boundary layer in the flow behind the incident shock. This type of shock nonideality is least restrictive in gas mixtures with high specific-heat ratios.

3. The "vessel," even for back wall explosions, does not have a constant volume in a shock-tube experiment, because gas dynamics allows expansion when a local pressure rise occurs.

4. Because of the extremely short heating times and relatively short observation times available in the tube, the bulk of the gas sample does not contact the walls of the tube during the experiment. Thus rate constants obtained from experiments in the shock tube are fundamentally more useful in the interpretation of overall explosion kinetics than those obtained from ordinary vessel experiments. This usefulness is augmented by the technique of working at high dilution. Quite frequently a reaction is studied in the presence of 95–99% argon as an inert carrier gas. The large excess of inert gas acts effectively as a thermostat bath for the reaction and allows one to study kinetics in even highly exothermic systems without risking the occurrence of detonation.

6-4. THE WELL-STIRRED REACTOR

The well-stirred reactor is a concept that may be approximated in the laboratory. It has application in studies of the recirculation stabilization of turbulent flames, and of the volumetric combustion chamber efficiency for continuous flow combustors. In this section, we wish to develop the concept and show the experimental verification; an application of the well-stirred reactor will be discussed in Chapter 8.

Consider a chamber containing inlet and outlet ducts. Let us add a mixture of fuel and oxidizer continuously at the inlet and remove the resultant products at the outlet. We assume that a combustion process is occurring inside the chamber and we observe that there must be some limit mass-flow rate above which the chemical processes can no longer be stabilized in the vessel during the residence time of the gas sample. The experimental values of this limit are actually dependent in a remarkedly complicated way on details of the flow situation inside the chamber. However, it is possible in the laboratory to construct chambers that nearly approach the theoretical efficiency of the "well-stirred" model; experiments in these devices have shown that their blowout limits are related in a rather fundamental way to the chemistry of the combustion reactions.

Because its blowout behavior represents a limiting efficiency for practical continuous flow engines, this type of device is useful in an engineering sense. However, because of the complex nature of the flow in the reactor (rapid reaction is obtained by high-intensity turbulent mixing), the apparatus does not lend itself to the study of the rates of fundamental kinetic steps. For these two reasons, studies in well-stirred reactors have been mainly limited to hydrocarbon–air mixtures, and the modeling of the reactor behavior has been used to determine kinetic parameters for overall rate equations similar to Eq. 6-3.

Consider a well stirred adiabatic chamber of volume V with the global kinetics of Eq. 6-3 occurring everywhere at the same rate throughout its internal volume under conditions which are identical to the output conditions. Since pressure is assumed constant, two balance equations, for mass and energy, must be satisfied. For a fuel lean mixture entering the reactor we write the balance equation for fuel, using Eq. 6-3 for the global rate expression

$$\frac{N(X_{f_o} - X_f)}{VP^{(n+m)}} = \frac{AX_{f_o}^{(n-1)}X_{o_o}^{m}}{RT^{(n+m)}} (1-\eta)^n(B-\eta)^m \exp(-E/RT) \qquad (6\text{-}10)$$

where N is the total throughput in moles/sec and $N(X_{f_o} - X_f)$ is the rate of consumption of fuel (moles/sec). The dimensionless quantity η equals $(1 - X_f/X_{f_o})$ and since the mixture is fuel lean $B > 1$. A similar relationship could be derived for a fuel rich system.

Defining $\theta = RT/E$, this equation may be written in the form

$$\eta = \tau_{k/s}\,(1-\eta)\,(B-\eta)\frac{\exp\,(-1/\theta)}{\theta^{\,(n+m)}} \tag{6-11}$$

where

$$\tau_{k/s} = \left(\frac{AX_{fo}^{(n-1)}X_{O_o}^{m}}{E^{(n+m)}}\right) \Bigg/ \left(\frac{N}{VP^{\,(n+m)}}\right)$$

and represents a dimensionless ratio of a chemical rate to a throughput rate. For any specific initial conditions $\tau_{k/s}$ is a constant and Eq. 6-11 may be solved to yield curved lines in Fig. 6-9.

If one assumes a constant molecular weight and heat capacity for the reactants and products, tlhe energy balance equation becomes

$$\eta = (\theta - \theta_o)/\omega \qquad \text{with } \omega = \frac{C_o R Q_r}{EC_p} = (\theta_m - \theta_o)$$

Here, Q_r is the heat of reaction and θ_m is the maximum temperature for complete reaction. This is the equation of a straight line and is plotted as such in Fig. 6-9.

For the system described above, as flow through the reactor is increased $\tau_{k/s}$ decreases and Eq. 6-11 can be progressively represented by the curves labeled A, B, C, D and E in Fig. 6-9. Curve A has only one solution, with high outlet temperature and nearly complete reaction; B has two, with the lower point a tangent; C has three with the midpoint being dynamically unstable; D has two with an upper tangency point; and E has only a low temperature solution. All of the low temperature solutions are physically unrealistic because the assumed kinetic law is unrealistic at low temperatures. All the high temperatures solutions represent combustion solutions and the tangency point on curve D represents the maximum throughput velocity at which high temperature combustion can occur in the reactor. Above this velocity theory predicts that the reactor will "blow out," under conditions where combustion is not complete (i.e., $\eta_b < 1$ and $\theta_b < \theta_m$ where the subscript *b* represents blowout conditions). Thus combustion efficiency at blowout is less than one. This behavior is observed experimentally.

Blowout conditions are usually determined as a function of equivalence ratio and correlated to the overall activation energy, *E*, and the overall order of the reaction, $\alpha = n + m$. Such a correlation for octane-air is shown in Fig. 6-10. Note that the overall reaction order, α, is very close to 2.

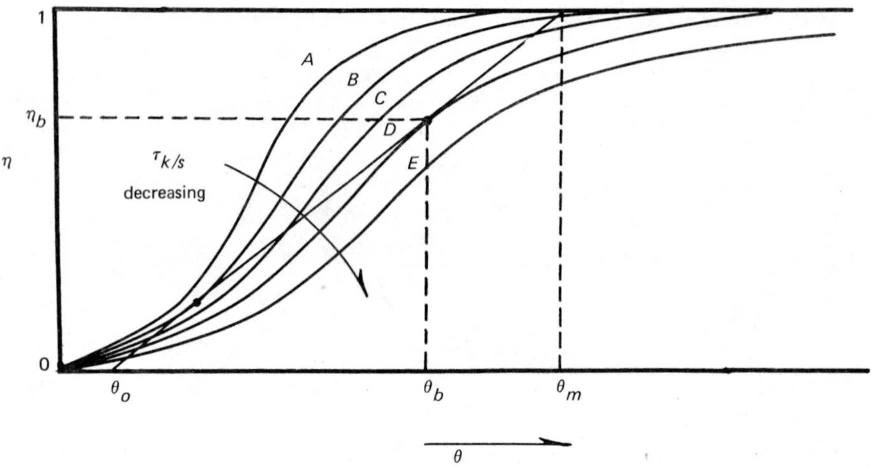

Fig. 6-9. Flow rate vs. efficiency in a typical well-stirred reactor.

Additionally, experiment has shown that the overall equation cannot model the process correctly for both lean and rich blowout. Specifically, for lean blowout very few hydrocarbons are present at incipient blowout and $CO \rightarrow CO_2$ oxidation is the rate controlling step. For rich blowout, on the other hand, hydrocarbon fragments and free hydrogen are present at blowout and earlier oxidation steps must be rate determining.

It is interesting to note that the volumetric heat release rate in well-stirred reactors approaches that of the very narrow hot-reaction zone of a premixed gas flame. Typical volumetric efficiencies are tabulated in Table 6-1.

TABLE 6-1*
VOLUMETRIC EFFICIENCY OF
VARIOUS COMBUSTION CHAMBERS

Apparatus	Heat Release Rate, Btu/Cu ft hr atm^2
Domestic boiler	$0.01–0.03 \times 10^6$
Navy boiler	$0.20–0.25 \times 10^6$
Industrial gas turbine	$0.70–3.00 \times 10^6$
Aircraft gas turbine	$4.0 – 11.0 \times 10^6$
Well-stirred reactor	$\approx 170 \times 10^6$
Premixed laminar flame zone	$\approx 300 \times 10^6$

*Data with permission from "Literature of Combustion of Petroleum," *Advances in Chemistry Series No. 20*, American Chemical Society (1958), p. 33.

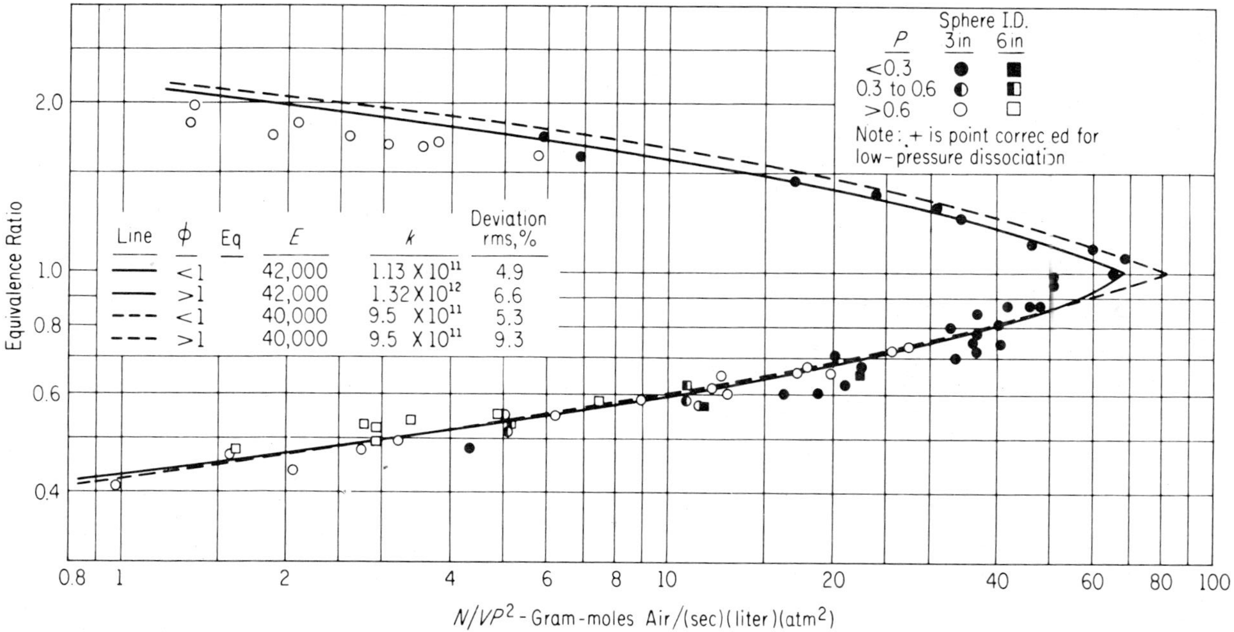

Fig. 6-10. Blowout data correlation for second-order reaction (α = 2.0), corrected to 400°K inlet temperature. [Adapted from P. A. Longwell and M. A. Weiss, *Ind. & Eng. Chem.*, *47*, 1634 (1955).]

6-5. GAS-PHASE DIFFUSION FLAMES

Diffusion flames have been known since antiquity. Natural fires are predominately a complex of diffusion controlled flames and the candle or oil-wick flames that early man used for artificial lighting are relatively simple examples of diffusion flames. The diffusion flame in its simplest form consists of an exothermic reaction zone separating relatively pure samples of oxidizer and fuel gases. As such it does not exhibit a characteristic propagation velocity. However, the characteristics of diffusion flames are markedly dependent on the aerodynamics of the particular flow situation, and the complete spectrum of diffusion-flame behavior cannot be discussed without invoking aerodynamic considerations.

In this section we will discuss two diffsuion-flame configurations that are particularly simple aerodynamically. These are the impinging-jet diffusion flame, and the cylindrical diffusion flame that occurs above a circular port. In addition, we will discuss the mechanism of soot formation in hydrocarbon–air diffusion flames, since ordinary convective diffusion flames of hydrocarbons in air almost invariably form soot in some portion of the flame zone.

The impinging-jet diffusion flame represents the simplest diffusion flame available for experimental study. Unfortunately this type of flame has been constructed in the laboratory only recently and studies of its behavior are not complete at this writing. Figure 6-11 is a schematic

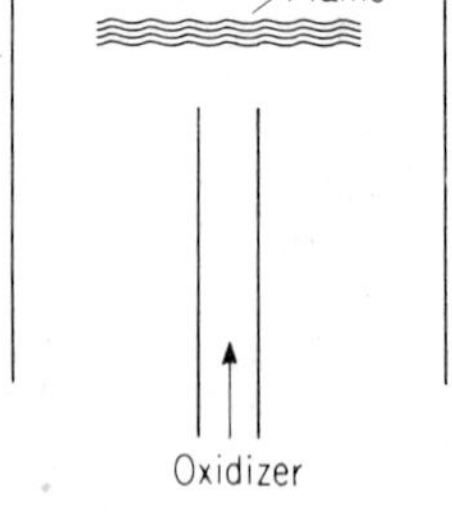

Fig. 6-11. Schematic diagram of the impinging jet diffusion flame burner.

diagram of the experimental apparatus used to produce and study the impinging-jet diffusion flame. In this apparatus, the diffusion flame is formed at the boundary between the fuel and oxidizer streams but not necessarily at the stagnation point of the flow. Its position relative to the stagnation point may be altered by diluting either the oxidizer or fuel streams and, because the apparatus is relatively small, the flame's behavior may be studied as a function of pressure. The most important quantity which has been studied using this flame is its "blowout" behavior. It is observed that as the fuel and oxidizer flow rates are increased, the flame becomes relatively thinner in the neighborhood of the axis of the flow, and at some critical flow rate the central portion of the flame will no longer be self-supporting and will "blowout." Under these conditions the flame will appear to have a hole in its center completely surrounded by an ordinary diffusion flame similar in character to those to be discussed in later paragraphs. This limit behavior of the flame is related in a complex way to the gross aerodynamics of the flow. However, in the simplest cases it has already been observed that the limit is essentially independent of diffusional control and represents a chemical reaction rate limit similar to that found in the well-stirred reactor or in a laminar premixed flame zone. We note that if the center of the flame is exactly at the stagnation point of the flow, the limit represents a maximum lateral expansion of the flame zone at which combustion may be maintained. Thus the fundamental "strength" of a diffusion flame may hopefully be defined, using this type of apparatus. Outside of the central thin reaction zone the lateral stretch decreases rapidly because of the cylindrical geometry, and the rate of the flame reaction is eventually limited by ordinary diffusion processes.

Before the invention of the impinging-jet burner, diffusion flames were studied mainly as attached flames in specific flow systems. A fairly simple example is the burning of a pure fuel jet of cylindrical section in a large enclosure under conditions where the flow remains laminar throughout. Experimental composition profiles at various heights above the port for a hydrogen jet are shown in Fig. 6-12. Figure 6-13 is a schematic model of the concentration profiles at any level made by assuming that the reaction zone is infinitely thin, and therefore that the rate of combustion at any level in this diffusion flame is controlled purely by diffusional processes. This assumption has been shown to be best in the main portion of the flame. It is probably not as valid near the port. In fact, one would expect that as the flow velocity is increased toward blowout, the region near the port would exhibit the character of the flame which is seen near the flow axis in the impinging-jet experiment. This is true and almost all attached diffusion flames clearly show a thin hot-reaction zone at the point of attachment. The base of a candle flame is a good example.

Unfortunately, the situation in a diffusion flame is relatively complex because the flame's behavior is intimately tied to aerodynamic considerations. These flames are really very slow burning at the local level and buoyancy effects, for example, contribute markedly to their overall behavior. Because of this type of difficulty, diffusion flames have not been studied as intensively or quantitatively as premixed gas flames. One may conclude, however, that in the thicker regions of the diffusion flame, diffusion processes are actually controlling the local conversion rate, while

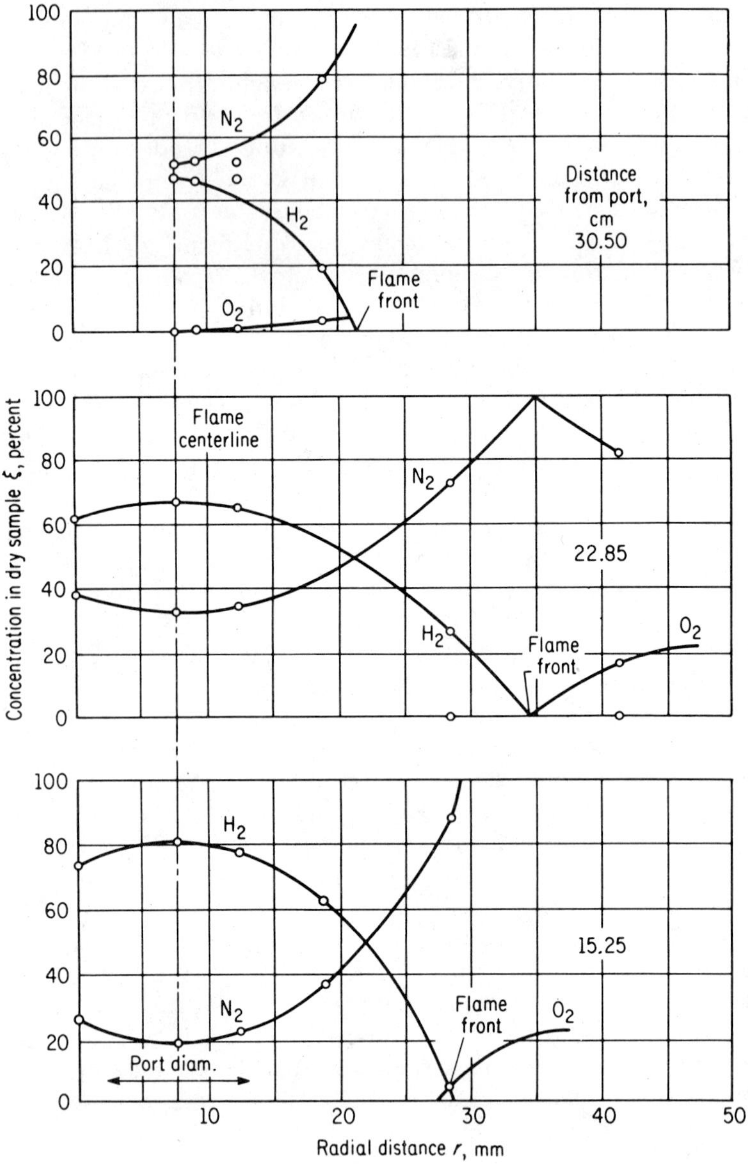

Fig. 6-12. Composition of hydrogen diffusion flame. Nozzle diameter, $^1/_4$ in.; nozzle velocity, 108 ft/pr sec. [Adapted with permission from H. C. Hottel and W. R. Hawthorne, *Third Symposium (International) on Combustion*, Williams and Wilkins Co., Baltimore, 1949, p. 258.]

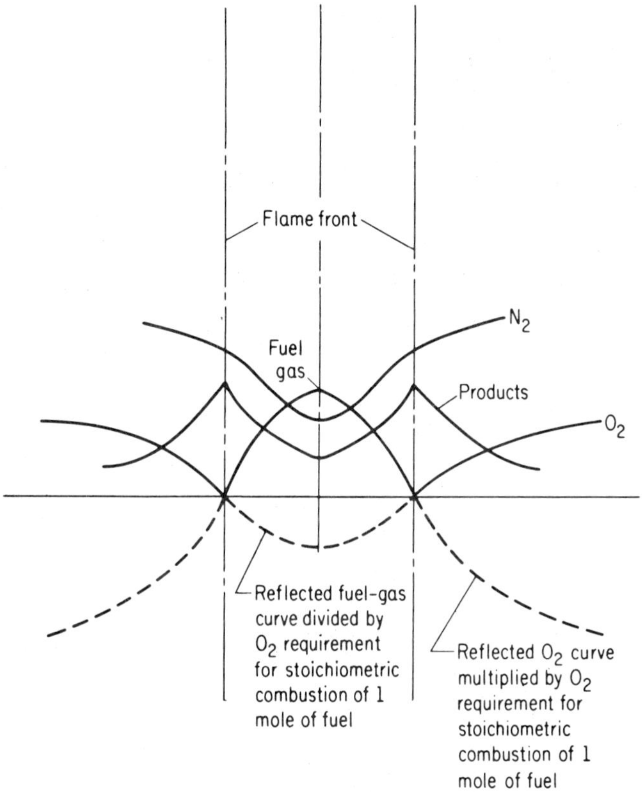

Fig. 6-13. Simplified diagram of concentration profiles in typical laminar diffusion flame. [With permission from H. C. Hottel and W. H. Hawthorne, *Third Symposium (International) on Combustion*, Williams and Wilkins Co., Baltimore, 1949, p. 258.]

in the thin-flame regions the gradients may become steep enough so that the chemical reaction rate becomes the controlling step.

The most striking characteristics of ordinary hydrocarbon–air diffusion flames are their predominantly yellow-white color, their relative opaqueness and their ability to produce a lampblack deposit when contacting a cold object. All these effects are due to soot particles forming in the flame zone of the diffusion flame. The formation of soot is due to the fact that high in the diffusion flame the relatively pure fuel core becomes quite hot by thermal conduction, and pyrolysis of the fuel occurs locally. If the flame is overventilated, a hot oxidation zone will completely surround the soot region and the flame will burn without discharging a sooty wake. However, in many cases the hot flame "opens" at the top and a soot stream is seen to leave the flame zone. This is particularly true if the flame is disturbed by external gusts of wind. Similarly, the insertion of a cool surface into the soot zone of the flame will lead to the formation of a lampblack deposit on the cool surface.

The primary mechanism of soot formation has been an open question for many years, and studies of the process are still continuing. However, recent evidence strongly indicates that soot actually consists largely of addition compounds whose structural formulas resemble higher cyclic unsaturated hydrocarbons of the type shown in Fig. 6-14. Soot shows the

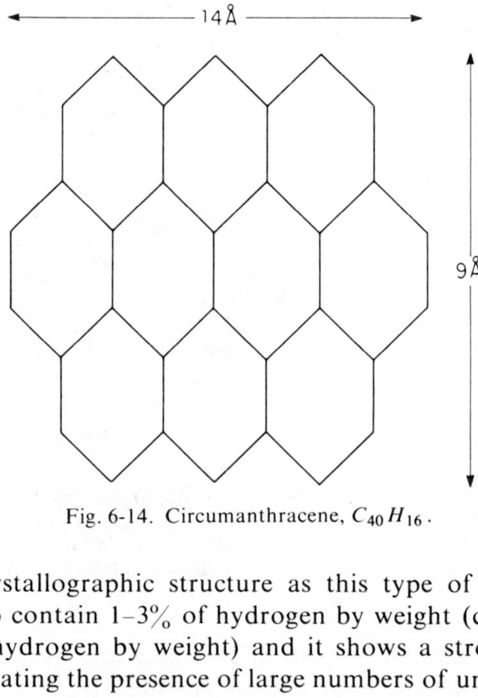

Fig. 6-14. Circumanthracene, $C_{40}H_{16}$.

same X-ray crystallographic structure as this type of compound; it is always found to contain 1–3% of hydrogen by weight (circumanthracene contains 3.2% hydrogen by weight) and it shows a strong electron-spin resonance, indicating the presence of large numbers of unpaired electrons. Thus it is highly probable that soot is simply an aggregate of very large polybenzenoid free radicals.

There is also evidence that the primary formation process involves addition reactions of relatively stable conjugated bond species. Thus the probable steps leading to soot formation are:

1. The formation of small hydrocarbon radicals by cracking.
2. The agglomeration of these small radicals into relatively stable larger conjugated bond radicals such as butadiene.
3. Addition reactions of the Diels-Adler type to form very large conjugated bond polybenzenoid free radicals from the simple conjugated bond molecules.

4. Subsequent agglomeration of these larger radicals to form the actual soot particles.

Practical studies of soot formation are quite important to industry, because fouling by soot represents an important problem in the operation of combustion power plants. There have been many empirical indices developed to aid the designer in his evaluation of the probable occurrence of soot in a specific situation. A detailed discussion of this work is outside the scope of this text and the reader is referred to the bibliography for references to this work.

6-6. SINGLE-DROPLET AND SOLID-PARTICLE COMBUSTION

The combustion of fuel sprays or suspensions of solid particles are important even though quite complex practical problems. In an attempt to understand the mechanism of this type of combustion process, there have been many investigations of the combustion of single droplets or single solid particles. Since the combustion of a single particle of either liquid or solid represents a well-defined limiting case which is primarily controlled by the behavior of a diffusion flame, we include its discussion in this general chapter. Further brief discussions of the behavior of spray or dust cloud combustion is left to Chapters 7 and 8.

The simplest case of small-particle combustion is the combustion of a liquid drop which leaves no solid residue on evaporating, and has no solid combustion products. The theory and experimental work on this type of droplet combustion has shown that there are three phases in the life of a fine droplet after its exposure to an infinite reservoir of hot gaseous oxidizer. First there is the ignition-delay time in which surface evaporation raises the concentration of the fuel vapor to the point where spontaneous ignition may occur. Second, after ignition there is the burning phase during which the droplet size decreases until all the liquid fule is evaporated, and third there is the postdroplet period during which the residual fuel is consumed.

The ignition period has not been studied extensively, but it is known that the diameter of an isolated drop changes only slightly during this portion of its life. In this connection it should be pointed out that we are discussing, in the main, the behavior of small droplets and therefore the motion of the gas stream relative to the drop is always small except in a flow situation containing regions of high acceleration. This is because viscous drag on a small droplet causes it to relax to the local flow velocity quite rapidly. Thus the preflame evaporation process is primarily controlled by conductive rather than convective heat transfer, and the evaporated fuel is ordinarily not markedly separated from the droplet by convective mechanisms. Not only is the data on the subject of ignition delay meager, but it is questionable if this type of single-droplet ignition data is applicable to the prediction of flame position in a multidroplet spray in, say, a turbulent-flow situation.

The second phase of the combustion process has been studied extensively and the results present a rather clear picture of the process. Once again, because of the small size of the droplet, large relative velocities do not occur, and the burning droplet may be assumed to be completely surrounded by a diffusion flame similar to the ideal flame shown in Fig. 6-13. The only change from Fig. 6-13 is that the system now has spherical symmetry and the center line of Fig. 6-13 represents the evaporating fuel surface. The theory based on this simple model predicts that the drop diameter should decrease with time according to the relation

$$d^2 = d_0^2 - kt \tag{6-12}$$

because heat transfer to the droplet surface controls the evaporation rate. The constant k in Eq. 6-12 is called the characteristic evaporation constant of the fuel and ranges from 3.6×10^{-3} cm^2/sec to 35.6×10^{-3} cm^2/sec for a large variety of fuel and oxidizer combinations. For most hydrocarbons burning in air, the constant is nearly invariant at $10 \pm 2 \times 10^{-3}$ cm^2/sec. These predictions have been verified by extensive experimental observations on single burning particles. However, we must again make the comment that during the combustion of dense fuel sprays the assumption that the burning droplet is indeed immersed in an infinite quantity of gaseous oxidizer is no longer valid, and the evaporation rate of burning droplets in this case might be expected to be controlled by different mechanisms than those that control single-droplet evaporation.

The third or postdroplet combustion period can be important because its duration is probably longer than the total evaporation period for a small drop. However, it is primarily of importance in dense-spray combustion because the attainment of complete combustion would be expected to be slowest in this situation. We will discuss dense spray combustion in Chapter 7.

We indicated earlier that droplet or dust combustion was more complex if the substance does not evaporate completely during the second stage of the process. The complexity here takes many forms and research is still quite active in this area. We will therefore simply list a few examples and qualitatively discuss their behavior.

1. Solid reactants—gas-phase products.

This system is best exemplified by the combustion of coal dust in air. Also, during the late phases of the combustion of higher hydrocarbons, pyrolysis may form a core of porous, solid residues which continue to burn. In this type of combustion the system behaves as though diffusion of oxygen to the burning solid was the rate determining step.

2. Liquid reactants—solid products.

The combustion of aluminum in air is an excellent example of this system. Here the aluminum liquifies before burning and forms solid aluminum oxide on combustion. In certain cases the aluminum oxide accumulates on the surface of the sphere as a solid skin. It may then develop a small hole and jet aluminum vapor into the air, causing the drop to have an erratic velocity during burning. Hollow aluminum oxide spheres have been recovered from these experiments.

3. Solid reactants—solid products.

The combustion of high-melting-point metal powders, such as zirconium, is an example of a solid-solid system. Very little quantitative work has been performed on these systems, and their detailed combustion mechanism is not understood at present.

BIBLIOGRAPHY

Vessel Explosions

1. A. S. Sokolik, "Self-Ignition Flame and Detonation in Gases." Israel Program for Scientific Translations. NASA TT F-125 OTS 63-11179 (Jerusalem, 1963) pp. 1–165.

2. A.C.S. No. 20, *Literature of the Combustion of Petroleum*, American Chemical Society, pp. 84–103.

3. Combustion Processes, Princeton Series on High-Speed Aerodynamics, Vol. 2, pp. 160–200.

4. P. A. Longwell, H. H. Reamer, N. P. Wilburn, and B. H. Sage, *Ind. and Eng. Chem.*, 50, 603, 1958.

5. E. F. Greene and J. P. Toennies, *Chemical Reactions in Shock Waves.* (New York: Academic, 1965.)

6. J. N. Bradley, *Shock Waves in Chemistry and Physics.* (New York: Wiley, 1962.)

7. A. G. Gaydon and I. R. Hurle, *The Shock Tube in High Temperature Chemical Physics.* (New York: Reinhold, 1963.)

Well-stirred Reactors

8. J. P. Longwell and M. A. Weiss, *Ind. Eng. Chem.*, *47*, 1634 (1955).

9. R. H. Essenhigh, "An introduction to stirred reactors theory applied to design of combustion chambers" in Combustion Technology, Some Modern Developments, H. B. Palmer and J. M. Beér, eds. (New York: Academic Press, 1974), pp. 373-415.

Diffusion Flames

10. NACA Report 1300, pp. 207–228, 1959.

11. A. G. Gaydon and H. G. Wolfhard, *Flames.* (London: Chapman and Hall, 1960.)

12. *Combustion Processes*, Princeton Series on High-Speed Aerodynamics, Vol. 2, pp. 365–444.

13. A. Thomas, *Combustion and Flame*, *6*, 46 (1962).
14. D. B. Spaulding, *ARS J.*, *31*, 763 (1961).
15. F. A. Williams, *Combustion Theory*. (Reading, Mass.: Addison-Wesley, 1965.)

PROBLEMS

6-1. Rework the equations for a thermal explosion in a vessel to illustrate an experimental technique for directly determining the orders of the overall rate equations, m and n.

6-2. Calculate and plot the lifetime of a typical hydrocarbon droplet while burning in air, using Eq. 6-12. Allow the initial diameter to range from 1 mm to 1 μ.

7

Premixed Gas Flames

7-1. INTRODUCTION

Even though flames in one form or another have been known and used since antiquity, a particularly simple form of combustion, the premixed gas flame, has been recognized and studied for only about 150 years. Its discovery was a side result of the researches of Berthollet, Dalton, Volta, and others (1776–1810) concerning the composition of combustible gases from different sources, particularly marsh gas (methane) and olefiant gas (ethylene). In their analyses, a combustible mixture of these gases with air was sparked in a bulb and the products of the premixed flame (explosion) were analyzed to determine the composition of the original combustible gas. During a later period (1805–1819) Davy's observation that a premixed flame would not propagate through a fine-mesh screen led to the invention of the safety lamp for mines. The "discovery" of premixed flames was essentially completed in 1855 by Bunsen's invention of a burner which allowed the stabilization of a premixed flame in a flow system.

We now know that a large collection of exothermic mixtures and pure substances which are capable of sustaining a homogeneous reaction will also support a subsonic "reaction wave" (or flame) which is self-propagating and has definite properties that are dependent only on the initial conditions in the gas. Unfortunately, these reaction waves are always rather thick and thus can be markedly influenced by curvature effects. Furthermore, they can be inherently unstable as a one-dimensional wave phenomena, and because of their low subsonic velocity they usually interact very strongly with the flow aerodynamics. In this chapter we wish to divorce our discussion from these complicating aerodynamic aspects as much as possible—reserving our discussion of that facet of flame behavior for Chapter 8. Thus in the following sections we will consider only structure and inherent stability of a laminar flat flame, the point ignition of a flame, and the cold-wall quenching of flames.

7-2. FLAME TEMPERATURE AND BURNING VELOCITY

Assume that Fig. 5-2 represents a strictly one-dimensional steady flame with a relatively low subsonic velocity. If this flame is to exist as a steady phenomena it must satisfy the equations derived in Chapter 5 at the two stations (1) and (2), where (2) is far enough upstream so that thermal and chemical equilibrium is reached in the hot gases. Therefore it must have two properties that are definable irrespective of the details of the processes occuring in the reaction zone—the burning velocity $S_u = u_1$ and the flame temperature $T_b = T_2$. In addition, we note from Chapter 5 that the momentum equation requires that $P_2 \neq P_1$ for a flow process with steady heat addition, and we first determine an upper limit for the pressure change associated with flames by solving Eq. 5-25a to yield

$$P_2 - P_1 = P_1 \gamma (1 - \epsilon) M_1^2 \qquad (7\text{-}1)$$

where $\epsilon = V_2/V_1$ and $M_1 = S_u/a_1$. Since for all observable flames $M_1 < 0.02$, $V_2/V_1 < 15$ and $\gamma \leq 1.667$, Eq. 7-1 predicts that a maximum pressure change of $P_1 - P_2 \leq +0.01\ P_1$ may exist across any flame. We thus note that flame propagation is always accompanied by a slight pressure drop, and that ordinary one-dimensional flames with $M_1 \simeq 0.001$ may be considered as essentially isobaric processes. This observation justifies the usual approximation of equating the flame temperature to an isenthalpic explosion temperature for the mixture. Also, for any specific system the isobaric assumption means that the flame temperature T_b is relatively independent of the observed burning velocity and may be determined with reasonable accuracy without knowledge of the burning velocity. Measurements of flame temperatures have shown that the translational gas temperature rises monotonically through the flame and eventually reaches a value comparable to the isenthalpic values calculated using thermodynamics. Ordinarily the accepted value of a flame temperature is the calculated isenthalpic value because difficulties with the experimental measurement of flame temperature and the relatively high accuracy of present-day thermodynamic calculations cause one to place more confidence in the calculated value. Figure 7-1 illustrates the effect of mixture stoichiometry on the calculated flame temperature for a few common flames.

In addition to flame temperatures, an isenthalpic calculation always yields the equilibrium composition of the mixture some distance downstream from the flame. We note that this equilibrium composition may not be simply related to the composition in or near the reaction zone of the flame. Section 7-3 discusses internal structure in more detail.

It has been observed that the flame temperature of the hotter flames

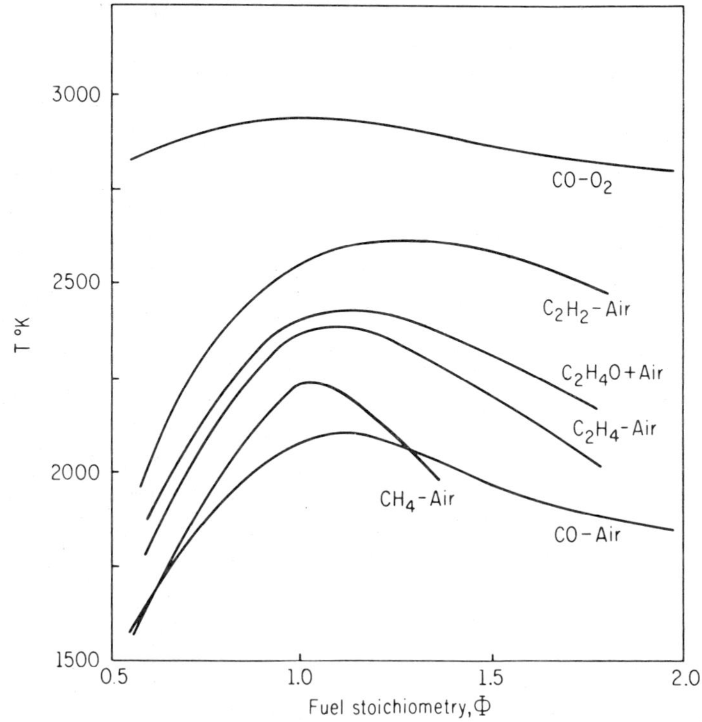

Fig. 7-1. Flame temperature for various fuel-air and fuel-oxygen mixtures.

is relatively insensitive to the initial temperature of the mixture because dissociation reactions cause the flame gases to have a large effective heat capacity. Low-temperature flames are relatively more sensitive to changes in initial temperature because dissociation reactions are not as important at these temperatures. However, the flame temperature of a high-temperature flame is more sensitive to changes in the ambient pressure, because pressure shifts the dissociation equilibria at high temperatures but can cause only relatively small composition changes in low-temperature flames.

We now turn our attention to the question of burning velocity. First, we note that our assumption that a flame is a steady-flow process in no way legislates the behavior of real flames. On the contrary, the justification for this assumption arises solely from our observations of flames under well-controlled conditions in the laboratory. It has been observed that in a variety of laminar-flow situations the flame behaves in a way best explained by assuming that it is a steady three-dimensional phenomena. The extrapolation of this three-dimensional real behavior to the strictly one-dimensional behavior required for the definition of a burning velocity is primarily a problem of careful experimental measurement and of properly assessing the effects of curvature on flame structure. Because of the difficulty of correcting for flame curvature in flows where the flame is positioned at an oblique angle to the flow, the most accurate burning velocities are obtained in situations where either the flame is strictly flat, even though oriented obliquely to the flow, or where the flame, even though slightly curved, is oriented strictly normal to the flow-velocity vector.

There are three measurement techniques which satisfy these requirements and which may be considered as the primary experimental techniques for determining burning velocity: (1) the slot-burner technique, (2) the soap-bubble technique, and (3) the flat-flame burner technique.

In the slot-burner technique a flame is stabilized above a rectangular nozzle (ratio of side lengths > 3), which has either a flat velocity profile generated by a two-dimensional Mache–Hebra nozzle, or a fully developed laminar pipe-flow profile generated by using a long entrance pipe of the same dimensions as the port. In this situation the flame is attached to the rim of the burner and completely burns the combustible mixture issuing from the port. The flame takes the appearance of a tent above the port and is therefore held at an oblique angle to the primary flow direction. It has been observed that a relatively flat oblique laminar flame may be transformed to a strictly one-dimensional flame by imposing a constant velocity transformation as shown in Fig. 7-2. Therefore the advantage of the slot burner is simply that the flame has rather large flat areas (both

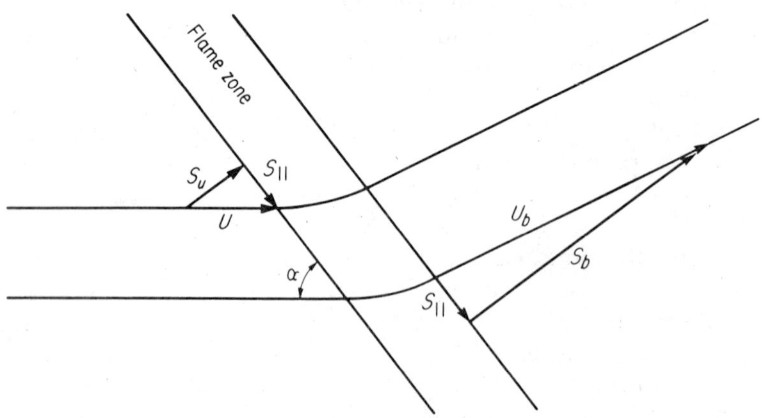

Fig. 7-2. The oblique one-dimensional flame. The flame appears as an oblique flame because of a simple velocity transformation at the velocity $S_{\|}$ along itself. The stream tube area parallel to the flame front does not change in this diagram.

sides of the tent) in which curvature effects are minimized. Furthermore, since the flame completely encloses the incoming flow the incoming streamlines are not deflected until they enter the heat conduction zone. Thus the angle that the flat-flame surfaces make to the incoming flow can be measured accurately and used to evaluate the normal burning velocity by applying the simple equation

$$S_u = U \sin \alpha \tag{7-2}$$

Experiments have shown that for this geometry the measurement of the local flow velocity and local flame angle at a point in the flame yields a burning velocity which is relatively constant over a large portion of the burner width. A typical evaluation is shown in Fig. 7-3. We will see later that the apparent increase in velocity at the tip of this flame is caused by an aerodynamic interaction related to the positive flame curvature in this region, and the apparent decrease of burning velocity near the rim of the burner is caused by heat losses to the wall and the negative flame curvature in this region.

The soap-bubble technique is another method which is known to yield accurate burning velocities without the application of an excessive number of corrections. In this technique a soap bubble is blown with a combustible mixture and ignited centrally by means of a capacitance spark. The flame burns as an outwardly propagating spherical flame, and since the bubble expands as combustion proceeds, the process occurs at constant pressure. In this case, after an initial propagation period the flame is only slightly curved and is oriented strictly normal to the flow direction. Its apparent propagation velocity is not S_u, however, but actually $S_b = u_2$ (as given in Fig. 5-2) since the gas behind the flame is quiescent. This measured velocity may be converted to the normal burning velocity S_u by noting that the continuity equation requires that $\rho_1 S_u = \rho_2 S_b$, and therefore that

$$\begin{aligned} S_u &= \left(\frac{V_1}{V_2}\right) S_b \\ &= \left(\frac{r_1}{r_2}\right)^3 S_b \end{aligned}$$

where r_1 and r_2 are the initial bubble radius and final flame-ball radius respectively. Measurement of burning velocity using this technique thus requires the measurement of both the apparent flame velocity and the ratio of the initial to final radius of the ball of combustible gases. Since a relatively exact final radius is required, the best measurements are obtained when the initial bubble is surrounded by an inert atmosphere to prevent afterburning. Figure 7-4 is a typical example of an experimental

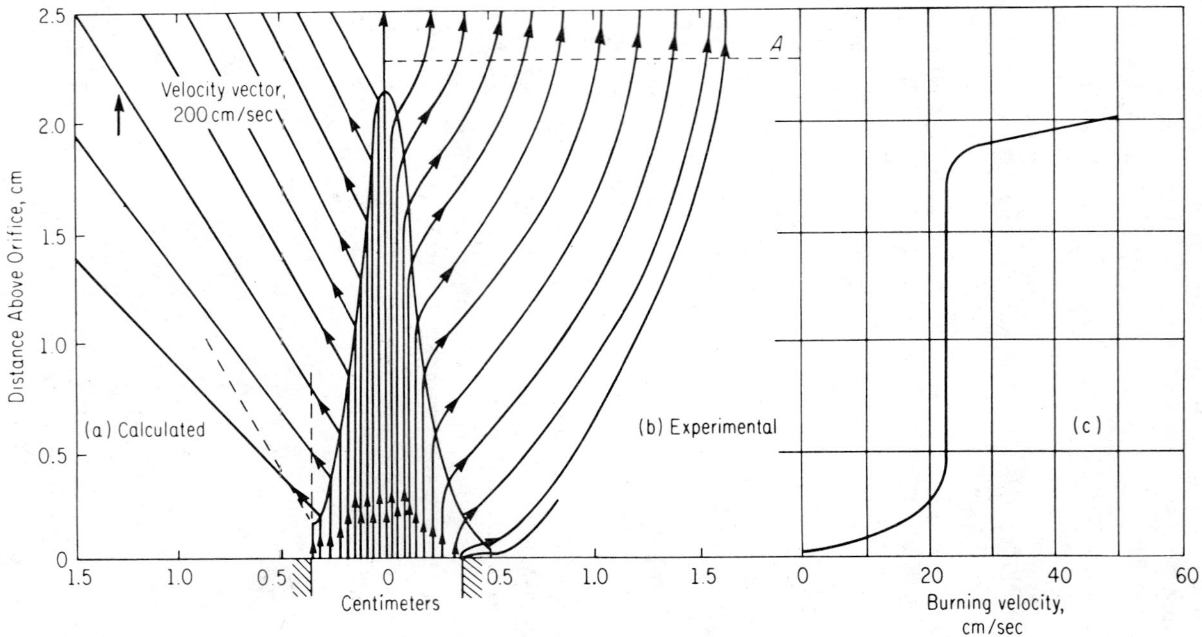

Fig. 7-3. Diagram of vertical center plane of natural gas–air flame on rectangular burner tube of 0.755 × 2.19 cm cross section. Mixture composition 7.5% natural gas; flow 204 cc/sec. (a) and (b) Combustion zone and flow pattern; (c) burning velocity. (With permission from B. Lewis and G. von Elbe, *Combustion Flames and Explosions of Gases*, 2d ed., Academic, New York, 1961.)

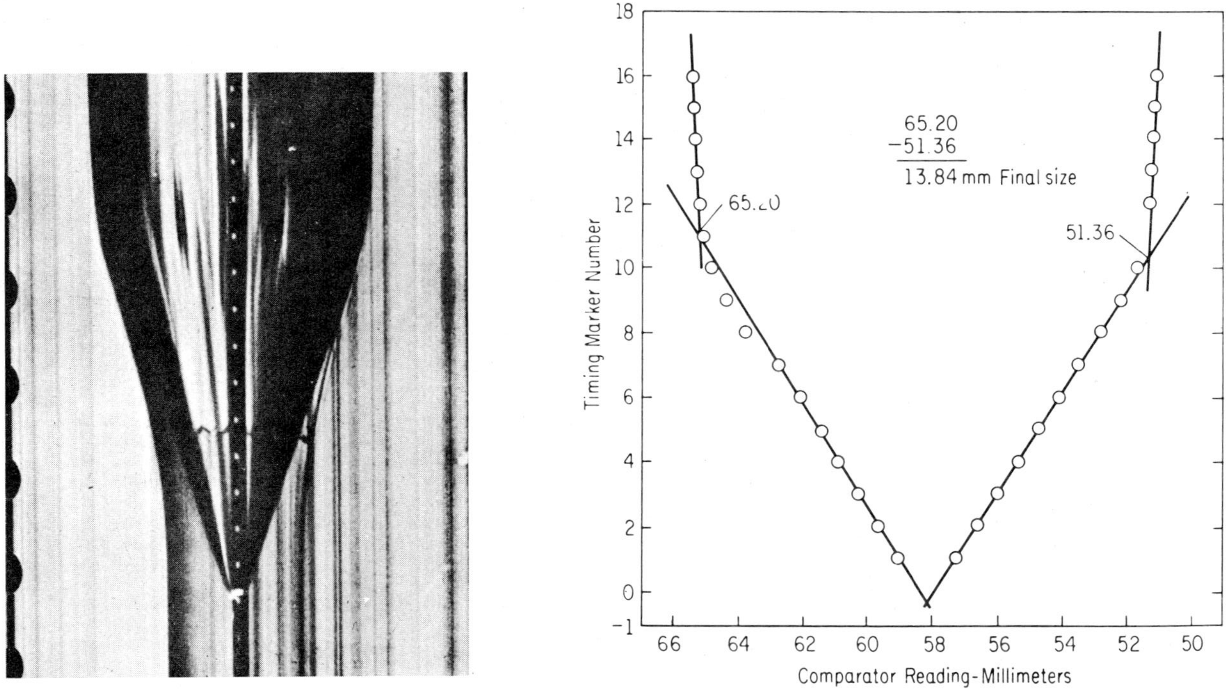

(a) (b)

Fig. 7-4. A 6.53% ethylene-air mixture; argon atmosphere; timing light = 1000 flashes per second; standard distance, 17.788 cm at bubble equals 23.38 mm comparator reading; initial diameter = 6.91 mm. $(r_2/r_1)^3$ = $(13.84/6.91)^3$ = 8.033; space velocity S_b (from graph) = $\Delta D/2\ \Delta t$ = 12.87 × 17.788/23.38 × .02 = 489.5 cm/sec; S_u = 489.5/8.033 = 60.9 cm/sec. (a) Bubble explosion. (b) Graph of comparator reading. [By permission from R. A. Strehlow and J. Stuart, *Fourth Symposium (International) on Combustion,* Williams & Wilkins, Baltimore, (1953), p. 332.]

measurement using a strip-film schlieren camera to observe the growth of the flame ball through a thin-slit aperture focused on the diameter of the flame ball. This technique is somewhat more difficult and less generally applicable than the slot-burner technique. It has the advantage, however, of requiring only small samples of the combustible mixture. Its disadvantages are:

(1) Because the final radius is required rather accurately and because it is assumed that the final value of this radius represents gas which is completely in equilibrium, the technique is limited to soap bubbles of rather large initial diameter.

(2) Soap bubbles are moist, and also are permeable to or react with certain gases. Thus mixtures whose burning velocity is markedly dependent on water-vapor content or which interact with the bubble material cannot be studied using this technique.

(3) The accuracy of this technique is limited most seriously by the requirement that the final radius be determined accurately.

However, measurements of flame-ball velocity versus time for sufficiently large flame balls have shown that the apparent velocity approaches S_b asymptotically as the diameter increases. Thus an alternative approach is to measure the asymptotic S_b value for large flame balls and divide it by the isobaric volume ratio calculated for a homogeneous constant-pressure process.

(4) Mixtures with very high burning velocities are not amenable to measurement using this technique. As burning velocity increases, the expansion process association with flame growth sends out finite pressure waves which, on reflection from the walls of the chamber, cause a disturbance at the flame ball and complicate the measurement of final size.

The third precise technique, using the flat-flame burner, is strictly limited to very low-burning-velocity flames ($S_u < 20$ cm/sec). In fact, flames observed using this technique usually cannot be observed as stable flames in other apparatus because of their low burning velocity. A schematic diagram of a flat-flame burner is shown in Fig. 7-5; Fig. 7-6 is a photograph of a flame stabilized on a flat-flame burner taken with illuminated particle tracks to show the velocity field ahead of the flame. In this type of apparatus, rather large-diameter, low-velocity combustible stream is produced by passing the flow through an appropriate series of screens and honeycomb filters. The flame is stabilized as a flat flame by using a capping screen and a surrounding low-velocity, inert-gas flow. Experimentally, it is observed that a flame will be stabilized in this burner over only a very small flow range. Once the flame is stabilized at a reasonable distance from the port, the burning velocity may be determined by simply dividing the input volumetric flow rate by the total flame area. This apparatus has also been used to stabilize flames for structure studies.

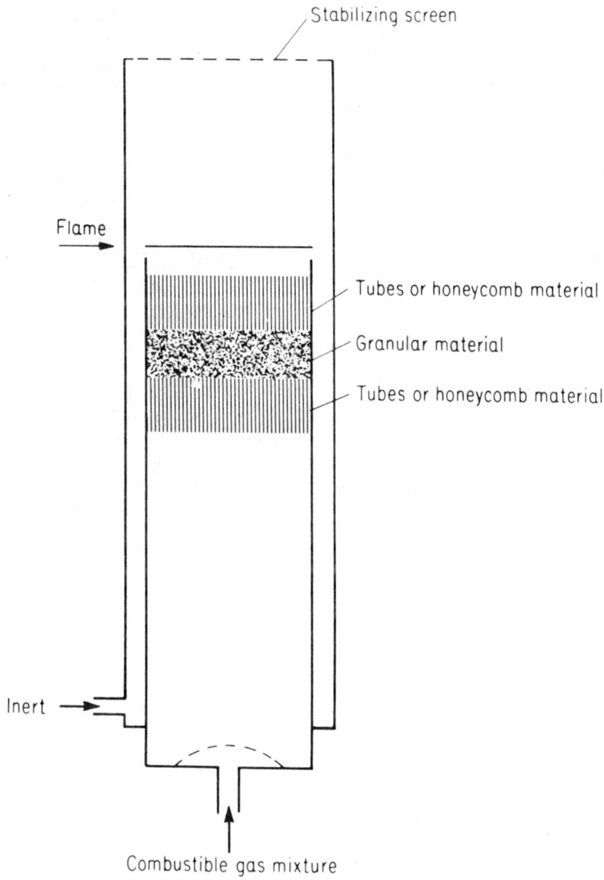

Fig. 7-5. Apparatus used for producing large flat flames. [Adapted from J. Powling, *Fuel*, *28* 25(1949).]

In addition to the above relatively precise techniques for determining burning velocity, there are many other techniques which have been used extensively. Five of these will be discussed briefly.

1. The spherical-bomb technique. A flame is ignited by a spark at the center of a spherical bomb. The flame diameter is recorded as a function of time, usually using the visible flame front, and the pressure in the bomb is recorded with a fast-response pressure gauge. The flame velocity is then determined from these two pieces of information. The most striking difficulty with this technique is the requirement in the analysis that the flame represent a thin discontinuity separating equilibrium products and unreacted reactants, and the equating of the visible image of the flame to this location. Unfortunately, slight errors in this location yield large errors in the calculated burning velocity.

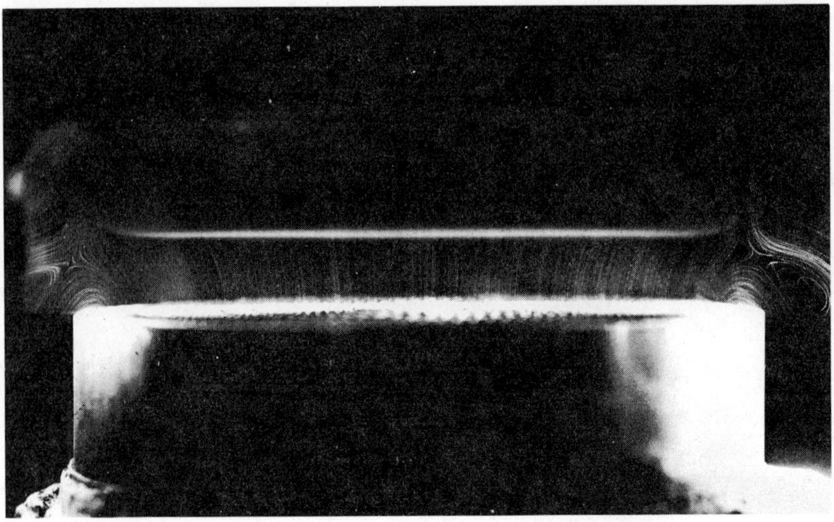

Fig. 7-6. Particle tracks in a flat ethylene–air flame. [Courtesy of F. J. Weinberg, Imperial College, London. Originally from A. Levy and F. J. Weinberg, Seventh Symposium (International) on Combustion, Butterworth, London (1959), p. 296.]

2. The burner method, using a flame-height measurement. It has been found that mixtures with very high burning velocity can be stabilized as laminar flames on relatively small burners. One technique for estimating the burning velocity of these small flames is to measure the total height of the inner cone and calculate the flame area, assuming that the flame has a simple conical shape. With this assumption the burning velocity of a flame on a circular port of radius r is given by the equation

$$S_u = \frac{2V}{\pi r \sqrt{h^2 + r^2}}$$

where h is the flame height and V the volumetric flow rate of the combustible gas mixture. It is known that this technique does not yield an exact result, and in fact errors of 15–20% may be expected using this method.

3. The burner method, using the actual flame shape to calculate a flame area. This is a refinement of the above technique. In this case the flame, which is stabilized on a circular port, is photographed and the area is calculated from the photographic trace by numerical integration assuming cylindrical symmetry. Even though it would seem that this technique should yield a more precise burning velocity than the cone-height method, it is subject to many errors which do not adequately cancel each other. Aside from end effects, associated with curvature at the tip and

heat losses at the rim, the technique suffers from the fact that the apparent flame position is different when measured using schlieren, shadowgraph, or visible-light photography, and the measured area therefore depends on the choice of optical technique.

4. Pressure-drop technique. Equation 7-1 states that a slight static pressure drop should be associated with flow through a steady flame. Therefore, a measured value of this pressure drop allows a calculation of the burning velocity, using Eq. 7-1. This technique has been attempted in flames with a low enough final temperature so as not to melt a static-pressure probe. The results indicate that the burning velocity and pressure drop correlate well if the flame is sufficiently flat. However, the technique has never been used for the routine measurement of burning velocity.

5. The tube method. It has been observed that with the correct set of end openings certain flames will propagate down a horizontal tube at a relatively constant velocity. With proper calibration, a measurement of this steady velocity coupled with photographs of the shape of the propagating flame yields a relative burning velocity which is reasonably accurate and reproducible. Because of its simplicity and the fact that it requires relatively small quantities of combustible mixture, this technique has been used extensively to evaluate the relative burning velocity of various hydro-

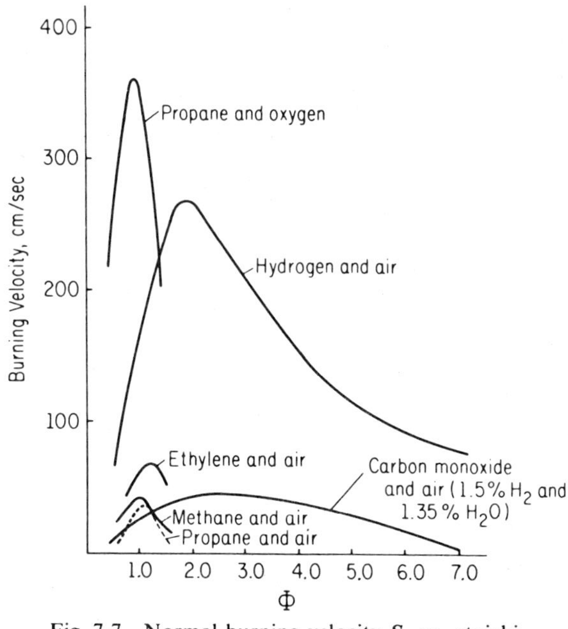

Fig. 7-7. Normal burning velocity S_u vs. stoichiometry for a few representative systems.

carbon fuels in air. However, the detailed behavior of a flame in a duct depends rather strongly on various aerodynamic considerations and is also greatly dependent on the presence of the walls. This technique has never been proposed as an absolute method of burning-velocity measurement.

The effect of mixture composition on the burning velocity for a few typical flames is shown in Fig. 7-7. It is interesting to note that flame temperature and flame velocity roughly parallel each other for any specific flame system, and that the maximum flame temperature ordinarily occurs in a mixture which also exhibits the maximum flame velocity.

In this section we have seen that the flame temperature of a gaseous premixed flame may be determined without knowing the burning velocity or mechanism of propagation, and that, in general, the burning velocity is quite difficult to measure accurately. In this discussion we have ignored a possible complication—namely, that certain flat flames are dynamically unstable and prefer to travel in other than the steady and strictly one-dimensional manner that we assumed. The behavior of these dynamically unstable flames will be discussed in Sec. 7-4.

7-3. MECHANISM OF FLAME PROPAGATION

Soon after laminar premixed flames were stabilized in the laboratory, it was recognized that they represent a unique case of wave propagation; the steady self-sustaining propagation of an "exothermic reaction wave" at a low subsonic velocity in a direction normal to itself. The question of the propagation mechanism of this unique wave has been discussed extensively during the past 90 years. Early in this period, heat conduction was postulated as a primary mechanism of propagation and the concept of an "ignition temperature" for the hot-flame reaction was introduced. As mentioned in Chapter 6, this led to many vessel studies of reaction kinetics in exothermic systems. After these studies showed the prevalence of exothermic chain-branching reactions in combustion systems, forward diffusion of highly reactive chain carriers was postulated as an alternate mechanism of propagation. The subsequent proliferation of "simple" propagation theories greatly stimulated experimental flame research, and in particular led to many studies of burning velocity as a gross property which could possibly be correlated to the various theoretical predictions. However, it is only recently that precise experimental probing techniques in conjunction with a correct and complete steady-flow theory have yielded a good descriptive model of the highly coupled processes that cause stable flame propagation. We will now develop this complete model in some detail, compare it to recent experimental results, and discuss a few useful approximations concerning flame structure.

Accumulated experimental evidence indicates that an ordinary one-dimensional flame propagates as a steady wave and that heat conduction, diffusion, and a coupled set of rapid exothermic chemical reactions all play important roles in the propagation mechanism. In the following development, we follow the approach of Hirschfelder and Curtiss[1] and include these three processes, but neglect the effect of radiative heat transfer, and second-order (i.e., coupled) transport effects, such as thermal diffusion. We also ignore nonstationary effects, such as flame acceleration or the possible occurrence of three-dimensional instabilities. That is to say, we will not consider the question of the inherent stability of a combustion wave at this time, but simply assume that the steady model represents a stable flame. One important type of observed flame instability will be discussed in Sec. 7-4.

Since the flame is essentially isobaric, we dispense with the momentum equation and discuss only mass and energy balances. As in Chapter 5 we must discuss the mass balance of each specie that appears in the flame. The velocity of each specie at any plane in the flame is simply $S + V_i$, where V_i is the average velocity due to diffusion of the ith specie at that location, and S is the average flow velocity at that point. Since the total mass flow is constant at each plane, and since each specie by definition has the same convective velocity S, we find that the global mass balance equation may be written as

$$\sum_{i=1}^{s} \rho_i (S + V_i) = \rho S = M$$

which yields an equation relating the diffusion velocities of the individual species,

$$\sum_{i=1}^{s} [I] m_i V_i = 0 \tag{7-3}$$

where a lower case m is used for molecular weight to distinguish it from the mass flow M. We may also define the diffusion velocity in terms of a diffusion coefficient,

$$[I] V_i = -N D_i \frac{dX_i}{dx} \qquad V_i = -D_i \frac{d \ln X_i}{dX} \tag{7-4}$$

where N is moles of mixture per cc, X_i is the mole fraction of the ith specie, and D_i is the diffusion coefficient of the ith specie in the local mixture subject to the constraint of Eq. 7-3. In general, these coefficients are not simply related to the binary diffusion coefficients discussed in Chapter 3. The negative sign appears in Eq. 7-4, because a positive concentration gradient yields a negative diffusion velocity.

[1] J. O. Hirschfelder and C. F. Curtiss, *J. Chem. Phys.*, *17*, 1076 (1949).

If we now consider a balance between two neighboring planes x and $x + dx$, we see that the net rate of production of the ith specie in the distance dx per square centimeter of flame area is simply

$$[I](S + V_i)\big/_{x+dx} - [I](S + V_i)\big/_{x}$$

However,

$$[I](S + V_i)\big/_{x+dx} = [I](S + V_i)\big/_{x} + \frac{d}{dx}\{[I](S + V_i)\}\, dx$$

and we may write

$$\left(\frac{\partial[I]}{\partial t}\right)_{\text{chem}} = \rho\left(\frac{dn_i}{dt}\right)_{\text{chem}} = \frac{d}{dx}\{[I](S + V_i)\}$$

where $(\partial n_i/\partial t)_{\text{chem}}$ is defined in Eq. 4-11 and is the net rate of production of the ith specie at a station in the flame due to all the chemical reactions which are occurring at that location. Note that if $(\partial n_i/\partial t)_{\text{chem}}$ is not identically zero throughout the flame, concentration gradients will exist and cause diffusion of individual species to be superimposed on the net motion of the flame. Thus the i mass-balance equations become

$$\rho \sum_{j=1}^{l} \nu_{ij} \frac{d\lambda_j}{dt} = \frac{d}{dx}\{[I]S\} - \frac{d}{dx}\left(ND_i \frac{dX_i}{dx}\right) \tag{7-6}$$

where there are l chemical reactions occuring, each with their reaction coordinate λ_j and rate $d\lambda_j/dt$.

We now construct an energy balance in the flame. Since the pressure is essentially constant and the flame is a low speed subsonic wave, we write an enthalpy balance and therefore implicitly include the external work performed at constant pressure and *exclude* the flow energy. The enthalpy of the ith specie is

$$H_i = H^{\circ}_{0_i} + \int_0^T C_{p_i} dT$$

where $H^{\circ}_{0_i}$ is the heat of formation at 0°K and C_{p_i} the heat capacity per mole. The total enthalpy flux through a plane located in the flame is therefore

$$\sum_{i=1}^{s} [I](S + V_i)H_i$$

while the flux through a plane located in the equilibrium region downstream of the flame zone is given by the equation

$$\sum_{i=1}^{s} [I]_{\text{eq}} S_b H_{i_{\text{eq}}}$$

since no diffusion is occurring in this region. The enthalpy gain of this volume element due to the mass flow is therefore

$$\sum_{i=1}^{s} [I](S + V_i)H_i - \sum_{i=1}^{s} [I]_{\text{eq}} S_b H_{i_{\text{eq}}}$$

Since the flow is steady, heat must be transported out at the rate of accumulation indicated in the above equation and since heat is transported only in the flame zone we obtain the equation

$$\kappa \frac{dT}{dx} = \sum_{i=1}^{s} [I](S + V_i)H_i - \sum_{i=1}^{s} [I]_{\text{eq}} S_b H_{i_{\text{eq}}}$$

where κ is the thermal conductivity of the mixture at x and the first term on the right is evaluated at x. This equation may also be written as

$$\sum_{i=1}^{s} [I]_{\text{eq}} S_b H_{i_{\text{eq}}} = \sum_{i=1}^{s} [I]SH_i - N\sum_{i=1}^{s} H_i D_i \frac{dX_i}{dx} - \kappa \frac{dT}{dx} \qquad (7\text{-}7)$$

The three terms on the right in Eq. 7-7 are energy fluxes due respectively to convection, diffusion, and conduction in the flame zone. Equation 7-7 may also be written in the differential form

$$\frac{d}{dx}\sum_{i=1}^{s} ([I]SH_i) - \frac{d}{dx}\left(N\sum_{i=1}^{s} H_i D_i \frac{dX_i}{dx}\right) - \frac{d}{dx}\left(\kappa \frac{dT}{dx}\right) = 0 \qquad (7\text{-}8)$$

Equation 7-7, the mass-flow Eqs. 7-6, and relationships defining reaction rates, diffusion coefficients, thermal conductivities and enthalpies may in principle be used to determine the burning velocity and detailed structure of any premixed gas flame. In practice, integrating this equation set to obtain concentrations, state properties, and fluxes in the flame zone requires a knowledge of the burning velocity S_u commensurate with specific assumptions for the reaction rates and transport properties. Since this information is not available, we are left with the solution of an eigenvalue problem in which the burning velocity is determined by the characteristic or eigenvalue.

We now wish to discuss the solution of these flame equations for a particularly simple model to illustrate both the solution technique and the coupled nature of the processes in an ordinary flame. We follow the approach first suggested by Friedman and Burke.[2] Consider the flame reaction to be an irreversible first order decomposition of pure A yielding only B as product:

$$A \rightarrow B$$

with the molecular weights $m_A = m_B = m$ and the enthalpies $H_A = C_p(T - T_u)$, $H_B = C_p(T - T_u) - q$, where C_p is taken to be a constant

[2] R. Friedman and E. Burke, *J. Chem. Phys.*, *21*, 710 (1953).

and $q = C_p(T_b - T_u)$; therefore, $H_b = C_p(T - T_b)$. Thus the convected enthalpy is zero at both stations 1 and 2 in this flame. We write the first-order rate equation as

$$\frac{d(NX_B)}{dt} = k_1 N X_A \exp(-E/RT_b\tau)$$

where

$$\tau = \frac{T - T_u}{T_b - T_u}$$

and X_A and X_B are mole fractions.

The introduction of τ in the rate equation forces the chemical rate to be zero at T_u without significantly altering it in the hotter flame regions. Thus this model is fundamentally different from earlier models which did not require a zero reaction rate at the initial temperature of the mixture but instead forced the effect of the reaction to be zero by withdrawing a small quantity of heat from the flow at a plane ahead of the flame. However, since most real flames propagate at a steady velocity into mixtures which have a vanishingly small decomposition rate at their initial temperature, the artifice of requiring a zero reaction rate at T_u is more satisfactory than the artifice of requiring an artificial heat sink. Friedman and Burke's model, in fact, leads to the conclusions that (1) no cold boundary is actually required for an eigenvalue to exist, and (2) the "ignition temperature" is really a difficultly defined dynamic property which depends on a coupling of reaction kinetics, diffusion, and thermal conduction in the flame.

We now define a simplified binary diffusion coefficient $D = cT^2$ and thermal conductivity $\kappa = fT$ and thus find that the Lewis number Le (defined with Eq. 3-14) is a constant throughout the flame. With this set of assumptions, we are ready to transform the general flame equations to a dimensionless form. We first note that in this binary gas mixture, Eq. 7-3 yields

$$V_A = -V_B\left(\frac{X_B}{1 - X_B}\right)$$

and Eq. 7-7 therefore becomes

$$\frac{d\tau}{dx} = \frac{NC_pS}{\kappa}(\tau - z) \tag{7-9}$$

where

$$z = \frac{(S + V_B)X_B\rho}{S_u\rho_u}$$

Equation 7-5 for the specie B may be written as

$$\frac{dz}{dx} = \frac{k_1\rho}{M}(1 - X_B)\exp(-\epsilon/\tau) \tag{7-10}$$

where $\epsilon = E/RT_b$, is a dimensionless activation energy. Furthermore, the definition of a diffusion coefficient yields the equation

$$\frac{dX_B}{dx} = -\frac{X_B V_B}{D} \tag{7-11}$$

Since the flow is steady, these equations are parametric in x and may be written in terms of τ, z and X_B by dividing Eq. 7-9 into Eq. 7-10 and Eq. 7-11 to eliminate x as a variable. This yields the equations

$$\frac{dz}{d\tau} = \beta_c \left(\frac{1 - X_B}{\tau - z}\right) \exp(-\epsilon/\tau) \tag{7-12}$$

and

$$\frac{dX_B}{d\tau} = \frac{1}{\text{Le}} \left(\frac{X_B - z}{\tau - z}\right) \tag{7-13}$$

with the boundary conditions $X_B = z = \tau = 0$ at station 1 (the cold boundary) and $X_B = z = \tau = 1$ at station 2 (the hot boundary). The constant β_c is the dimensionless eigenvalue of the physically realizable solution to this equation set, and is given by the relation

$$\beta_c = \frac{k_l \kappa \rho m}{S_u^2 \rho_u^2 C_p}$$

The value of β_c must be determined by solving Eqs. 7-12 and 7-13 simultaneously. Notice that β_c defines the burning velocity in terms of the reaction kinetic and transport properties of the combustible mixture. Also note that for our assumptions, the temperature dependence of ρ and κ cancel so that $\rho\kappa = \rho_u \kappa_u$.

Therefore we find that β_c is truly a constant throughout this flame. Thus we obtain the equation

$$S_u = \sqrt{\frac{mk_1 \kappa_u}{\rho_u \beta_c C_p}} \tag{7-14}$$

which defines the burning velocity.

Again, following Friedman and Burke's treatment, we note that Eq. 7-13 is directly integrable for two cases; Le = 1 and Le = 0, yielding respectively $X_B = \tau$ or $X_B = z$. In the case where we assume Le = 0 we do so by setting the diffusion coefficient to zero, thereby producing a flame with $V_A = V_B = 0$. With these assumptions we may rewrite Eq. 7-12 to yield

$$\frac{dz}{d\tau} = \beta_c \frac{1 - \tau}{\tau - z} \exp(-\epsilon/\tau) \tag{7-15}$$

for Le = 1, or

$$\frac{dz}{d\tau} = \beta_c \frac{1 - z}{\tau - z} \exp(-\epsilon/\tau) \tag{7-16}$$

for Le = 0. At $\tau = z = 1$, both of these equations are indeterminant. Furthermore, the cold boundary ($\tau = z = 0$) for both these equations must be approached along the direction $dz/d\tau = 0$, in order for the solutions to be physically meaningful. Therefore the solution of these equations may be obtained by numerical integration from the hot boundary, using an expansion of Eqs. 7-15 and 7-16 to determine the initial τ, z relation at this boundary. The proper eigenvalue β_c for an assumed ϵ and for either Le = 1 or Le = 0 is determined from this numerical integration by assuming a series of values for β until z remains positive and approaches zero with a slope $dz/d\tau = 0$. Friedman and Burke found that, for flames having Le = 1 or Le = 0, β_c may be quite accurately represented by the equation

$$\beta_c = [0.514 \text{ Le } \epsilon^2 + (1.0 + 0.35 \text{ Le})\, \epsilon + 0.20 + 0.05 \text{ Le}] \exp(\epsilon) \qquad (7\text{-}17)$$

Figures 7-8 and 7-9 illustrate the structure of simple first order decomposition flames for Le = 1 and Le = 0. These figures were obtained by integrating Eqs. 7-15 and 7-16 using the following assumptions: $m = 32$; $P = 1$ atmosphere, $T_u = 300°K$ (which means that $\rho_u = 1.348 \times 10^{-3}$ gm/cc), $C_p = 3.5$ cal/mole°K, $\kappa_u = 4.75 \times 10^{-5}$ cal/cm sec°K, $k_1 = 10^{13}$, $E_{act}/R = 30{,}000$ and $T_b = 2000$°K. The calculated burning velocity using these assumptions are: $S_u = 85.1$ cm/sec for Le = 1 and $S_u = 254.6$ cm/sec for Le = 0. The plots of mass fluxes, concentrations, energy fluxes, and temperature distribution shown in Figs. 7-8 and 7-9 should

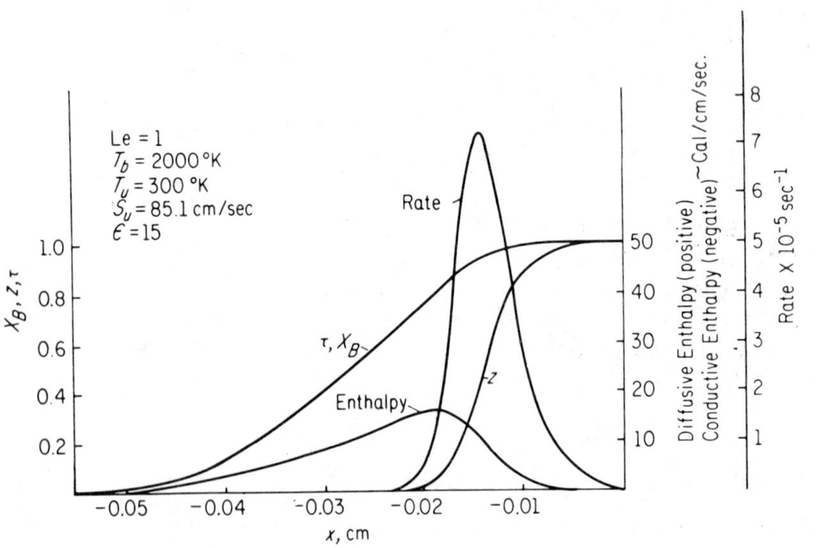

Fig. 7-8. Flame structure for a flame with Lewis number = 1, supported by a first-order reaction. Rate is the rate of disappearance of the specie A, and equals $k_1(1 - X_B)\exp(-\epsilon/\tau)$.

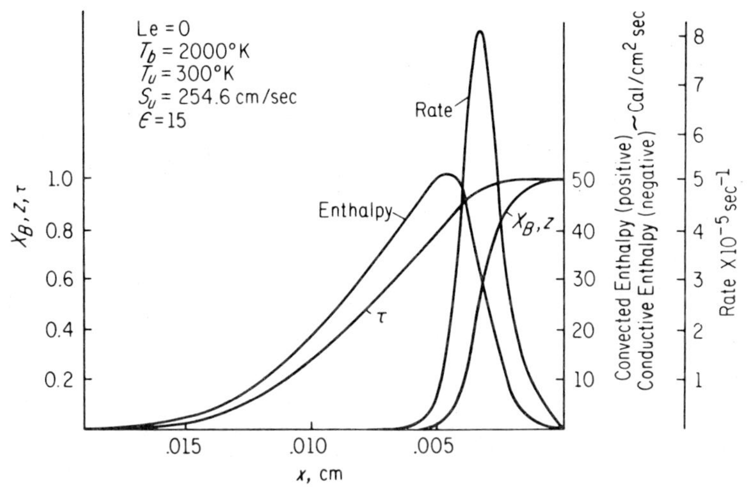

Fig. 7-9. Flame structure for a flame with a Lewis number which is equal to zero because the diffusion coefficient is assumed to be zero. Reaction is assumed to be first-order in the specie A. Rate is the local rate of disappearance of the specie A and equals $k_1(1 - X_B) \exp(-\epsilon/\tau)$.

be compared to Fig. 7-10 through 7-13, which were reported by Fristrom and Westenberg[3] and were obtained from actual experimental measurements in $^1/_{10}$th or $^1/_{20}$th atmosphere lean methane-oxygen flames. The methane–oxygen system is known to react with a chain-branching mechanism which is, of course, kinetically quite different from the first-order reaction assumed in obtaining Figs. 7-8 and 7-9. However, many striking similarities are evident in these figures. In the first place, as Fristrom and Westenberg point out, actual flames are usually dominated by one primary reaction which determines the flame velocity. This is because it is not very likely that two parallel reactions will occur with exactly balanced rates in a flame. Although different reactions can be important in, say, different composition or temperature ranges, in a specific flame it is ordinarily found that a single reaction is rate-controlling. Second, in real systems one might always expect the final approach to equilibrium to be a slow process, because recombination reactions are usually slow even at high temperatures. Thus, experimental work on flame structure shows three distinct zones—the preheat zone, where there is very little chemical reaction, the fast or primary reaction zone which is usually very thin, and the final slow-reaction zone. In contrast, our simple first-order theory shows only a distinct preheat zone with very slight reaction, followed by a fast reaction zone. Third, a comparison of the

[3] R. M. Fristrom and A. A. Westenberg, *Flame Structure*, McGraw-Hill, New York, 1965.

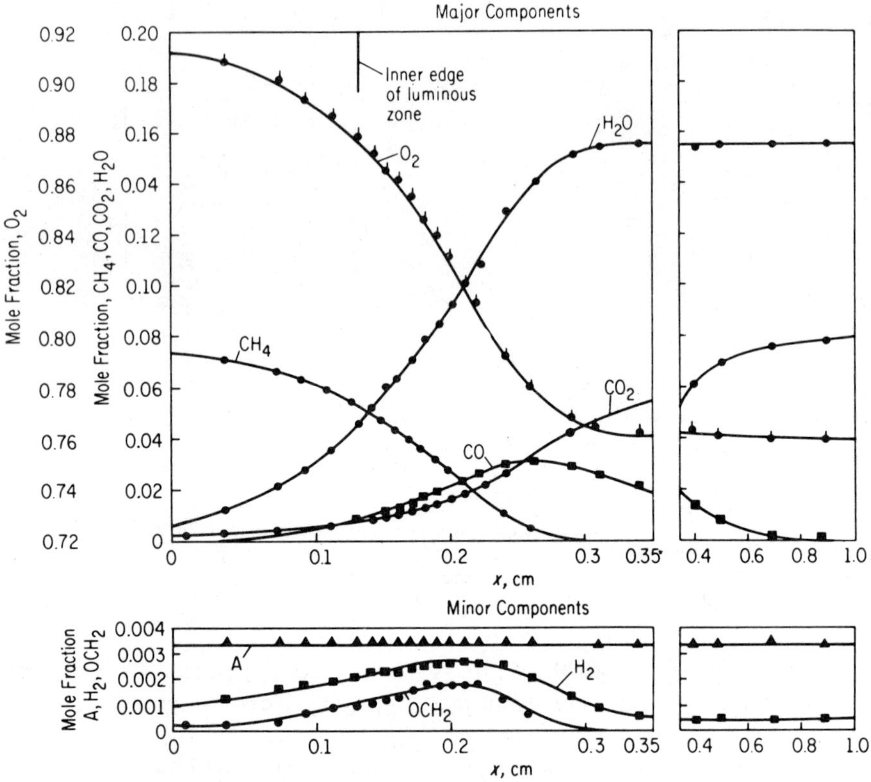

Fig. 7-10. Specie concentrations in a lean methane–oxygen flame. Mole fraction CH_4 = 0.078; mole fraction O_2 = 0.922, pressure = 0.1 atm. Flat flame on 3.2 cm screen burner. (By permission from R. M. Fristrom and A. A. Westenberg, *Flame Structure*, McGraw-Hill, New York, 1965.)

figures show that in a particular real flame, mass diffusion has an important effect, and that conducted sensible enthalpy is balanced in the main by a diffusion-controlled conducted chemical enthalpy, and, therefore, that the convected enthalpy (i.e., the first term on the right in Eq. 7-7) is relatively constant throughout this flame. This behavior may be generalized to almost all real flames because the Lewis number is approximately unity in most gases, and therefore diffusion must make an important contribution to the enthalpy balance.

The theory and experiment therefore show that flame propagation is primarily controlled by a coupling of the processes of thermal conduction and chemical reaction. Diffusional control in a one-dimensional flame is essentially a secondary effect, and is important only as it alters the rate of the chemical reaction. For example, the first-order reaction flame dis-

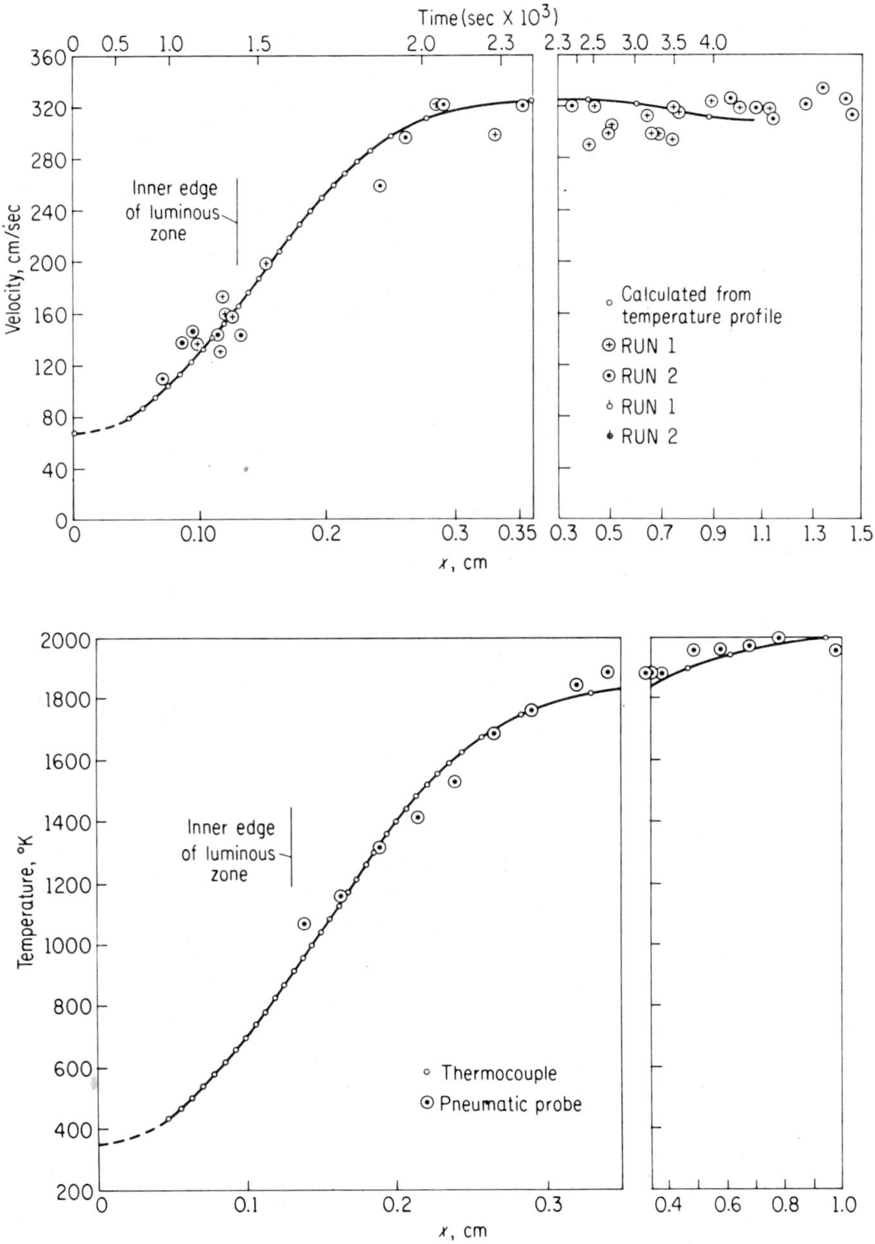

Fig. 7-11. Velocity and temperature profiles in the methane–oxygen flame of Fig. 7-10. (By permission from R. M. Fristrom and A. A. Westenberg, *Flame Structure*, McGraw-Hill, New York, 1965.)

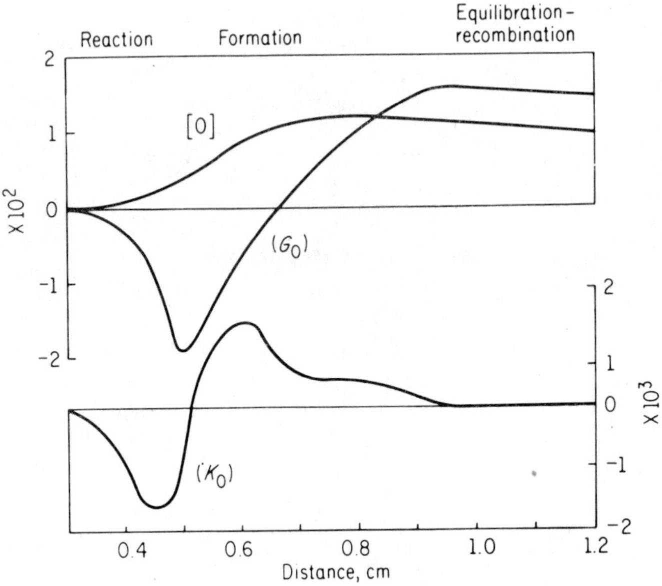

Fig. 7-12. Oxygen radical concentration [O], flux (G_0) and net rate of production (k_0) in the methane–oxygen flame. Same initial mole fractions as the flame of Fig. 7-10, P = 0.05 atm. (By permission from R. M. Fristrom, "Flame Structure," in A. F. Scott (ed.) *Survey of Progress in Chemistry*, Vol. 3, Academic, New York, 1966.)

cussed above has a higher burning velocity when Le = 0 since for this case back diffusion does not cause dilution of the reactants before reaction occurs. It should be pointed out, however, that this "secondary effect" can sometimes have a primary importance. In those cases where a reactive chain carrier diffuses downstream from the hot reaction zone and starts a chain-branching reaction in the cooler regions of the preheat zone, the burning velocity can show a strong dependence on the diffusivity of this chain carrier, and the flame will appear to be propagated primarily by a diffusional mechanism.

The fact that most flames are propagated by a coupled chemical reaction-conduction mechanism led to the postulate of the existence of an "excess enthalpy" in the flame zone by Lewis and von Elbe.[4] However, our simple flame example shows that flames which have a Lewis number equal to one have no excess total enthalpy in the flame zone even though they do have an excess conducted sensible enthalpy in the preheat zone. Since this conducted enthalpy is always present when the flame propa-

[4] B. Lewis and G. von Elbe, *Combustion Flames and Explosions of Gases*, 2d ed., Academic, New York, 1961.

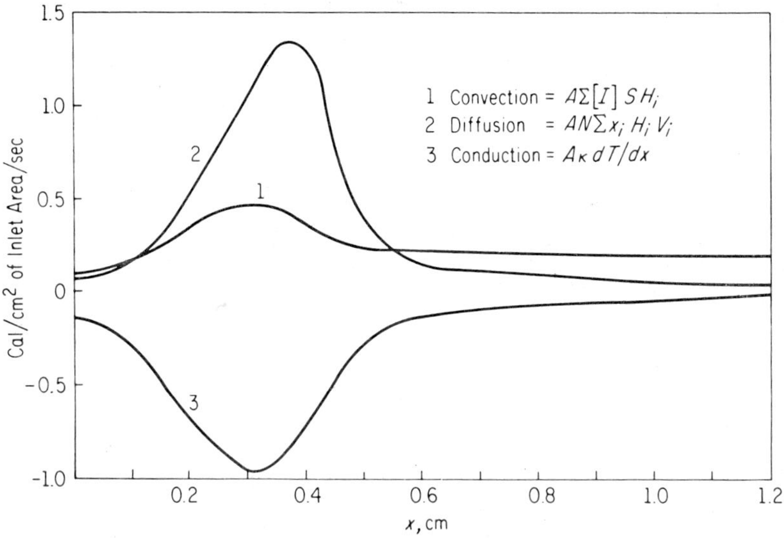

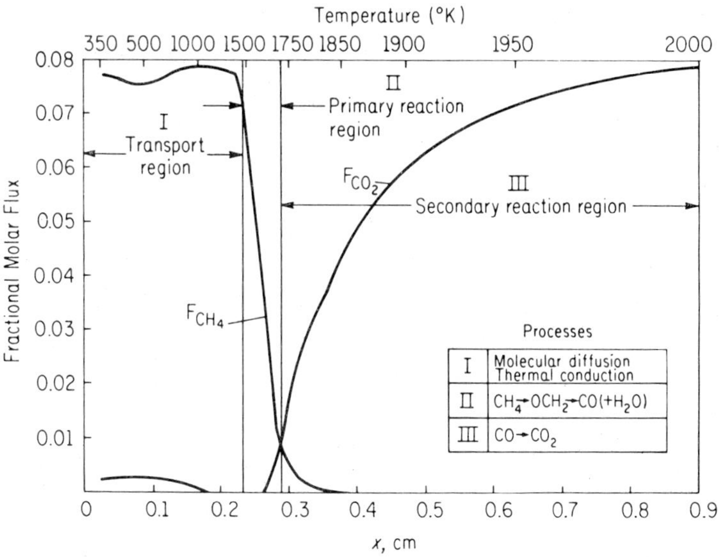

Fig. 7-13. Enthalpy and molar fluxes in the methane–oxygen flame of Fig. 7-10.

gates, it is the quantity which must represent a fundamental property of the flame. The fact that it is offset by either a convective enthalpy or by the diffusion of species-carrying chemical enthalpy in no way detracts from its existence in the flame, or its importance to the propagation of the flame. The importance of this property will be discussed further in Sec. 7-6.

We now wish to discuss the relationship between the theory and some of the gross properties of flames using an extended theoretical model. We restrict ourselves to a flame with Le = 1 and investigate the effect of the order of the primary chemical reaction on the flame parameters. The rate equation therefore becomes

$$\frac{d(NX_B)}{dt} = k_\alpha (NX_A)^\alpha \exp(-E/RT_b\tau) \tag{7-18}$$

where α is the reaction order and k_α has appropriate units. This yields a new conservation equation (replacing Eq. 7-10)

$$\frac{dz}{dx} = \frac{k_\alpha \rho^\alpha}{Mm^{(\alpha-1)}} (1 - X_B)^\alpha \exp(-\epsilon/\tau) \tag{7-19}$$

Since Le = 1 we again find from Eq. 7-13 that $X_B = \tau$ and therefore arrive at the final flame equation,

$$\frac{dz}{d\tau} = \Lambda_c \left(\frac{1}{1 + \tau\phi}\right)^{\alpha-1} \frac{(1 - \tau)^\alpha}{\tau - z} \exp(-\epsilon/\tau) \tag{7-20}$$

where

$$\Lambda_c = \frac{k_\alpha \kappa_u \rho_u^\alpha}{M^2 C_p m^{(\alpha-2)}}$$

and

$$\phi = \frac{T_b - T_u}{T_u}$$

Therefore

$$S_u = \rho_u^{[(\alpha/2)-1]} \sqrt{\frac{k_\alpha \kappa_u}{\Lambda_c C_p m^{\alpha-2}}}$$

In this flame of order α, Λ_c also shows a primary dependence on the activation energy and may be assumed to be roughly proportional to exp (ϵ). See Eq. 7-17. Therefore the theoretical model predicts that to a first approximation

$$S_u \simeq \text{const.} \cdot P^{[(\alpha/2)-1]} \exp(E/RT_b)$$

for any specific flame. Figure 7-14 is an experimental determination of the exponent $n = \alpha/2 - 1$ for a number of hydrocarbon flames and yields the simple result that these flames are propagated by a primary reaction which is predominantly second order. It must be pointed out that this is only an approximate statement because the simple theory has neglected the effect of pressure on flame temperature, and for high-temperature flames we might expect that this effect would become important.

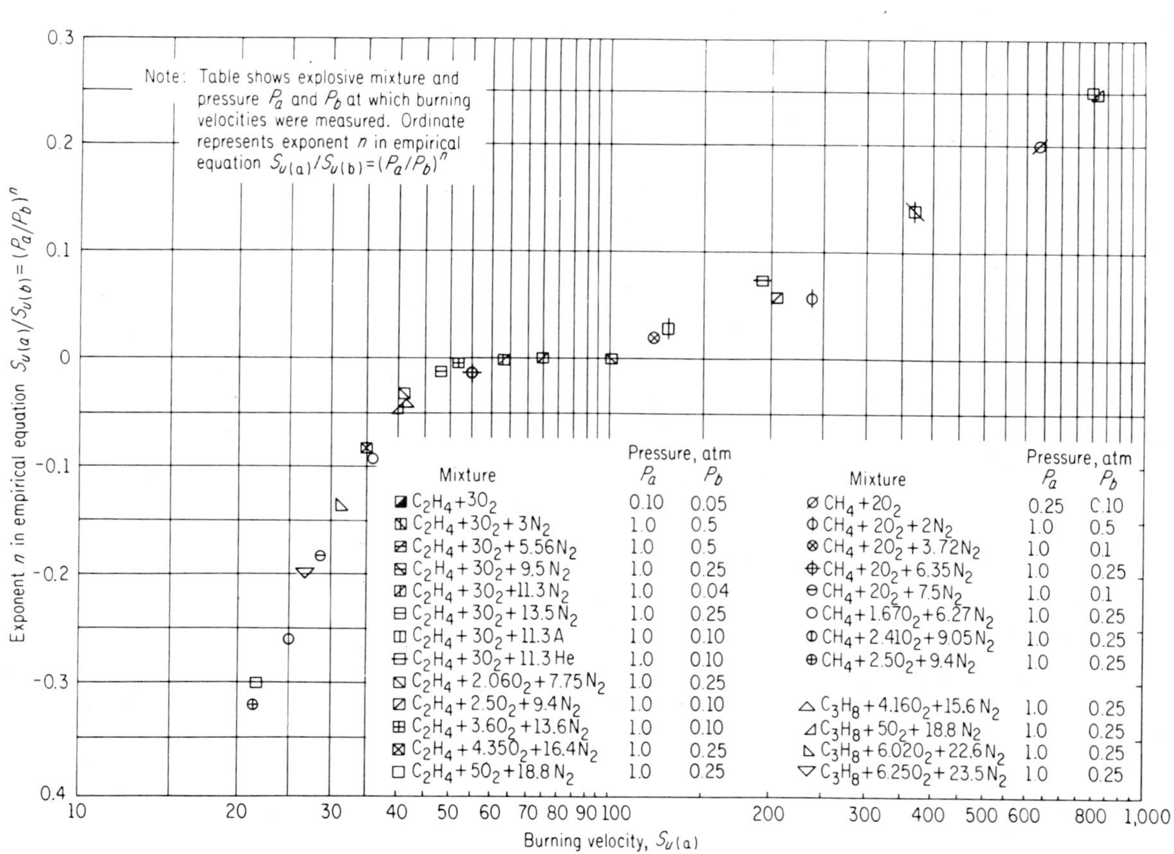

Fig. 7-14. The pressure exponent n for a variety of hydrocarbon flames. [Courtesy Dr. Bernard Lewis, Combustion and Explosions Research, Inc., Pittsburgh, Pa.; originally appeared in AGARD, *Selected Combustion Problems*, Butterworth, London, (1954), p. 177.]

Flame thickness may be investigated by rewriting Eq. 7-9 to yield

$$\frac{d\tau}{dx} = \frac{\rho_u S_u C_p}{m\kappa_u} \cdot \frac{(\tau - z)}{(1 + \tau\phi)} \tag{7-21}$$

We note from Fig. 7-8 that z drops more rapidly than τ as we leave the hot boundary and we see that Eq. 7-21 predicts a relative maximum in $d\tau/dx$ at the point where z first becomes negligible. This is the point which formally defines the boundary between the preheat zone and the beginning of the hot-reaction zone. We let $\tau = \tau_{\text{ign}}$ at this point and define a preheat-zone thickness η_0 by evaluating Eq. 7-21 at $\tau = \tau_{\text{ign}}$

$$\left(\frac{d\tau}{dx}\right)_{\max} = \frac{\rho_u S_u C_p}{m\kappa_u} \cdot \frac{\tau_{\text{ign}}}{1 + \tau_{\text{ign}}\phi}$$

This yields

$$\eta_0 = \frac{\tau_{\text{ign}}}{\left(\frac{d\tau}{dx}\right)_{\max}} = \frac{\kappa_u}{\rho_u S_u (C_p/m)} (1 + \tau_{\text{ign}}\phi) \tag{7-22}$$

(See Fig. 7-15.) The preheat-zone thickness η_0 defined in this manner is larger (by the factor T_{ign}/T_u) than the usually defined preheat-zone thickness. However the usual definition of a preheat-zone thickness is made under the assumption that κ is independent of the temperature. Since we have assumed that $\kappa = fT$ in our flame model and experimentally $\kappa \simeq gT^{0.7}$, the above definition of η_0 may be expected to be more realistic than the usual simple theory definition.

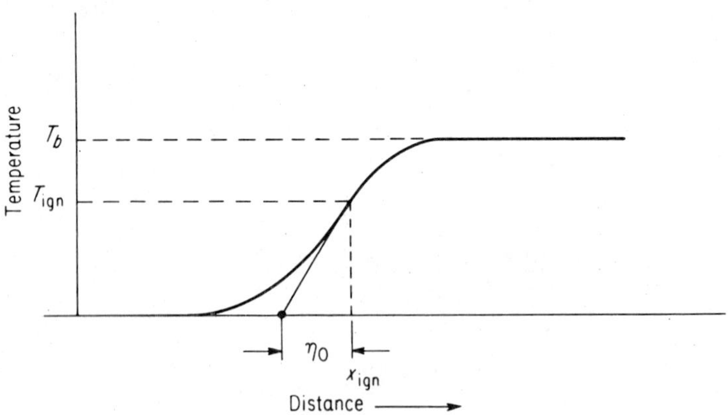

Fig. 7-15. Schematic diagram showing the definition of η_0. Here x_{ign} is the point in the flame where the rate of chemical reaction is much less than the maximum rate. In this text we will assume that $(dn_i/dt)_{\text{chem}} = 1/100\,(dn_i/dt)_{\text{chem max}}$ defines this point.

In summary we should mention again that the majority of flames actually appear to propagate by a second-order mechanism, and that in most flames the rate-controlling reaction is actually the slow step in a chain-branching reaction similar to that discussed in Sec. 4-6. Thus most flames (especially the cooler ones) contain a nonequilibrium excess of reactive radicals in the hot-reaction zone (see Fig. 7-10, for example) and back diffusion of these radicals enhances the reaction rate markedly. The simple, direct decomposition flame which we used as a theoretical model cannot be used to discuss the effects of diffusional control by the back diffusion of these radicals. However, it does illustrate other pertinent facts concerning the structure of flames, including the fact that unreacted fuel and oxidizer are diffusing toward the flame zone throughout the entire preheat zone. The theory also reintroduces the old concept of an ignition temperature in a new context; pointing out that the ignition temperature in a flame is related to the dynamics of the situation and therefore is not directly comparable to the "vessel" ignition temperatures discussed in Chapter 6.

7-4. CELLULAR FLAMES

The theory of flame propagation developed in the previous section is based on the *a priori* assumption that the flame is indeed a one-dimensional steady and self-propagating reaction wave of low subsonic velocity. While many observed flames are known to behave in this manner, experiments have shown that there is a class of flames which prefer to propagate with a characteristic two- or three-dimensional structure rather than as a strictly one-dimensional phenomena. Markstein has pioneered in this field and has recently published an extensive review[5] of the existing experimental and theoretical work on this so-called "cellular" instability.

In this section, we will discuss only the behavior of spontaneously cellular flames. Externally driven flame instability will be discussed in Chapter 8. Figure 7-16 contains a set of photographs of typical three-dimensional cellular flames in fuel rich hydrocarbon–air–nitrogen mixtures propagating downward against a nearly balancing flow in a tube of 10-cm diameter. The photographs were taken using the light emitted by the flame. Clearly defined cellular structure may also be observed and studied in a rectangular slot burner, which has been specially designed to hold a flame on only one of its long edges. In this case, flame curvature occurs primarily along the length of the slot, and the cellular structure may become strictly two-dimensional. Figure 7-17 is an edge-on photograph of a flame front for this case. Spontaneous cellular instability has also

[5] G. H. Markstein, *Non-Steady Flame Propagation*, Macmillan, New York, 1964.

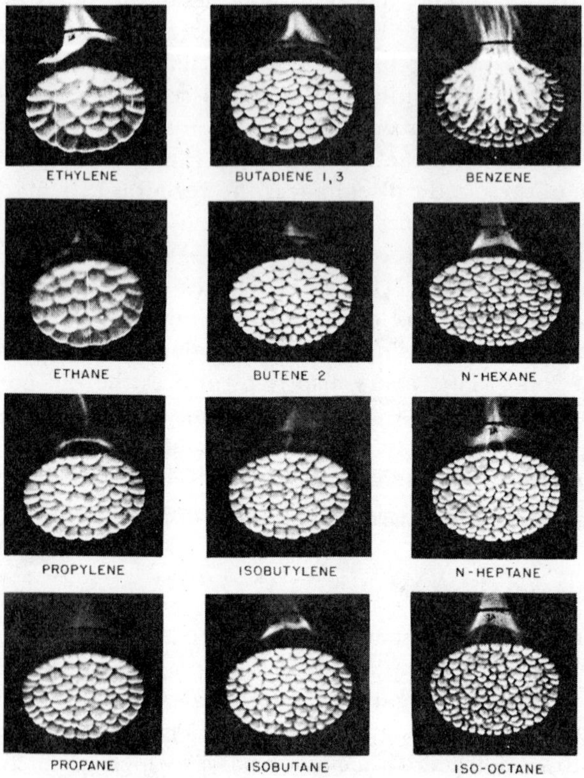

Fig. 7-16. Cellular flames in fuel-rich hydrocarbon-air-nitrogen mixtures at atmospheric pressure. Tube I.D. 10.0 cm. [Courtesy G. H. Markstein, Cornell Aeronautical Laboratory, Buffalo, N.Y.; originally appeared in *J. Aero. Sci. 18*, 199 (1951).]

been observed in premixed gas flames having other geometries. For example, certain spherically expanding flames show the spontaneous formation of cellular structure, and certain mixtures when burned as attached flames on circular ports are known to exhibit a fluted appearance due to the occurrence of cellular instability.

If we consider only the simplest geometric situations, it is evident that we may observe either a two- or three-dimensional instability in a propagating combustion wave. The two-dimensional wave shown in Fig. 7-17 will be used to illustrate the nature of the phenomenon. We note from this figure that the cellular flame contains relatively long flame sections that are convex toward the flow, separated by very short sections that are concave toward the flow. We also note that the convex sections have a

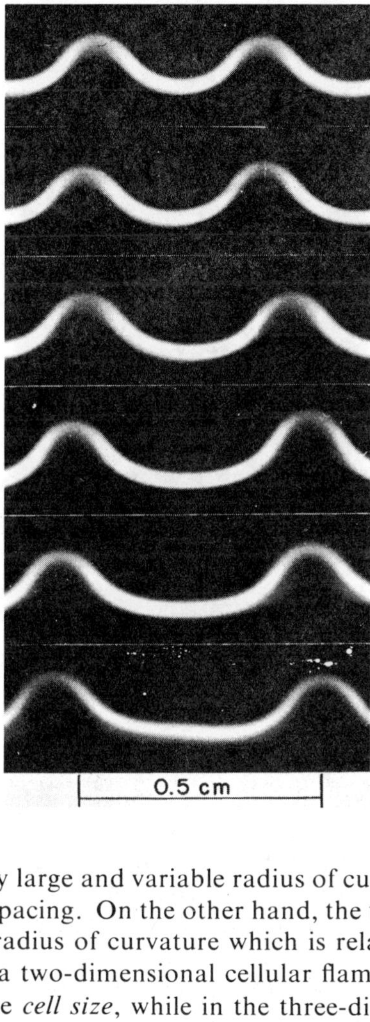

Fig. 7-17. Profiles of steady cellular slot-burner flames, *n*-butane-air mixture, equivalence ratio 1.47. Flow is upward. (Courtesy H. G. Markstein, Cornell Aeronautical Laboratory, Buffalo, N.Y.; originally appeared in H. G. Markstein, *Non-Steady Flame Propagation,* Macmillan, New York, 1964.)

relatively large and variable radius of curvature which is a function of the trough spacing. On the other hand, the troughs or concave sections have a shorter radius of curvature which is relatively independent of their spacing. In a two-dimensional cellular flame, the spacing between troughs is called the *cell size*, while in the three-dimensional structure illustrated in Fig. 7-16, the average cell diameter is usually defined as the cell size. These two- and three-dimensional cell sizes are not the same; the minimum size in two dimensions always being smaller than the minimum cell diameter in three dimensions.

Cellular structure is usually in a constant state of flux during flame propagation. High-speed photographs of three-dimensional flames have shown that large cells grow larger and then divide, while small cells are absorbed. Figure 7-18 is a streak photograph of the behavior of a typical two-dimensional cellular flame obtained on a slot burner. The dark regions are the troughs. Notice (1) that large cells split to form new cells and sometimes these disappear entirely after their formation; (2) that the

Fig. 7-18. Streak-camera record of nonsteady cellular slot-burner flame. Time increases from top to bottom. *n*-butane–air mixture, $\Phi = 1.56$ (Courtesy G. H. Markstein, Cornell Aeronautics Laboratory, Buffalo, N.Y.; originally in G. H. Markstein and Schwartz, *Proc. Gas Dynamics Symposium on Aerothermochemistry*, Northwestern University, Evanston, Ill., 1956.)

cells can propagate along the burner direction; and (3) that a fluctuation in one cellular region can propagate to other neighboring regions at a well-defined velocity. Other observations have shown that there is a characteristic minimum cell size for either two- or three-dimensional cells in any mixture, and that for two-dimensional cells a single cell spontaneously breaks into two cells when it reaches approximately twice the minimum size.

The most interesting and significant piece of information concerning spontaneous cellular structure is its dependence on the stoichiometry of the combustible mixture. Experimentally it has been found that all the fuels illustrated in Fig. 7-16 propagate spontaneously cellular flames in rich mixtures. In fact, the appearance stoichiometric ratio has been found to be 1.00 ± 0.04 for all these hydrocarbons. It is also known (see Fig. 7-19) that for hydrocarbon–air flames, the minimum cell size is dependent on the fuel's molecular weight, and that for a specific fuel the cell size is relatively insensitive to the stoichiometry and quite dependent on the ambient pressure.

For a rich butane–air flame, propagating at low velocity, Markstein found that $DP^{3/4}$ = constant best correlated the experimental data. It

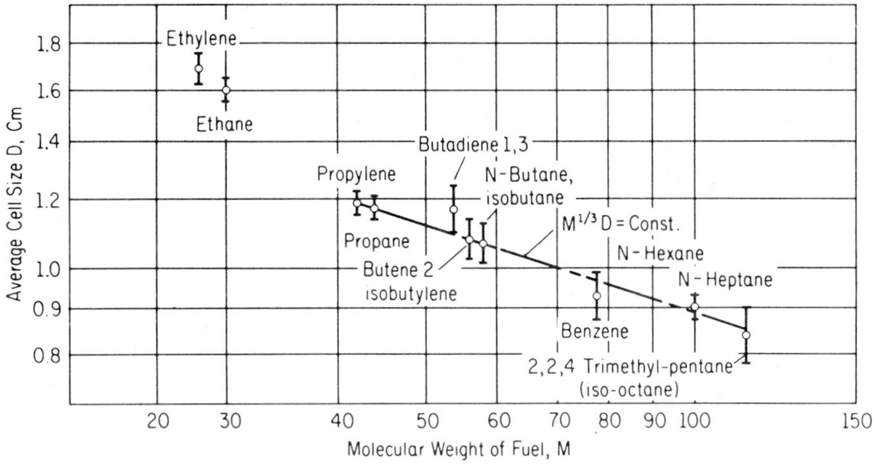

Fig. 7-19. Three-dimensional cell size vs. fuel molecular weight for a series of hydrocarbon fuels burning in air at 1 atm. pressure. [G. H. Markstein, *J. Aero. Sci.*, *18*, 199 (1951).]

has also been observed that very light fuels exhibit cellular structure in lean mixtures and burn as strictly one-dimensional flames in rich mixtures.

The theory of cellular formation, stability, and its other characteristics, such as cell size, trough depth, etc., is not entirely adequate at present. For one thing, a flame is a subsonic wave phenomenon and is, therefore, quite easily influenced by external conditions which are not obviously connected to the local flame conditions. On the other hand, any nonlinear or linearized theory for the instability or the structure must necessarily be based on a model of the flame zone, and it has been found that the results of the analysis can be quite dependent on the assumptions of the model.

However, the experimental observations suggest that the relatively simple mechanism of preferential diffusion of the reactants in the preheat zone is the primary cause of cellular instability in premixed gas flames. Figure 7-20 is a schematic drawing illustrating a perturbation of an otherwise laminar flame front. The behavior of isoconcentration lines of C in the preheat zone are shown for the flame $A + B \rightarrow 2C$. If there were no preferential side diffusion these lines would also be lines of constant composition for the species A and B. We see that for case (a) the concentration gradients will cause preferential diffusion of the lighter reactant away from the centerline of the perturbation, while the reverse will be true for case (b). Since this preferential diffusion will change the stoichiometry of the mixture along the centerline of the perturbation, the flame velocity will also be altered in this region. If the change in flame velocity due to

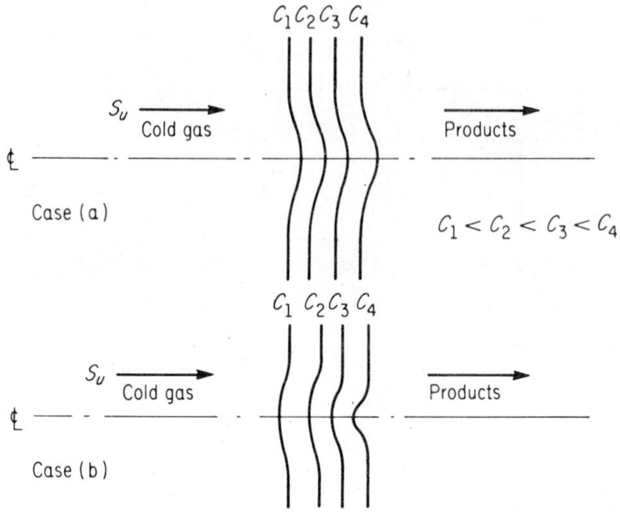

Fig. 7-20. Isoconcentration lines in a single line perturbation of an $A + B \rightarrow C$ flame.

this composition change causes the perturbation to grow relative to the rest of the flame, a laminar flame will be unstable to this type of perturbation. Thus, any fuel-oxidizer mixture should cause perturbations of type (a) to grow if the heavy reactant is in excess and those of type (b) to grow if light reactant is in excess. Experimentally only type (a) perturbations are observed. This may be explained by noting that preferential diffusion of the reactants occurs only toward the hot-reaction zone from the preheat zone, and therefore, that the curvature in case (b) causes relatively small composition changes ahead of the flame, whereas that of (a) can cause large local composition changes along the centerline. Thus the effect of lateral thermal conduction, which always tends to flatten the flame front, could be expected to override the preferential diffusion effect in case (b) but not in case (a).

As one might expect, this very simple theory for the formation of perturbations does not adequately explain all the observed facts concerning spontaneous instability. For example, since ethane and ethylene have molecular weights quite close to but less than that of oxygen, it is surprising that they show spontaneous instability at all, and even more remarkable that their instability appears in rich mixtures. Also, even though methane is considerably lighter than air, its lean mixtures are only weakly unstable. However, the theory does point out that preferential lateral diffusion of the reactants in the preheat zone is probably the primary cause of the trough formation which is a characteristic property of cellular flames.

Unfortunately, there is as yet no adequate theory to explain the observation that a definite cell size is preferred by a propagating cellular flame. However, the experimental cell-size relation $DP^{3/4}$ = const. found for butane–air flame illustrates that the minimum cell size is simply related to the preheat-zone thickness of the flame. This relation was obtained in flames with a burning velocity of about 20 cm/sec. From Fig. 7-14 we see that flames of this velocity have a pressure dependence approximately given by the equation

$$S_u = \text{const.} \cdot P^{-1/4}$$

Therefore, since Eq. 7-22 states that the preheat-zone thickness is related to the pressure in a constant-temperature flame by the expression

$$\eta_0 \cong \frac{\text{const.}}{P^{3/4}}$$

we find that the minimum cell size for a butane–air flame is directly proportional to the preheat-zone thickness of the equivalent laminar flame.

$$\frac{\eta_0}{D} = \text{const.} \tag{7-23}$$

This relation indicates that the mechanism which controls minimum cell size is very probably related to processes occuring in the preheat zone of the flame.

7-5. INFLAMMABILITY LIMITS

The one-dimensional flame theory discussed in Sec. 7-3 predicts a velocity and structure for a one-dimensional flame irrespective of the reaction's exothermicity or its absolute rate. However, it is well known that continued dilution with either inert gases or with an excess of fuel or oxidizer causes the burning velocity and flame temperature of the mixture to decrease until at some limit concentration propagation ceases to be spontaneous.

Experimentally this point is not well defined, but depends to a great extent on the apparatus and criteria used in its determination. Empirically it has been found that the extinction of a flame propagating upward from the open end of a 2-in.-diam. tube doward the closed end of the tube yields the widest inflammability limits. Most of the limits quoted in the literature have been determined in this manner. Table 7-1 lists some representative samples. However, mixtures whose compositions are outside these limits can usually be burned as a very low-velocity flame on a flat-flame burner, and therefore the limits ordinarily quoted in the literature do not represent absolute limits of propagation but simply reasonable practical limits.

TABLE 7-1.
INFLAMMABILITY LIMITS OF SOME COMMON FUEL-OXIDIZER MIXTURES (IN MOLE %)*

Fuel	Oxidizer	Lean limit	Rich limit
Hydrogen	Air	4.0	75.0
Carbon monoxide (moist at 18°C)	Air	12.5	74.0
Ammonia	Air	15.0	28.0
Cyanogen	Air	6.0	32.0
Methane	Air	5.3	15.0
Ethane	Air	3.0	12.5
Propane	Air	2.2	9.5
Butane	Air	1.9	8.5
Ethylene	Air	3.1	32.0
Acetylene	Air	2.5	80.0
Benzene	Air	1.4	7.1
Methyl alcohol	Air	7.3	36.0
Ethyl alcohol	Air	4.3	19.0
Diethyl ether	Air	1.9	48.0
Carbon disulphide	Air	1.2	44.0
Hydrogen	Oxygen	4.0	94.0
Carbon monoxide (moist at 18°C)	Oxygen	15.5	94.0
Ethane	Oxygen	3.0	66.0

*Data from H. F. Coward and G. W. Jones, *Limits of Flammability of Gases and Vapors*, U.S. Bureau of Mines Bulletin 503(1952).

The limit of propagation may also be controlled by a combination of preferential diffusion and buoyancy effects in certain flames under the correct circumstances. For example, lean hydrogen–air flames exhibit cellular instability and also have markedly different limits for upward propagation (4% hydrogen) and downward propagation (9% hydrogen) in a tube. An examination of the propagation process, however, shows that the upward-propagating wave is not a simple flame but consists of a series of small-radius-of-curvature fronts separated by quite large distances. Also, after propagation, combustion in the tube has been found to be relatively incomplete. In this case, during combustion, buoyancy and preferential diffusion effects are causing the flame to produce a local increase in the hydrogen concentration ahead of each of a series of highly curved fronts, and this causes their preferential propagation in a mixture whose composition is normally outside the mixtures inflammability limit. The correctness of this mechanism has been verified by passing a flame through a lean combustible mixture containing a hydrogen–deuterium mixture as the fuel. In this case the residual gas after flame propagation is found to be relatively enriched in deuterium, because the hydrogen had preferentially diffused into regions ahead of the curved flamelets where it reacted to form water.

Theoretically, the question of the inflammability limits for an infinite one-dimensional flame may be related to the question of transient one-dimensional stability or to the question of heat loss in a steady system. For flames of the type discussed in Sec. 7-3, transient one-dimensional stability could be investigated by rewriting the fundamental balance equations to include time-dependent terms, and then using these equations to discuss stability for specific flame models. This has yet to be done. Heat-loss models show that for a sufficient loss of heat a flame cannot exist as a steady one-dimensional phenomenon.

Without going through this theoretical development, some insight into the differences between the structure of limit and ordinary flames may be obtained by examining the one-dimensional flame discussed in Sec. 7-3. Let us consider the Le = 1 flame discussed on page 212, and determine the effect of changing the flame temperature, holding all other parameters constant. This is equivalent to diluting an ordinary combustible mixture with an inert such as nitrogen or argon. Figure 7-21 illustrates the effect of the flame temperature for this flame on the burning velocity, on the thickness of the hot-reaction zone when compared to the preheat-zone

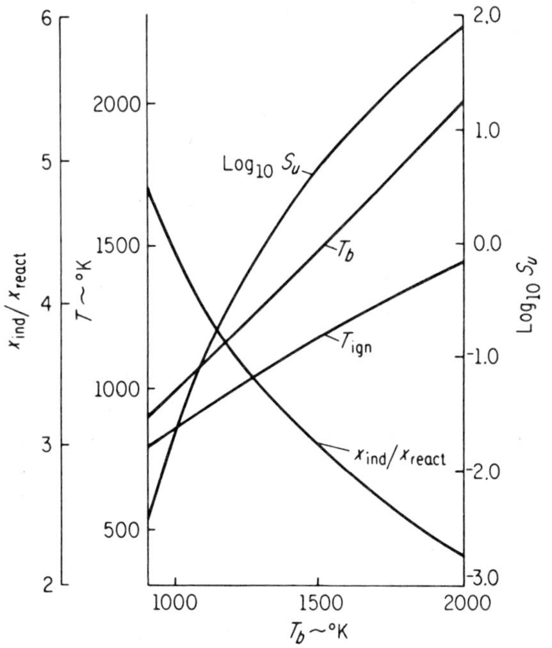

Fig. 7-21. Relative behavior of various flame properties for an Le = 1, ϵ = 15 flame plotted vs. flame temperature. The plotted quantities are the $\log_{10} S_u$, T_{ign}, and X_{ind}/X_{react} for this first-order flame. S_u units are cm/sec.

thickness and on the "effective" ignition temperature (defined as the temperature where the reaction rate is $^1/_{100}$ of the maximum reaction rate or that flame). We note two things: (1) the ignition temperature is dropping much more slowly than the flame temperature, and (2) as the flame temperature drops, the hot-reaction zone is becoming thinner, relative to the preheat-zone thickness. Thus the character of the coupling between the chemical reaction and thermal conduction is changing as the flame temperature drops.

Qualitatively, one might expect this shift in the details of the flame structure to lead to a dynamically unstable situation for just these two reasons. We would expect that if the effective ignition temperature were above $(T_b - T_u)/2$ the flame would become unstable, and we also might expect that a flame which has an extremely thin reaction zone supporting and coupled with an extensive preheat zone, would also represent a dynamically unstable situation.

Recent experimental work on a large flat-flame burner has shown that as the burning velocity and flame temperature decrease with dilution the downstream heat loss increases, and thus this heat loss could become sufficient to cause the one-dimensional extinguishment predicted by theory. However, the experiments also show that three-dimensional phenomena are extremely important at the limit. In fact, the primary mechanism of extinguishment was found to be the spontaneous development of "holes" in the flat flame. These holes were found to be caused either by a buoyancy effect which caused recirculation, or in another case by a complex three-dimensional effect in which the flame prefers to propagate tangentially into the mixture passing through the hole. When hole formation was inhibited in a portion of the flame, catastrophic downstream excursions of the flame were observed at the limit. This effect first appeared as a "fluttering" of the flame and probably represents another type of three-dimensional instability.[6]

7-6. MINIMUM IGNITION ENERGY AND QUENCHING DISTANCE

These two flame properties are quite useful for characterizing practical combustion systems. The minimum ignition energy is the smallest quantity of energy that needs to be added to a system to start flame propagation. Its value is quite dependent on the local rate and method of heat addition, and on the geometry of the heat source. Quenching distance is defined as the largest channel dimension that will just keep a flame from propagating throughout the channel. Values of the quenching distance are also quite dependent on experimental geometry.

[6] P. H. Kydd and W. I. Foss, *Combustion and Flame*, *8*, 267 (1964).

Combustible mixtures may be exploded "homogeneously" in "vessel" experiments like those discussed in Chapter 6, or they may be ignited by rapid heat addition at a localized region in the gas. In the latter case, successful ignition leads to the propagation of either a flame or detonation. As the total quantity of energy is reduced for any one ignition technique, direct detonation ceases to occur, if it did at all (see Chapter 9 for a discussion of detonation initiation), and flames are observed to propagate from the source. Further reduction of the total quantity of energy that is added to the fluid continues to produce ordinary flame propagation until the minimum ignition energy for that source configuration is reached. At this point only a small quantity of the combustible gas is converted to products and the energy is dissipated harmlessly by thermal conduction.

A multitude of different energy sources and source configurations have been used to study flame ignition experimentally. However, an ordinary capacitor discharge spark has consistently yielded the lowest ignition energy for any specific combustible mixture and has a geometry and temporal history which are quite amenable to mathematical analysis.

For an experimental determination of an ignition energy, high-performance (usually air-gap) condensers are used so that the majority of the stored energy will appear in the spark gap. The stored energy may be calculated using the equation $E = CV^2/2$, where C is the capacity of the condenser and V the voltage just before the spark is passed through the gas. The spark must always be produced by a spontaneous breakdown of the gap because an electronic firing circuit or a trigger electrode would either obviate the measurement of spark energy or grossly change the geometry of the ignition source. It has been found experimentally that for this type of spontaneous spark up to 95% of the stored energy appears in the hot kernel of gas in less than 10^{-5} sec. The losses are thought to be due primarily to heat conduction to the electrodes. Since the total stored energy can be varied by changing either the capacity or the voltage, the electrode spacing (which is proportional to the voltage at breakdown) may be varied as an independent parameter. Two problems now arise: If the electrode spacing is too small, the electrodes will interfere with the propagation of the incipient flame and the apparent ignition energy will increase. If, however, the spacing is too large, the source geometry will become essentially cylindrical and the ignition energy will again increase because the area of the incipient flame is greatly increased in this geometry. This condition is shown schematically in Fig. 7-22a. The fact that the increase in minimum ignition energy is due to quenching at small electrode spacings may be confirmed by using electrodes flanged with electrically insulating material. Figure 7-22b shows the effect of electrode flanging on the ignition energy at small separations. Note that below a certain spacing, as the spacing is reduced the spark

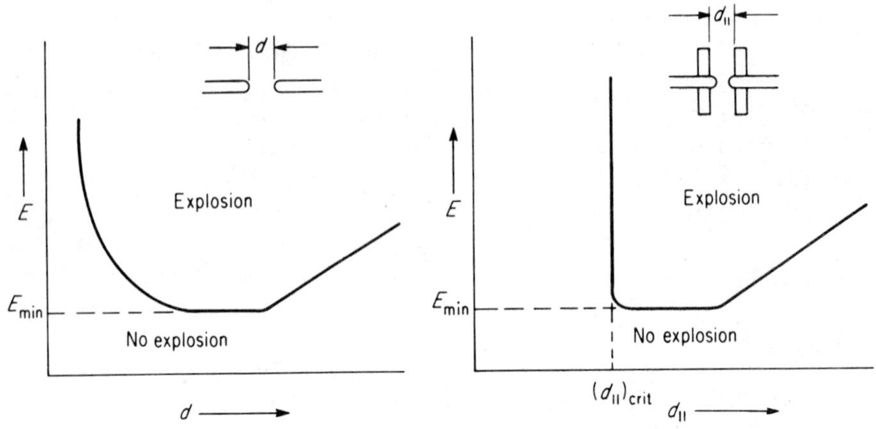

(a) Unflanged electrodes (b) Flanged electrodes

Fig. 7-22. Quenching distance and minimum-ignition-energy measurement with a capacitor spark.

energy required to ignite the bulk of the gas sample rises very much more rapidly with flanged tips. This critical flange spacing is defined as the flat-plate quenching distance of the flame. Therefore this simple experiment may be used to determine the flat-plate quenching distance and the minimum ignition energy for spherical geometry, and may also be used to evaluate whether the ignition source is essentially spherical or cylindrical. The latter piece of information is obtained by determining the effect of electrode separation on E_{min} at large separations.

Quenching distances may also be measured by quickly stopping the flow through a tube of the desired geometry when a flame is seated on the exit of the tube. If the flame flashes down the tube, the minimum dimensions are above the quenching distance for that mixture. Quenching distances that have been measured between two flat plates using this technique agree quite well with the quenching distances measured using the flanged electrodes in the spark-ignition experiment described above. Parallel-plate quenching distances and minimum ignition energies determined by the spark method for propane, oxygen, and nitrogen mixtures at various initial pressures are given in Fig. 7-23. It should be noted that, in general, minimum ignition energies are extremely small—in some cases, corresponding to the passage of a barely audible spark through the mixture. In fact, it is well known that static electric sparks can cause the ignition of many explosive mixtures.

The relation between the characteristic quenching distance and the geometry of the quenching tube has been developed theoretically and confirmed experimentally in a number of instances. The theory can be based

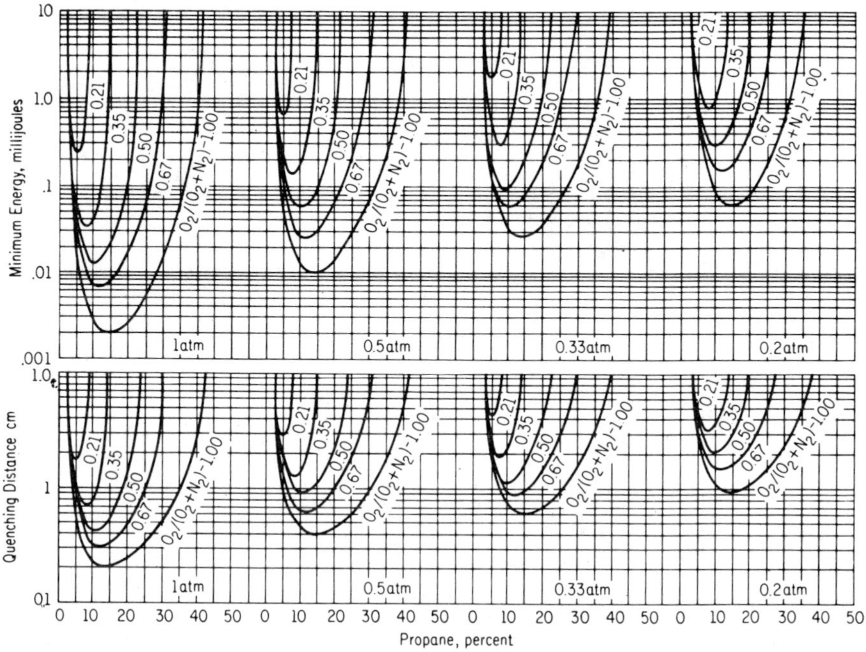

Fig. 7-23. Effect of pressure, nitrogen dilution, and stoichiometry on propane-oxygen flat-plate quenching distances and spark minimum-ignition energies. (With permission from B. Lewis and G. von Elbe, *Combustion Flames and Explosions of Gases*, 2d ed., Academic, New York, 1961.)

on the assumption either that (1) wall capture of reactive species, or (2) heat transfer to the wall controls the quenching. Both these approaches yield the same theoretical relations relative to geometry effects, because of the similarity between diffusion and thermal conduction. Quenching measurements for a variety of geometries show quite good relative agreement ($\pm$ 10%) with the theoretical predictions for a number of hydrocarbon systems. The theory predicts that the circular tube quenching diameter should be related to the parallel plate quenching distance by the equation

$$d_{\parallel} = 0.65 d_0$$

Experimentally, one may show that the quenching distance is simply related to the pressure (for any one flame (Pd = const.) and also that it is related to the minimum ignition energy of the system for virtually all hydrocarbon–air flames, through the equation $E_{\min} = \text{const.} \times d^2$ (see Fig. 7-24).

This last relation is a rather interesting one and bears some further comment. First, however, we will look at the details of incipient flame

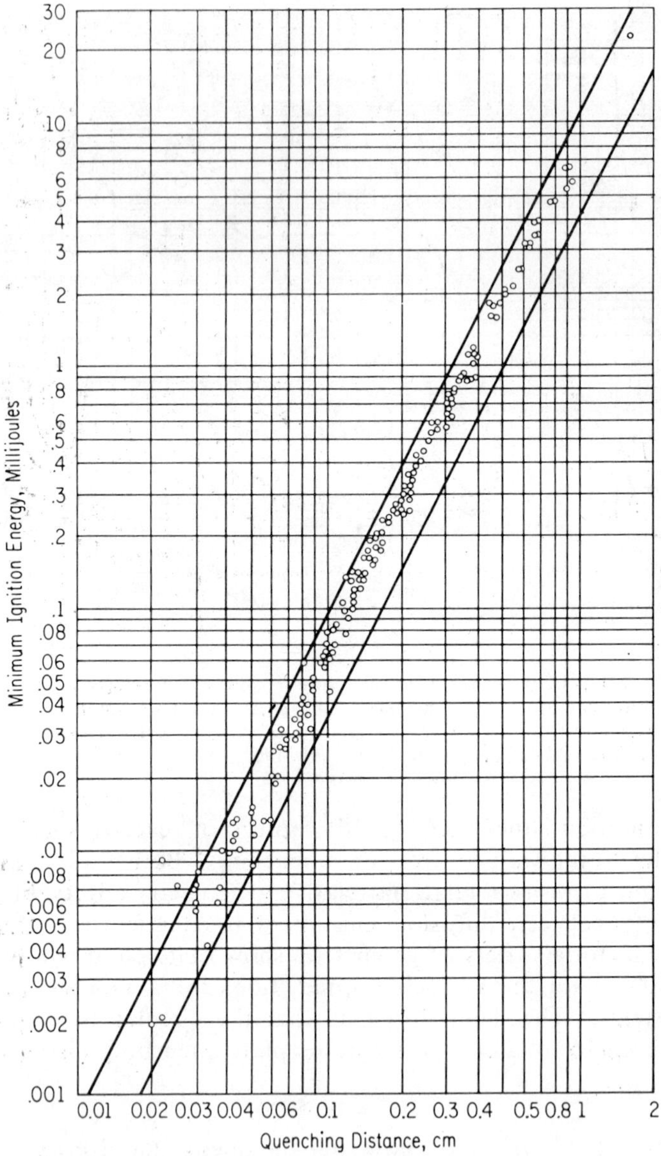

Fig. 7-24. The relation between flat-plate quenching and spark minimum ignition energies for a number of hydrocarbon–air mixtures. (With permission from B. Lewis and G. von Elbe, *Combustion Flames and Explosions of Gases*, 1st ed., Academic, New York, 1951.)

growth from a small spark. Figure 7-25 contains repetitive schlieren photographs illustrating the appearance of the hot kernel of gas for an inert mixture and for a reactive mixture with both a subcritical and supercritical energy addition. In the reactive case, the spark kernel initially contains temperatures well above the ignition temperature of the mixture,

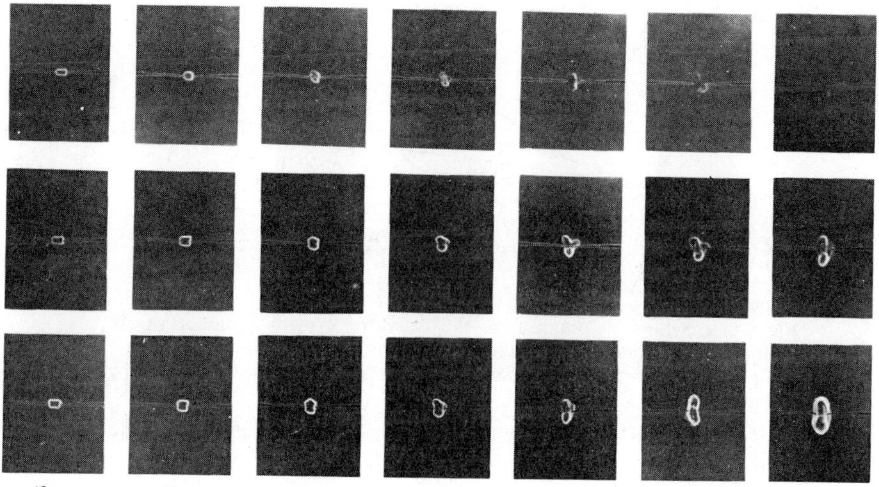

Fig. 7-25. Effect of change in ignition energy upon flame propagation. Times are in microseconds (Courtesy H. Lowell Olsen, Applied Physics Laboratory, Silver Spring, Md. Originally in AGARD, *Selected Combustion Problems*, II, Butterworth, London, 1956.) Read down:

	Gas	$\frac{1}{2}CV^2$
	Air only	1.33 mJ
Subcritical energy	Propane–Air	1.14 mJ
Supercritical energy	Propane–Air	1.23 mJ

and an incipient flame is seen to form. This behavior should be contrasted with the inert case where the initially high gradient simply flattens with time and eventually disappears. However, as the two reactive cases show, the formation of an incipient flame does not insure subsequent propagation. At some point, well after spark passage (ca 100 μsec in Fig. 7-25) the incipient flame becomes a true flame in the supercritical case and starts to decay by thermal conduction in the subcritical case. This behavior is shown schematically in Fig. 7-26. Thus there is evidently a minimum ignition volume which will lead to stable flame propagation for spherical geometry.

Theoretically, ignition and quenching are difficult to discuss and complete theories are not as yet available for all geometries. However, Rosen[7]

[7] G. Rosen, *J. Chem. Phys.*, *30*, 298 (1959).

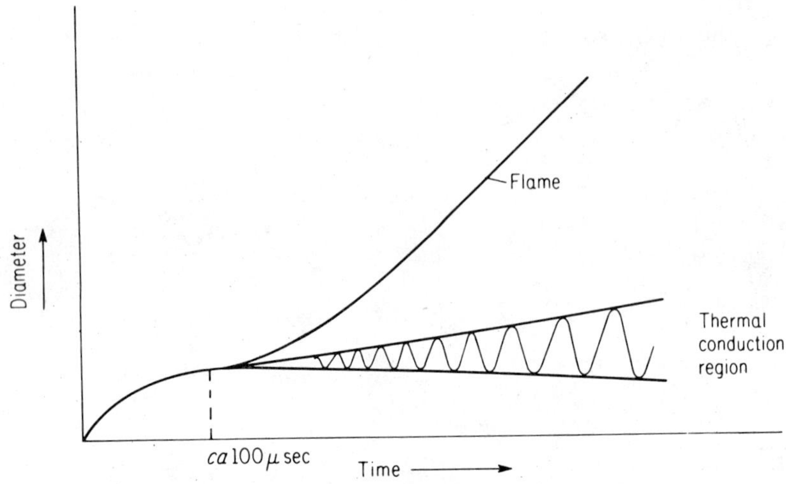

Fig. 7-26. Schematic diagram of the decay or growth behavior of an ignition kernel (incipient flame ball) near the threshold value of energy for ignition of a propagating flame.

has justified Lewis and von Elbe's[8] empirical relation for a plane source (i.e., one-dimensional) minimum ignition energy by applying the non-steady equations to a simple first-order $A \rightarrow B$ flame with Le = 1, similar to the steady flame discussed in Sec. 7-3. He assumed that heat was added in a very short time at an infinite flat plane in the mixture, and determined that the critical quantity of heat addition to cause flame propagation was related to the steady-flame properties by the equation

$$E_{\min} \cong \frac{\kappa_u q}{(C_p/m)S_u} = \frac{\kappa_u}{S_u}(T_b - T_u) \tag{7-24}$$

where $E_{\min}$ has the units of cal/cm^2 and $q = (C_p/m)(T_b - T_u)$ = heat of reaction at constant pressure in calories per gram. We note that q is defined as an enthalpy, while $E_{\min}$ is an energy. However, this equation is correct because the energy added to the system appears as an enthalpy in the system, since the process is taking place at constant pressure. We now use Eq. 7-22 for the preheat zone thickness to determine the approximate sensible enthalpy storage in the preheat zone of the flame in terms of flame parameters. First we integrate over the temperature distribution in the preheat zone, assuming that it is linear from T_i to T_u over a distance η_0 defined by Eq. 7-22. With this assumption the equation becomes

$$H_{p_h} = \frac{C_p}{m}\rho_u T_u \int_0^{\eta_0}\left(1 - \frac{T_u}{T}\right)dx$$

[8] B. Lewis and G. von Elbe, *Combustion Flames and Explosions of Gases*, 2d ed., Academic, New York, 1961.

where $T = T_u + \frac{x}{\eta_0}(T_i - T_u)$ and H_{p_h} is the sensible enthalpy stored in the preheat zone. This equation integrates to yield

$$H_{p_h} = \frac{C_p}{m} \rho_u T_u \eta_0 \left[\frac{T_i - T_u - T_u \ln \frac{T_i}{T_u}}{T_i - T_u} \right]$$

However, $\ln (T_i/T_u) \simeq 1$ and therefore,

$$H_{p_h} \simeq \frac{C_p}{m} \rho_u \eta_0 T_u$$

Substituting for η_0 from Eq. 7-22 and $E_{\min}$ from Eq. 7-24 yields the equation

$$H_{p_h} = \frac{E_{\min} T_i}{T_b - T_u} \simeq E_{\text{mim}} \qquad (7\text{-}25)$$

Equation 7-25 states that the minimum ignition energy of a flat flame is indeed related to the sensible enthalpy stored in the preheat zone. This relationship has never been checked for a flat ignition source, however, because it is difficult to add heat instantaneously at a plane in a gas without the presence of a wall or other disturbances.

A semiquantitative argument may now be used to relate these concepts to the experimental observations of cylindrical tube quenching and the minimum ignition energy in spherical geometry. Each unit area of a flat semi-infinite flame must heat by thermal conduction a unit area of gas in the region ahead of the flame. However, a small spherical flame propagating outward is always increasing its area. Therefore in this case each element must heat a larger area of cold gas in order to sustain its propagation. Consider a flame ball of diameter d and preheat-zone thickness η_0, where d is taken to be the diameter of the spherical ignition surface in the flame front and is assumed to be at least somewhat larger than η_0. When the flame propagates a distance equal to the thickness of the undisturbed preheat zone, its diameter increases from d to approximately $d + 2\eta_0 \frac{\rho_u}{\rho_b}$. Therefore the flame's relative area increase when it consumes the gas initially trapped in the preheat zone is equal to the quantity

$$1 + \frac{4\eta_0}{d} \frac{\rho_u}{\rho_b}$$

The quantity

$$K = 4 \frac{\eta_0}{d} \frac{\rho_u}{\rho_b} \qquad (7\text{-}26)$$

is called the *Karlovitz number* for an outwardly propagating spherical flame.[9] It is the percentage increase in flame area or the relative lateral stretch of an outwardly propagating spherical flame when the flame travels a distance equal to its preheat-zone thickness for any particular flame diameter. We note that K approaches zero as d approaches infinity, i.e., as the flame becomes a flat flame.

Since smaller flame balls exhibit a larger lateral stretch and therefore have a larger value of K, one might expect that a limit stretch exists which defines the smallest flame ball which will spontaneously propagate without being quenched by excessive heat loss to the gas ahead of the flame front. This concept may be applied to the problem of spherical ignition by rewriting the equation for K using the definition of η_0, Eq. 7-22.

$$K = \frac{4\kappa_u(1 + \tau_{\text{ign}}\phi)}{(C_p/m)dS_u\rho_b}$$

or

$$K = 4\,\frac{\kappa_u(T_b - T_u)\pi d^2/S_u}{(C_p/m)\rho_b(T_b - T_u)\pi d^3}\cdot\left(\frac{T_i}{T_u}\right) \tag{7-27}$$

In Eq. 7-27 the quantity $(C_p/m)\rho_b(T_b - T_u)\pi d^3$ is closely related to the total sensible enthalpy contained in a flame ball of diameter d. This may be determined by noting that

$$H' = \int_0^{d/2+\eta_0} 4\pi r^2(C_p/m)\rho(T_b - T_u)\,dr$$

Since $\rho \cong (T_b/T)\rho_b$, we may rewrite this equation to yield

$$H' = 4\pi(C_p/m)\rho_b \int_0^{d/2+\eta_0}\left(T_b - \frac{T_bT_u}{T}\right) r^2\,dr$$

However, the temperature distribution is close to T_b in a large portion of the flame ball and if we assume $T = T_b$ over the range of the integral to $d/2$ and $T = T_u$ from $d/2$ to $d/2 + \eta_0$ we obtain

$$H' \cong \frac{\pi}{6}(C_p/m)\rho_b(T_b - T_u)\,d^3$$

We may therefore rewrite Eq. 7-27 to obtain

$$K = \frac{2}{3}\,\frac{\kappa_u(T_b - T_u)\pi d^2/S_u}{H'}\cdot\left(\frac{T_i}{T_u}\right) \tag{7-28}$$

for a flame ball of diameter d. We note that the quantity $\kappa_u(T_b - T_u)\cdot\pi d^2/S_u$ is simply the total ignition energy required at the surface of a sphere of radius d and is approximately equal to the total energy stored in the preheat zone of a flame of radius d. This relation explains the

[9] B. Karlovitz, et al., *Fourth Symposium on Combustion*, Williams & Wilkins, Baltimore, 1953, p. 613.

experimental correlation $E_{\min} = cd_0^2$, where d_0 is the quenching distance and shows that a minimum-diameter propagating flame has a quantity of energy roughly equal to its ignition energy stored in its preheat zone.

In addition, Eq. 7-28 states that the Karlovitz number of any specific spherical flame will give an indication of the relative quantity of energy stored in the preheat zone and in the central flame region; that is,

$$K = \frac{2}{3} \cdot \frac{T_i}{T_u} \cdot \frac{\text{(conducted enthalpy)}}{\text{(hot-gas enthalpy)}}$$

We therefore see that as the flame diameter d gets smaller K should increase because the conducted enthalpy decreases more slowly than the hot-gas enthalpy, and it is rather instructive to investigate the magnitude of K for an incipient spherical flame. Since studies of the type illustrated in Fig. 7-24 have not been performed on a large number of systems, we use instead the minimum ignition energy and cylindrical-tube quenching distance for our correlation. This is a reasonable approximation because one would expect the total enthalpy in an incipient flame ball to be somewhat greater than, and simply related to, the minimum spark ignition energy of the flame, while the minimum flame diameter for propagation as used in the equations should be somewhat less than and simply related to the quenching distance of the flame. Furthermore, because of the many assumptions used in deriving this concept, we should expect no better than an order-of-magnitude agreement between prediction and experimental measurements. The preheat thickness η_0 for any flame may be calculated from normal burning velocities and cold-gas values of heat capacity, density, and thermal conductivity. The only other variable that we need to determine the conducted and total enthalpy at ignition and also to independently calculate the Karlovitz number at ignition is the minimum flame diameter d, and we equate this to the tube-quenching distance of the flame. We thus find (as shown by Lewis and von Elbe)[10] that for a large number of common flame systems K_{ign} as given by Eq. 7-27 ranges from 0.5 to 1.5 and the ratio of the minimum ignition energy to the quantity H' (for $d = d_0$) ranges from 0.1 to 0.65 for methane–oxygen–nitrogen mixtures and is approximately 0.3 for hydrogen–air, 0.5 for propane–oxygen, and 0.2 for propane–air. This is quite reasonable, since the ignition temperature of these flames ranges from about 900–1200°K. Thus we obtain an order of magnitude agreement between the Karlovitz number calculated simply from the quenching distance and that calculated from the ratio of conducted to total enthalpy in these incipient flames, and we arrive at the conclusion that a Karlovitz number of approximately unity represents a threshold flame diameter for an incipient

[10]B. Lewis and G. von Elbe, *Combustion Flames and Explosions of Gases*, 2d ed., Academic Press, New York, 1961, pp. 336–346.

flame ball. We therefore have experimental verification that ignition is accomplished by constructing a correctly shaped preheat zone in a local region of the gas sample whose size is given by $K \cong 1$.

One difficulty which has not been discussed is the problem of preferential diffusion associated with curved flame fronts. Evidence that preferential diffusion influences flame development in systems where the fuel and oxidizer have different diffusivities is shown in Fig. 7-27, which is a plot of minimum ignition energy vs. composition for various fuels in air. Since from the elementary theory one would expect the minimum ignition energy to occur at the maximum burning velocity of these mixtures, these curves should all show a minimum in the neighborhood of $\Phi = 1.1$ (the approximate stoichiometry for the maximum laminar burning velocity for all these systems). Notice that for methane, which is

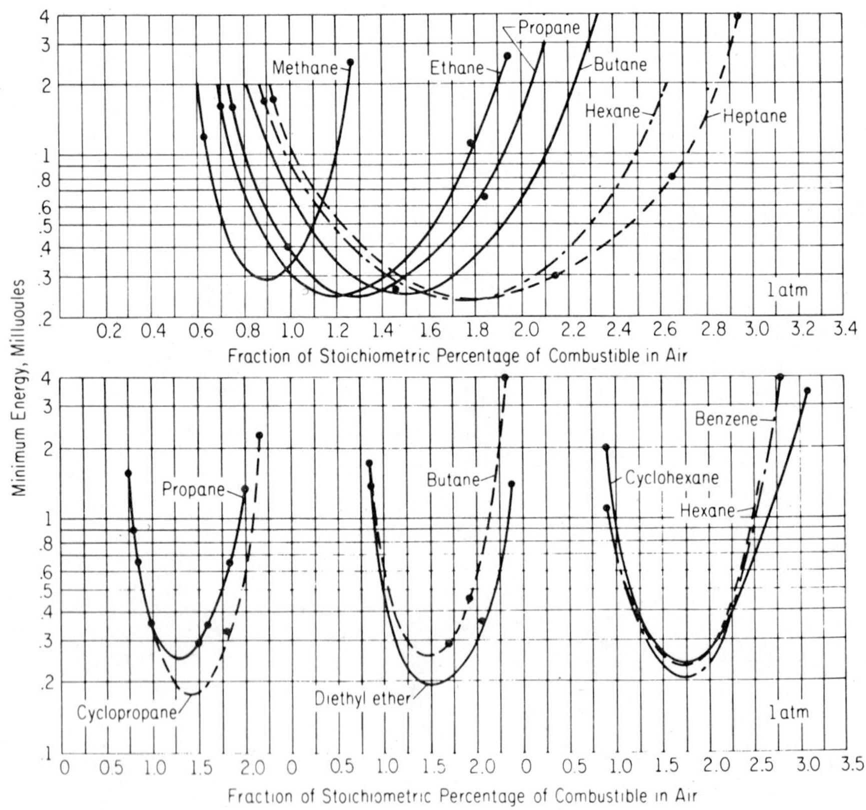

Fig. 7-27. Minimum ignition energy vs. stoichiometry for various fuel–air mixtures showing the effects of preferential diffusion. (with permission from B. Lewis and G. von Elbe, *Combustion Flames and Explosions in Gases*, 2d ed., Academic, New York, 1961.)

lighter than oxygen, Φ_{min} is 0.8, while for those fuels whose diffusivity is less than oxygen Φ_{min} is always greater than one and increases as the fuel molecular weight increases. Evidently preferential diffusion of the lighter species in these highly curved flames changes the composition in the flame region markedly during the ignition process. When this composition approximates the stoichiometric composition the minimum quenching distance is observed because the incipient flame has the maximum burning velocity. The larger the difference in relative diffusivity, the larger the deviation of Φ_{min} from stoichiometric.

Another indication that preferential diffusion plays an important role in incipient flame behavior is obtained from the measured burning velocities of outwardly propagating flames. For example, as shown in Fig. 7-28, in propane–air mixtures (i.e., for a heavy fuel and light oxidizer mixture), rich mixtures exhibit a higher space velocity at small radii while lean mixtures show a lower space velocity at small radii. In all

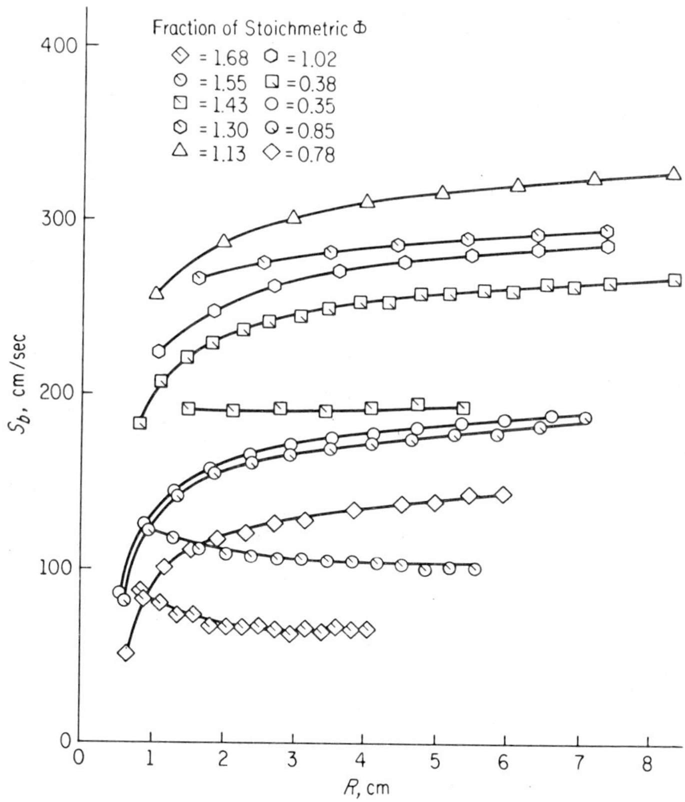

Fig. 7-28. Space velocity S_b vs. radius for a laminar flame ball in propane–air mixtures.

cases the space velocity of the flame relaxes to the theoretically expected laminar space velocity after a few centimeters of flame travel. The effect is somewhat masked by two additional effects which cannot be easily discussed quantitatively: (1) The slow relaxation of the burned products to the equilibrium composition adds a slow initial increase in velocity to the experimental curves, and (2) any excess ignition energy at the spark source can cause a higher propagation velocity of the early flame. However, the preferential diffusion effect is the only effect which shifts sign as the stoichiometry passes through unity and is therefore quite definitely operating in this system.

7-7. UNUSUAL FLAMES

A few of the more important examples of unusual flame systems are given in Table 7-2 along with their maximum burning velocity. We note from Table 7-2 that there are many exothermic compounds which may be made to decompose in a spontaneously propagating flame front to yield less exothermic products. Certain of these decomposition flames have

TABLE 7-2*
UNUSUAL FLAME SYSTEMS

	Method of Velocity Determination	Burning Velocity, S_u cm/sec	Flame Temperature, °K
Decomposition Flames			
Acetylene, C_2H_2	Tube	20	—
Ethylene oxide, C_2H_4O	Flat flame	5	1200
Methyl nitrate, CH_3NO_3	Bomb	250	1800
Ethyl nitrate, $C_2H_5NO_3$ $T = 200°C$	Bomb	10	—
Methyl nitrite, CH_3NO_2	Flat flame	7.5	1300
Hydrazine, N_2H_4 $P = 40$ mm	Bomb	110	1900
Hydrogen azide, HN_3 $P = 50$ mm	Bomb/burner	950	3000
Other Flame Systems			
Hydrogen – bromine, $H_2 + Br_2$	Burner	200	1650
Diborane–hydrazine, $B_2H_6 - N_2H_4$	Burner	5,000	—
Hydrazine – nitrous oxide, $N_2H_4 - N_2O$ $P = 40$ mm	Bomb	164	2650
Hydrazine – nitric oxide, $N_2H_4 - NO$ $P = 40$ mm	Bomb	245	2740
Hydrogen–chlorine, $H_2 - Cl_2$ (60% H_2)	Burner	235	2100
Perchloric acid–methane, $HClO_4 + CH_4 + 1.5N_2$	Burner	122	2460

*Data from *Combustion Symposium* volumes—see Index in Volume X, The Combustion Institute (1965), for individual experimental values.

been studied quite extensively because they represent kinetically simple systems. This is particularly true of the ozone-decomposition flame whose burning velocity has both been studied as a function of oxygen dilution and calculated, using various approximations of the theory developed in Sec. 7-3. Another decomposition flame, that of nitric oxide, is unusual because it represents a flame that was first observed after its possible existence was predicted theoretically. Nitric oxide is a relatively stable exothermic molecule and it decomposes slowly even at 1500°K. Furthermore, its heat of reaction is sufficiently low so that its ordinary flame temperature is not far above 1500°K. In 1948 Henkel, Hummel, and Spaulding[12] predicted that a nitric oxide decomposition flame should be obtainable if the gas were preheated to at least 1000°K, and in 1952 Parker and Wolfhard[13] produced such a burner flame in preheated nitric oxide. They obtained the flame by first burning a nitric oxide–hydrogen flame and then slowly reducing the quantity of hydrogen until they obtained the pure decomposition flame. Of the other flame systems listed in Table 7-2 the hydrogen–bromine flame has been studied extensively because the primary flame reactions represent an example of a simple chain reaction without branching.

Flames which contain either nitrogen dioxide or an excess of nitric oxide as the oxidizer hold a special interest because they will exhibit two distinct luminous flame zones under the proper circumstances. This is caused by the relative inertness of the nitric oxide, which is either present in the initial mixture or appears as a consequence of the decomposition of nitrogen dioxide in the flame. These double flames are actually a fuel-oxidizer flame closely followed by a simple nitric oxide decomposition flame. This phenomenon has been observed in a number of other systems all of which have the property that two independent decomposition reactions with properly different rates are occurring in a tandem fashion in the flame. (The second reaction must occur at approximately the same velocity as the first at the flame temperature of the first.)

7-8. IONIZATION IN FLAMES

Most flames exhibit ion concentrations well above those theoretically available on the basis of equilibrium thermodynamic calculations. Since an investigation of ions in flames must be performed in situ the exact determination of the composition of the ions is quite difficult. Three techniques have been used. These are (1) the Langmuir probe, (2) the mass spectrometer, and (3) microwave absorption. Of these, the third

[12] M. J. Henkel, H. Hummel, and W. P. Spaulding, *Third Symposium on Combustion*, Williams & Wilkins, Baltimore, 1949, p. 135.

[13] W. G. Parker and H. G. Wolfhard, *Fourth Symposium (International) on Combustion*, Williams & Wilkins, Baltimore, 1952, p. 420.

produces the least quantitative data since with it one observes the whole flame. Microwave absorption has been found to be most useful in observing the relaxation of ion concentrations in the postflame gases or studying the effect of external parameters on the ion concentration in the reaction zone of thick, low-pressure flames.

The Langmuir probe technique measures the current to a fine probe held at a particular voltage at a point in the flame zone. Figure 7-29

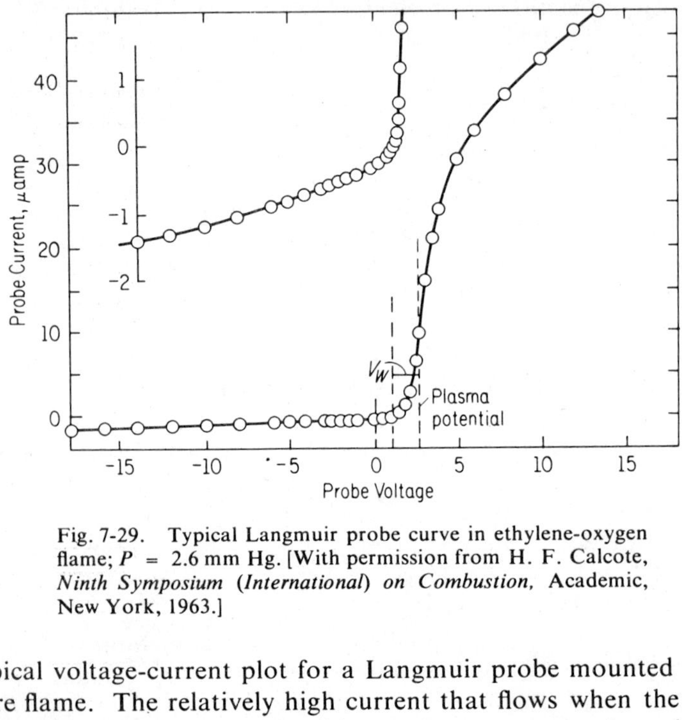

Fig. 7-29. Typical Langmuir probe curve in ethylene-oxygen flame; $P = 2.6$ mm Hg. [With permission from H. F. Calcote, *Ninth Symposium* (*International*) *on Combustion*, Academic, New York, 1963.]

is a typical voltage-current plot for a Langmuir probe mounted in a low-pressure flame. The relatively high current that flows when the probe is charged positively and the net positive current at zero probe voltage may be explained by the very high mobility of one specific negative particle, the electron. Since the current at positive voltage is predominantly caused by electron capture, the voltage-current curve in this neighborhood may be used to determine the free-electron temperature in the flame. Values determined in a variety of low-pressure hydrocarbon-oxygen-diluent flames indicate that this temperature is always quite close to the translational gas temperature.

The Langmuir probe is not an appropriate tool for detailed studies of the positive ions in the flame because it cannot distinguish species. Also, because of the extremely high mobility of electrons the probe cannot be used to observe the presence of the negative ions in a flame. For the quantitative observation of ion concentrations a mass spectroscopic technique is necessary. With this technique both positive and negative ions have been observed in hydrocarbon flames and their concentrations throughout the flame has been mapped in certain cases. At present it is thought that the primary ion-producing reaction in hydrocarbon flames is the reaction

$$CH + O \rightarrow CHO^+ + e$$

The remaining ions are thought to be formed by a series of exchange reactions with either this ion or the free electrons. This mechanism of primary ion formation also explains the observation that hydrogen–oxygen flames have extremely low ion concentrations relative to the hydrocarbon flames of much lower temperature.

7-9. DUST AND SPRAY FLAMES

In Chapter 6 we briefly discussed the combustion of individual dust particles or liquid drops and pointed out that in the case of an isolated particle the combustion process was not controlled by a wave-propagation mechanism but simply by diffusion and heat transfer at the drop, coupled with chemical reaction. We now wish briefly to discuss the other "pure" process available in dust or droplet combustion—i.e., the propagation of a "flame" through a highly concentrated cloud of fine combustible particles which are either solid or liquid.

It has been found experimentally that highly concentrated dust or mist clouds of sufficiently small particles of a fuel in an oxidizing atmosphere are capable of supporting laminar "flames" whose properties are quite similar to the properties of ordinary premixed gaseous flames. In particular, these flames have been found to exhibit characteristic ignition energy and burning velocities which are functions of the stoichiometry. However, in the case of dust or spray flames the particle size and relative volatility of spray or dust material are both quite important in determining the flame structure and burning mechanism. For very small particles of relatively high volatility the particles completely evaporate in the preheat zone of the flame and, since for this case concentration inhomogeneities would exhibit a scale considerably smaller than the preheat-zone thickness, the flame appears to burn as an ordinary premixed gas flame. At the other end of the particle-size–volatility range, larger and less volatile particles will not completely evaporate before entering the hot heterogeneous reaction zone. Thus in this case the reaction zone consists of numerous particles in various stages of diffusive burning—i.e., surrounded by small local diffusion flames. The apparent "laminar" flame thickness increases markedly and the flame-propagation mechanism becomes somewhat dependent on the heat-transfer and diffusion processes occurring at the individual particles.

The reader is referred to the bibliography for more detailed experimental and theoretical information on the problem of flame propagation through concentrated dust or spray clouds.

BIBLIOGRAPHY

Flame Velocity and Temperature

1. NACA Report 1300, pp. 127–162 (1957).

2. A. S. Sokolik, *Self-Ignition Flame and Detonation in Gases.* Israel program for scientific translation. (Jerusalem, 1963). NASA TTF125 OTS 63-111 79, pp. 165–274.

3. W. Jost, *Explosions and Combustion Processes in Gases.* (New York: McGraw-Hill, 1946).

4. A. G. Gaydon and H. G. Wolfhard, *Flames.* (London: Chapman and Hall, 1960.)

5. B. Lewis and G. von Elbe, *Combustion Flames and Explosions of Gases.* (New York: Academic, 1961.)

Flame Structure

6. R. M. Fristrom and A. A. Westenberg, *Flame Structure.* (New York: McGraw-Hill, 1965.)

7. F. A. Williams, *Combustion Theory.* (Reading, Mass.: Addison-Wesley, 1965.)

8. C. P. Fenimore, *Chemistry in Premixed Flames.* (New York: Pergamon, 1964.)

9. F. J. Weinberg, *Optics of Flames.* (London: Butterworth, 1963.)

10. G. H. Markstein, *Non-steady Flame Propagation.* (New York: Macmillan, 1965.)

Ignition Quenching and Inflamibility Limits

11. B. Lewis and G. von Elbe, *Combustion Flames and Explosions of Gases.* (New York: Academic, 1961.)

12. A. E. Potter, Jr., *Progress in Combustion Science and Technology*, J. Ducarme, M. Gerstein, and A. H. Lefebvre, eds. (New York: Pergamon, 1960.)

13. NACA Report 1300, pp. 83–126 (1957).

Unusual and Dust or Spray Flames

14. *The Combustion Symposium*, Vols. 3–10 (index in Vol. 10).

15. ACS Monograph No. 20, pp. 104–165 (1958).

PROBLEMS

7-1. Calculate the isobaric flame temperature of a stoichiometric hydrogen-oxygen flame for an initial temperature of 300°K and an initial pressure of one atmosphere. Assume (a) that water is the only product formed, (b) that the flame products are an equilibrium mixture of H_2, O_2, and H_2O, (c) that the flame products are an equilibrium mixture of H_2, O_2, H_2O and OH, and (d) that H_2, O_2, H, O, OH, and H_2O are present at equilibrium.

7-2. Calculate the isobaric flame temperature of a stoichiometric carbon monoxide–oxygen mixture for an initial temperature of 300°K and an initial pressure of one atmosphere. Assume (a) only CO_2 is formed as a product, (b) that CO, O_2, and CO_2 are present at equilibrium, and (c) that CO, O_2, O, and CO_2 are present at equilibrium.

7-3. Determine the effect of the reactions activation energy on burning velocity for the $A \rightarrow B$ flame if all other properties of the flame are held constant. Do the calculation for both an $\mathrm{Le} = 0$ and $\mathrm{Le} = 1$ flame.

7-4. Calculate the ratio of the minimum three-dimensional cell size to the flame preheat-zone thickness for a slightly rich propane–air flame.

8

Aerodynamics of Flames

8-1. LAMINAR FLAME HOLDING

Premixed gas flames may be observed as steady waves in a variety of laminar-flow situations, and except for the case of the flat flame on a flat-flame burner, they will exist as a steady-wave phenomena only if the flow velocity of the main stream is well above the normal burning velocity of the mixture. This prevents the flame from propagating downstream into the apparatus. Thus all attached laminar flames are oriented obliquely to the flow, as indicated in Fig. 7-2 or 8-1. We see from these

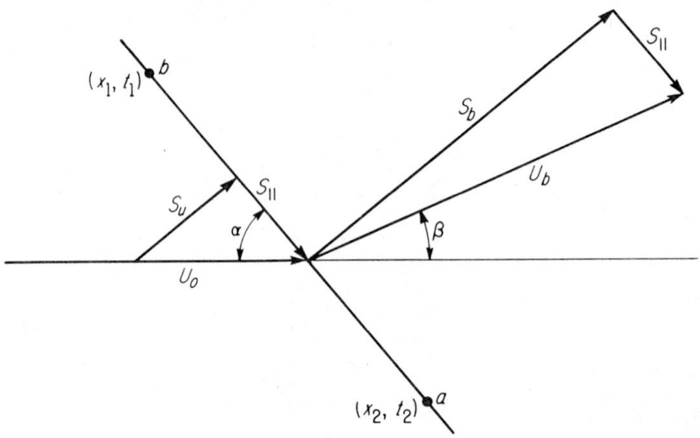

Fig. 8-1. Oblique flame orientation and flow-velocity relations if the flame is considered as an infinitely thin flat flame. (Compare to Fig. 7-2.)

figures that even though a flame of this local geometry will appear steady to an observer, an element of this flame is, in reality, propagating along the flame at a velocity $S_{\parallel}$. Thus, as shown in Fig. 8-1, the flame can exist only as an apparently steady flame at some time t_2 at the point a, if at some earlier time t_1 it appeared steady at a point b whose distance from point a (along the flame in the upstream direction) is given by the expression $x_a - x_b = (t_2 - t_1)S_{\parallel}$. This implies that all steady flames must have an attachment region which continually reignites the oblique flame sheet, and that attachment must itself be steady if the flame is to appear steady.

From the vector relations in Fig. 8-1 we note that any holding region must necessarily always be at the furthest upstream point of the steady flame, and we further see that at this point the local flow velocity vector and the local flame velocity vector must be equal in magnitude, coincident in direction, and of opposite sign. Furthermore, as we travel away from the attachment point into the bulk of the combustible mixture, the local flow velocity must always exceed the local burning velocity. If for any particular geometry and flow velocity these conditions are not met, the flame will either "flash back" i.e., travel upstream, "blow off" i.e., be washed downstream by the flow, or exist as some type of nonsteady repetitive phenomena. All three of these nonsteady behaviors have been observed.

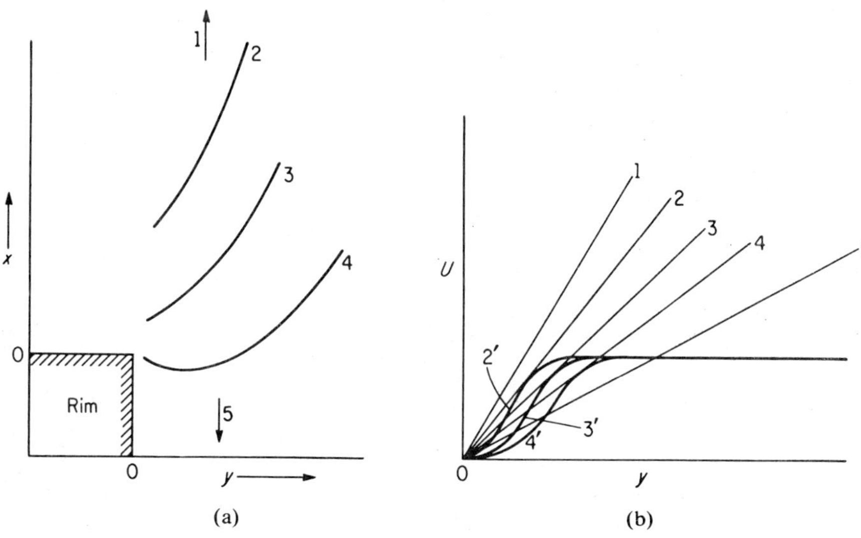

Fig. 8-2. Schematic diagram of holding behavior: (a) physical plane showing various stable flame positions, (b) U-y plane showing the relationship between the local flow and flame velocity.

The mechanism of steady attachment may be discussed by considering the flow in the neighborhood of an idealized burner rim as illustrated in Fig. 8-2. Experimentally, it is observed that laminar flame attachment usually occurs within approximately 1 mm of the rim lip. The attachment point is therefore in the viscous boundary layer region for these low subsonic flows. For simplicity we assume that the velocity gradient of the flow near the attachment point is a constant that is a monotonically increasing function of the bulk-flow velocity. Typical velocity profiles just inside the rim are plotted as the straight lines 1 through 5 in Fig. 8-2b. The normal component of the burning velocity for three arbitrarily chosen flame positions and quenching behaviors are also plotted as the curves 2′, 3′, 4′. The differences in these curves are due to both the presence of the cold wall and the effect of diffusive mixing with the surrounding gas after the combustible jet issues from the tube. These are typical burning velocity curves for the flame positions shown in Fig. 8-2a. Curve 4′ represents the burning velocity for a laminar flame which is positioned entirely inside the rim of the burner. This curve is drawn tangent to the straight-line curve 4 which, therefore, represents a critical velocity gradient at the rim for flashback. Note that a lower velocity gradient, such as 5, allows the local burning velocity to exceed the local flow velocity, and therefore, will allow flashback to occur in the neighborhood of the rim. The velocity gradient represented by curve 3 will cause a flame in position 4 to blow off, but this flame will be stabilized in a new position, because, above the burner, the effect of rim quenching is diminished before the effect of diffusion causes excessive dilution of the peripheral combustible gases. Thus as the flow velocity is increased the flame will "seat" at progressively higher positions. However, its local burning velocity will eventually fall below the local flow velocity at all points in the flow and blowoff will occur. This blowoff limit behavior is illustrated with curve 2′ in Fig. 8-2.

This rather simple conceptual approach predicts that blowoff and flashback should correlate well with the velocity gradient in the neighborhood of the holder. This correlation has been verified experimentally and Fig. 8-3 is a typical summary of flashback and blowoff behavior (in this case for a natural gas flame) plotted against the velocity gradient dU/dy with units of $\sec^{-1}$. Of these two critical behaviors, blowoff is the more complex. This is because at flashback the flame is always attached in the immediate neighborhood of the rim and its attachment is therefore controlled primarily by the nature of quenching at the rim or holding body. Blowoff, on the other hand, may occur when the flame is a considerable distance above the attachment body, and may therefore be controlled by other factors than simple quenching.

Two alternate processes that affect blowoff have been identified. These are (1) jet mixing at the edge of the jet, and (2) flame stretch. The first behavior occurs in those situations where jet-edge mixing produces a more nearly stoichiometric mixture such as the case of a fuel-rich jet exhausting into air. If the circumstances are right, the mixing process in this case can lead to attachment well outside the ordinary expected limits,

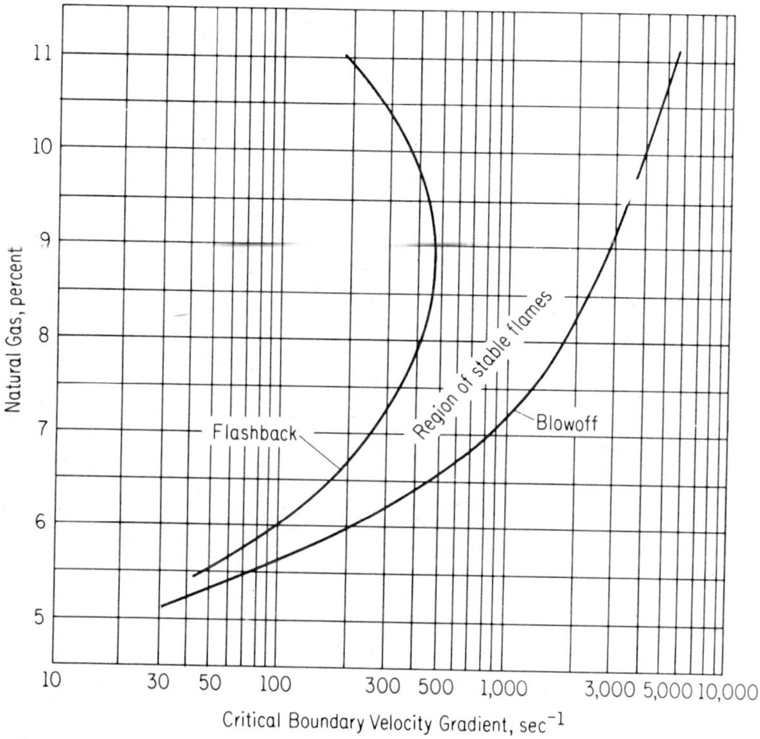

Fig. 8-3. Flashback and blowoff critical velocity gradient for a natural-gas air flame. (With permission from B. Lewis and G. von Elbe, *Combustion Flame and Explosions in Gases*, 2d ed., Academic, New York, 1961.)

and in fact can lead to stable lifted flames. These flames are always turbulent and quite complex aerodynamically. Their behavior will be discussed in Sec. 8-3.

The other mechanism, flame stretch, is observed in steady oblique flames when the approach flow contains a velocity gradient. This situation is shown schematically in Fig. 8-4 for a case of positive flame stretch. Since a flame is a wave, it can only propagate into the cold gas in a direction that is everywhere normal to its orientation. Thus, as we mentioned earlier, each element of such an oblique flame sheet is slipping along the sheet at the velocity $U_{\parallel}$. When the flow gradient of Fig. 8-4 exists ahead of the flame, $U_{\parallel}$ is not a constant with time and the frontal area of the flame increases as it propagates. We now define the Karlovitz number which was first introduced on page 238 as the logarithmic rate of increase of flame area with time, nondimensionalized by the characteristic time it takes the flame to burn through its preheat zone.

$$K = \frac{\partial(\ell n\, \Delta A)}{\partial t}\,\frac{\eta_o}{S_u}$$

This definition is compatible wiш the simple derivation on pages 237-238.

In the geometry of Fig. 8-4 an element of the flame front is slipping along the flame front at the velocity $U_{\|} = S_u \tan \delta$. Since the geometry is two-dimensional the rate of area increase may be derived as follows. Consider two points on the flame front which are separated by a distance $\Delta s = s_2 - s_1$.

The area of a unit of depth D of this flame section is

$$\Delta A = \Delta s D$$

and

$$\frac{\partial(\Delta A)}{\partial t} = \frac{\partial \Delta s}{\partial t} = \left((U_{\|})_o + \frac{\partial U_{\|}}{\partial s}\,\Delta s - (U_{\|})_o\right) D$$

Therefore in this case

$$\frac{\partial \ell n(\Delta A)}{\partial t} = \left(\frac{\partial U_{\|}}{\partial s}\right)$$

or

$$K = \frac{\eta_o}{S_u}\left(\frac{\partial U_{\|}}{\partial s}\right)$$

Since $U_{\|} = U \sin \delta$ we obtain

$$K = \frac{\partial U}{\partial y}\frac{\eta_o}{U} \sin \delta + \frac{\partial U}{\partial y}\frac{\eta_o}{U} \cot \delta \cos\delta \qquad (8.1a)$$

or

$$= \frac{\partial U}{\partial y}\frac{\eta_o}{U}\frac{1}{\sin \delta} \qquad (8.1b)$$

In another approach one can substitute $U_{\|} = S_u \tan \delta$ to obtain

$$K = \frac{\eta_o}{R} \sec^2 \delta$$

where R is the local radius of curvature of the flame. the first term on the right hand side of Eq. 8-1a is the result obtained by Karlovitz *et al.** However, we obtained Eq. 8-1b which, though markedly different for small δ, approaches the same limit value for large δ. The reason for this difference is that the original

*Ref 9, pg 238

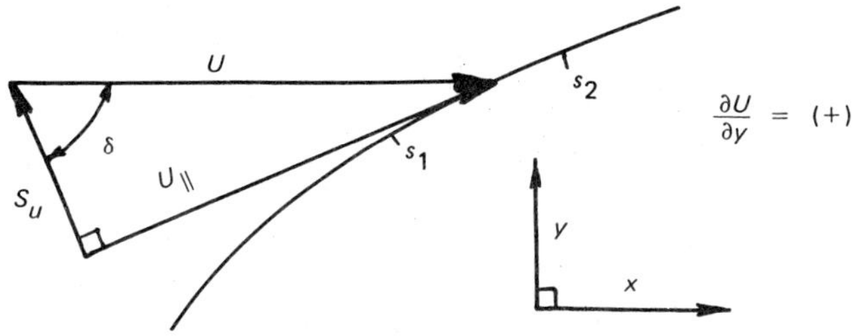

Fig. 8-4. Vector diagram for a stretched oblique flame in steady flow.

derivation does not recognize that the flame angle δ also changes as the oblique flame sheet slips along itself at the velocity $U_{\parallel}$.

At "stretch blowoff" the quantity K should be a constant of order unity if we consider this behavior to be analogous to the stretch criterion for extinction in the neighborhood of an ignition source (Chapter 7). Unfortunately, even though the stretch concept is valid at stretch "blowout" above a pilot flame, there has been no direct test of the concept because of the difficulty in evaluating the value of S_u and U in the neighborhood of the extinction point. We note that even though S_u was assumed to be constant in the above derivation, its value should be changing in the neighborhood of the hot rim gases, and this complicates the application of the simple theory.

In certain cases of buff-body holding in a free laminar stream nonsteady oscillatory flame holding is observed. This type of holding usually occurs at relatively low flow velocities and is characterized by flame oscillation between a forward and aft portion of the body. In a typical sequence, a flame seated near the aft end of the body will propagate toward the forward end through the developing boundary layer (from the previous cycle) and then increase its angle to the flow α until it can no longer maintain its forward position. The entire flame is then washed downstream. After some time the flame again repropagates up the newly established boundary layer and the cycle is repeated. This nonsteady attachment can lead to rather violent oscillations of the entire oblique flame sheet if the circumstances are correct. The effect of this process on the behavior of the flame will be discussed in the next section.

8-2. LAMINAR FLAME SHAPES

In this section we consider the aerodynamics of either an attached or detached flame sheet in a laminar approach flow. Detached flames will always be nonsteady, but an attached flame may be either steady or nonsteady depending upon the circumstances. For steady attached flames only curvature or quenching effects are important; however, if the attached flame is nonsteady we must differentiate between driven instabilities caused by the nonsteady holding behavior discussed above, and spontaneous instabilities triggered by the cellular behavior discussed in Chapter 7.

The only important detached flames in a laminar stream are those that occur when ignition is effected in a region devoid of flame holders and the stream velocity is sufficient to cause the entire flame to translate downstream as it propagates to encompass an ever-increasing volume of gas. Observations of these outwardly propagating flames in a predominantly uniform laminar stream show some interesting effects triggered by aerodynamic interactions. The two flame balls shown in Fig. 8-5 were both obtained in a laminar flow of about 20 ft/sec by passing a spark between two neighboring 0.001-in. tungsten wires placed mutually perpendicular to each other and to the flow direction. In this case the wires are sufficiently fine to prevent attachment and are also fine enough to yield a laminar wake. Figure 8-5a illustrates the large effect of this very weak

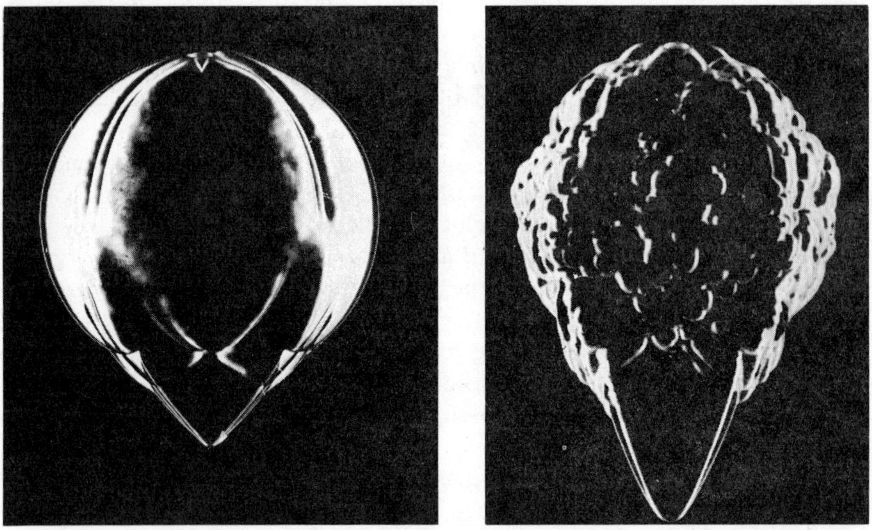

(a) (b)

Fig. 8-5. Free flame balls obtained in a low-speed laminar flow by spark ignition. (a) Lean butane air flame which is inherently stable. Note the surface features caused by the wake of the crossed ignition wires upstream in the flow. (b) Lean hydrogen–air flame which is inherently unstable. Note the extremely rough flame surface caused by spontaneous cellular instability and the very pronounced lower tip caused by the enhancement of the burning velocity at the point where the wake of the crossed wires in the upstream flow produces and initially low-radius-of-curvature section in the flame.

wake on the flame-front shape in a situation where cellular instabilities are not present. The presence of the wake in the flow has caused a deep depression in the upper surface of the flame ball and a corresponding extension in the lower portions of the flame ball, and the effect is most pronounced in the region where the wakes of both the wires are present. Figure 8-5b was obtained in an extremely lean hydrogen–air mixture which tends to propagate with a highly developed cellular structure. The very pronounced lower tip of this flame is in reality a simple example of the interaction of aerodynamic and flame-curvature effects in an inherently unstable flame system. In this case the strong curvature of the tip coupled with the extremely high preferential diffusion of the light hydrogen into the preheat region ahead of the tip yields a sufficient increase of the local burning velocity to cause the tip to dominate the flame shape throughout the flame-ball life. The smooth conical section in this flame is an oblique flame generated by this tip, and this oblique surface is seen to spontaneously develop cellular instabilities only in the region near the main flame ball. Thus this flame illustrates an enhancement of the burning velocity in a strong curvature region, and in addition shows the transient slow development of cellular instability on the initially smooth conical section generated by the tip motion.

Steady attached flames may be observed in a variety of laminar-flow situations. The multiple exposure of a series of Bunsen flames in Fig. 8-6 illustrates typical behavior for a flame stabilized on a cylindrical nozzle that produces a flow having a constant-velocity profile. This flame is burning at low pressure and is dominated by the tip and rim behavior. We have already discussed the attachment mechanism at the rim and now wish to discuss the aerodynamics of the tip. This is a region of negative stretch, in the Karlovitz sense, because the flame's radius of curvature is such that the stream-tube area at the start of the preheat zone is less than at the start of the reaction zone. Fristrom[1] has recently shown that in well-behaved (i.e., noncellular) flames, the flow velocity profile through the hot-reaction zone of the flame is not affected by flame curvature for curvatures ordinarily encountered at the flame tip. However, as he indicates, this is not true for the preheat region, since it is always much thicker than the hot-reaction region (see Fig. 7-21). Thus in the typical enclosed flame front as produced on a Bunsen burner, divergence of the central stream tube as it passes through the preheat zone causes the flow velocity to lower until it matches the burning velocity locally at the start of the reaction zone. For highly curved tips, however, one might also expect an enhancement of the burning velocity at the tip due to preferential preheating in its neighborhood. The effects of preferential diffusion are also

[1] R. M. Fristrom, *Phys. Fluids*, *8*, 273(1965).

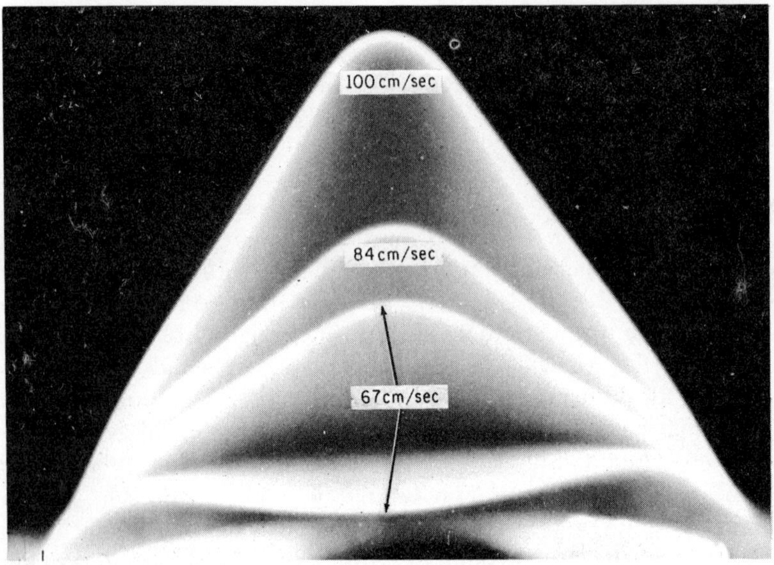

Fig. 8-6. Visible-light photograph of a low-pressure methane-air flame on a Mach-Hebra nozzle (constant velocity across duct) at various flow velocities, showing the effect of velocity on flame front shape. [Courtesy of R. M. Fristrom, Applied Physics Laboratory, Silver Spring, Md.; originally in *Phys. Fluids*, *8*, 273 (1965).]

evident in this region in certain cases. For example, in a rich benzene flame preferential diffusion of the oxygen out of the central region lowers the burning velocity sufficiently to cause the flame to open at the tip and allow a thin stream of unburned gas to escape.

In general, the problem of the interaction of aerodynamically induced curvature (as at a flame tip) and the processes of thermal conduction and preferential diffusion have not been studied in sufficient detail to allow a calculation of, say, flame-tip radius from basic principles and the downstream flame geometry. It has been observed however that, in regions of slight flame curvature, flame shapes may be quite well approximated by the simple subsonic aerodynamic construction in which the flame is assumed to be an infinitely thin region of heat release whose front locally obeys the equation $S_u = U \sin \alpha$, when U is the approach-flow velocity. If it is further assumed that there is no flame stretch (i.e., that each stream tube retains its one-dimensional geometry) the local-flow velocity and flow deflection downstream of the flame may be calculated using the following expressions: For the flow-deflection angle

$$\tan(\beta + \alpha) = \frac{\rho_u}{\rho_b} \tan \alpha \cong \frac{T_b \tan \alpha}{T_u}$$

which may be written explicitly for β as the equation

$$\beta \cong \tan^{-1}\left[\frac{\left(\dfrac{T_b}{T_u} - 1\right)\tan\alpha}{1 + \dfrac{T_b}{T_u}\tan^2\alpha}\right]$$

while for the flow velocity,

$$U_b = U\frac{\rho_u}{\rho_b}\frac{\sin\alpha}{\sin(\alpha+\beta)} \cong U\frac{T_b}{T_u}\frac{\sin\alpha}{\sin(\alpha+\beta)}$$

The quantities β and U_b are defined in Fig. 8-1. Unfortunately, the calculation of a steady-flame shape using these assumptions in no way insures that the flame will behave in a steady manner, and in general one must still resort to experiment to determine the flame shape and inherent stability in any specific environment. It is also true that a steady flame will distort the upstream velocity field in those cases where the flame does not completely enclose the upstream flow. Therefore, while the above approximations may be used to estimate flame shape for an enclosed flame, such as a Bunsen burner flame, distortion ahead of an open flame, such as a rod held flame, will negate any simple calculation of its steady shape.

If an open attached flame in a strictly laminar approach flow appears "unstable" or nonsteady, the cause can be related to either one or both of the two effects, cellular instability or nonsteady attachment as mentioned above. A quantitative assessment of the effect of these two sources of instability on the behavior of a particular unstable flame is usually quite difficult to make. However, a recent definitive study of this problem by Petersen and Emmons[2] has shed considerable light on the mechanism of instability growth, and the properties of these unstable flames. They studied V-flames in a laminar propane–air flow held on fine wires that could be *vibrated* normal to the flow direction with very small amplitudes at frequencies up to 1,700 cps. They were able to observe the effects of both natural instabilities as shown in Fig. 8-7a and "attachment" instabilities produced artificially by vibrating the wire, as shown in Fig. 8-7b. The instability patterns shown in Fig. 8-7 are traveling along the flame sheet away from the holder at approximately the velocity $S_{\parallel}$ which is defined in Fig. 8-1. Their work strikingly shows the importance of the concept that an element of the flame front always propagates normal to itself, and that flame-shape perturbations in the laminar stream are attached to the flame front and only secondarily affect or interact with the flow field. In addition, their results show that

1. Flame-front stability regions are intimately tied to the stoichi-

[2] R. E. Petersen and H. W. Emmons, *Phys. Fluids*, *4*, 456(1961).

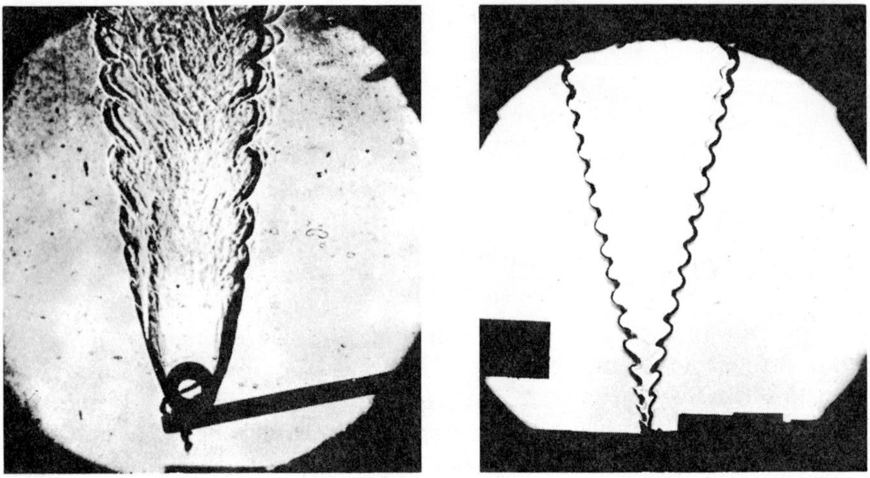

(a) (b)

Fig. 8-7. Instabilities on a flame sheet. (a) Natural instabilities in a system showing cellular instability. Note the symmetrical nature of the fluctuations. (b) Instability in same mixture caused by holder vibrations. Note that the fluctuations on the opposing sheets are out of phase. [Reproduced from H. E. Petersen and H. W. Emmons, *Phys. Fluids*, *4*, 456 (1961).]

ometry of this propane–air flame, and driven instabilities can be made to grow outside of the region of appearance of spontaneous instabilities. They also found stoichiometric ranges where the flame refused to exhibit gross unstable behavior even though the holder was oscillated at relatively high amplitudes.

2. Both the spontaneous and driven instabilities exhibit a characteristic minimum wavelength or cell spacing which, when coupled with the flame's oblique angle, determine the upper limit on the wire frequency that will yield a growing flame-front perturbation. These driven instabilities were observed to have a slightly shorter characteristic wavelength than the spontaneously occurring instabilities.

3. The initial rate of growth of the instabilities was found to be exponential until an amplitude of about 0.3 λ was reached, where λ is the wavelength of the disturbance. This initial growth period is followed by a negative rate of growth due to cutoff of the deeply extended cusps of the flame by ordinary propagation. However, distortion of the flow field ahead of the flame, coupled with preferential diffusion effects, was found to prevent the return of the flame to simple planar geometry, and once they started to grow the cylindrical disturbances were always observed to equilibrate at an amplitude of about 0.2 λ. Furthermore, these

"steady" cusped flames always had their cusps oriented downstream, i.e., as in the spontaneous cellular flames of Sec. 7-4.

The authors also summarize data which show that the preferred wavelength of the disturbance in a flame varies with the stoichiometry for any particular fuel–oxidizer mixture and is also dependent on the fuel type. They further point out that even in this simple geometry the two sides of the V-flame do not behave in a truly independent manner. Thus, as general as their results are, they are still quite definitely dependent on the gross flame geometry and cannot be easily extrapolated to other geometric situations.

We therefore see that an attached flame held in a laminar approach stream can exhibit steady behavior or a nonsteady cellular type of behavior. In addition, we find that gross aerodynamic interactions across the entire flame can sometimes determine the type of behavior that will occur. However, we also see quite clearly that all the observed flame-front instabilities are tied to the flame's previous surface shape by the requirement that each element of the flame front must propagate in a direction normal to its instantaneous orientation. We have thus incidentally introduced the concept of the "life" of an element of an attached flame as the time after leaving the attachment region for such an element to completely consume the gas available to itself through the normal propagation process. We note, in addition, that the overall effect of local flame-front curvature is quite complex, since it leads to the production of induced flow disturbances ahead of the flame, to preferential diffusion in the preheat zone of the flame and, possibly in certain circumstances, to "stretch" in the Karlovitz sense.

8-3. OPEN TURBULENT FLAMES

In this section we discuss the behavior of either free or attached flames burning above a port exhausting a turbulent flow of premixed gases. However, before coming to those flames that have been studied experimentally, we define the more important characteristics of both the isotropic turbulence that appears downstream of a mesh or grid placed in a uniform flow and the turbulent field which occurs in a fully developed turbulent pipe flow. We restrict ourselves to these two turbulent fields since they have been frequently used as approach flows to study the "turbulent burning" of open flames in the laboratory. The behavior of enclosed turbulent flames (i.e., flames burning inside a duct of chamber) will be discussed in the next two sections.

A turbulent flow is one whose instantaneous velocity at a point is fluctuating in a random manner about the mean stream velocity.[3] If we orient a coordinate system so that the flows mean velocity vector (of magnitude U = const.) lies in the $+x_1$ direction, the instantaneous velocity may be represented by the vector $(U + u_1, u_2, u_3)$ at any point of the flow with average values $\overline{u}_1 = \overline{u}_2 = \overline{u}_3 = 0$. Here a bar over any quantity represents the time average value of that quantity. With this notation, the intensity of a particular component of turbulence is defined as

$$I_x = \sqrt{\overline{u_1^2}}/U$$
$$I_y = \sqrt{\overline{u_2^2}}/U$$
$$I_z = \sqrt{\overline{u_3^2}}/U$$

and the turbulence is said to be isotropic if the average velocity U is a constant over any x_2, x_3 plane and

$$I_x = I_y = I_z = I$$

The relative frequency spectrum of the fluctuating components in the turbulent field is usually expressed in terms of the correlation of the fluctuating velocities measured at two points a known distance apart in the flow. While these correlation coefficients have relatively complex definitions and interrelationships in a nonisotropic turbulent flow, they reduce to a single definition of "turbulent scale" in an isotropic turbulent field. Consider the two correlation quantities around the point $(0, 0, 0)$ defined by the equations

$$g(x_2) = \frac{\overline{u_1(0) \cdot u_1(x_2)}}{I_x(0) \cdot I_x(x_2) \cdot U^2}; \quad x_1 = x_3 = 0$$
$$f(x_1) = \frac{\overline{u_1(0) \cdot u_1(x_1)}}{I_x(0) \cdot I_x(x_1) \cdot U^2}; \quad x_2 = x_3 = 0$$

when the bars again refer to time average values. Here both $g(x_2)$ and $f(x_1)$ approach unity as either x_2 or x_1 approach zero and for a random field $g(x_2)$ and $f(x_1)$ could be expected to approach zero as either x_2 or x_1 approach infinity. However, we note that these definitions allow for the occurrence of a negative correlation at some separation if either of the products $\overline{u_1(0) \cdot u_1(x_2)}$ or $\overline{u_1(0) \cdot u_1(x_1)}$ is negative at any x_1 or x_2 value.

The theory of isotropic turbulence (as summarized by Hinze, for example) may be used to show that $g(x_2) = g(x_3)$, and that for small

[3]Our discussion of turbulent flow and the definitions we introduce are based on the treatment of J. O. Hinze, *Turbulence*, McGraw-Hill, New York, 1962.

values of either x_2 or x_1 both $g(x_2)$ and $f(x_1)$ decrease with the co-ordinate separation in a manner given by the quadratic functions

$$g(x_2) \cong 1 - \frac{x_2^2}{\lambda_g^2}$$

and

$$f(x_1) \cong 1 - \frac{x_1^2}{\lambda_f^2}$$

where λ_g and λ_f are called the microscale and dissipative scale of isotropic turbulence. In addition, it may be shown that for the case of isotropic turbulence

$$\lambda_f = \sqrt{2}\,\lambda_g$$

Another scale, the macro or integral scale of turbulence, is defined with the equation

$$\Lambda_g = \int_0^\infty g(x_2)\,dx_2$$

where Λ_g represents the base length of a rectangle of unit height which has the same area as that under the $g(x_2)$ correlation curve. A schematic representation of a correlation coefficient $g(x_2)$ vs. x_2 and the defined correlation lengths λ_g, λ_f, and Λ_g are shown in Fig. 8-8.

With the use of these first- and second-order correlations we may adequately define the isotropic approach flow turbulence that appears at least 20-mesh diameters downstream of a screen or grid. In the case of fully developed pipe turbulence in a cylindrical pipe, it has been found that the turbulence is isotropic only along the centerline of the pipe. Away

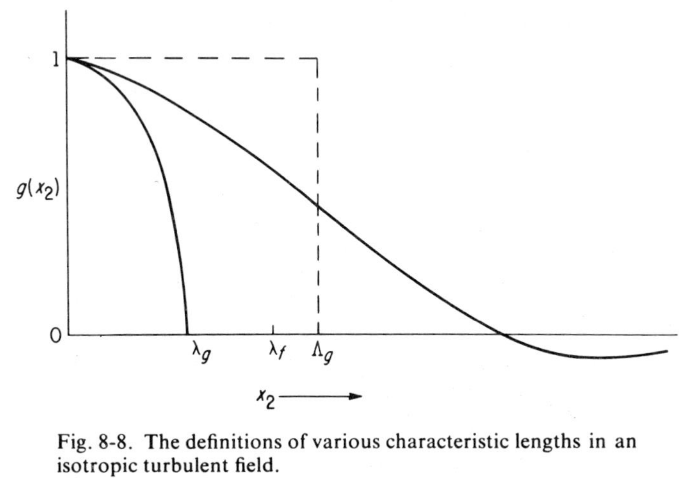

Fig. 8-8. The definitions of various characteristic lengths in an isotropic turbulent field.

from the centerline the turbulent intensity is found to increase to a maximum before decreasing to zero at the pipe wall. One additional complication in using a pipe flow for the study of open turbulent flames is the effect of the shear turbulence generated in the mixing zone along the jet edge. This usually influences the flow at the centerline about 5 diameters downstream of the pipe lip. Therefore, for a rim-held flame burning a jet of combustible mixture exhausting into quiescent air, one must distinguish between "core-zone" flames burning entirely inside the conical region of influence of the turbulent mixing region and "mixing-zone" flames which are burning in the region influenced by the shear-generated turbulence of this mixing zone.

We now define the "equilibrium isotropic turbulent burning velocity" of a flame in terms of the parameters defined above, and the spirit used in the definition of the laminar burning velocity. Consider a semi-infinite isotropic turbulent field of intensity I and scale λ_g flowing in the x_1 direction at a velocity U and continuously supporting a flame zone of infinite lateral extent lying in the neighborhood of $x_1 = 0$. If we find that we can define a value of $U = S_T$ at which such a flame zone exists as a time-average steady phenomena, we will say that we have defined the turbulent-burning velocity of the system. If such a flame velocity is definable (i.e., if such a flame truly exists) we should be able to determine the dependence of S_T on the initial mixture composition and the static properties of the flow (which allow one to define the laminar-burning velocity, S_u) as well as its dependence on the approach-flows turbulent intensity I and scale λ_g at $x_1 =$ some small negative value.

Unfortunately, even though this represents a valid definition of a turbulent-burning velocity, it is doubtful that such a definition can ever be used to interpret an experimentally determined "turbulent-burning velocity." The two main reasons for this are (1) most observed turbulent flames are, in reality, transitional flames, and (2) the behavior of a turbulent flame in most geometries is governed to a great extent by a strong coupling of neighboring flame sheets because of the highly subsonic nature of these flows.

Experimentally, we may observe two types of open turbulent flames, either attached or free. The attached flames that are usually observed are either simple "Bunsen burner" flames attached at the rim of the burner, or V-flames attached at a rod holder perpendicular to the flow across the port. The most readily obtainable free flame is the outwardly propagating "spherical" flame which may be obtained by spark ignition, using a set of thin crossed wires placed in the turbulent flow. This type of free flame is most useful for comparison to the theoretical model described above. However, there is no *a priori* reason to expect even this spherical flame to exist as a fully developed "equilibrium" turbulent flame immediately after its inception. In fact, high-speed photographs of flame balls taken in relatively low-intensity turbulent flows show that during the initial phase of flame-ball growth the flame appears as an ordinary laminar flame ball similar to that shown in Fig. 7-25. The flame then slowly relaxes until it contains a roughened or turbulent-flame surface as illustrated in Fig. 8-9.

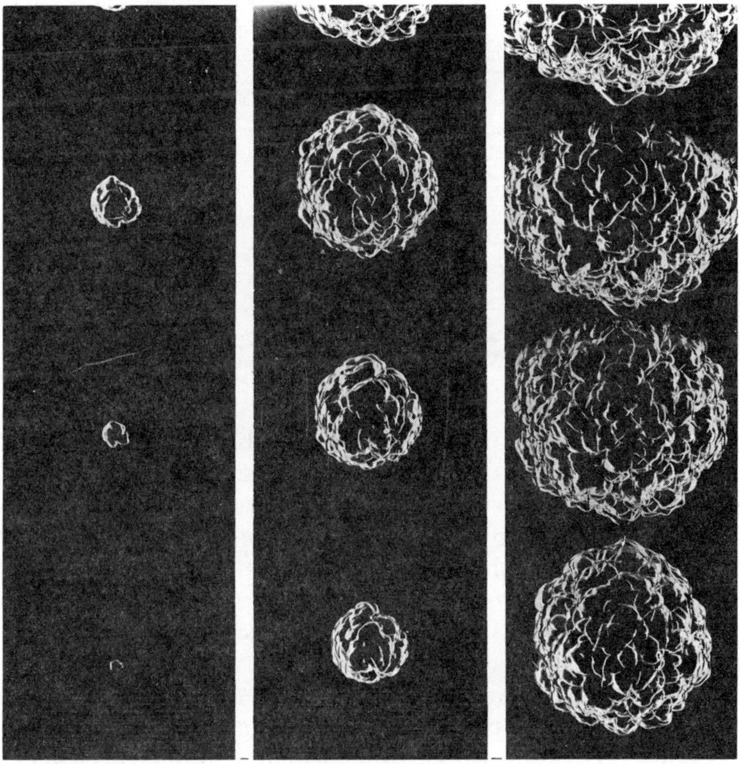

Fig. 8-9. Flame ball burning in an isotropic turbulent stream. Note that surface roughness increases with time but that characteristic roughness spacing or size is remarkably constant. Frame width = 5 in.; time between exposures = 1/300 sec.

Measurements of the turbulent burning velocity S_T, made by measuring the average local $(S_b)_T$ as a function of flame-ball radius and dividing this by the calculated isenthalpic expansion ratio yields the interesting results that S_T is increasing with the flame-ball radius over the entire experimental range. Figure 8-10 is such a plot for a series of stoichiometries in a propane–air mixture. The implication of these results is quite clear. A turbulent approach flow presents a continuous source of velocity perturbations

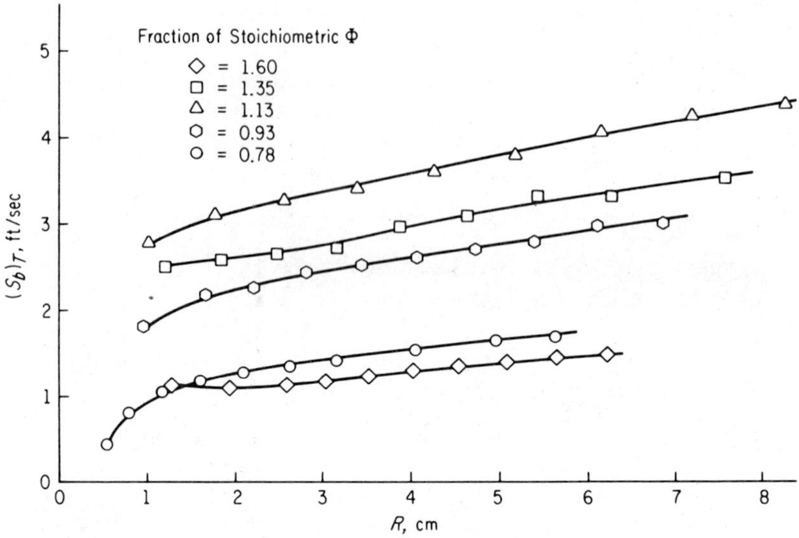

Fig. 8-10. $(S_b)_T$ vs. radius of a turbulent flame ball.

to the advancing flame front, and an equilibrium turbulent flame, as defined, can exist only if the flame front has had sufficient time to reach a relaxation equilibrium with these perturbations, if it is indeed capable of relaxing to an equilibrium state. Thus an important parameter in turbulent-burning studies should be the characteristic relaxation time of each "element" of the flame front for a known approach turbulence. Here an element is again defined as a section of the flame front propagating normal to itself into the flow. In determining the data of Fig. 8-10 the flame was observed for a travel distance of 6 in. This corresponds to a travel of 3 in. relative to the gas ahead of the flame, and thus the observation that this flame is still accelerating is consistent with the results of Petersen and Emmons, who observed that for a simple driven perturbation the flame required a considerable time to reach an equilibrium cusped perturbation shape. The current spherical flame results show therefore that the existence of a truly steady "turbulent burning velocity" is as yet unproved.

Unfortunately, this problem cannot be resolved by observing attached turbulent flames because (1) in these flames each flame element has a finite life which is dependent upon the value of $S_{\parallel}$ and the burner size, and (2) attached turbulent flames are all transitional since they are always held in a laminar holding region and therefore start their life as laminar flames. In addition, studies of turbulent-burning velocity using Bunsen or "closed" flames show that these flames exhibit a violent interaction at the tip which is not present at the downstream extremities of V-flames nor in the later stages of free spherical flames. These effects all conspire to make the measured value of the "turbulent-burning velocity" extremely apparatus dependent, no matter how carefully it is determined. In fact, the apparatus dependence of measured flame velocities is so striking that extrapolation from one experiment to another with a different geometric arrangement is essentially impossible at the present time. Also, because of the nature of the flame-relaxation process, attached flame velocities cannot be profitably compared to any of the current theories of turbulent propagation since unfortunately all the current theories are "equilibrium turbulent-burning velocity" theories and do not include the relaxation behavior. For this reason they are not summarized in this text.

Two additional properties of open turbulent flames should be mentioned. The first is that at relatively low turbulence levels the flame appears to be simply a highly wrinkled laminar flame, and the wrinkle size or characteristic surface roughness of virtually all these turbulent flames is sufficiently regular to merit comment. Note Fig. 8-10. It has been noted in fact that this characteristic size for the turbulent-flame structure is relatively independent of the scale of the approach turbulence and is most closely related to the characteristic size of the cellular structure found in unstable laminar flames. The reader will recall that this size was found to be directly proportional to the preheat-zone thickness of a laminar flame. This result reinforces our earlier-conclusions concerning the transitional nature of turbulent flames; for if the flame were truly transitional with a relatively long time constant its roughness should be related to the cellular instability size. If, however, the roughness was generated by the flame's instantaneous response to the turbulent flow, one might expect the turbulent-flame roughness to be characterized by the turbulent scale in some manner.

The above statements concerning flame roughness and surface features apply only to those cases where the flame is not violently perturbed by the presence of neighboring flame sheets. The other possible effect of turbulence, a bulk effect in the preheat zone for small-scale–high-intensity turbulence, would be expected to enhance the local "laminar" flame velocity. This effect has been postulated but not as yet proven. The reader is referred to the bibliography for current controversies concerning the behavior of open turbulent flames. The author feels that a considerable revision of approach is needed to describe turbulent combustion adequately. Possible parameters for this new description are (1) the flames relaxation rate, (2) the equilibrium turbulent flame velocity if such velocity exists, and (3) an aerodynamic interaction parameter which is dependent on gross flame geometry to describe the effect of the cross-coupling of neighboring flame sheets during the later portions of flame life.

The lifted flame represents another type of turbulent flame which is really a quite complex flame system. In this case, turbulent mixing of a fuel-rich stream exhausting into an oxidizing atmosphere will produce attachment, at some point above the burner rim, where the burning velocity is sufficiently high (because of dilution with the surrounding oxidizing stream) and the flow velocity is sufficiently low. As can be seen from Fig. 8-11, lifted flames may be stabilized over a considerable range of flow velocities and rich stoichiometries, indicating that this type of balance occurs over a considerable range of parameters. The most spectacular example of such a lifted flame of practical importance is the disastrous flame which seats, after ignition, above the port of an accidentally opened casing leading to a high-pressure gas well.

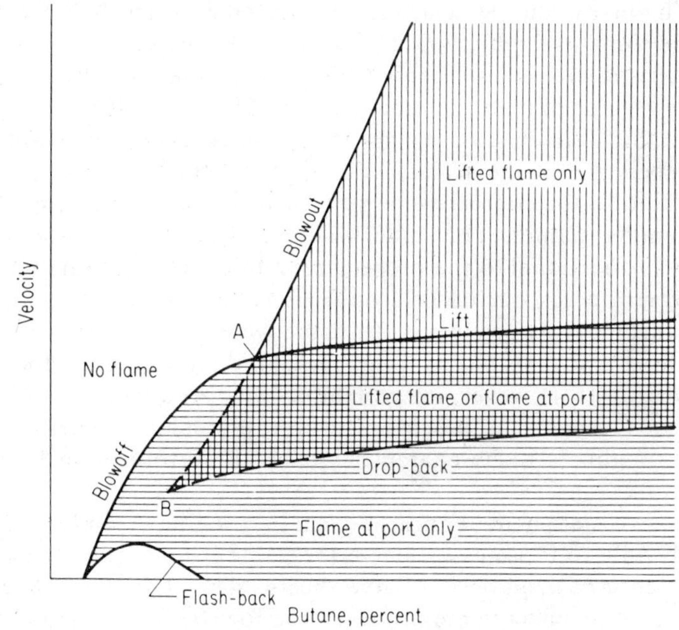

Fig. 8-11. Stability limits for attached and lifted flames for a port exhausting into the atmosphere. [With permission from K. Wohl, N. M. Kapp, and C. Gazley, *Third Symposium (International) on Combustion*, Williams and Wilkins, Baltimore, Md., 1949.]

8-4. STEADY ENCLOSED TURBULENT FLAMES

With the proper apparatus one may stabilize a flame inside a duct in a premixed gas flow which is either laminar or turbulent. For a steady laminar approach flow the same principles that apply to open flames may be used to discuss holding and/or flame shapes, with the added boundary condition imposed by the presence of the duct walls. However, such low-speed phenomena have little practical importance and therefore have not been studied extensively. On the other hand, flame holding and combustion behind a holder in a high-speed turbulent flow have extensive application in the design of various high-performance air-breathing jet engines, and there have been many fundamental studies of these processes in flow situations similar to those that occur in actual engines.

The flame holder in a high-speed (and incidentally turbulent) flow must be a buff body which is followed by a separated flow. Experiments have shown that when a flame is held on such a buff body the Kármán vortex street of shed eddies that commonly follows the body in a cold flow is replaced by a single recirculation zone. A typical example is shown in Fig. 8-12. The gas trapped in this recirculation zone is known

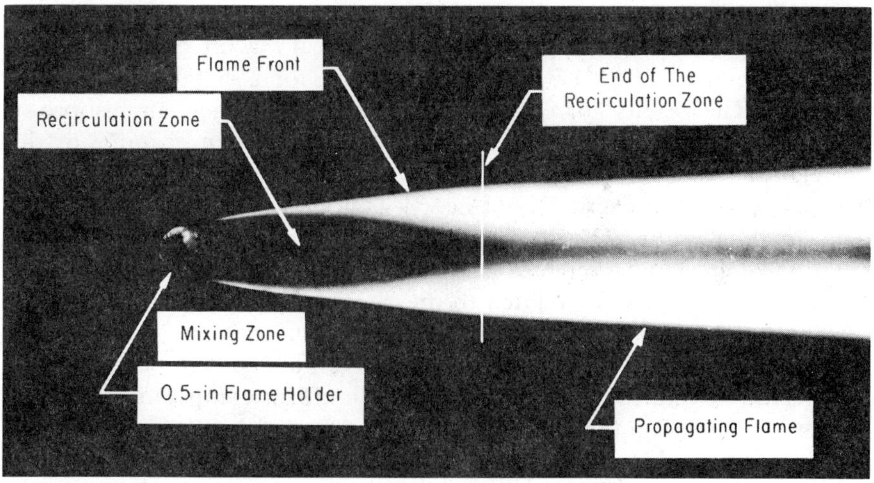

Fig. 8-12. Side view of rod-stabilized flame in gasoline-air mixture. (Reproduced from E. E. Zukoski and F. E. Marble, *Combustion Researches and Reviews, 1955* AGARD, Butterworth, London, 1955.)

to be quite hot (temperatures of approximately 0.9 of the theoretical flame temperature have been reported) and the flame is evidently reignited and held at the edge of this zone. Many parametric studies have been made to correlate the blowoff limits of such buff bodies to the aerodynamic and chemical effects that might influence the holding process and while these studies have shown that both chemistry and aerodynamics influence holding, at present they have not yielded any unambiguous and simple mechanism for this type of flame holding. Zukoski and Marble[4] have pointed out that for a large variety of low-blockage flame-holder shapes and sizes the residence time,

$$t = L/U$$

where L = length of the recirculation zone and U = mean stream velocity past the recirculation region, is a function of only fuel type and stoichiometry for simple buff-body holding. However, in practical applications the use of this correlation requires a knowledge of L at blowout, which is normally not available. Weiss, Roherer, and Longwell[5] have discovered an interesting correlation with the fuel's combustion chemistry by studying the relative lean blowout limits for a variety of fuels in a system in which all variables except the fuel type and concentration were fixed during the experiment. They determined the lean blowoff stoichiometric ratio for these fuels at the flow conditions which corresponded to blowoff for a $\Phi = 0.60$ isooctane mixture. They obtained a very good correlation by comparing the measured blowoff stoichiometry of a specific holder to the lean blowoff stoichiometry of a specific spherical reactor operating at the loading factor ($N/VP^{1.8}$) which gave blowout for a $\Phi = 0.60$ isooctane mixture. The results correlated well even for hydrogen and isopropyl chloride which have markedly different properties. An example of this correlation is shown in Fig. 8-13. Note that the correlation with burning velocity or ignition energy is not nearly as good as with the spherical reactor loading factor. The implication of these results is that the hot recirculation zone behind a buff body has a maximum holding efficiency when its composition is simply related to the composition inside a well-stirred reactor operating at maximum efficiency. The problem is not simple however, because, as many investigators have pointed out, the recirculation region behind the buff body is not of uniform composition but is a region of incomplete turbulent mixing. It is doubtful if this problem will ever yield to a truly fundamental analysis; direct experimentation on flame holders in the desired engine geometry will probably always be a part of the engine-design procedure. However, the interesting chemical correlation discussed above indicates that relatively inexpensive well-stirred reactor experiments may be useful for comparative fuel evaluation.

The study of turbulent flame propagation downstream of the holder in a high-speed turbulent duct flow is quite difficult, and even though rather extensive measurements have been made no really fundamental progress has been attained. The flame sheet in this case is propagating

[4] E. E. Zukoski and F. E. Marble, *Combustion Researches and Reviews, 1955*, AGARD, Butterworth, London, 1955.

[5] M. A. Weiss, J. C. Roherer, and J. P. Longwell, *Sixth Symposium (International) on Combustion*, The Combustion Institute, Pittsburgh, Pa., (1957).

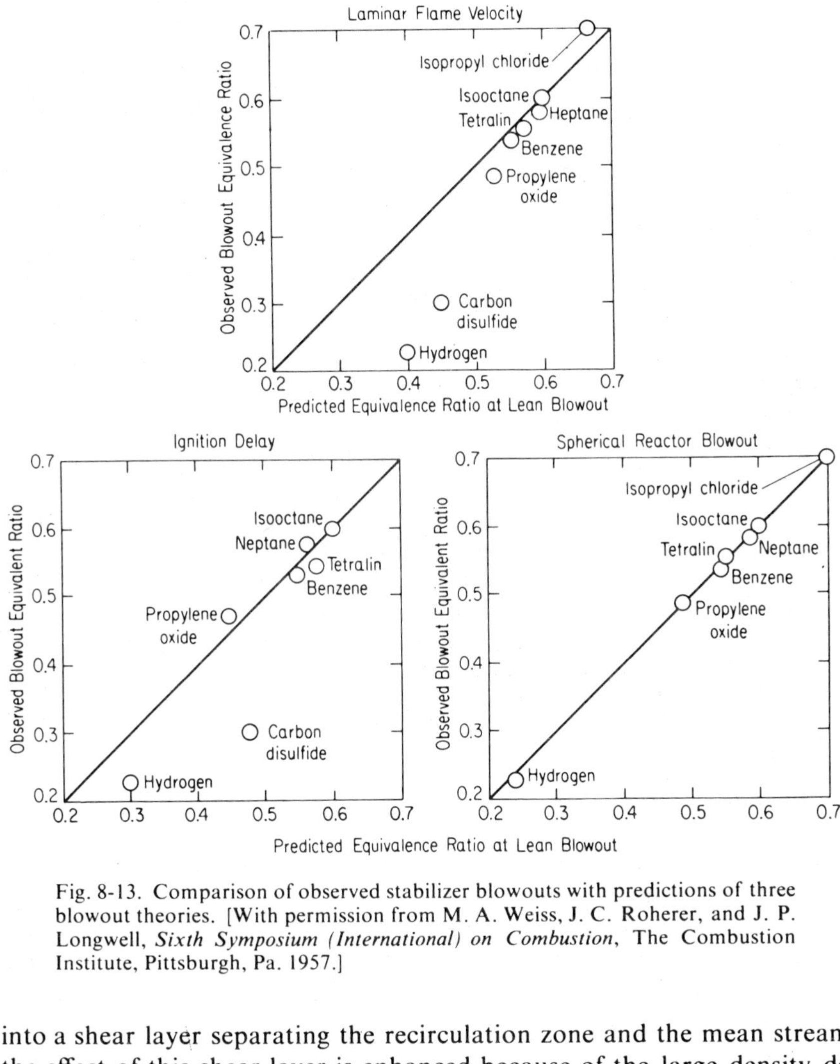

Fig. 8-13. Comparison of observed stabilizer blowouts with predictions of three blowout theories. [With permission from M. A. Weiss, J. C. Roherer, and J. P. Longwell, *Sixth Symposium (International) on Combustion*, The Combustion Institute, Pittsburgh, Pa. 1957.]

into a shear layer separating the recirculation zone and the mean stream; the effect of this shear layer is enhanced because of the large density decrease that occurs during combustion. Experimental observations have shown that, as a time average, the flame "fills" the duct some distance from the holder, even for small three-dimensional holders, and there is evidence that islands of unburned gas completely surrounded by an inwardly propagating flame exist during the later stages of burning. Measurements of "burning velocity" tend to be quite dependent on the local geometry, and theories for this type of propagation are still quite controversial. The reader is referred to the current literature for further discussion of the behavior of attached flames in ducts.

8-5. NONSTEADY PROPAGATION IN VESSELS

Transient propagation of a flame from a point ignition source through the entire contents of a closed vessel of arbitrary shape represents an interesting nonsteady propagation problem that has been studied rather extensively. By far the simplest closed vessel explosion is that following central ignition in a spherical bomb. In this case, if buoyancy may be neglected, gross aerodynamic effects are absent and the course of the entire process may be discussed in some detail. This simplification occurs for two reasons. First, at all times during its propagation the flame front is orientated strictly normal to the local-flow velocity vector and second, there is never any gas motion along the vessel wall and therefore never a boundary layer with which the flame may interact.

Lewis and von Elbe[6] have presented an extensive analysis of this system and have reached the following conclusions:

1. Throughout the propagation process the local flame may be assumed to be burning as though in a constant-pressure environment. This assumption is valid, since pressure changes due to the formation of hot products are relatively slow during the combustion of a gas layer of thickness equivalent to the preheat-zone thickness of the flame.
2. Because of spherical geometry the early burning takes place at essentially constant pressure (when the flame radius is one-half of the bomb radius, the hot gas ball occupies only one-eighth of the bomb's volume).
3. The early pressure rise is a cubic function of time.
4. Burning velocity is not constant during the entire propagation process. This is the result of adiabatic compression which significantly raises the temperature and pressure of the outer shell of gas before it is reached by the combustion wave.
5. The combustion process produces a gas sample which is significantly hotter at the center of the vessel. We note that the thermodynamic path for a sample at the center is different than that for a sample in the outer shell because the center sample is first burned irreversibly and then compressed isentropically, whereas the outer shell sample is first compressed isentropically and then converted to products by the irreversible combustion process. Thus the specific entropy at the center will be significantly higher than that near the wall, and temperature will be higher in the center of the bomb after combustion is complete. This result has been verified experimentally.

The above results mean that with sufficient information concerning the burning velocity and the thermodynamic properties of the gas one

[6] B. Lewis and G. von Elbe, *Combustion Flames and Explosions of Gases*, 2d ed., Academic, 1961, pp. 367–381.

may follow, in detail, the combustion process as it occurs in a spherical vessel with central ignition. This is not true for a vessel of arbitrary shape, however, because secondary aerodynamic effects, primarily produced by the expansion process associated with flame propagation, lead to complex flame behaviors which are not easily generalized.

There are however, two types of important secondary behaviors involved in these nonsteady processes. These are (1) boundary layer-flame interactions, and (2) the gross interaction of the flame with compression or expansion waves propagating in the system. Figure 8-14 illustrates typical behavior of flame propagation in a long rectangular tube in which ignition occurred at a spark source placed on the centerline very close to one end. We note that the apparent flame speed relative to an observer first increases slowly and then starts to decrease, but that near the end of the record it again accelerates. However, in the neighborhood of the last acceleration the flame appears to occupy a larger length of the tube and its previously smooth character in this (x,t) plot disappears. This flame is now propagating into a turbulent field produced by the gas motion ahead of the flame. That there is gas motion is evident from the presence of shock waves which appear as a series of black lines propagating away from the ignition source ahead of the flame. Thus transient flames ignited at the closed end of a long channel are characterized by the presence of a shock wave ahead of the flame, and after propagating some distance always become turbulent as they enter the turbulent-flow regime produced by this flow. Since the strength of a weak one-dimensional shock wave is proportional to the velocity of the piston producing it, and since for a system ignited at the closed end the effective piston velocity has a minimum value given by $S_b - S_u$, one might expect that the entire acceleration process is markedly dependent on the flame's laminar burning velocity— and indeed this is exactly what is observed.

The process discussed above is approximately a centered-wave process leaving the ignition source, and there are no major compressional waves traveling to the left in Fig. 8-14. If, however, the chamber were to be closed on the right, reflected waves would be produced which could then overwrite the rightwardly propagating combustion front. The primary result of such a process appears to be a gross interaction due to a Taylor instability of the surface (i.e., flame front) separating the light and heavy gases in the presence of an adverse acceleration field. This effect has been observed independently by passing a shock wave over a growing flame ball or even over a soap bubble filled with helium.

The result of this relatively pure interaction is shown in Fig. 8-15. The top surface of a hot (light) bubble of gas is unstable to the acceleration produced by the downwardly propagating shock in this photo, and the sphere very quickly becomes a torus. Eventually the upper flame surface

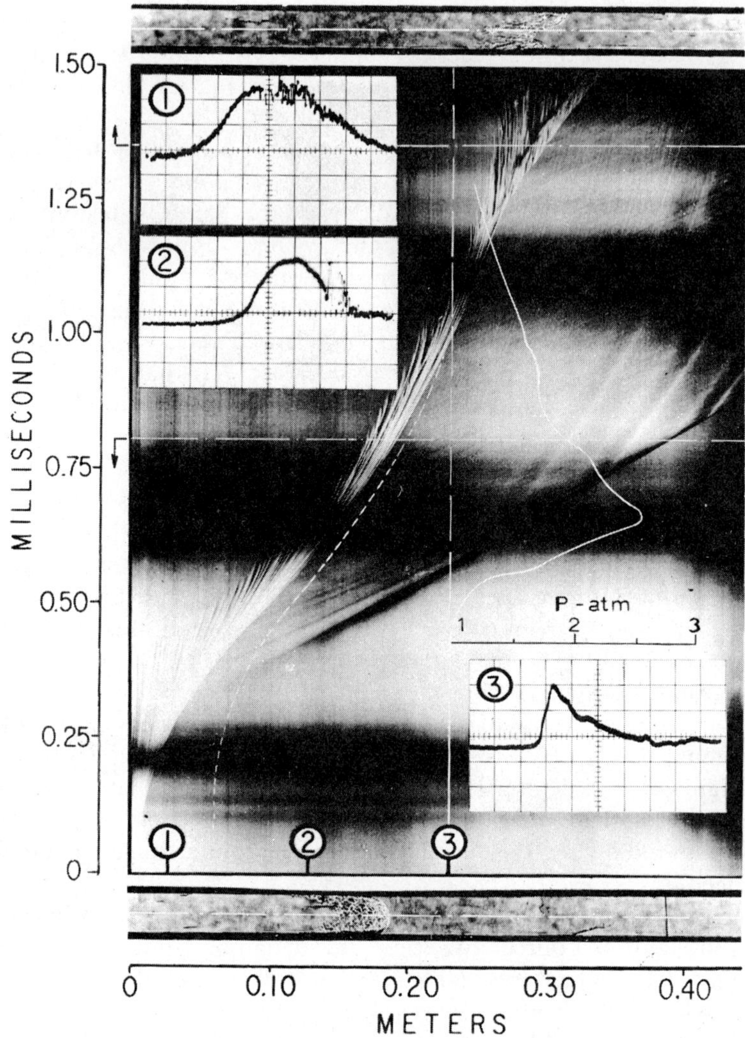

Fig. 8-14. Flame propagation from the closed end of a long rectangular duct. Note that the flame is becoming highly turbulent near the end of the record and that compression waves ahead of the flame are coalescing to produce shock waves. Pressure records taken from stations 1, 2, and 3 are shown as well as two pulsed laser photographs of the flame and shock wave development. (Courtesy A. K. Oppenheim, University of California, Berkley; photograph originally in paper presented at the OAR Research Applications Conference, 1966.)

Fig. 8-15. Interaction of a shock wave and a flame of initially roughly spherical shape. Pressure ratio of incident shock wave 1.3; stoichiometric butane–air mixture ignited at center of combustion chamber, 8.70 millisec before origin of the time scale. (Courtesy G. H. Markstein, Cornell Aeronautical Laboratory, Buffalo, N.Y.; originally in *Combustion and Propulsion, Third AGARD Colloquium*, p. 153, Pergamon, 1958.)

completely breaks through the lower flame surface to produce a highly extended turbulent-flame surface. That this mechanism is operative during transient propagation in a closed vessel may be seen from Fig. 8-16 which shows the flame position for equal time intervals after the start of explosion in a variable length cylindrical tube. Note that in the shortest

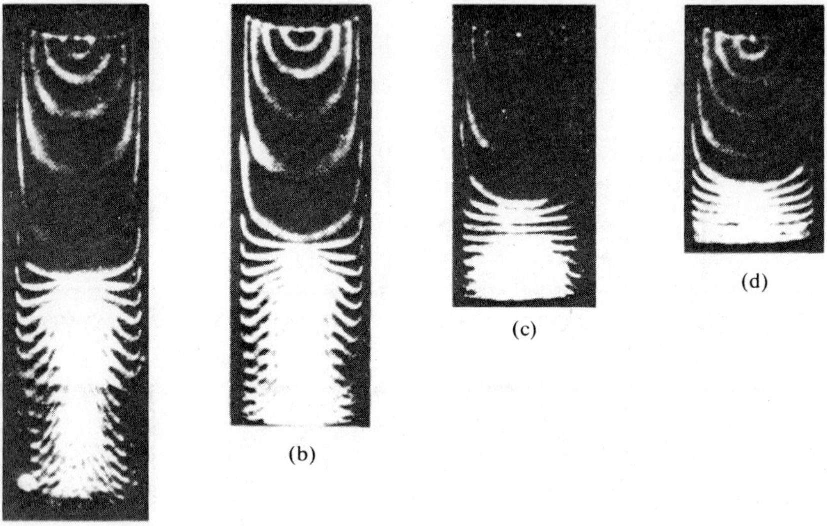

(a) (b) (c) (d)

Fig. 8-16. Stoboscopic flame records. Mixture: 10 parts CO + 1 part O_2, saturated with water vapor at 15°C. Tube closed at both ends; diameter 5 cm; length: (a) 19.5 cm; (b) 17 cm; (c) 12 cm; (d) 9.5 cm. Ignition at top center of each vessel. [From O.C. de C. Ellis, *Fuel and Science in Practice, 7,* 502 (1928).]

tube, where distinct compression waves do not have time to form, the flame is well behaved throughout the process. However, as the chamber becomes longer the central flame region becomes more and more inverted by this gross interaction as time progresses.

In long tubes there is also the possibility that the coupled-flame and flow-acceleration processes will eventually produce a detonation wave in the undisturbed gases ahead of the leading shock. The detonation initiation phase of this process is discussed in Chapter 9.

8-6. OSCILLATORY BEHAVIOR

In the study of combustion and the development of practical combustion devices, system instabilities frequently occur. These instabilities manifest themselves in a variety of ways. We have seen examples of cellular instabilities which appear spontaneously on the propagating laminar flame front, as well as examples of externally driven flame-front instabilities caused by the presence of a nonsteady holding region in the flow.

In addition to these "free" instabilities which occur in the open atmosphere there are many examples of enclosed-flame instability which arise when one attempts to carry out a combustion process inside an enclosure. These can be important to the design of combustion devices because the majority of practical power plants require that combustion take place in an enclosure, and because these instabilities usually produce large transient pressure fluctuations which may damage the vessel.

The primary cause of such combustion-chamber instability is the fact that in a combustion chamber the flame system is supplying energy to the flow at a very rapid rate, and the fact that if only a small portion of this energy is properly supplied to a stable oscillatory mode of the chamber the system may resonate in this mode. In a great many cases this mode is simply a particular acoustic mode of the chamber, but it need not be (see, for example, Chapter 10).

In this section we wish to develop the theory of acoustic instability in terms of a very simple model which yields a criterion for instability that is fundamental to the occurrence of all self-excited instabilities. In addition, a few applications of the theory to gas-phase systems will be discussed. For our simple example we consider the case of a nonsteady, one-dimensional flow in a tube of length L whose ends may be either open or closed. We follow the approach of Boa-Teh Chu.[7] The equations of motion for this case are obtained by writing Eqs. 5-1, 5-2, and 5-11 for the case of the one-dimensional nonsteady flow of a constant gamma gas with heat addition:

$$\frac{\partial \rho}{\partial t} + \rho \frac{\partial u}{\partial x} + u \frac{\partial \rho}{\partial x} = 0 \tag{8-2a}$$

$$\frac{\partial u}{\partial t} + u \frac{\partial u}{\partial x} + \frac{1}{\rho} \frac{\partial P}{\partial x} = 0 \tag{8-2b}$$

$$a^2 \left[\frac{\partial \rho}{\partial t} + u \frac{\partial \rho}{\partial x}\right] = \frac{\partial P}{\partial t} + u \frac{\partial P}{\partial x} - \rho(\gamma - 1)Q \tag{8-2c}$$

where the Q in Eq. 8-2c replaces the term $\partial q/\partial t$ as defined in Eq. 5-11. Q therefore has units of erg/gm sec. Since we are specifically interested in

[7]Boa-Teh Chu, "Stability of Systems Containing a Heat Source—The Rayleigh Criterion," NACA RM 56D27 (1956).

a cyclic process we may simplify these equations by using the following assumptions:

(a) No mean flow in the tube; i.e.,

$$\int_{\text{cycle}} u(x,t)dt = 0$$

(b) No net heat addition in the tube; i.e.,

$$\int_{\text{cycle}} Q(x,t)dt = 0$$

(c) The standard acoustic assumptions, i.e.,

$$\rho = \rho_0 + \delta\rho, \quad P = P_0 + \delta \mathrm{P}$$

where

$$\delta P/\gamma P_0 = s \ll 1 \text{ and } u/a_0 \ll 1$$

with

$$a_0^2 = \gamma P_0/\rho_0 = \text{const.}$$

We note here that the acoustic condensation s is defined in terms of pressure rather than density because the pressure and flow velocity are the important balance parameters along a particle path in the flow. This yields the equations

$$\frac{\partial(\delta\rho)}{\partial t} + \rho_0 \frac{\partial u}{\partial x} = 0 \tag{8-3}$$

$$\frac{\partial u}{\partial t} + \frac{1}{\rho_0}\frac{\partial(\delta P)}{\partial x} = 0 \tag{8-4}$$

$$\frac{\partial(\delta\rho)}{\partial t} = \frac{1}{a_0^2}\frac{\partial(\delta P)}{\partial t} - \frac{\rho_0}{a_0^2}(\gamma - 1)Q \tag{8-5}$$

Cross-differentiation of Eqs. 8-3 and 8-4 and substitution of 8-5 then yields the acoustic equation in terms of the dimensionless pressure:

$$\frac{\partial^2 s}{\partial t^2} - a_0^2 \frac{\partial^2 s}{\partial x^2} = \frac{\partial(Q/\mathfrak{C}_p T_0)}{\partial t} \tag{8-6}$$

where the right-hand side represents a forcing function. This equation may be solved for any arbitrary forcing function if initial conditions and boundary conditions are given for the specific problem.

The requisite boundary conditions for the case of one-dimensional flow in a tube of length L are

(a) at an open end $s = 0$ and $\partial s/\partial t = 0$ for all time

(b) at a closed end $u = 0$ and $\partial u/\partial t = 0$ for all time

Using Eq. 8-4, this last boundary condition is equivalent to the requirement that $\partial s/\partial x = 0$ at a closed end for all time. We note that if

$(Q/\mathcal{C}_p T_0)$ is harmonic in time with a period $4L/na_0$ where $n = 2, 4, 6, \ldots$, for both ends open or closed, or $n = 1, 3, 5, 7, \ldots$, for one end open and the other closed, the solution that we obtain will represent a simple acoustic mode of the tube whose amplitude is growing linearly with the time. This is the case of an externally forced oscillation at resonance and is not generally of interest in combustion systems.

Observable instabilities in combustion systems ordinarily arise through an internal feedback mechanism which is independent of any external forcing function. Mathematically, the simplest type of such feedback mechanism is that in which the forcing function is directly proportional to the perturbation quantity throughout the entire enclosure. This causes Eq. 8-6 to become homogeneous. A homogeneous differential equation also arises if the forcing function is assumed to be proportional to the first derivative of a perturbation quantity. However, we will assume only the simplest case. Therefore let us assume that

$$\frac{Q}{\mathcal{C}_p T_0} = ks \tag{8-7}$$

throughout the tube. With this assumption Eq. 8-6 becomes

$$\frac{\partial^2 s}{\partial t^2} - a_0^2 \frac{\partial^2 s}{\partial t^2} = k \frac{\partial s}{\partial t} \tag{8-8}$$

Equation 8-8 has a solution of the form

$$s = \mathrm{Re}\left\{\sum_{n=1}^{\infty} C_n \exp\left[n(\Lambda t + i\omega x)\right]\right\} \tag{8-9}$$

where C_n and Λ may be imaginary and ω is real. We choose this form because we note that Eq. 8-8 is separable and has a strictly harmonic solution in x. Substituting Eq. 8-9 into Eq. 8-8 for a particular value of n and solving for Λ yields

$$\Lambda = \frac{k}{2n} \pm \sqrt{\frac{k^2}{4n^2} - a_0^2\omega^2}$$

If we assume that $k/2n \ll a_0\omega$ (this is a good assumption for weak coupling) we obtain

$$\Lambda = \frac{k}{2n} \pm i\omega a_0$$

Therefore the solution is of the form

$$s = \exp(kt/2) \cdot \mathrm{Re}\left[\sum_{n=1}^{\infty} \{C_{n_+} \exp\left[i n \omega(x + a_0 t)\right] + C_{n-} \exp\left[i n \omega(x - a_0 t)\right]\}\right] \tag{8-10}$$

Without resorting to any further simplifications, we note that if k is a positive quantity this solution states that s will grow exponentially with time, and under this condition the system will be dynamically unstable to any small-amplitude perturbation. Thus a positive value of k for this assumed functional dependence of the heat-release rate represents a positive feedback mechanism for the system. Equation 8-10 may now be solved for any set of boundary and initial conditions. We note in particular that if we have a tube with both ends open, the real part of Eq. 8-10 becomes

$$s = \exp(kt/2) \sum_{n=1}^{\infty} \sin \frac{n\pi x}{L} \cdot \left(A_n \sin \frac{n\pi a_0 t}{L} + B_n \cos \frac{n\pi a_0 t}{L} \right) \qquad (8\text{-}11)$$

where the coefficients A_n and B_n are functions of the initial conditions. This solution is harmonic in x and quasi-harmonic in t, and it predicts the exponential growth (at the same rate) of all the standing organ-pipe modes of the tube. However, Eq. 8-11 may be applied only to the beginning phases of the growth process because only in these phases are the acoustic assumptions valid.

One of the assumptions used in deriving Eq. 8-11 makes the application of this equation to combustion-driven instabilities rather difficult. In an actual combustion system it is much more common to find the region of heat release (which is subject to perturbation) concentrated in a small region of the tube rather than distributed throughout the tube as is assumed in the derivation. We therefore approach the problem again, using a different model which is more closely related to physical reality than the homogeneous model discussed above. The result of this new analysis is a formal mathematical proof of the criterion for instability which Rayleigh first stated without proof in 1878.[8]

Consider a tube of length L with either open or closed ends containing a homogeneous heater in the region $h - \Delta x$ to $h + \Delta x$ as shown in Fig. 8-17. We assume as before that the acoustic approximations hold, that there is no mean flow, and that $\int_{\text{cycle}} Q dx = 0$ at the heater. We also assume that $2\Delta x \ll L$.

For this analysis we divide the tube into three control volumes as labeled in Fig. 8-17 and calculate the rate of change, with time, of the energy stored as acoustic energy in volumes 1 and 2 so that we may relate this to a fluctuating heat input at the heater through a set of compatible boundary conditions at the interfaces 1–3 and 2–3. We note from the

[8] J. W. S. Rayleigh, *Nature, 18*, 319 (1878).

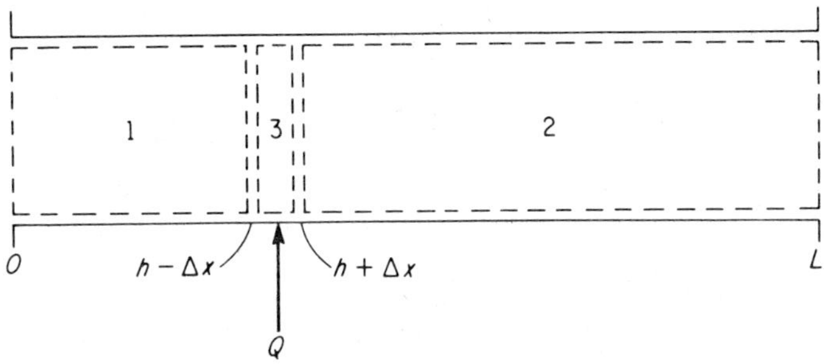

Fig. 8-17. Tube geometry for acoustic model. Ends (at O and L) may be on either open, as shown, or closed.

theorem of Gauss that

$$\int_A \phi_i n_i dA = \int_V \frac{\partial \phi_i}{\partial x_i} dV$$

and we apply this to the quantity Pu for one-dimensional flow occurring in the control volumes 1 and 2 for a tube of unit cross-sectional area. For the first control volume we obtain

$$(P_1 + \Delta P_1)_{h-\Delta x}(u_1 + \Delta u_1)_{h-\Delta x} - (P_1 + \delta P_1)_0(u_1 + \delta u_1)_0$$
$$= \int_0^{h-\Delta x} \frac{\partial}{\partial x}[(P_1 + \delta P_1)(u_1 + \delta u_1)]dx$$

Since P_1 = constant and $u_1 = 0$ by definition, this equation may be written as

$$(\delta P_1)_{h-\Delta x}(\delta u_1)_{h-\Delta x} - (\delta P_1)_0(\delta u_1)_0$$
$$= \int_0^{h-\Delta x}\left[\delta P_1 \frac{\partial}{\partial x}(\delta u_1) + \delta u_1 \frac{\partial}{\partial x}(\delta P_1)\right]dx \qquad (8\text{-}12)$$

However, for an open end $(\delta P_1)_0 = 0$ while for a closed end $(\delta u_1)_0 = 0$. Therefore, the second term on the left-hand side of Eq. 8-12 becomes zero for either of these cases. Furthermore, the coordinate derivatives under the integral on the right-hand side of Eq. 8-12 may be replaced by time derivatives using Eqs. 8-3 and 8-4 to yield the equation

$$(\delta P_1)_{h-\Delta x} \cdot (\delta u_1)_{h-\Delta x} = -\frac{\partial}{\partial t}\int_0^{h-\Delta x}\left(\frac{1}{2}\rho_1 a_1^2 s_1^2 + \frac{1}{2}\rho_1 u_1^2\right)dx \qquad (8\text{-}13a)$$

Here we have defined $s = \delta P_1/\gamma P_1 = \delta\rho_1/\rho_1$, since $Q = 0$ in this control volume. In a similar manner Gauss' theorem when applied to the second control volume yields the equation

$$(\delta P_2)_{h+\Delta x} \cdot (\delta u_2)_{h+\Delta x} = \frac{\partial}{\partial t}\int_{h+\Delta x}^{L}\left(\frac{1}{2}\rho_2 a_2^2 s_2^2 + \frac{1}{2}\rho_2 u_2^2\right)dx \qquad (8\text{-}13b)$$

The integrand in Eqs. 8-13a and b represents the total acoustic energy contained in these two control volumes at any instant of time. The second term under the integral represents the kinetic energy per cubic centimeter, while the first term is the internal or potential energy of the gas relative to the energy of a fluid element at rest at a pressure P_0 and a density ρ_0. To prove this we consider an elemental volume of fluid ΔV whose volume at $s = 0$ is ΔV_0 where $\Delta V = \Delta V_0(1 - s)$. The potential energy (P.E.) is defined as either the work which must be performed on the system to reversibly reach the desired state from the reference state or the work that the fluid element does on its surroundings when it is reversibly returned to the reference state. For our case this definition may be summarized with the equation

$$\text{P.E.} = \int_{\Delta V}^{\Delta V_0} \delta P d(\Delta V)$$

However, $d(\Delta V) = -\Delta V_0 ds$ and $\delta P = \rho_0 a_0^2 s$. Therefore

$$\text{P.E.} = -\Delta V_0 \int_s^0 \rho_0 a_0^2 s ds = \frac{\rho_0 a_0^2 s^2}{2} \Delta V_0$$

Thus in the framework of acoustic theory the total potential energy of the gas in a control volume is

$$\text{P.E.} = \int_V \frac{\rho_0 a_0^2 s^2}{2} dV$$

and Eq. 8-13 may be written as

$$(\delta P_2)_{h+\Delta x} \cdot (\delta u_2)_{h+\Delta x} - (\delta P_1)_{h-\Delta x} \cdot (\delta u_1)_{h-\Delta x} = \frac{\partial}{\partial t}(E) \qquad (8\text{-}14)$$

Where E is the total energy stored in the acoustic modes of the tube.

We now turn our attention to the third control volume, where the heater is located. We assume that $2\Delta x$ is sufficiently small so that $\delta P_2 = \delta P_1$ at any instant of time (i.e., the region has uniform static properties throughout) and we write the specifying equation for this region as

$$\frac{\partial(\delta\rho)}{\partial t} = \frac{1}{a^2}\frac{\partial(\delta P)}{\partial t} - \frac{\gamma - 1}{a^2} Q$$

This equation may be integrated over the control volume to yield

$$\int_{h-\Delta x}^{h+\Delta x} \left[\frac{a^2}{\gamma - 1}\frac{\partial(\delta\rho)}{\partial t} - \frac{1}{\gamma - 1}\frac{\partial(\delta P)}{\partial t}\right] dx = -\int_{h-\Delta x}^{h+\Delta x} \rho_0 Q\, dx = -\delta Q$$

where we now define δQ to be independent of Δx. This will allow us to reduce the extent of the heating region without varying the magnitude of the heat-addition term. Substituting from Eq. 8-3 we obtain

$$\int_{h-\Delta x}^{h+\Delta x} \left[\frac{a^2 \rho}{\gamma - 1}\frac{\partial u}{\partial x} - \frac{1}{\gamma - 1}\frac{\partial(\delta P)}{\partial t}\right] dx = -\delta Q \qquad (8\text{-}15)$$

(it should be noted that $\delta\rho/\rho_0 \neq s$ in this region because of the heat addition). Equation 8-15 may be integrated to yield

$$\frac{a^2}{\gamma - 1}\left[(\delta u_2)_{h+\Delta x} - (\delta u_1)_{h-\Delta x}\right] + \int_{h-\Delta x}^{h+\Delta x} \frac{1}{\gamma - 1}\frac{\partial(\delta P)}{\partial t}\,dx = \delta Q \qquad (8\text{-}16)$$

Combining Eqs. 8-14 and 8-16 and recalling that

$$(\delta P_1)_{h-\Delta x} = (\delta P_2)_{h+\Delta x} = \delta P$$

yields the equation

$$\frac{\partial}{\partial t}(E) = \delta P[(\delta u_2)_{h+\Delta x} - (\delta u_1)_{h-\Delta x}]$$

$$= \frac{\delta Q \delta P(\gamma - 1)}{a^2 \rho} - \delta P \int_{h-\Delta x}^{h+\Delta x} \frac{\gamma P_0}{\gamma - 1}\frac{\partial s}{\partial t}\,dx \qquad (8\text{-}17)$$

We now assume that $\Delta x \to 0$ and integrate Eq. 8-17 over a single cycle of a pure mode of the tube to obtain the equation

$$\Delta E_{\text{cycle}} = \frac{(\gamma - 1)}{\gamma P_0}\int_{\text{cycle}} \delta P \cdot \delta Q\,dt \qquad (8\text{-}18)$$

Equation 8-18 states that the local-energy addition or abstraction must be positively correlated with a naturally available pressure fluctuation in the system to produce a spontaneous amplification of that particular pressure fluctuation. While this theory was developed for an acoustic fluctuation, its application is not limited to acoustic-mode amplification. An example of a nonacoustic mode which is amplified by this mechanism is given in Sec. 10-4.

The criterion developed above has been found to be generally applicable to the discussion of combustion instabilities in a variety of systems. We note, however, that it places no limitation on the nature of the coupling which leads to positive correlation (i.e., to amplification). Thus while this theory may be used to discover if a particular coupling mechanism is viable in a system, it cannot be used to rule out all other possible types of coupling that may occur in the system. In other words, even though the theory cannot predict the stability of a real system it can be used to discuss possible mechanisms of coupling in systems which are known to be unstable.

We now apply Rayleigh's criterion, Eq. 8-18, to a particular case of coupled instability, the Rijke tube. In 1859 Rijke[9] discovered that an open vertical tube of length L containing a heated gauze at a position $L/4$ from the bottom spontaneously sounded an extremely loud tone. If the gauze is heated by an external burner he found that the tone becomes audible approximately one second after removal of the burner, and that

[9] P. L. Rijke, *Phil. Mag.*, *17*, 419 (1859).

after this time lag the amplitude of the acoustic vibration grows to a maximum and then decays. He also observed that an electrically heated gauze produced a continuous tone whose amplitude decreased slowly as the surrounding tube was warmed by the heater.

To discuss this system we consider an open vertical tube of length L containing a heated gauze mounted at a station a distance h from the lower end as shown in Fig. 8-18. We first determine the relationship between the condensation s and the local perturbation velocity u for a standing acoustic mode in this tube with no heat addition at the gauze. The exact form of this relationship is dependent on initial conditions. If we assume that we are interested only in standing modes of the tube and that $s = 0$ throughout the tube at $t = 0$, we obtain the relation

$$s = \sum_{n=1}^{m} A_n \sin \frac{n\pi x}{L} \sin \frac{n\pi a_0 t}{L} \tag{8-19}$$

for the first m modes of the tube by writing Eq. 8-11 with $k = 0$ and the boundary conditions for an open-ended tube. We may differentiate this equation with respect to x and use Eq. 8-4 to obtain the equation

$$\frac{\partial u}{\partial t} = -\frac{a_0^2 n\pi}{L} \sum_{n=1}^{m} A_n \cos \frac{n\pi x}{L} \sin \frac{n\pi a_0 t}{L}$$

This expression may be integrated to yield an equation for the flow velocity in the tube:

$$u = a_0 \sum_{n=1}^{m} A_n \cos \frac{n\pi x}{L} \cdot \cos \frac{n\pi a_0 t}{L} + B \tag{8-20}$$

Where the constant $B = 0$ since we assume that

$$\int_{\text{cycle}} u\, dt$$

is zero at every point in the tube. Equations 8-19 and 8-20 represent the coupled pressure and velocity perturbations that occur during a resonant oscillation in this tube. These equations may be applied to the Rijke tube if the drift velocity is low and if the average temperature rise of the gas sample due to heat transfer from the gauze is low. The behavior of these coupled oscillations is plotted in Fig. 8-18 at various times as a function of position and at two positions as a function of time for the fundamental mode of the tube.

We now return to Rayleigh's criterion, Eq. 8-18, noting in addition that heat can be transmitted to the gas from the gauze only by conductive or convective heat transfer, and that the convective heat-transfer rate is a function of the local flow velocity u. We see from Fig. 8-18 that if the rate of heat transfer Q is simply proportional to u, as it is for steady flow, there will be no net amplification for any heater position because

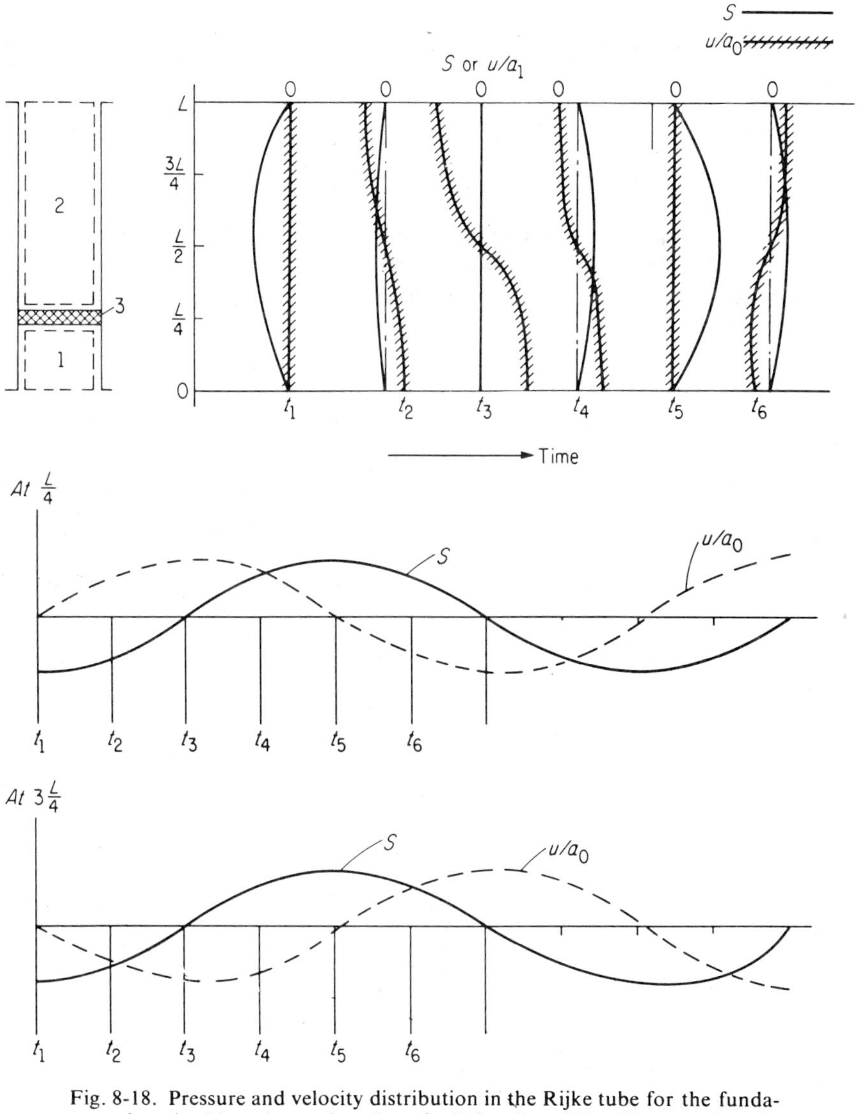

Fig. 8-18. Pressure and velocity distribution in the Rijke tube for the fundamental mode. Note that a phase lag of $\pi/2$ for Q at $L/4$ will cause Q and s to be positively correlated throughout the cycle.

there is no position in the tube which will yield the required positive correlation. However, it has been shown that heat transfer to a fluctuating flow from a transverse fine wire lags the flow velocity with a phase angle which approaches $\pi/2$ at reasonably high frequencies. As Fig. 8-18 shows, such a lag will cause a positive correlation in the fundamental mode if the heater is placed in the bottom quadrant of the tube. Figure 8-18 also predicts that a cooled gauze placed in the upper quadrant with a forced upward convection should cause the tube to sound. This second effect was first observed experimentally by Bosscha and Riesse[10] in 1859.

It has also been observed that a premixed gas flame seated on a gauze flame holder in the tube will cause the tube to sound if the flame holder's position is the same as that for the heated gauze. Here there is a marked temperature difference across the gauze, but the theory may be applied by simply writing the equations for the upper and lower portions separately and connecting them with compatible boundary conditions at the flame's location. A careful investigation by Bailey[11] of both this and the Rijke tube effect has been reported recently. In his work the frequency of the first tone to sound was determined as a function of heater position and compared to a rather detailed theoretical analysis which included system losses. Bailey assumed that the coupling for the flame case was caused by a perturbation of the flame's burning velocity due to its nearness to the gauze. The agreement between experiment and theory was found to be quite good.

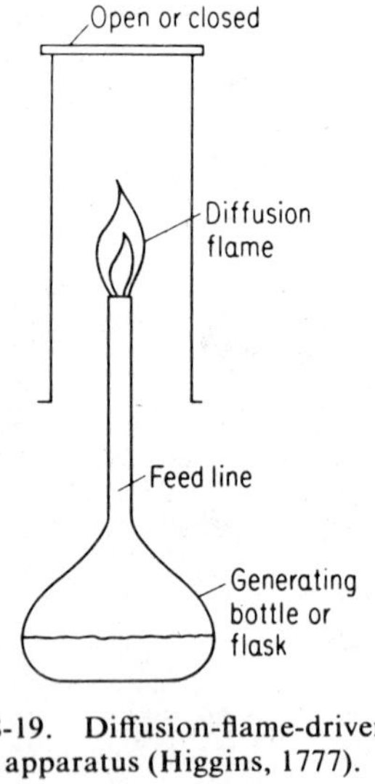

Fig. 8-19. Diffusion-flame-driven resonance apparatus (Higgins, 1777).

Other coupling mechanisms which cause instability behavior have been observed. One of the more interesting examples is the instability that may occur when a diffusion flame is located in a pipe as shown in Fig. 8-19. This resonance was first observed by Higgins[12] in 1777, using

[10] H. Bosscha and P. Riesse, *Pogg. Ann.*, *108*, 653 (1859).
[11] J. J. Bailey, *J. Appl. Mech.*, *24*, 333 (1957).
[12] B. Higgins, *Nicholson's Journal*, *1*, 130 (1802).

a hydrogen diffusion flame. In this case the coupling is with a resonant oscillation of the flow in the fuel feed line such that the size of the diffusion flame is caused to change with the proper phase relationship. Here the active coupling agent is pressure and the oscillation may be suppressed by simply changing the length and therefore the resonant frequency of either the feed line or the surrounding pipe. Kaskan[13] has observed and described another instability mechanism which requires that a flat flame propagating down a tube extend into the acoustic boundary layer before amplification is effected. This is undoubtedly the mechanism which causes tone production when a hydrogen flame flashes down a test tube after the flame has been ignited at the open end.

Solid-propellant, liquid-propellant, and detonation systems also show instabilities which may be related to Rayleigh's criterion. The discussion of these systems will be deferred to Chapters 9 and 10.

8-7. AERODYNAMICS OF DIFFUSION FLAMES

Aerodynamics plays an important part in the behavior of virtually all diffusion flames. In fact, as we saw earlier, it is quite difficult to design a system for the study of a diffusion flame in which aerodynamic effects are absent. On the other hand, in the majority of practical combustion systems, heat release is primarily effected with the use of a highly turbulent diffusion flame. Unfortunately at the present time there is a very large gap between the fundamental studies of diffusion-flame aerodynamics, which are always made in relatively simple systems, and empirical studies, which are concerned with the nature of combustion in an actual engine geometry. In fact, the application of true fundamentals to engine design has yet to be done, and the engine designer must rely almost entirely upon empirical test results to improve or alter existing engines or design a completely new engine.

Such phenomena as

(1) the nature of holding in a lifted diffusion flame
(2) the length of the diffusion flame on holders of various geometries
(3) the effect of swirl in the entering air on the flame's behavior
(4) the effect of recirculation on the nature of the flame
(5) the effect of geometry on the threshold stoichiometry for soot formation
(6) the effect of nozzle design on liquid-fuel combustion and flame holding

have all been extensively studied in an empirical manner. While the observed behavior of these systems has sometimes been summarized with

[13] W. E. Kaskan, *Fourth Combustion Symposium*, Williams & Wilkins, Baltimore, Md., 1953, p. 575.

empirical mathematical relationships, fundamental notions have yet to be applied extensively. The reader is referred to the literature for further information on this subject.

BIBLIOGRAPHY

1. G. Markstein (ed.), *Non-Steady Flame Propagation*, Agardograph No. 75. (New York: Macmillan, 1964.)

2. A. G. Gaydon and H. G. Wolfhard, *Flames.* (London: Chapman and Hall, 1960.)

3. B. Lewis and G. von Elbe, *Combustion Flames and Explosions of Gases.* (New York: Academic, 1963.)

4. K. I. Shchelkin and Ya. K. Troshin, *Gas Dynamics of Combustion.* (Baltimore: Mono Book, 1965.)

5. *The Combustion Symposium*, Vol. 3–10; Index in Vol. 10.

6. E. S. Shchetinkov, *Physics of the Combustion of Gases.* (Moscow, 1965.)

7. J. O. Hinze, *Turbulence.* (New York: McGraw-Hill, 1962.)

PROBLEMS

8-1. Consider a flame in which a fixed amount of heat Q is added to a constant-gamma gas. Calculate the entropy and temperature difference ΔT between the center and edge region for combustion in a centrally ignited spherical chamber. Determine the temperature difference for a few different values of Q and show that

$$\Delta T \rightarrow 0 \quad \text{as} \quad Q \rightarrow 0$$

8-2. Write the conservation equations for a steady flow stabilized as either a spherical or cylindrical flame surrounding either a point or line source or sink. Note that the flame is burning inward for a source and outward for a sink flow. Show that an oblique flame held in a gradient region cannot be transformed with a single velocity transformation to everywhere produce the purely cylindrical source or sink flame described above. In particular, show that this is true for a flame held in a linear velocity-gradient region.

8-3. Using the linear approach discussed in Sec. 8-6, discuss the preferential excitation of higher modes of the Rijke tube by the proper placement of the heater. Assume that $A_1 = A_2 = A_3 = \ldots = A_n = \ldots$ in Eqs. 8-19 and 8.20 and that the phase lag is exactly $\pi/2$. Discuss the effect of the phase lag being somewhat less than $\pi/2$ on your results.

9

Detonation

9-1. INTRODUCTION

The word *detonation* has been in common usage for at least 175 years. In earlier times it referred to the sudden decomposition of certain chemicals and mixtures with the production of considerable noise "like thunder," even though these materials were unconfined when they decomposed. For example, there is the report[1] that in 1793, a Dr. Wurzer was mixing $1\frac{1}{2}$ grains of three parts sodium chlorate and one part sulfur at Bonn in a mortar and pestle when "he obtained a detonation which rendered him deaf for several days." It was, in fact, well known at that time that detonation is the rapid decomposition of certain substances which occurs upon percussion or with frictional heating, and that these same substances can be decomposed quietly under the proper conditions.

At the present time the term *detonation* should, in the author's estimation, be applied only to processes where a shock-induced combustion wave is propagating through a reactive mixture or pure exothermic compound. The phenomenon of detonation is known to be similar for solids or liquids and for gaseous substances and we will therefore discuss only gas-phase detonation in detail. The nature of detonation for a condensed phase and its similarities to and differences from gas-phase detonation are summarized in Sec. 9-7.

Since a shock wave is always observed to precede the reaction front in a detonation, the propagation velocity of such a combustion wave is always supersonic relative to the undisturbed media. The first experimental evidence for this supersonic nature of gaseous detonation waves was obtained in 1881, when Berthelot and Vieille and independently Mallard and Le Chatelier observed supersonic combustion waves while studying flame propagation in tubes filled with gaseous combustible mixtures. A detailed study of this phenomenon by these investigators in France and by Dixon in England produced the interesting observation that the propagation velocity of this newly discovered wave was very high, and that it was remarkably constant for any mixture composition and initial pressure and temperature. Specifically, it was observed that

[1] *Nicholson's Journal*, *2*, 473 (1798).

the tube's geometry had little effect on the propagation velocity of a detonation wave. This, of course, contrasts very strikingly with the propagation behavior of ordinary combustion waves, in which the tube geometry is extremely important. It was also observed that detonation could be initiated in a tube from a spark or other low-energy ignition source after a highly variable and quite geometry-dependent delay time during which an accelerating flame propagated down the tube.

The observation of a new phenomenon as unusual as detonation led to a number of models and attempted explanations. In 1899 Chapman[2] essentially stated the one-dimensional flow requirements summarized in Chapter 5 and compared the minimum theoretical velocity with the experimental results of Dixon. In 1905 Jouguet,[3] in a long theoretical monograph, independently restated Chapman's rule that the tangency point of the Rayleigh line and Hugoniot curve should represent the proper end state for experimental detonations and therefore should allow a calculation of the detonation velocity from *thermodynamic* considerations only. As we have seen, this point is now called the upper Chapman-Jouguet (CJ) point in the P, V plane. It is shown in Fig. 5-3. Close quantitative agreement between the theory and experimentally measured velocities was not obtained immediately, however, because of the paucity of thermodynamic and chemical data. For example, the existence of atoms and radicals such as H, O, N, and OH at equilibrium was not suspected and, in fact, the variation of heat capacity with temperature was only known qualitatively at the time of the first calculations. Chapman even suggested that detonation studies might be useful for the determination of high-temperature heat capacities.

We now know that the CJ criterion when applied to detonations that are well-removed from limit behavior predicts velocities which agree with the experimentally measured propagation velocities within a few percent. A comparison of a few recently measured and calculated detonation velocities are given in Table 9-1.

TABLE 9-1.
DETONATION VELOCITIES (m/sec)

System	Calculated	Measured
80% H_2 20% O_2	3,427	3,390
66.7% H_2 33.3% O_2	2,852	2,825
25% H_2 75% O_2	1,747	1,763
$CH_4 + O_2$	2,637	2,528
$CH_4 + {}^3\!/_2 O_2$	2,531	2,470
41.2% C_2N_2 + 58.8% O_2	2,528	2,540
50.5% C_2N_2 + 49.5% O_2	2,762	2,768

[2] D. L. Chapman, *Phil. Mag.*, *213*, Series 5, *47*, 90 (1899).
[3] E. Jouguet, *J. Pure Appl. Math.*, *70*, Series 6, *1*, 347 (1905); *2*, 1 (1906).

The theoretical justification for the stability of CJ flow is, in general, lacking even though it predicts correct velocities, and the local flow conditions near the shock front are now known to be far from one-dimensional in any self-sustaining detonation. Interestingly enough, the first evidence that detonations might prefer to travel with a configuration which is locally three-dimensional and nonsteady was obtained in 1926 when Campbell and Woodhead[4] first observed spin in limit mixtures in circular tubes. At the time of its discovery (and for at least 25 years thereafter), spinning detonation was thought to be a phenomenon peculiar to limit mixtures. A typical open-shutter photograph of a spinning detonation in an acetylene–oxygen mixture is shown in Fig. 9-1. Its detailed structure will be discussed in Sec. 9-3.

Fig. 9-1. Open-shutter photograph of a single-spin detonation in an acetylene-oxygen mixture diluted with argon. Photograph was taken through a 1-in.-diameter glass tube. [Courtesy G. L. Schott, Los Alamos Scientific Laboratory, Los Alamos, N.M.; originally appeared in *Phys. Fluids*, *8*, 850 (1965).]

The modern era of detonation research was opened in the years 1940–1945 when Zel'dovich,[5] von Neumann,[6] and Döring[7] each independently formulated essentially the same one-dimensional model for the structure of detonation waves. The true three-dimensional nature of all self-sustaining detonations was not suspected until the late 1950's, however, and a reasonable model for detonation initiation did not appear until the early 1960's. In the next sections we summarize these modern results.

9-2. STRUCTURE OF A ONE-DIMENSIONAL STEADY DETONATION WAVE

To discuss the internal structure of a one-dimensional detonation wave we further manipulate the equations derived in Chapter 5 for the

[4]C. Campbell and D. W. Woodhead, *J. Chem. Soc.*, (1926), p. 3010; (1927), p. 1572.
[5]Y. B. Zel'dovich, *J. Exptl. Theor. Phys.* (USSR), *10*, 542 (1940).
[6]J. von Neumann, O.S.R.D. Rep. No. 549 (1942), Ballistic Research Laboratory File No. X-122, Aberdeen Proving Ground, Md.
[7]W. Döring, *Ann. Physik.*, *43*, 421 (1943).

constant-gamma heat-addition model. We first rewrite the Hugoniot equation (5-23) in terms of a dimensionless pressure ratio $Y = P_2/P_1$ and volume ratio $\epsilon = V_2/V_1$, noting that $a_1^2 = \gamma P_1 V_1$. This yields the equation

$$\frac{\gamma}{\gamma - 1}(Y\epsilon - 1) - \frac{\gamma q}{a_1^2} = \frac{1}{2}(Y - 1)(\epsilon + 1) \tag{9-1}$$

where q is the heat added between stations one and two in erg/gm. The Rayleigh line is given by rewriting Eq. 5-25a to yield

$$M^2 = \frac{1}{\gamma}\frac{(Y - 1)}{(1 - \epsilon)} \tag{9-2}$$

Here we have dropped the subscript on M_1, since in this chapter we will always refer to M as the approach-flow Mach number. The pressure ratio Y may be eliminated from Eqs. 9-1 and 9-2 and the resulting relationship rearranged to obtain the following explicit relation for the volumetric (or reciprocal density) ratio between stations 1 and 2:

$$\epsilon = \frac{1 + \gamma M^2 \pm \sqrt{(M^2 - 1)^2 - 2(\gamma + 1)(\gamma - 1)M^2 q/a_1^2}}{(\gamma + 1)M^2} \tag{9-3}$$

We consider only flows with $M > 1$ and observe that there are either 2, 1, or no possible end states for each value of M, and that the number of such end states depends on the magnitude of q, since q occurs under the radical sign in the expression. The condition for only one end state may be determined by setting the radical equal to zero to obtain the equation

$$\frac{(\gamma - 1)q_{max}}{a_1^2} = \frac{(M^2 - 1)^2}{2(\gamma + 1)M^2} \tag{9-4}$$

This equation allows one to calculate the maximum amount of heat, q_{max}, which may be added to the flow for any specific incoming Mach number. It represents choked flow heat addition for a one-dimensional steady flow and we may therefore define $q_{max} = q_{CJ}$. We now wish to examine flows for which $q \leq q_{max}$ and thereby determine the nature of possible intermediate flow states for various heat addition models. We therefore define a heat-addition parameter $\phi = q/q_{max}$, where $0 < \phi < 1$ for any steady exothermic flow. With this definition, Eq. 9-3 reduces to the expression

$$\epsilon = \frac{(1 + \gamma M^2) \pm (M^2 - 1)\sqrt{1 - \phi}}{(\gamma + 1)M^2} \tag{9-5}$$

We note from Eq. 9-5 that for $\phi = 0$ the plus-sign solution yields $\epsilon = 1$ (i.e., the trivial case of no change between the stations) and the minus-sign solution yields the expression

$$\epsilon = \frac{2 + (\gamma - 1)M^2}{(\gamma + 1)M^2}$$

which is the equation for the volumetric change across a nonreactive shock in a constant-gamma gas.

These equations, relating heat addition to the flow properties, were first applied to the discussion of one-dimensional detonation structure by Zel'dovich, von Neumann, and Döring in the period 1940–1945. In its simplest form, their model (the ZND model) considers a steady one-dimensional detonation to consist of a nonreactive shock discontinuity followed by a region of heat addition (due to chemical reaction) which continues until the flow becomes sonic at $\phi = 1$ for a CJ detonation or until ϕ reaches a value of

$$\phi_{\max} < 1$$

for an overdriven detonation wave. The question of accessibility of various flow regimes in the ZND model for cases of unusual heat-addition processes has been discussed rather extensively, but we will content ourselves here with representing the detonation as a $\phi = 0$ (adiabatic) shock transition followed by a flow in which ϕ increases monotonically with time until a value of $\phi_{\max} \leq 1$ is reached. Thus if we know a value of ϕ at some plane in this one-dimensional wave we will be able to calculate all the properties of the flow at that plane. The density may be obtained from Eq. 9-5 (using the negative sign on the radical) and the pressure from the Rayleigh equation. Using these, we may also calculate the temperature ratio $\delta = Y\epsilon$, the velocity of sound ratio $\mathfrak{A} = \sqrt{\delta}$ and the dimensionless flow velocity $\mathfrak{U} = M\epsilon$. The actual physical structure of this one-dimensional detonation is not given by simply listing these properties, however, but may now be obtained by noting that at any x position in this steady reactive flow

$$\frac{\partial \phi}{\partial x} = \frac{1}{u q_{\mathrm{CJ}}} \cdot \frac{Dq}{Dt} \tag{9-6a}$$

where $u = \mathfrak{U}\, a_1$ and the quantity Dq/Dt is the local rate of heat addition at the plane of interest. The actual structure therefore depends on the reaction kinetic model assumed for the gas behind the shock front, and one may construct any appropriate detonation structure in the framework of the constant-gamma heat-addition model by using these equations and various simple or complex reaction schemes. Ordinarily the structure is built by starting the calculation at the shock front, where $\phi = 0$ and integrating to the plane ϕ by determining the appropriate upper limit on the integral in the equation

$$\phi = \int_0^x \frac{1}{u q_{\mathrm{CJ}}} \cdot \frac{Dq}{Dt}\, dx \tag{9-6b}$$

Numerical integration is usually required to determine $\phi(x)$ from this equation. The structure of such a typical detonation wave is shown in Fig. 9-2 for a particular heat-release model.

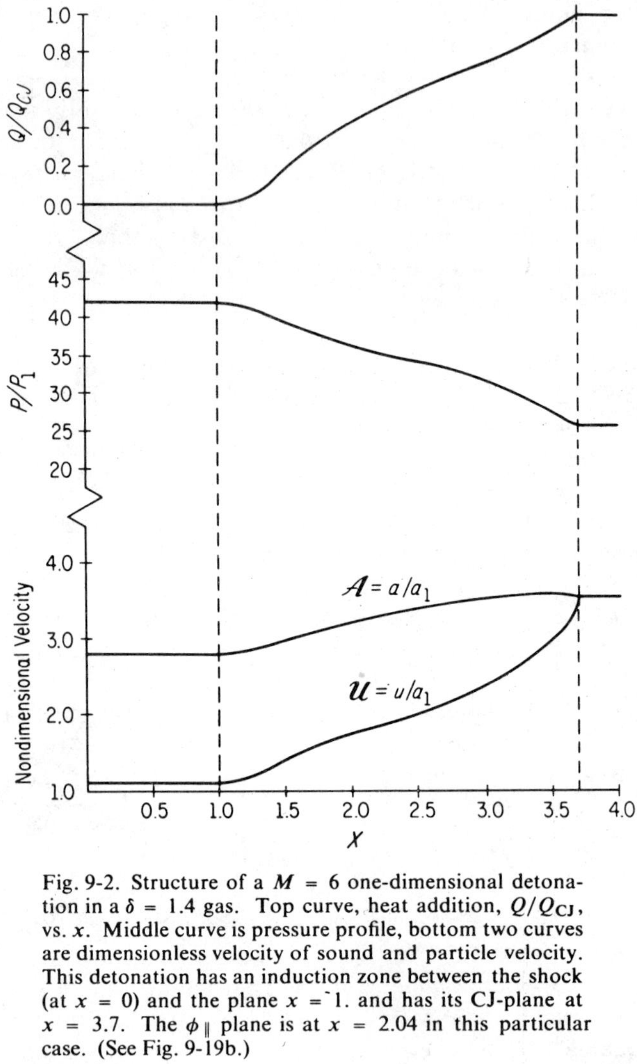

Fig. 9-2. Structure of a $M = 6$ one-dimensional detonation in a $\delta = 1.4$ gas. Top curve, heat addition, Q/Q_{CJ}, vs. x. Middle curve is pressure profile, bottom two curves are dimensionless velocity of sound and particle velocity. This detonation has an induction zone between the shock (at $x = 0$) and the plane $x = 1$. and has its CJ-plane at $x = 3.7$. The $\phi_{\parallel}$ plane is at $x = 2.04$ in this particular case. (See Fig. 9-19b.)

This one-dimensional steady approach is not limited to the case of heat addition in a constant-gamma, constant-molecular-weight fluid. In fact, it may be easily extended to the general case of a one-dimensional steady wave in a real gas with realistic caloric and thermal equations of state, realistic transport properties, and complex kinetics. In this case one first calculates a CJ detonation velocity by simultaneously solving the Rayleigh and Hugoniot equations for tangency, assuming thermodynamic and chemical equilibrium at the CJ plane. This yields the lowest possible value for the steady detonation velocity. A knowledge of this velocity allows one to construct a shock-wave reaction-zone structure of any desired complexity by forward integration from the cold flow. In this type of calculation, one need not even assume a shock discontinuity but may integrate through the shock wave with the transport equations. One may thus produce a steady one-dimensional structure whose complexity is dependent on the complexity of the starting assumptions. One may also construct an overdriven wave using this technique by simply assuming that the cold gas flow velocity is above the calculated CJ detonation velocity of the mixture.

In conclusion, we note two important properties of our simple ZND model. First, as shown in Fig. 9-2, the model predicts a pressure maximum at the shock transition. This maximum is commonly called the *von Neumann spike*, and its prediction by the ZND model stimulated a considerable amount of experimental work on structure in the 1950's. Second, we observe that the ZND model is strictly a steady one-dimensional model and therefore cannot be used to discuss the question of stability. For this purpose the original equations must be modified to include time-dependent terms. However, we defer our discussion of detonation stability until we describe the experimental results concerning the actual three-dimensional nonsteady nature of self-sustaining and overdriven detonation waves.

9-3. THREE-DIMENSIONAL STRUCTURE

As mentioned eariler, spin was discovered in 1926 and was thought to be a property of limit mixtures. However, it is now known that all self-sustaining detonation waves exhibit significant three-dimensional structure in a region which extends from the leading shock front through the reaction zone of the detonation. This structure is characterized by a nonplanar leading shock wave which at every instant consists of many curved shock sections which are convex toward the incoming flow. The lines of intersection of these curved shock segments are propagating in various directions at high velocity across the front and actually consist of triple-shock interactions (i.e., Mach-stem configurations). The third shock of these interactions extends back into the reactive flow regime and is required for the flow to be balanced at the intersection of the two convex leading shock waves. In general, the flow in the neighborhood of the shock front is therefore quite complex. However, it has been observed that these propagating transverse waves exhibit a "preferred" spacing which is primarily related to the "thickness" of the reaction zone. Furthermore, even though in spherical detonations these waves appear in a completely random manner and are disoriented relative to each other, in those cases where a detonation tube is used they have been observed to couple with the geometry of the tube. It is true, of course, that a finite-amplitude transverse shock wave cannot exist behind a leading shock front without being attached to a transverse pressure wave propagating in the hot-gas column well behind the front. Therefore, coupling might be expected in those cases where the detonations preferred spacing corresponds to the spacing required for a transverse standing or traveling mode in the hot-gas column downstream of the front. In fact, it has been found that the coupling phenomenon is so general that virtually every possible mode of coupled transverse-wave phenomena may be observed in self-sustaining detonations.

In describing the structure of real detonation waves we must therefore contend with three important non-one-dimensional properties of the detonation, each of which must be discussed in detail. These are (1) the nature of the coupling with the tube geometry, (2) the detailed structure of transverse-wave phenomena at the shock front and in the reaction zone, and (3) the relationship between the preferred spacing exhibited by the transverse phenomena and the initial conditions in the detonatable mixture.

It is interesting that the nature of the coupling was essentially explained before any details concerning the structure of the front were observed or surmised. Experimentally it was noted quite early that a particular limit detonation not only prefers to travel in a round tube as a spinning wave but that the pitch ratio (i.e., the ratio the length of the tube required for the phenomenon to traverse one circumference to the diameter of the tube) is a constant for any initial conditions and tube geometry. In 1946 and 1952 Manson[8] and Fay[9] independently presented an acoustic theory of spin which produced excellent agreement with experimental P/d values. In this theory an acoustic equation is solved to find the allowable transverse modes for tubes of various cross-sectional geometries. We summarize Fay's analysis here. Consider a long column of quiescent hot gas whose properties are given by the static CJ properties of the particular detonation wave under consideration. Write the three-dimensional acoustic equation (similar to Eq. 8-6, only with no source term) in a coordinate system which is appropriate to the tube's geometry and solve this equation for either propagating or standing transverse waves, using the proper boundary conditions at the walls. For the case of single spin in a circular tube, Fay looked for a solution which corresponds to a single rotating pressure antinode which may be wrapped along the length of the tube as a long helix at any instant of time. He determined the wavelength of this helix from slope measurements on experimental streak photographs which showed the rotating pressure head

[8]N. Manson, *Compt. Rend.*, *222*, 46 (1946).
[9]J. A. Fay, *J. Chem. Phys.*, *20*, 942 (1952).

as a luminous band behind the front. With this information he was able to determine the characteristic rotational frequency of the first spin mode in the quiescent hot-gas column.

To determine the spin pitch, P/d, at the detonation front from the rotational frequency of this antinode, a velocity transformation is required. This is illustrated in Fig. 9-3. Here we have drawn a stationary coordinate system relative to the hot-gas column (whose velocity of sound

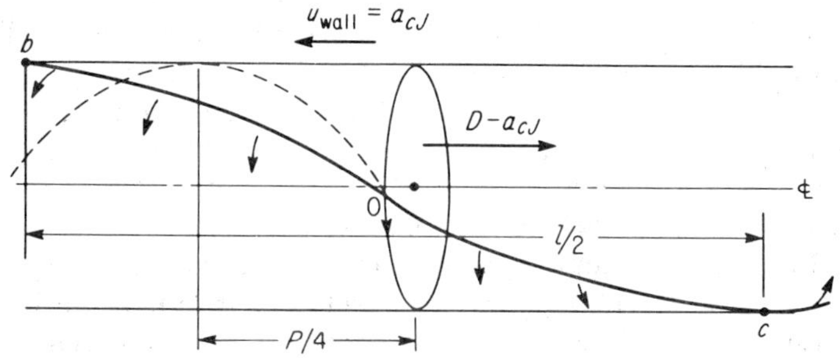

Fig. 9-3. Single-spin writing for an antinode of wave length l. P/d is the pitch of the spin in a tube diameter d.

is a_{CJ}) and have schematically represented an instantaneous position of the positive acoustic-pressure antinode at the wall by the line *b*-0-*c*. This antinode is propagating along the wall in the direction of the arrows at the velocity specified by the acoustic theory. In this case, since we assume that the detonation is CJ, we may assume that the walls of the tube are traveling to the left at the velocity $u_{wall} = a_{CJ}$ and therefore that the detonation front is traveling to the right at the velocity $(D - a_{CJ})$. Note that the detonation front is always uncovering new material as it propagates. If a coupling of the transverse phenomena at the detonation front and the acoustic signal in the gas column following the front is assumed, the trajectory of the point 0 on the wall will produce the spin writing indicated schematically by the dashed line. Thus one complete rotation of this trajectory will define a pitch which may be compared to experiment. Fay also applied his theory to rectangular tubes, triangular tubes, and circular annular spaces, and showed that the application of acoustic theory predicted the correct pitch in all cases. However, he also showed that true spin occurs only as the helical transverse mode of a round tube, and that the acoustic modes which appear in rectangular or triangular geometries are simple standing transverse modes of the tube.

Fay also calculated the spin pitch for the case of a multiheaded, single-handed (i.e., either right or left) rotating acoustic mode in a circular tube. We now know, however, that the only purely rotational mode which is observable is the single-spin mode of a round tube. In fact, this mode of detonation propagation is unique for two reasons. First, since it is a traveling and rotating mode, it may be rendered three-dimensional steady by applying the proper rotational and translational velocity transformation. On the other hand, all of the observed higher-order modes in round tubes and all the acoustic modes observed in rectangular tubes involve the simultaneous occurrence of wave heads propagating in both directions. Therefore, for these cases there is no transformation that will render the flow steady in a new coordinate system. Second, since the solution for a rotating mode in a cylindrical tube requires the use of a Bessel function for the construction of the radial pressure distribution, the peripheral velocity of the mode is not sonic but is determined by the radial and rotational mode numbers of the desired mode. For the case of a first radial, first rotational mode (which is the observed single-spin mode) the peripheral velocity of the wave head is $1.84 a_{CJ}$. In the case of higher mode spin, the transverse acoustic velocities are always nearer to a_{CJ}, while for rectangular modes the transverse velocity is simply a_{CJ}. Thus single-spin in a round tube represents a rather violent transverse phenomenon and may be viewed as a steady flow in the proper coordinate system, while all other types of observable transverse wave motion contain weaker transverse waves and are truly nonsteady phenomena.

We now wish to discuss the detailed shock structure that occurs at the detonations shock front. The fundamental principles of gas dynamics place some very strong restrictions on the character of the interactions observed and on the flow properties associated with these interactions. Therefore, before discussing the properties of the actual interactions that occur, we will briefly review the general problem of unreactive steady-shock interactions.

Consider a flow of velocity $w = Ma$ where $M > 1$ and require that two steady oblique shocks intersect at a point in this flow (labeled T in Fig. 9-4a). We may represent all possible behaviors of these two shocks (each considered individually) by assuming that shock AT has an angle β which lies in the range $\pi/2 > \beta > \mu$ where $\mu = \sin^{-1}(1/M)$ and that shock BT has an angle β' which lies in the range $\pi - \mu > \beta' > \mu$. Individually, the flow properties behind these shocks are determined by the value of the incident velocity vector normal to the shock (i.e., by specifying $u_1 = w \sin \beta$ and $u_1' = w \sin \beta'$) and using these values to calculate state property changes across the shocks.

Also note that these oblique shocks are "traveling" along their length relative to the coordinate system at the velocity $v_{\|} = w \cos \beta$ or $v_{\|}' = w \cos \beta'$. Therefore all the AT shocks (except for $\beta = \pi/2$) "originate" at point A. However the BT shocks with $\beta' > \pi/2$ originate at B, while

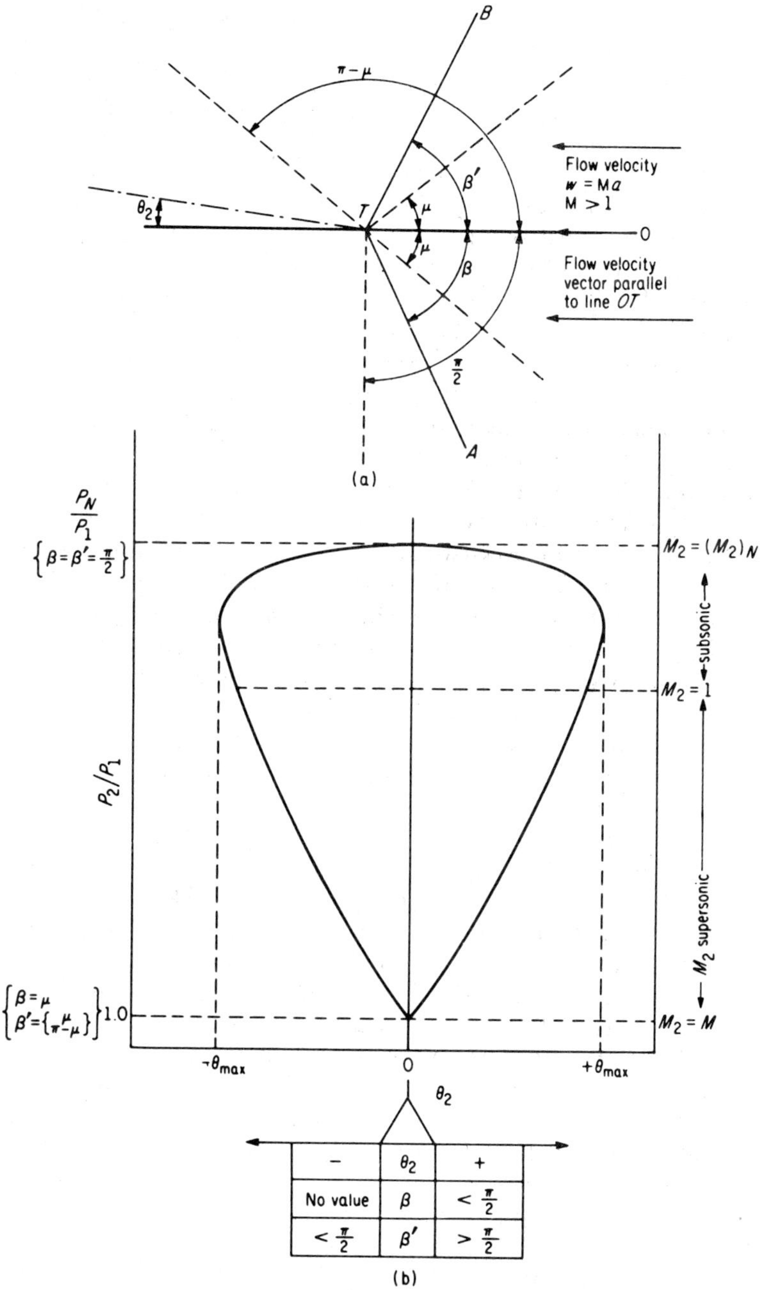

Fig. 9-4. Possible oblique shock configurations in a supersonic flow and (P, θ) shock polar for all these configurations. (a) The configurations. (b) The (P, θ) polar (schematic).

those with $\beta' < \pi/2$ originate at T and the shock with $\beta' = \pi/2$ is normal to the flow and has $v'_{\parallel} = 0$.[10]

If we now require that the two shocks produce a steady flow relative to the point T, we restrict the number of possible values of β and β' in a relatively complex way. For such a steady flow to exist, the equations of motion require that the flow deflection angle θ and the pressure P be balanced across the boundary streamline which separates gas that has been processed by the two shocks, TA and TB (this streamline is usually called the slipstream). We call this the (P,θ) balance requirement for a shock interaction. The P,θ behavior of the flow for all the oblique shocks passing through T is shown in Fig. 9-4b. In this diagram θ is taken to be a positive flow-deflection angle, when drawn in the direction shown in Fig. 9-4a. An examination of Fig. 9-4b leads to the conclusion that the slipstream balance requirement may be satisfied (without further flow adjustments) for only the trivial case $\beta = \pi - \beta'$. However, we also note that for lower values of β the flow remains supersonic after the transition, thus leading to the possibility that another steady wave, originating at T, may traverse the flow downstream of T to produce a nontrivial steady structure. The investigation of all the possible configurations which satisfy this type of requirement is rather lengthly and will not be covered here.[11] We will simply state that for the shock intersections observed on propagating detonations, shock TA is called the *incident shock* and always produces a supersonic flow ($M_2 > 1$), whereas shock TB is called the *Mach-stem shock* and, since it always occurs at an angle slightly greater than $\pi/2$, it has a subsonic flow behind it. We designate as region 4 the flow region that has been processed by the Mach-stem shock, and note that it must lie on the upper branch of the shock polar of Fig. 9-4b to the right of $\theta = 0$. It must lie on the same shock polar as the (P_2,θ_2) point because it is a region produced by a single shock transition, and it must lie to the right of $\theta = 0$ since it is observed that physically this shock is always produced by the shock interaction (i.e., it originates at point T). We also note that since in our diagram we restricted the incident shock to the range $\beta < \pi/2$, θ_2 is always positive.

Let us now consider a case when the properties of the incident shock are known. This fixes point 2 on the (P,θ) diagram (see Fig. 9-5e). We now ask, What kind of oblique waves may originate at point 2 and traverse only region 2 of the flow? There are only two possible waves that may occur—either an oblique shock wave or an oblique centered rarefaction fan, both originating at T. Furthermore, if these waves are

[10]Notice the similarity between this behavior and the behavior of the oblique flames discussed in Chapter 8.

[11]See, for example, D. C. Pack, *J. Fluid Mech.*, *18*, 549 (1963); L. F. Henderson, *J. Fluid Mech.*, *26*, 607 (1966).

to traverse region 2 they must be of such an orientation that (1) for a centered rarefaction fan, pressure is decreased while θ increases, or (2) for a shock wave, pressure is increasing while θ decreases. Since the rarefaction fan produces a locus of end states on the P,θ diagram which never intersects the original polar, it cannot by itself represent a valid solution to the problem. However, the (P,θ) locus for a single oblique shock transition (producing region 3 properties in Fig. 9-5d) will intersect the original polar as shown in Fig. 9-5e. The shock which is of correct strength to exactly produce a $P_3 = P_4$; $\theta_3 = \theta_4$ balance is called the *reflected shock* in the Mach-stem configuration. Thus for every set of initial velocities w and shock angles β there is the possibility of defining a unique Mach-stem structure on the basis of an unreactive-unreactive slipstream balance. However, for generality, we note that the simple act of specifying a value for w and β does not insure that a steady Mach stem is dynamically possible.

Experimentally two types of Mach-stem interactions have been observed at the wave front in a propagating detonation. The simplest interaction is a single Mach-stem configuration propagating across the detonation's main shock front at a velocity which causes it to couple with an acoustic mode of the tube. Figure 9-5 is an interferometric photograph of such a Mach stem (in this case isolated) and a schematic diagram of the shocks and reaction waves that occur in its neighborhood. Interestingly enough, the position and nature of the "reaction waves" that occur in this situation may be predicted by assuming that each particle of gas reacts, after shock passage, with the proper kinetics for the chemical system under consideration. Note that the "reaction wave" is not a "flame" in the true sense, since it is not propagating through the mixture by a conduction mechanism. Instead, it exhibits a unique velocity only because the shock wave is processing the gas locally at a unique velocity. It should be added that its velocity is highly subsonic and therefore the flow through such a reaction wave may not be adequately represented as a constant-pressure process. The local structure of such a wave differs from the one-dimensional waves discussed in Sec. 9-2 only because it may have been produced by a locally curved shock wave.

A slipstream balance in a coordinate system which renders the interaction point steady in Fig. 9-5 requires that P and θ be equal in regions 3 and 4 of this figure. The configuration of an interaction of this type may be calculated if it is assumed that it is a quasi-steady phenomenon by using a method of shock polars in the P,θ plane as shown in Fig. 9-5. However, this calculation must be performed with a real gas caloric equation of state to be realistic. Such a calculation has shown that the structure of such an isolated interaction in a reactive gas corresponds to an unreactive-flow–unreactive-flow slipstream balance in the neighborhood

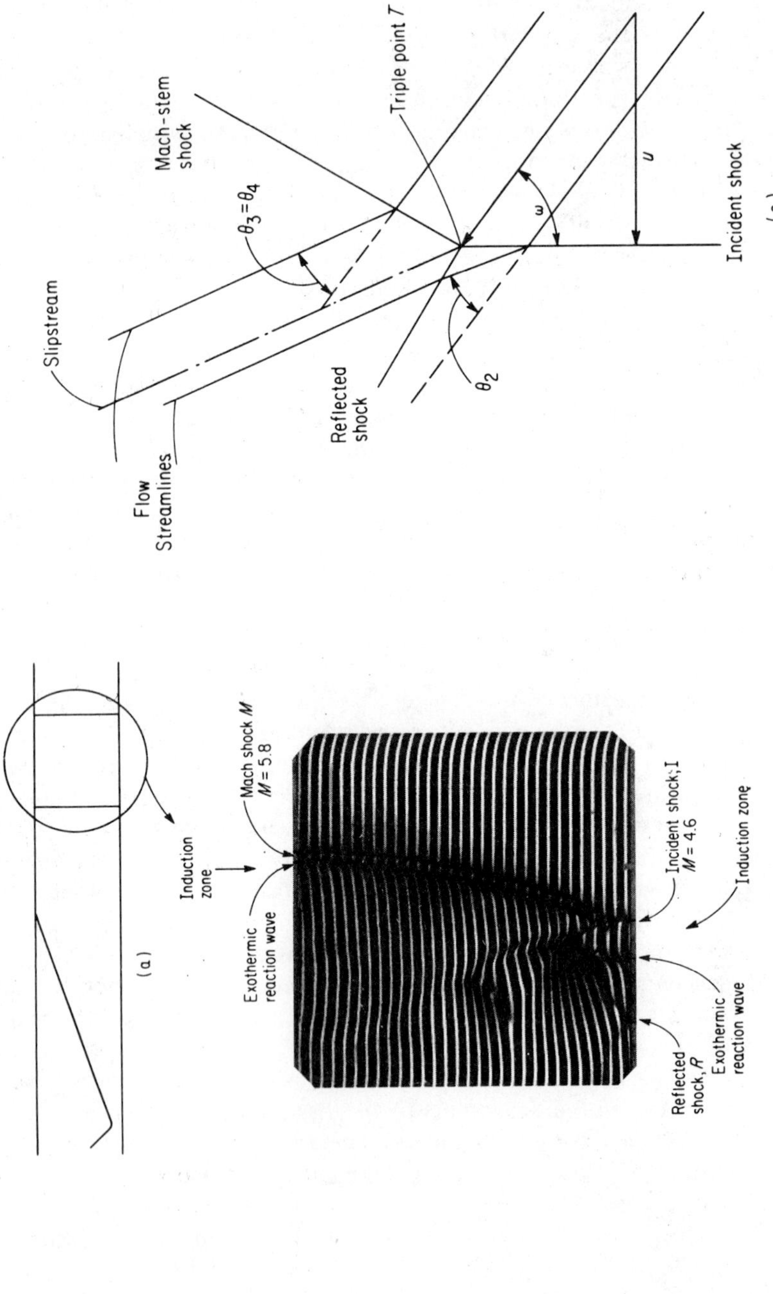
(a)
Induction zone
Mach shock M
M = 5.8
Exothermic reaction wave
Incident shock, I
M = 4.6
Induction zone
Reflected shock, R
Exothermic reaction wave
(b)
Slipstream
Flow Streamlines
Mach-stem shock
$\theta_3 = \theta_4$
Triple point T
Reflected shock
θ_2
ω
u
Incident shock
(c)

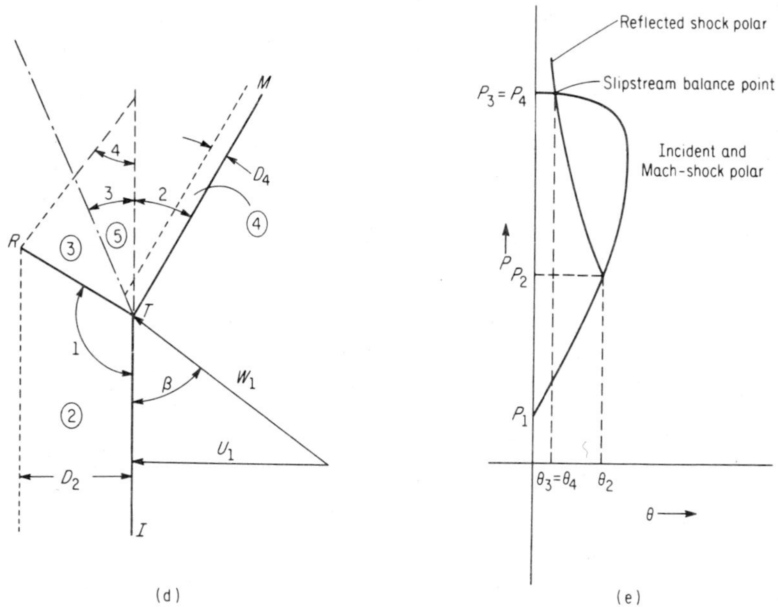

Fig. 9-5. Mach stem structure in a reactive gas. (a) Two-dimensional ramp geometry used to produce the mach stem illustrated in part b. (b) Spark interferogram of a reactive mach stem in a $2H_2 + O_2 + 2CO$ mixture, $P_1 = 15$ torr. (c) Single mach stem structure showing angles of flow deflection in a coordinate system with point T stationary (unreactive flow only). (d) Same structure as (c) showing location of reaction waves based on induction delay kinetics. Numbers in this figure are keyed to values tabulated in Table II, page 302. (e) Shock polars used to calculate mach stem structure shown in (c) and (d). Unreactive–unreactive slipstream balance assumed. [Photograph in (b) courtesy D. R. White, General Electric Research Laboratory, Schenectady, New York.]

of the triple point, even though a rapid exothermic reaction is occurring a few microseconds later or a few millimeters downstream of this point.

It should be emphasized that the (P,θ) criterion does not limit the interaction to a single configuration. On the contrary, for any specified incident shock strength there are an infinite number of allowable configurations, one for each incident specified shock angle over a rather large range of β. The angles summarized in Table 9-2 thus represent the best fit of a theoretical calculation to data obtained from Fig. 9-5, and the comparison therefore yields a determination of β for the experimentally observed interaction.

One interesting feature of these Mach stems is the ability of their triple point to write a line on a smoked surface. These thin lines were first

TABLE 9-2
REACTIVE MACH-STEM STRUCTURE

Quantity (See Fig. 9-5)	Theoretical Solution $\beta = 50°$	$\beta = 52°$	$\beta = 54°$	Experimental Values	Indicated Value of β
Angle 1	124.35	121.23	117.75	122.0 ± 1°	51.5 ± 0.5°
Angle 2	33.93	31.06	28.29	30.5 ± 1°	52.4 ± 0.6°
Angle 3	23.73	23.65	23.53	23.5 ± 1°	53.9 ± 4.3°
Angle 4	48.90	38.84	31.07	37.0 ± 1°	52.4 ± 0.2°
D_2/D_4	7.52	6.18	5.13	5 $^{+2.3}_{-1.3}$	54. ± 3°
					52.4 ± 1° average

observed by Antolik[12] in 1875 on soot-coated plates held near a spark discharge. Mach later interpreted them as being caused by the intersection of "sound waves," and in 1877 he observed the same type of writing when a bullet passed by obstacles placed near a coated surface. This work preceded by eight years his use of spark schlieren photography to observe the shock waves associated with a traveling bullet. In 1959, Denisov and Troshin[13] revived the technique and first applied it to the observation of transverse waves in detonations. Since that time it has been extensively applied to the study of detonation structure, mainly because of its simplicity. The current consensus is that wood smoke deposited in an almost opaque layer on the surface produces the best smoked-foil records. The foils may be "fixed" after firing by spraying with a clear lacquer.

A few examples of smoked-foil records obtained with propagating detonations are shown in Figs. 9-6 and 9-7 and a record obtained from an adaptation of this technique which allows the orientation of the leading

[12] K. Antolik, *Pogg. Ann. 230*, Series 2, *154*, 14 (1875).
[13] Y. H. Denisov and Y. K. Troshin, *Dokl. Akad. Sci. SSSR*, *125*, 110 (1959). (Translation *Phys. Chem. Sect*, *125*, 217, 1960.)

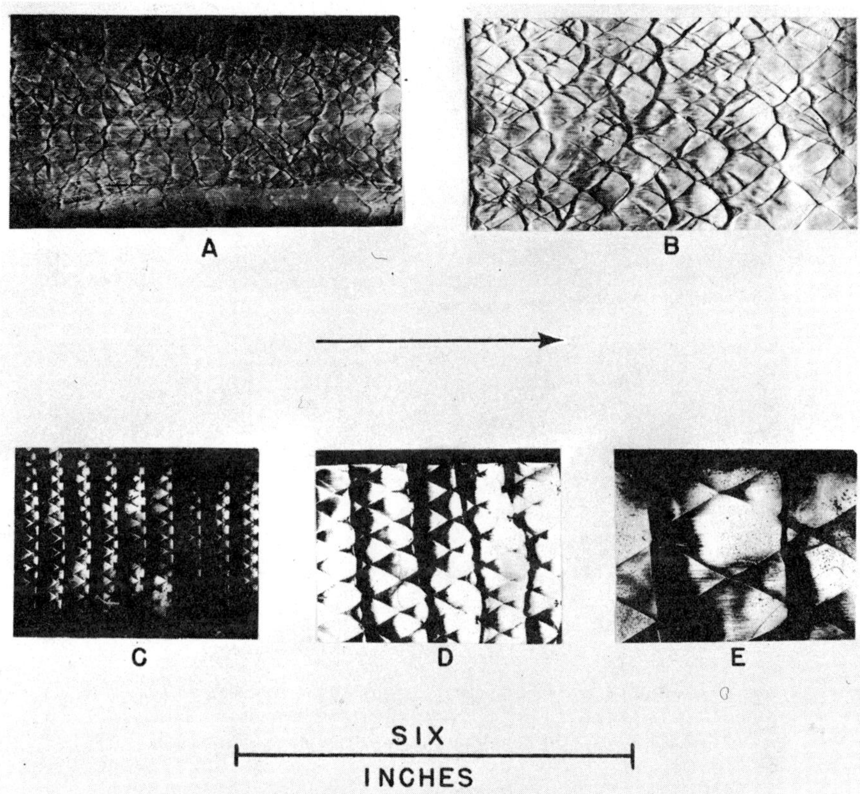

Fig. 9-6. Smoked foil inscriptions obtained under the following conditions: (a) $2H_2 + O_2$ in $3\frac{1}{4} \times \frac{1}{4}$ tube, P_0 = 150 torr. (b) $2H_2 + O_2$ in 2-in.-diameter tube, P_0 = 120 torr. (c) $\frac{2}{5} C_2H_2 + O_2$ + 75% argon, $3\frac{1}{4} \times 1\frac{1}{2}$ tube, P_0 = 100 torr. (d) $2H_2 + O_2$ + 70% helium, P_0 = 350 torr. (e) $2H_2 + O_2$ + 77.5% argon, P_0 = 100 torr.

shock wave to be observed is shown in Fig. 9-8. This figure was obtained by sprinkling sand grains on the film after smoking and before firing the detonation. Each sand grain produces a small local and symmetric interaction with the leading shock wave of the detonation and thereby flags its orientation at that location.

Figure 9-7 is an example of the smoked-foil writing produced on the wall of a round tube by a single-spin detonation. This complex pattern has been extensively analyzed by Voitsekhovskii and co-workers[14] in Russia and by Duff[15] and Schott[16] in this country. They find that this

[14]B. V. Voitsekhovskii, V. V. Mitrofanov, and M. E. Topchian, *Structure of Detonation Waves in Gases* (Moscow) 1963 (Translation AD 633821, Defense Documentation Center, Washington, D.C.).

[15]R. E. Duff, *Phys. Fluids*, *4*, 1427 (1961).

[16]G. L. Schott, *Phys. Fluids*, *8*, 850 (1965).

(a)

(b)

Fig. 9-7. Smoked foil writing produced by single-spin detonation in a round 1-in. diam. tube, using a $^2/_3$ acetylene–oxygen mixture diluted with argon. Detonation is propagating from left to right. (a) Spin with some bounce (i.e., with a nonconstant transverse velocity). (b) Constant-velocity spin. [Courtesy G. L. Schott, Los Alamos Scientific Laboratory, Los Alamos, N.M.; originally appeared in *Phys. Fluids*, *8*, 850 (1965).]

extremely violent transverse wave actually contains two triple-shock intersections, one on the primary front and one on the reflected shock of the primary interaction. Figure 9-9 contains a schematic diagram of the interaction, a summary of pressure-gauge records, and a time-resolved visible photograph of the peripheral transverse structure for a detonation wave traveling in the single-spin mode in a round tube. The second Mach-stem shock is oriented essentially normal to the flow in the region 2 and is an extremely strong shock wave. It heats the gas sufficiently to cause immediate reaction (< 1 μsec delay along its entire length). The high peripheral velocity of this phenomenon is dictated by the acoustic requirement for that mode and this in turn causes the unique structure of the transverse mode for this case. The broad writing band seen in Fig. 9-7 corresponds to the wall locus of this very strong transverse shock. We note that the characteristic cross-hatched structure in this band indicates that this transverse shock itself contains transverse propagating waves (which are not evidenced in Fig. 9-9).

Fig. 9-8. Smoked foil record showing orientation of the incident shock waves by use of sand grains. Propagation from left to right. $2H_2 + O_2 + 70\%$ argon. $P_0 = 100$ torr. The two white dots at the top are 1 in. apart on the original.

The nonsteady transverse structure which caused the smoked-foil writing shown in Fig. 9-6 is obtained most frequently and always contains transverse waves of different families which are continually intersecting. We note that the simplest type of transverse-wave intersection is that which occurs in a planar mode, and that this mode is purely two-dimensional nonsteady. A mode of this type may be produced by passing a detonation through a rather wide straight channel whose depth is less than the range of $a/6$ to $a/10$, where a is the preferential spacing of waves of the same family in that detonatable mixture. If one adjusts the initial pressure in such a narrow tube so that the tube's width is an integer multiple of the detonation's preferred spacing, we obtain the simplest two-dimensional nonsteady propagation behavior. An example of this ex-

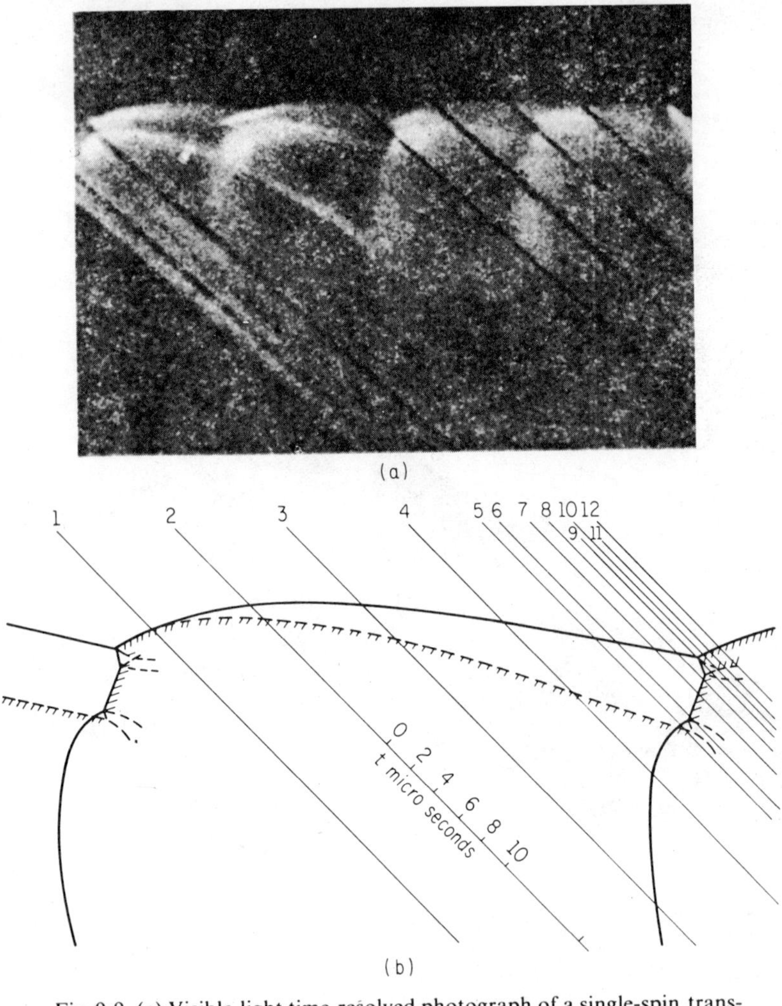

Fig. 9-9. (a) Visible-light time-resolved photograph of a single-spin transverse wave in an acetylene–oxygen mixture. Compare its structure with (b) A schematic diagram of structure of single-spin front at the wall. (c) Pressure records obtained at the gauge positions shown in (b). The highest pressure in this record is 160 P_1 and represents a streamline traversed by two very strong shocks before it reacted. [Adapted from B. V. Voitsekhovskii, V. V. Mitrofanov, and M. E. Topchian, *Structure of Detonation Waves in Gases*, Moscow, 1963 (translation AD 633821, Defense Documentation Center, Washington, D.C.).]

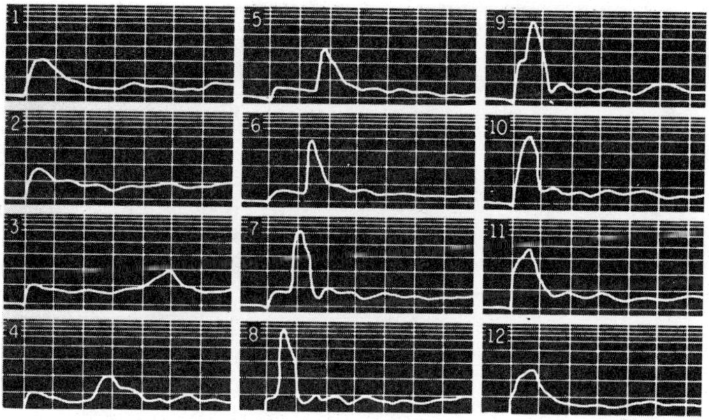

(c)

Fig. 9-9. (Continued)

tremely symmetric behavior is shown in Fig. 9-10 for the case of a stoichiometric hydrogen–oxygen mixture diluted with 70% argon. Here the Mach stems are always single Mach interactions, and the locus of a set of successive intersections of these Mach stems lies in exactly the direction of propagation of the detonation. The plane of each of these successive intersections defines a streamline of the flow for which the entire flow is nonsteady in the propagation direction, but along which the flow velocity in the transverse direction is always identically zero. The behavior of the Mach stem during the course of such an intersection is exhibited in Fig. 9-11. The dynamics of this type of intersection requires that the incident and reflected Mach-stem structures both have correctly balanced slipstreams and that the velocity and orientation of the shocks A–A', B–B', C–C', and D–D' remain constant during the interaction. In this calculation the angle ϕ is found to be a relatively insensitive function of the incident shock velocity u_A for a fixed value of the incident writing angle α. Furthermore, the velocity increment of the normal shock $(u_B - u_A)/u_A$ is also relatively insensitive to the value of u_A and has been found to remain relatively constant for a variety of different initial conditions. It ranges from 0.25 to 0.40 for detonations propagating in argon-diluted hydrogen–oxygen mixtures, argon-diluted hydrogen–acetylene mixtures, and carbon monoxide–oxygen mixtures containing a few percent of hydrogen. Thus, for writing patterns which everywhere appear as single lines, one concludes that the single interaction structure is the correct one and that the intersection of these interactions occurs as though they were unreactive–unreactive interactions. Furthermore, the analysis yields the simple schematic model shown in Fig. 9-12, in which it is obvious that all leading shocks of the detonation wave are continuously

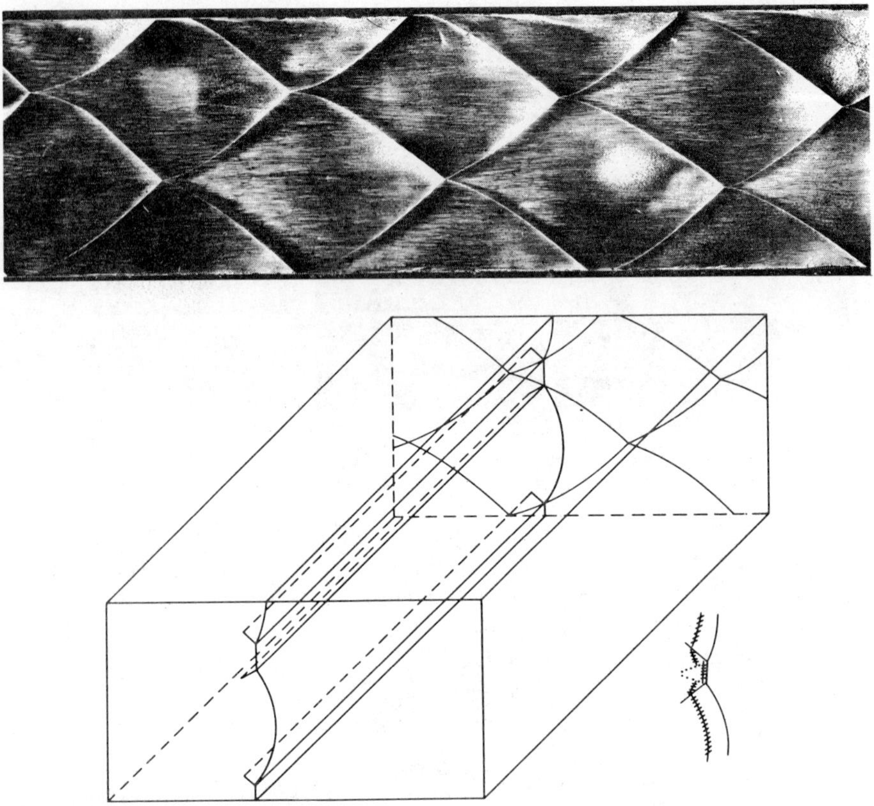

Fig. 9-10. Smoked foil record and schematic diagram of symmetric planar interaction. Note that the intersection of the interactions is on a centerline–streamline of the flow.

decelerating with time. Also the transverse velocity of the Mach-stem interactions themselves are constantly changing and the Mach stems are therefore only quasi-steady. The intersection of these Mach stems therefore represents the only mechanism which can cause self-sustenance of the wave as it propagates down the tube. Note that the entire gas sample must be converted to products by the detonation, however, since the leading shock wave is everywhere strong enough to cause complete reaction of each element of gas. Even though this is true, the wave pattern is quite inhomogeneous with almost immediate ignition of the gas sample just downstream of each intersection.

Returning for a moment to single-spin, we note that even in this case every element of the leading shock front always propagates in a direction normal to its instantaneous orientation and therefore is everywhere decaying with time at the tube's periphery. Thus for single spin, detonation is

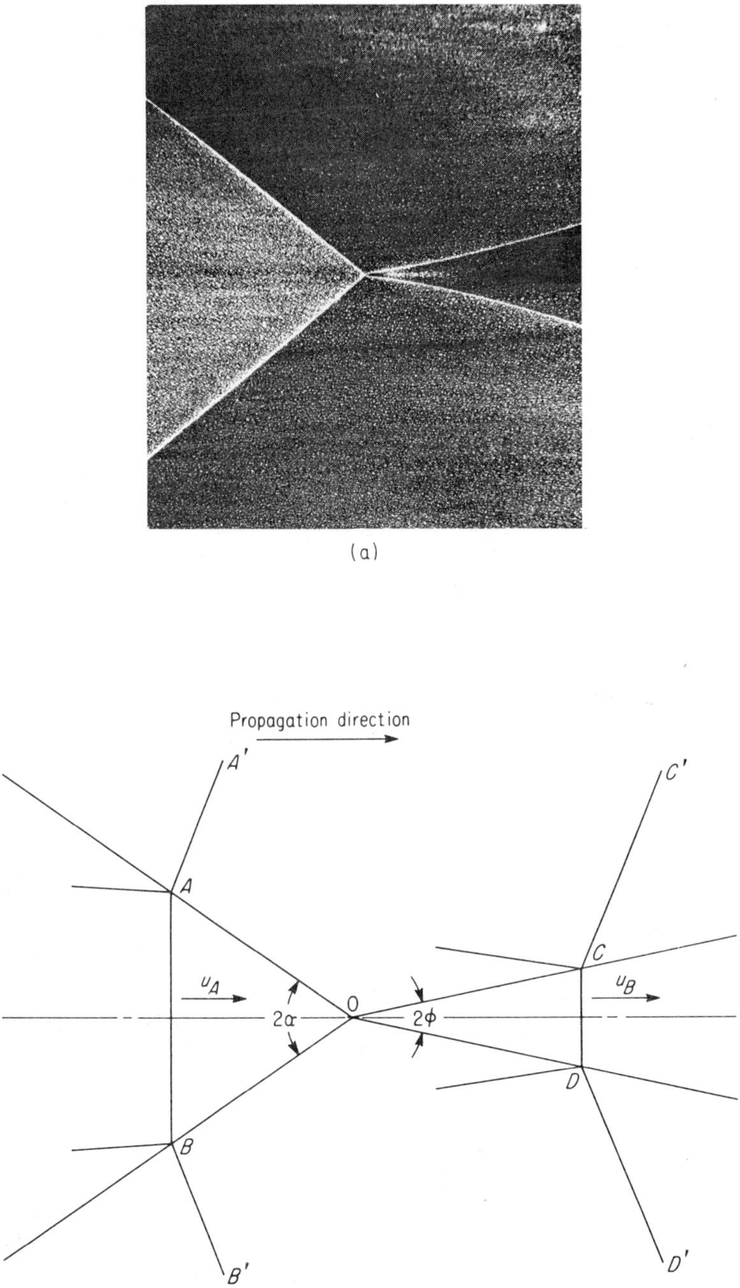

Fig. 9-11. (a) Interaction point in a $2H_2 + O_2 + 70\%$ argon mixture (76 torr); x15. (b) Shock structure at an interaction point. Line $A'\ A\ B\ B'$ represents shock at time t (before intersection). Line $C'\ C\ D\ D'$ represents shock at time $t + \Delta t$ (after intersection).

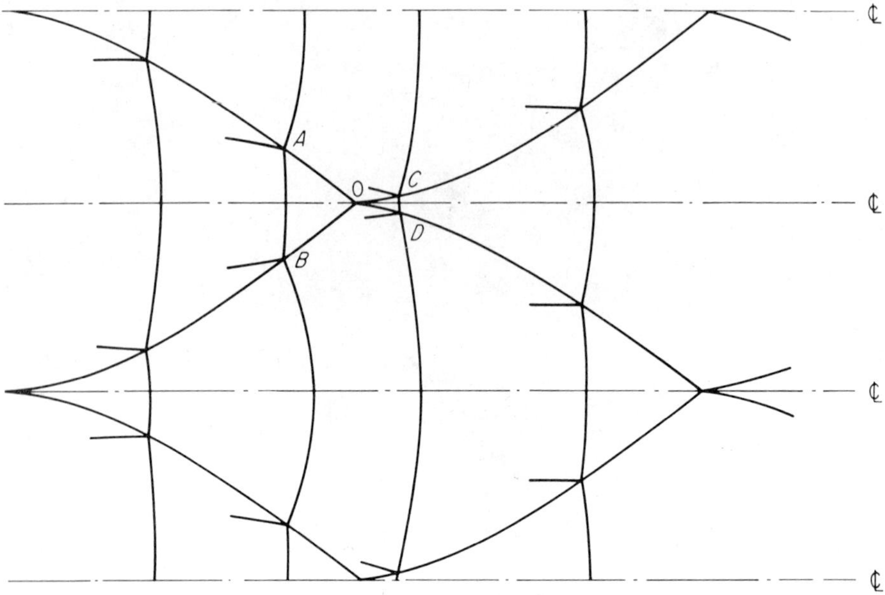

Fig. 9-12. Idealized two-dimensional detonation showing shock-wave structure at various times and associated smoked-foil pattern.

sustained by a rather violent Mach-stem interaction which continually overwrites this decaying shock but is steady in the proper coordinate system.

Relative to the occurrence of the single interaction or the double interaction, there is some evidence that the double interaction will occur in situations where the acoustic coupling requires a momentary high transverse velocity. There is also some evidence that the single structure occurs more readily in gases with a high heat-capacity ratio. For example, Soloukhin[17] mentions that the double structure appears just before the intersection of planar waves and that this intersection then produces a single interaction (which has a relatively low transverse velocity). However, the data of Figs. 9-6c, 9-6d, and 9-6e, 9-8, and 9-11 were obtained in hydrogen–oxygen or acetylene–oxygen mixtures having a high heat-capacity ratio (produced by argon dilution), and in this case the interaction does not show any evidence of double structure before the intersection. This regular behavior is not observed in all systems. In particular methane–oxygen mixtures exhibit a very unstable cellular behavior.

The last property of this transverse structure which should be discussed is its preferred spacing. Experimental results have shown that for

[17] R. I. Soloukhin, *Shock Waves and Detonations in Gases*, Mono Book Corp., Baltimore, Md., 1966.

a specific mode of propagation, the spacing of transverse waves of the same family is intimately related to the mixture composition and the initial pressure. Figure 9-13 summarizes data obtained for rectangular-mode propagation in a $3\frac{1}{4} \times 1\frac{1}{2}$-in. tube in stoichiometric hydrogen–

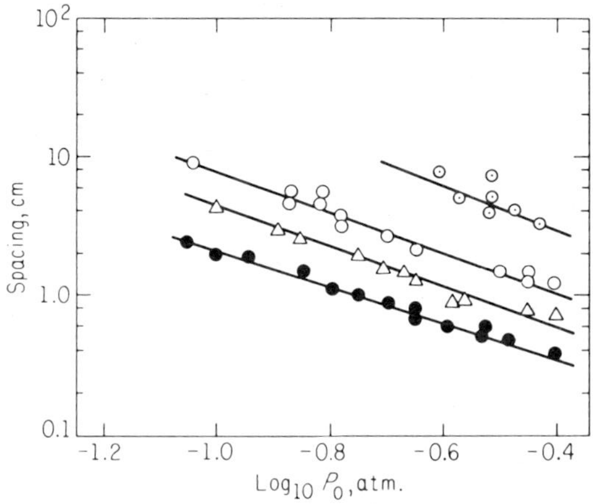

Fig. 9-13. Preferred spacing in rectangular mode ($3\frac{1}{4} \times 1\frac{1}{2}$ in. tube) $2H_2 + O_2 +$ diluents. ⊙, 85% argon, ○, 77.5% argon, △, 70% argon, ●, 50% argon.

oxygen mixtures diluted with various percentages of argon at various initial pressures. From these plots we see that the spacing is dependent on the pressure with qualitatively the same dependence as the reaction-zone thickness for the hydrogen–oxygen reaction. Thus the indications are that the spacing is in some way related to the reaction-zone thickness of the detonation. This will be discussed in further detail in the next section.

Figure 9-14 shows an example of the effect of mode type on the measured spacing. In this case, the planar-mode data were taken in a $3\frac{1}{4} \times \frac{1}{4}$-in. detonation tube, the rectangular-mode data in a $3\frac{1}{4} \times 1\frac{1}{2}$-in. tube and the round-tube data in a 2-in. diam. tube. For the planar mode, as pressure is lowered the transverse waves propagating in the $\frac{1}{4}$-in. direction become progressively weaker until at sufficiently low pressure they fail to appear on the record. Note that, as the pressure is decreased, the preferred spacing progressively deviates from the rectangular-mode spacing determined in a $3\frac{1}{4} \times 1\frac{1}{2}$-in. tube. However, there is evidence that in this case the detonation's velocity is progressively lowered by propagation in the narrow channel. This would affect the spacing in the observed manner.

Another interesting proof that the detonation does indeed have a preferred transverse spacing related to the initial conditions is given by the

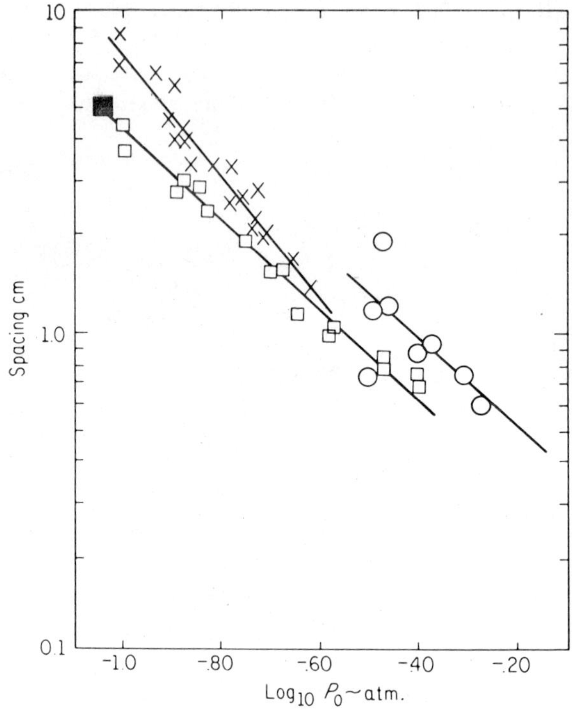

Fig. 9-14. Effect of tube geometry on preferred spacing $2H_2 + O_2 + 70\%$ argon. ○, round tube 2 in. diam.; □, $3\frac{1}{4} \times 1\frac{1}{2}$ in. rectangular tube, **X**, $3\frac{1}{4} \times \frac{1}{4}$ in. planar tube, ■, single planar mode data point in $3\frac{1}{4} \times 1\frac{1}{2}$ in. rectangular tube.

smoked-foil record shown in Fig. 9-15. Here the pressure was adjusted to lie between the correct pressure for first-mode and second-mode spacing, and the mode that occurred was asymmetric with an average spacing of about 2/3 of the tube width. This represents an interaction of the frontal spacing with a coupled first- and second-transverse acoustic mode in the hot-gas column behind the front.

Another piece of evidence which reenforces this conclusion concerning spacing is given by the composition and pressure limits of propagation for detonation. Single spin in a round tube or first-mode propagation in a rectangular tube are always observed just inside these pressure or composition limits, and it therefore appears that detonation will fail to propagate as a self-sustaining phenomenon when its preferred transverse spacing becomes sufficiently larger than the maximum cross-sectional dimension of the tube. Thus the accumulated evidence points to the requirement that some sort of transverse-wave structure must be present

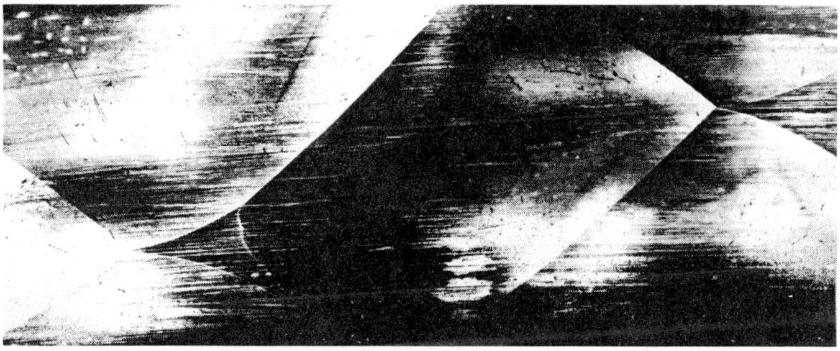

Fig. 9-15. Asymmetric mode in planar channel $3\frac{1}{4} \times \frac{1}{4}$ in. at intermediate pressure between first- and second-mode pressure. Full width of channel is shown; propagation is from left to right. $2H_2$ + O_2 + 70% argon dilution.

on all self-sustaining detonation waves. At present, however, observations of the nature of this transverse structure are strictly phenomenological and have not led to a theory which explains either its cause or its apparent "stability." In the next section we will discuss various theories of detonation stability and an acoustic model for transverse waves. These represent a first step in the direction of explaining the observed structure.

One qualitative statement concerning stability may be made at this point. If we examine the relative rate of heat release in various portions of a detonation wave which exhibits either steady spin or nonsteady transverse structure, we find that the region of maximum heat release always corresponds to the region of highest pressure and that, with acoustic coupling, this region also connects to a positive antinode of the condensation in the flow behind the leading front. Thus Rayleigh's criterion for combustion instability in a sense qualitatively justifies the existence of the observed transverse structure.

9-4. DETONATION STABILITY AND THEORY OF TRANSVERSE STRUCTURE

Early investigations of the stability of Chapman-Jouguet flow were concerned primarily with a comparison of the properties of this flow with the properties of neighboring steady flows which also satisfy the Hugoniot and Rayleigh equations. However, transient stability cannot be fruitfully investigated using this technique, since this type of analysis does not allow for the occurrence of the truly nonsteady flow which is a possible solution to the equations of motion. In other words, any analysis of stability to be realistic must specifically allow for nonsteady behavior.

Erpenbeck[18] has pioneered in the field of this nonsteady analysis and has shown, with a linearized treatment in Laplace transform space, that all detonations with realistic exothermic kinetics should be linearly unstable to transverse perturbation waves of the proper frequency for any arbitrarily overdriven detonation. In a sense, the results of his analysis are in agreement with the observation that transverse structure exists on all self-sustaining detonations. Unfortunately, because of the complexity of his analysis it is impossible to describe the nature of the flow that arises from an instability of his type. We therefore turn to simpler models and investigate the nature of the nonsteady flow for two very restrictive cases which have some relation to actual experimental observations.

First, we look at the effect of a purely longitudinal expansion wave on an overdriven one-dimensional detonation. This model quite closely represents the behavior of the leading shock that travels along the centerline of the Mach-stem intersections illustrated in Fig. 9-12. Therefore, we construct an overdriven steady detonation wave as shown in Fig. 9-16. In this case, the steady reaction zone has been modeled to be a simple zero-

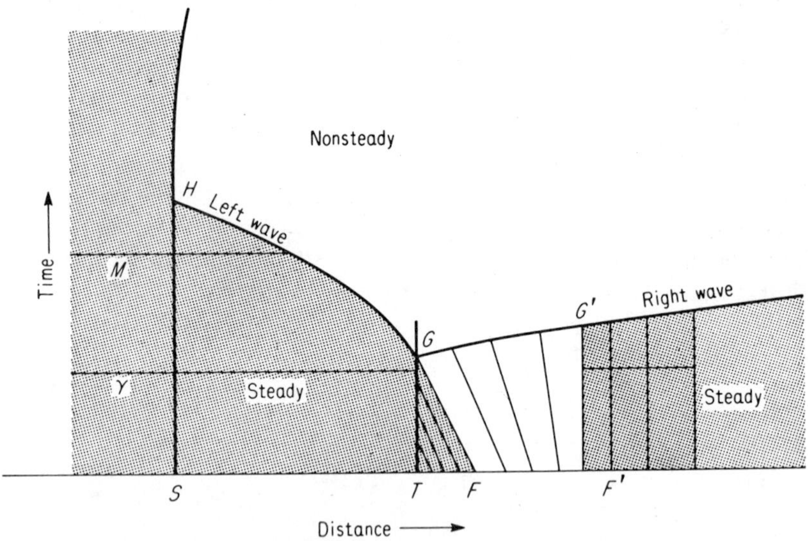

Fig. 9-16. Schematic x, t diagram of the one-dimensional model. SH is initially an $M = 6$ left-facing steady shock wave and the line TG represents the plane of last heat addition along any particle path. The stippled regions represent steady flow regions while the region $FF'G'G$ is a simple left-facing centered rarefaction fan. The truly nonsteady region originates in the reaction zone at the point G where the head of the rarefaction fan just intersects the terminating line.

[18] J. J. Erpenbeck, *Phys. Fluids*, *8*, 1192 (1965); *7*, 684 (1964).

order reaction with a constant activation energy in a constant-gamma gas as given by the equation

$$\frac{Dq}{Dt} = A \exp(-E/RT)$$

where $q_{max}/q_{CJ} = 0.9$ along any streamline. Therefore, the rate of the chemical reaction is the only variable and it is only dependent on the local gas temperature in the detonation's reaction zone. We now place a strong forward-facing rarefaction fan behind the front (as shown in Fig. 9-16) and use the method of characteristic to examine the nonsteady behavior of the flow for various values of the activation energy E. The resulting behavior of the leading shock wave is shown in Fig. 9-17. We note from this figure that for $E = 18$ or 50 kcal, the shock velocity quickly decays to less than the CJ detonation velocity, and the detonation does not asymptotically approach the CJ condition as a one-dimensional phenomenon. However, for zero activation energy—i.e., for a time-independent rate of heat addition behind the leading shock wave—the detonation velocity always remains above the CJ velocity. While this type of numerical analysis cannot be used to prove that a CJ detonation is stable for the case of zero activation energy, the very rapid decay of the shock velocity to below the CJ velocity for the case of high activation energy definitely proves the instability of CJ flow for these high activation energies.

Second, to investigate the nature of a more general type of instability, we again turn to the one-dimensional ZND model of the detonation

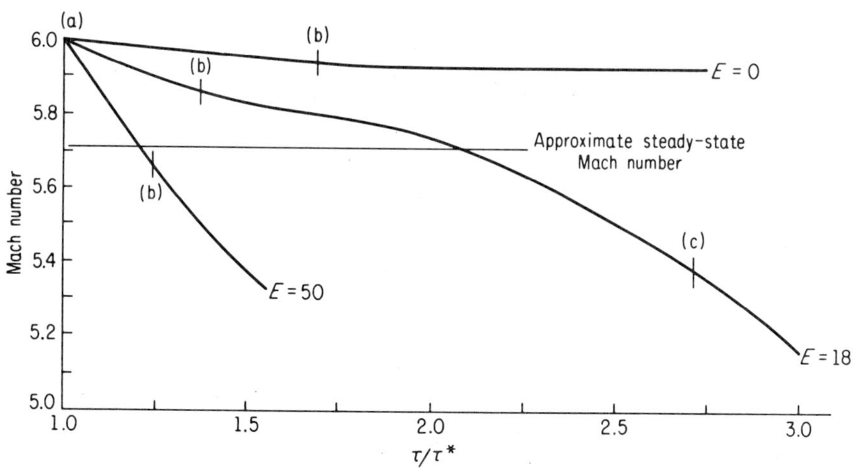

Fig. 9-17. Change of shock Mach number with dimensionless time τ/τ^*. τ^* is the transit time of the head of the left-facing rarefaction fan, i.e., the time between points G and H in Fig. 9-16. (a) expansion fan just contacts shock wave, (b) middle of expansion fan contacts shock wave, (c) end of expansion fan contacts shock wave.

introduced in Sec. 9-2. We note that the entire reaction zone of this detonation is subsonic and therefore that acoustic amplitude wave fronts may possibly propagate for long transverse distances in the reactive-flow region. We therefore consider in detail the propagation of high-frequency waves of acoustic amplitude and arbitrary initial orientation in this reactive flow. For our purposes, the frequency of the wave train is always assumed to be sufficiently high so that dispersion is not important. This frequency limit is determined by the requirement that the flow properties change only very slowly over a distance equal to a wavelength of the wave being considered. If this is true, the behavior of a coherent wave front may be approximated by writing the parametric equations for a sound ray, where a ray is defined as a curve whose direction at every point represents the direction of propagation of energy in a specific wave front passing through that point. Using the ZND detonation of Fig. 9-2 as a background flow for the acoustic treatment, we note that

$$a = a(x),\ u = u(x),\ v = w = 0,\ \frac{\partial a}{\partial x} = \frac{\partial a}{\partial x}(x) \text{ and } \frac{\partial u}{\partial x} = \frac{\partial u}{\partial x}(x)$$

while all other gradients are 0. Consider, therefore, a wave front whose orientation to the x-axis is given by the direction cosines l, m and n between its normal and the x-, y-, and z-directions as shown in Fig. 9-18a. Milne[19] in 1921 first derived the ray equations for the general case. For the one-dimensional flow considered here these equations reduce to the set

$$\frac{dx}{dt} = l\,a + u$$

$$\frac{dy}{dt} = ma \qquad (9\text{-}7)$$

$$\frac{dz}{dt} = na$$

and

$$\frac{1}{l}\left(\frac{dl}{dt} + \frac{\partial a}{\partial x} + l\frac{\partial u}{\partial x}\right) = \frac{1}{m}\frac{dm}{dt} = \frac{1}{n}\frac{dn}{dt} \qquad (9\text{-}8)$$

Equation 9-7 gives the direction of propagation at any instant and Eq. 9-8 gives the rate of change of direction of the normal to the front in terms of the gradients of the flow. We note that the second half of Eq. 9-8 reduces to

$$\frac{m}{n} = \text{const.}$$

which means that the projection of any ray on a y, z plane always yields a straight line. Thus, in complete generality, we may dispense with the direction cosine in the z direction and look at ray propagation in the x,y

[19] E. A. Milne, *Phil. Mag.*, *42*, 96 (1921).

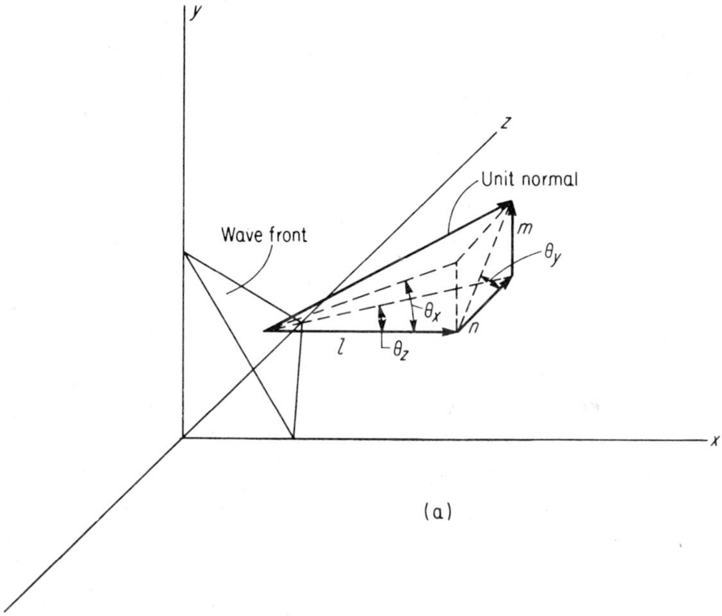

(a)

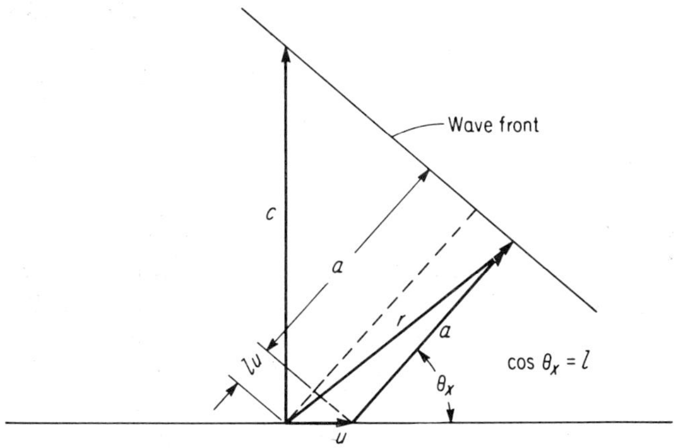

(b)

Fig. 9-18. (a) Direction cosine definitions for a three-dimensional orientation of the wave front. (b) Two-dimensional case. Unit normal is oriented in the direction of vector a but energy propagates in the direction of the vector **r**.

plane as shown in Fig. 9-18b. Since $n = 0$, we may write

$$m = (1 - l)^{1/2}$$

and the final parametric equations become

$$\frac{dx}{dt} = la + u \tag{9-9}$$

$$\frac{dy}{dt} = (1 - l)^{1/2}a \tag{9-10}$$

and

$$\frac{dl}{dt} = (l^2 - 1)\left(\frac{\partial a}{\partial x} + l\frac{\partial u}{\partial x}\right) \tag{9-11}$$

However, since all the flow variables and gradients are functions of x only, we may eliminate time by combining Eq. 9-9 and Eq. 9-11 to obtain the equation

$$(la + u)\frac{dl}{dx} = (l^2 - 1)\left(\frac{da}{dx} + l\frac{du}{dx}\right) \tag{9-12}$$

Equation 9-12 has the solution

$$c = \frac{a + lu}{(1 - l^2)^{1/2}} \tag{9-13}$$

where c is a constant along any ray as it propagates throughout the variable u, a field. By inspection of the vector diagram, Fig. 9-18b, we note that geometrically c is the apparent uncovery velocity of the wave front in the y direction at fixed x for any value of u,a and l. We also see that Eq. 9-13 may be solved for l to yield the equation

$$l_{\pm} = \frac{-au \pm c\sqrt{c^2 + u^2 - a^2}}{u^2 + c^2} \tag{9-14}$$

Thus at any position (and therefore fixed value of u and a) there are either 0, 1 or 2 values of l, depending on the value of c. A plot of l for a number of fixed values of c as a function of ϕ or x is shown in Fig. 9-19 for the detonation diagramed in Fig. 9-2. The arrows on the lines of constant c in this figure represent the behavior of the rays with time as given by Eqs. 9-9 and 9-11. The (l, ϕ) behavior is obtained by transforming the equations using Eq. 9-6a. We note that a horizontal line in Fig. 9-19 represents a coherent wave front which initially has a fixed constant inclination in the flow field, while a vertical line in this figure represents a wave front which is initially a sound circle at a point in the flow. We also note that there are two singular points in this flow, one being the saddle point at the intersection of the c = constant curves, the other being at the point $(-1, 1)$ in the (l, ϕ) plot or $(-1, x_{CJ})$ in the (l, x) plot.

We wish to investigate the nature of the saddle point in more detail. From Eq. 9-14 we note that for every value of x there is a minimum value

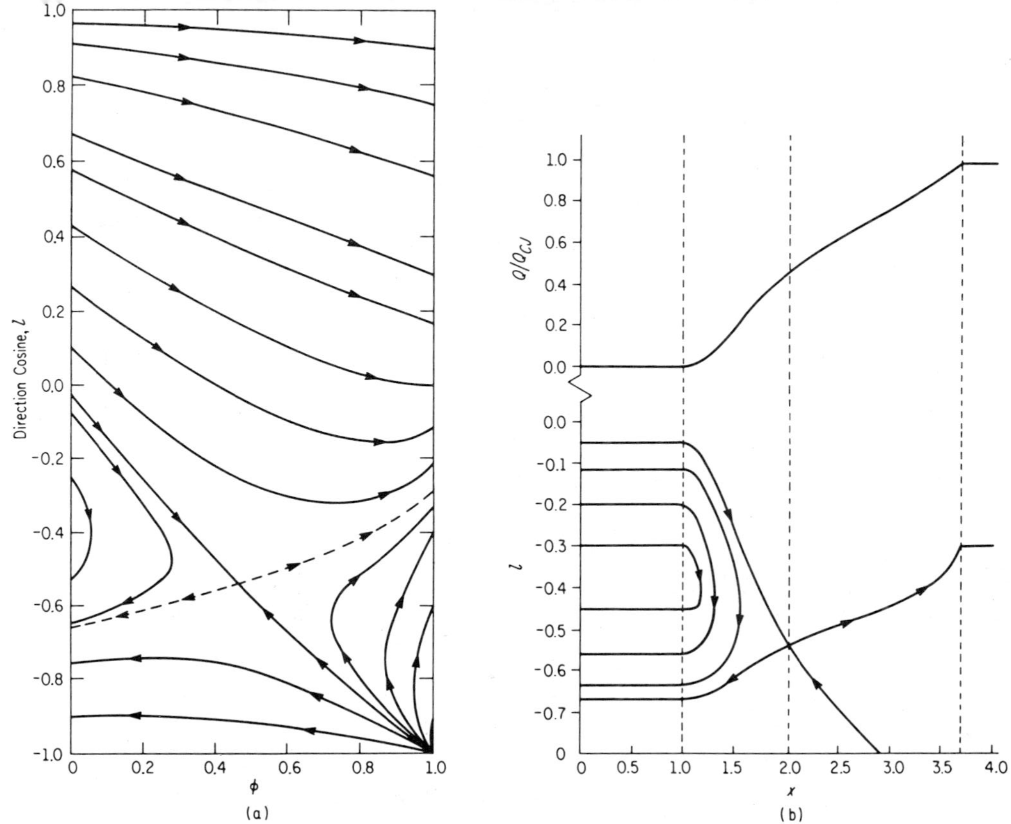

Fig. 9-19. $M = 6$, $\gamma = 1.4$ detonation of Fig. 9-2. (a) An (l, ϕ) plot of the ray behavior for a number of c values. (b) An (l, x) plot showing the mapping of the ray behavior into a physical plane for a specific heat addition model. $x = 1$ at the end of the induction zone, the $\phi_{\parallel}$ plane occurs at $x = 2.04$ and the CJ plane occurs at $x = 3.70$. Arrows on the curves indicate direction of ray propagation.

of $c = (a^2 - u^2)^{1/2}$ at which l has but one value. At this point

$$l = -\frac{u}{a}$$

In other words, the ray of critical orientation has a direction cosine equal to the negative of the local Mach number of the flow. From Eq. 9-9 we see that for this orientation of the wave front and precisely at this point in the flow

$$\frac{dx}{dt} \equiv 0$$

which means that the ray (or the energy trapped in the wave front) is propagating purely in the y direction. Furthermore, substituting this unique value of l into Eq. 9-11 yields the equation

$$\frac{dl}{dt} = \frac{(u^2 - a^2)}{2a^3} \frac{d}{dx}(a^2 - u^2)$$

which means that $dl/dt = 0$ at the point in the flow where

$$\frac{d}{dx}(a^2 - u^2) = 0$$

We note from Fig. 9-2 and from the nature of the equations that a one-dimensional detonation in which Dq/Dt is everywhere positive has a plane at which the quantity $(a^2 - u^2)$ has a simple maximum and therefore that this detonation has a plane in the flow for which a wave front of orientation $l = -u/a$ will have $dl/dt = 0$ and $dx/dt = 0$. Therefore in this case a ray will propagate parallel to the shock for an infinitely long time without changing its orientation. The behavior of a number of the rays connected with a specific wave front emanating from a sound source in this type of flow field are shown in x, y coordinates in Fig. 9-20.

Up until this point, even though we utilized the model of Fig. 9-2 to discuss the behavior of transverse waves, the resulting conclusions have been quite general and may be applied to any u,a field constructed for a particular kinetic scheme in a gas with a complex thermal and caloric equation of state. However, in order to discuss the condition for the existence of a saddle point in the flow and the amplitude behavior of the transverse waves propagating in its neighborhood, we must now reduce mathematical complexity by simplifying the detonation model. We therefore return to the model of a constant-gamma gas with heat addition developed in Sec. 9-2 to obtain the ray equations

$$\frac{d\phi}{dt} = (l\mathcal{A} + \mathcal{U})\frac{d\phi}{dx} a_1 \tag{9-15a}$$

and

$$\frac{dl}{dt} = (l^2 - 1)\left[\frac{\partial \mathcal{A}}{\partial \phi} + l\frac{\partial \mathcal{U}}{\partial \phi}\right]\frac{d\phi}{dx} a_1 \tag{9-15b}$$

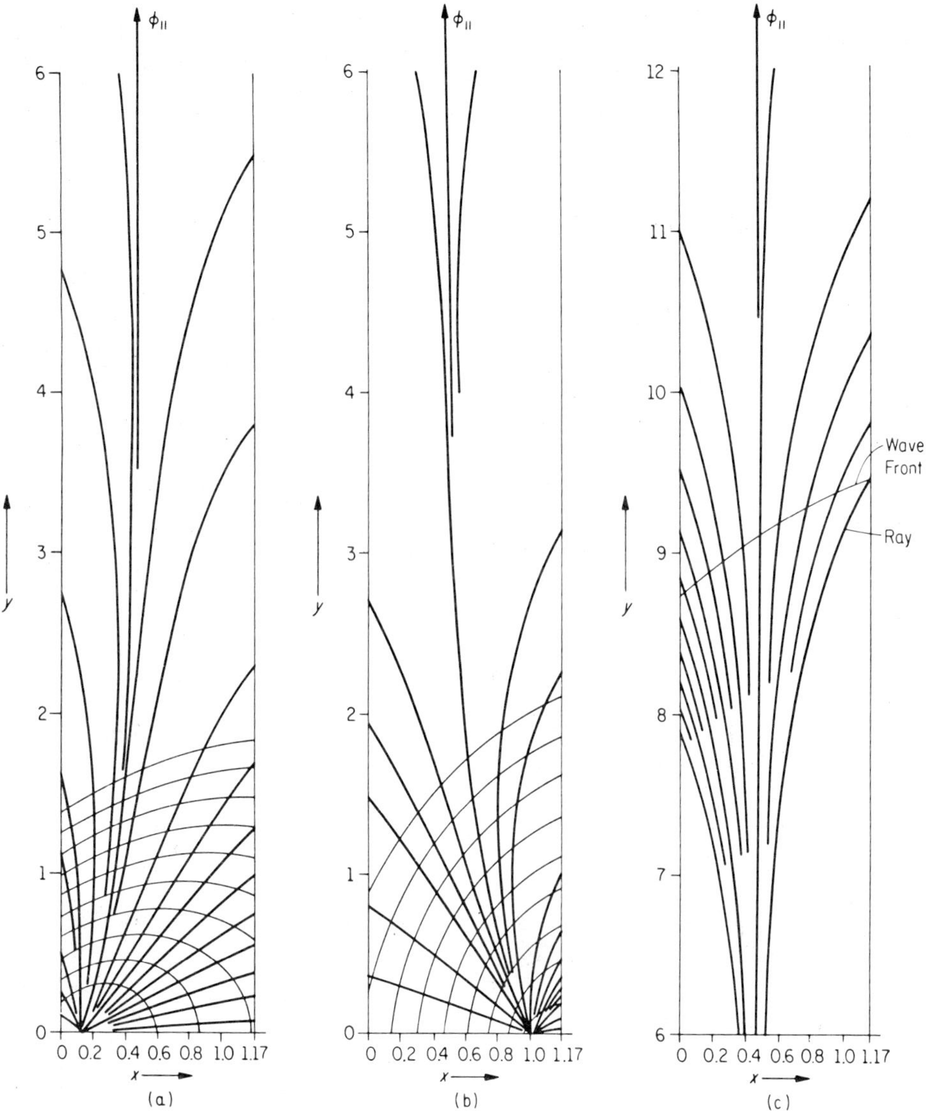

Fig. 9-20. The behavior of rays and wave fronts emanating from a sound source in the reactive region of a detonation with no induction zone. Reflection from the shock is not shown ______ wave fronts, ________ rays. Figure (c) is the late behavior of the rays from either (a) or (b).

where $\mathcal{A} = a/a_1$ and $\mathcal{U} = u/a_1$. By solving for the conditions that $dl/dt = 0$ and $d\phi/dt = 0$, we find that the ϕ value for the plane of a ray propagating in the y direction without curvature is

$$\phi_\parallel = 1 - \left[\frac{1 + \gamma M^2}{2(M^2 - 1)}\right]^{1/2}$$

and that the orientation of such a ray is given by the relation

$$l_\parallel = -1/(2 + \gamma)^{1/2}$$

Thus we see that there is a lower Mach number for the occurrence of $\frac{d}{dx}(a^2 - u^2) = 0$ in a constant-gamma detonation with strictly exothermic reactions and that this Mach number is always lower than $M = 3.0$. We also note that the local subsonic Mach number of the flow at the critical point is always between 0.5 and 0.6 for all realistic values of gamma. We now wish to use Eqs. 9-15a and 9-15b to discuss the behavior of rays in the neighborhood of $(\phi_\parallel, l_\parallel)$, and to this end we rewrite these equations in terms of the new variables

$$\xi = \phi - \phi_\parallel$$

and

$$\eta = l - l_\parallel$$

for small values of ξ and η. After dropping all quadratic and higher terms in ξ and η we obtain the relations

$$\frac{d\xi}{dt} = \left[\frac{(2 + \gamma)^{1/2}(1 + \gamma M^2)}{2(\gamma + 1)M}\eta + \frac{(M^2 - 1)^2}{(2 + \gamma)M(1 + \gamma M^2)}\xi\right]\frac{d\phi}{dx}a_1 \qquad (9\text{-}16)$$

$$\frac{d\eta}{dt} = \left[-\frac{(M^2 - 1)^2}{(2 + \gamma)M(1 + \gamma M^2)}\eta + \frac{2(\gamma + 1)^2(M^2 - 1)^4}{(2 + \gamma)^{5/2}M(1 + \gamma M^2)^3}\xi\right]\frac{d\phi}{dx}a_1$$

(9-17)

We note that neither of these equations is singular at $(\xi, \eta) = (0, 0)$ but that their ratio is.

We are interested in determining the behavior of a ray bundle (defined as the totality of rays contained in a short element of a coherent wave front propagating in the field) near the $(\xi, \eta) = (0, 0)$ point after it propagates for some time. That is, we want to look at the ray bundle surrounding the point $(\xi, \eta) = (0, 0)$. This represents the asymptotic behavior of the wave front traveling parallel to the shock. Therefore in the (ϕ, l) plane we wish to find the ray direction given by the equation

$$\frac{\xi}{\eta} = \frac{d\,\xi/dt}{\delta\,\eta/dt} = \lambda$$

where λ is a constant. Solving Eqs. 9-16 and 9-17 with this requirement yields two unique values for λ,

$$\lambda = \frac{(2+\gamma)^{3/2}(1+\gamma M^2)}{2(\gamma+1)^2(M^2-1)^2}[1 \pm (2+\gamma)^{1/2}] \tag{9-18}$$

substitution into Eq. 9-16 yields the relation

$$\frac{d(\ln \xi)}{dt} = \frac{\pm(M^2-1)^2}{M(1+\gamma M^2)}\frac{\partial \phi}{\partial x}a_1 \cdot \frac{1}{(2+\gamma)^{1/2}} \tag{9-19}$$

Equation 9-19 may be rewritten in x coordinates by substituting the variables from the ZND model to obtain the equation

$$\frac{d(\ln \delta x)}{dt} = \frac{\gamma - 1}{a_\parallel^2}\frac{dq_\parallel}{dt}(2+\gamma)^{1/2} \tag{9-20}$$

where $x_\parallel = x - \delta x$. In Eq. 9-20 we choose the positive sign, since this represents the spreading of a ray bundle leaving the point $(\phi_\parallel, l_\parallel)$ in the reactive-flow region.

Since a ray bundle which is properly oriented at the $\phi_\parallel$ plane remains at the $\phi_\parallel$ plane but is spreading with time its amplitude must be decreasing as time progresses because of this spreading. However, this acoustic signal is propagating in a reactive region which has other unique properties which also affect its amplitude behavior. If we assume that we can neglect the effects of the cross-flow, we find that we may discuss this behavior by following the behavior of an acoustic signal propagating in a truly homogeneous reactive flow with constant-background properties. Therefore the acoustic theory introduced in Chapter 8 may be applied. However, because the ray bundle is spreading we must derive the acoustic relation for a wave propagating down a quasi-one-dimensional stream tube whose walls are spreading at a rate given by the Eq. 9-20. Using the usual linearization assumptions we obtain the equation

$$\frac{\partial^2 s}{\partial t^2} - a_\parallel^2\frac{\partial^2 s}{\partial \nu^2} = \frac{\gamma - 1}{a_\parallel^2}\left(\frac{dq_\parallel}{dt}\frac{\partial s}{\partial t} + \frac{d^2(\Delta q_\parallel)}{dt^2}\right) - \frac{\partial s}{\partial t}\cdot\frac{\partial(\ln \delta x)}{\partial t} \tag{9-21}$$

where the direction ν has a direction cosine $l_\parallel$ to the x direction. In order to apply Eq. 9-21 to the case of a wave propagating in the reactive region of the detonation we must assume that dq/dt is known for the region of interest. We therefore assume that the reaction is either a simple first-, second-, or third-order reaction in the neighborhood of the $\phi_\parallel$ plane. For a first- or second-order reaction we assume that the reaction has an activation energy and that its rate is given by the relation

$$\frac{Dq}{Dt} = A\,\rho^{(k-1)}\exp(-E/RT) \tag{9-22}$$

where k = the order of the reaction. Since third-order recombination reactions can also occur in detonations and are exothermic, we model their rate, using the equation

$$\frac{Dq}{Dt} = B\rho^2 T^{\zeta} \tag{9-23}$$

Thus if we let $\zeta = E/RT$ in Eq. 9-22 we find that

$$\frac{d(\Delta q)}{dt} = \frac{dq}{dt}[\zeta(\gamma - 1) + (k - 1)]$$

for both types of reaction and that Eq. 9-21 becomes

$$\frac{\partial^2 s}{\partial t^2} - a_{\parallel}^2 \frac{\partial^2 s}{\partial \nu^2} - K\frac{\partial s}{\partial t} = 0 \tag{9-23}$$

where

$$K = \frac{\gamma - 1}{a_{\parallel}^2}\frac{dq_{\parallel}}{dt}[\zeta(\gamma - 1) + k - (2 + \gamma)^{1/2}] \tag{9-24}$$

Equation 9-23 is of the same form as Eq. 8-8 and has a solution of the form

$$s = \exp(Kt/2) \cdot C_{-} \exp[i\omega(\nu - a_{\parallel}t)] \tag{9-25}$$

when the frequency ω is high. The negative sign is used in the exponential since the wave is always propagating in the plus ν direction. Since we have assumed a functional form for dq/dt we note that we may analytically determine the characteristic growth behavior of such a $(\phi_{\parallel}, l_{\parallel})$ wave front. If we assume that the chemical reaction is exothermic throughout the detonation, K will be positive if the bracket in Eq. 9-24 is positive and under these conditions Eq. 9-25 states that the amplitude of the wave will increase with time. The threshold value of ζ for positive amplification may be determined as a function of the reaction order and the gamma of the gas by setting $K = 0$ in Eq. 9-24. Figure 9-21 is a plot of this threshold value for the case of first-, second-, and third-order exothermic reactions occurring at the $\phi_{\parallel}$ plane in the detonation. Note that a second-order reaction with a positive activation energy will always yield amplification and that recombination reactions with realistic temperature sensitivities will also cause the amplitude of the acoustic signal to grow with time. Thus, all one-dimensional detonations are inherently unstable to the propagation of infinitesimal high-frequency transverse signals.

While we have not proved instability in general terms we have shown that two types of perturbations in a propagating detonation will produce the collapse of the one-dimensional steady ZND structure. This, of course agrees with the observation that in the reactive region of a detonation, one-dimensional behavior does not obtain. However, we have still not justified the existence of the finite amplitude structure that is observed.

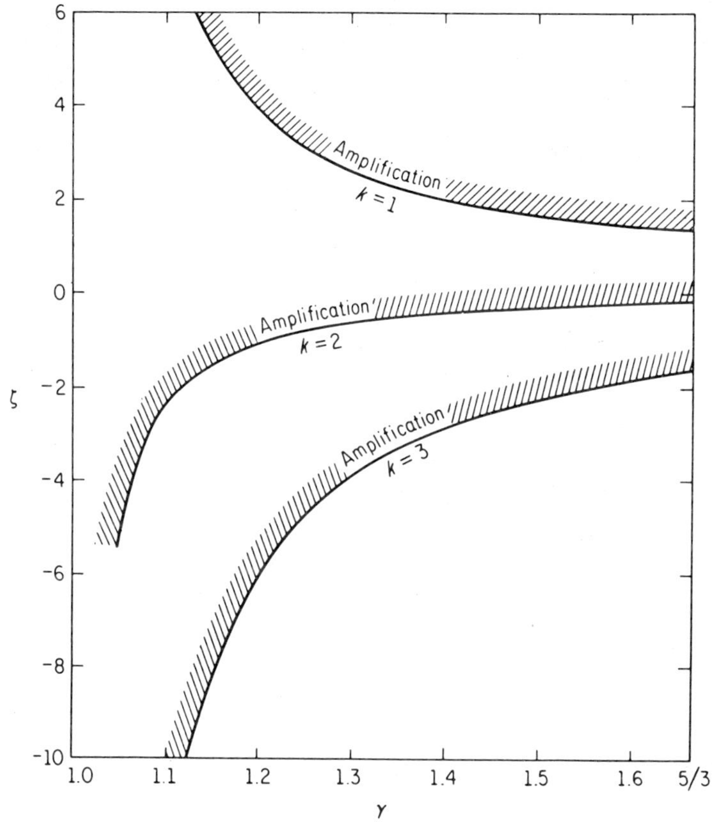

Fig. 9-21. Threshold values of ζ for positive amplification plotted for first-, second-, and third-order reactions vs. the gamma of the gas.

Even though at the present time this must remain an open question, one extension of the acoustic theory may be made which predicts transverse structure similar to that which is observed. In Fig. 9-19a consider the region between the shock and the $\phi_{\parallel}$ plane bounded by the c = constant curves that intersect at the point $(\phi_{\parallel}, l_{\parallel})$. Note that all rays in this region always propagate to the shock wave and recall that upon reflection from the shock wave the quantity c must be conserved because of the geometric relationship illustrated in Fig. 9-18b. Therefore, a ray of this orientation in this region of the flow will propagate away from the shock for some distance and then return, reflect and repeat its performance. The transient time between shock contacts for the different rays are, however, different. Therefore an initially smooth wave front trapped in this region will convolute with time and eventually produce multiple-shock contacts as shown in Fig. 9-22. Here we assume that the wave starts as a sound circle at the shock front (as might be produced by the shock wave over-writing a line bump on the wall of the detonation tube) and we numerically follow the wave-front behavior with time for a long period of time.

We see from this diagram that the active convolution point for this

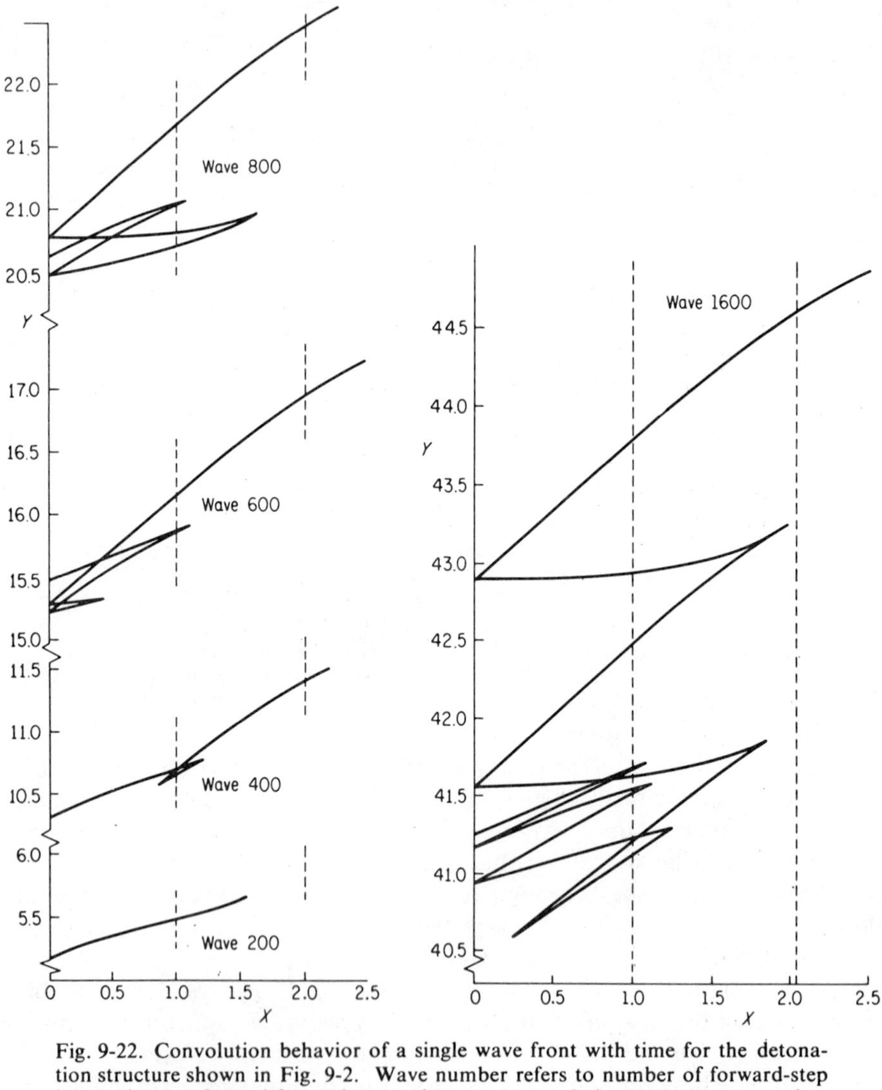

Fig. 9-22. Convolution behavior of a single wave front with time for the detonation structure shown in Fig. 9-2. Wave number refers to number of forward-step integrations performed for each wave drawn. At $t = 0$ the wave was assumed to be a sound circle located at the shock front at $y = 0$. (Adapted from H. O. Barthel and R. A. Strehlow, *Phys. Fluids*, *9*, 1896 (1966).)

wave front is located just inside the reactive region of the detonation, and that this active convolution point separates the wave front into two different types of convolution regions. One of these is asymptotically folding upon itself at the initial shock contact point. The reflection coefficient for this orientation of the wave front is -1. This branch of the convoluted wave front is therefore asymptotically destroying itself with time. The other branch, however, is asymptotically approaching a spacing which is uniform and which may be calculated from the behavior of the ray propagating through the point $(\phi_{\|}, l_{\|})$. The reflection coefficient for a wave of this orientation is approximately $+0.25$ and this wave will therefore retain its amplitude for long propagation distances. This is because as it asymptotically approaches the $(\phi_{\|}, l_{\|})$ behavior, a ray element will spend an increasingly longer time in the neighborhood of $\phi_{\|}$ and will be sufficiently amplified before it travels to the shock wave and suffers another reflection. Even though this qualitative statement may be made, the quantitative amplification behavior has yet to be discussed in this case because of the mathematical complexity of the problem. However, since the asymptotic behavior corresponds to a wave propagating through the point $(\phi_{\|}, l_{\|})$ acoustic spacing may be easily calculated for any detonation once the kinetics are known. Furthermore, this may be done without recourse to the constant-gamma heat-addition model as long as transport phenomena are neglected and the shock is assumed to be a true discontinuity. For this purpose one assumes any degree of complexity of the reaction kinetics, the equation of state and the caloric equation of state for the detonation, integrating the ZND equations to obtain the true $u(x)$, $a(x)$ field out to the point where $a^2 - u^2$ attains its maximum value. This allows a calculation of $c^2 = a^2 - u^2$ for the point $\phi_{\|}$ and therefore allows a calculation of the orientation of this critical wave front at every point between the shock and the $\phi_{\|}$ plane (using Eq. 9-13). The spacing may then be obtained from the equation

$$Z = \int_{x_s}^{x_{\phi\|}} \frac{l_+}{(1 - l_+^2)^{1/2}}\, dx - \int_{x_s}^{x_{\phi\|}} \frac{l_-}{(1 - l_-^2)^{1/2}}\, dx \tag{9-26}$$

Spacings calculated for hydrogen-oxygen detonations using Eq. 9-26 are compared to observed finite-amplitude spacings in Fig. 9-23. For these calculations, the true structure was modeled as realistically as possible using the best kinetic data available. The disagreement that is observed is undoubtedly due to the finite amplitude of the transverse waves that actually occur in these detonations.

Thus one-dimensional CJ flow has been shown to be nonexistent, both experimentally and theoretically, if one looks with sufficient detail. The reason why self-sustaining detonations tend to propagate at a velocity

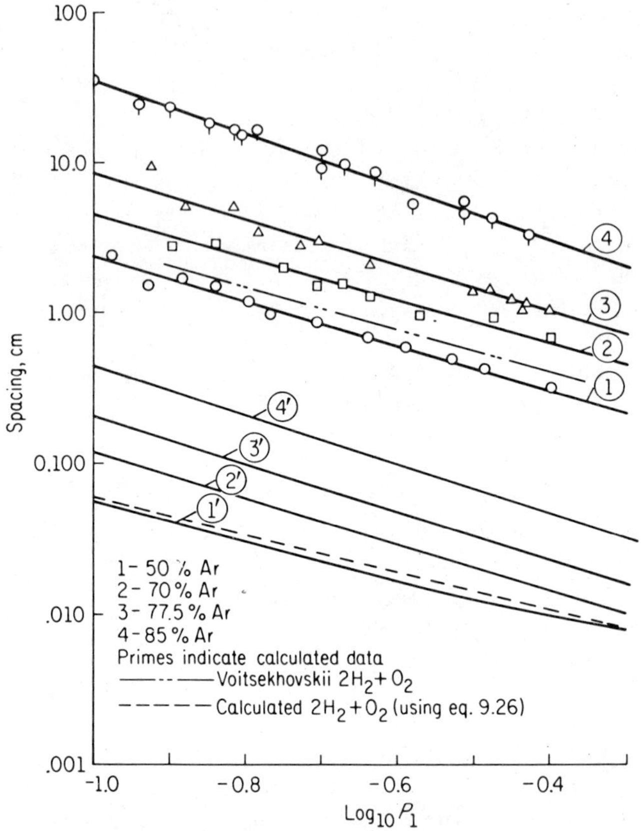

Fig. 9-23. Experimental and calculated transverse-wave spacing for dilute stoichiometric H_2–O_2 systems.

near the calculated CJ velocity of the mixture is therefore open to fresh interpretation. In the case of a detonation whose reaction-zone thickness is much less than the tube dimensions, there is no problem in applying the steady concepts between two stations, one ahead of the "steady" shock front and another sufficiently downstream of the front to be considered "one-dimensional equilibrium" flow. The small amount of energy trapped in the transverse wave motion in this case will not contribute significantly to the energy balance, and the CJ detonation solution therefore may be used as a valid representation of the flow. However, a justification for this observed velocity on the basis of a one-dimensional flow argument is no longer valid. Furthermore, the extensive recent observations of transverse structure point to the fact that further investigation of the stability must be based on quantitative investigations of this structure as a nonlinear and finite amplitude phenomena.

9-5. INITIATION OF DETONATION

Detonation initiation is fundamentally different from the processes which lead to either the formation of a flame or a homogeneous adiabatic explosion in a gas. Detonation may be initiated by a flame propagating from the closed end of a long tube, by a spark of sufficient energy, or by the passage of either an incident or reflected shock wave in a conventional shock tube. In all these cases a strong shock wave is observed to process the gas before initiation actually occurs, and the initiation process itself is a truly nonsteady wave process.

Initiation behind the reflected shock wave in a conventional shock tube produces the simplest one-dimensional initiation that has been observed experimentally. This process has been analyzed rather thoroughly by Strehlow and co-workers[20] and their results are summarized here. In mixtures containing a high percentage of argon, the shock-reflection process is quite ideal because the reflected shock is not grossly distorted by its interaction with the boundary layer (as it is in gas mixtures having a higher heat capacity). In this ideal case, the region between the reflected shock and the end wall of the shock tube is relatively quiescent and has quite uniform properties initially. Furthermore, because the reflected shock must be reasonably strong to produce high temperatures, the temperature jump across the reflected shock will be sufficiently large to cause a short delay to adiabatic explosion behind the reflected shock, whereas the concurrent delay behind the incident shock is quite long. Thus one may easily find a range of incident Mach numbers over which the first explosion process always occurs behind the reflected shock. This range of Mach number is bounded on the lower end by the weakest shock strength which will just cause initiation behind the reflected shock during the available run time, and on the upper end by an incident shock strength which is sufficiently high to cause initiation before reflection. In this short Mach-number range, initiation is observed behind the reflected shock.

Four types of initiation have been observed behind the reflected shock in these experiments. The first type is spurious and is due to the presence of bumps or crevasses on the tube walls or in the end plate. If large enough, these irregularities will destroy the one-dimensional nature of the flow and lead to local shock reenforcement and non-one-dimensional initiation. The second and third types of initiation are strictly one-dimensional processes and strikingly illustrate the nonsteady wave nature of initiation. Figure 9-24 contains schlieren and schlieren-interferometric photographs illustrating these two forms of initiation.

[20] R. A. Strehlow and A. Cohen, *Phys. Fluids*, *5*, 97 (1962); R. B. Gilbert and R. A. Strehlow, *AIAA J.*, *4*, 1777 (1966).

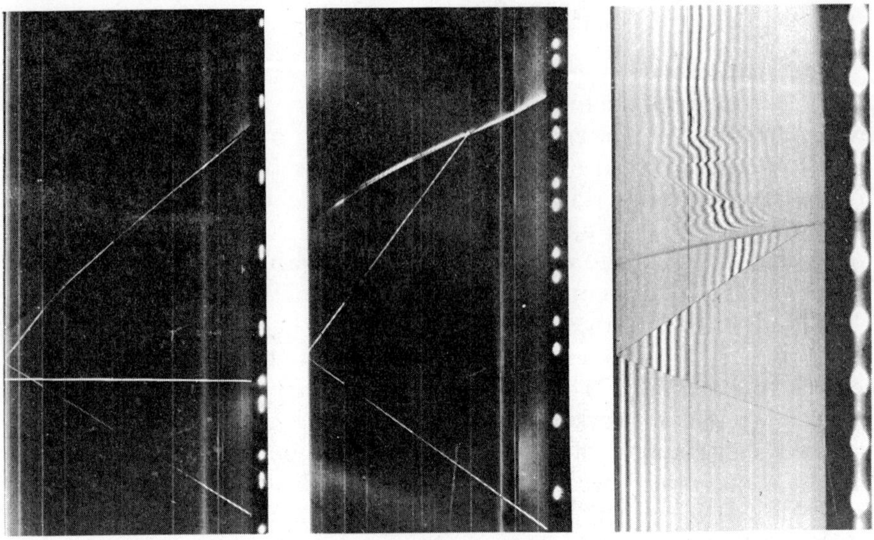

A B C

Fig. 9-24. (x, t) photographs obtained behind the reflected shock in a conventional shock tube containing $2H_2 + O_2 + 70\%$ argon mixtures. Time increases upward, back wall to the right, timing dots are 100 μsec apart on left. (a) Schlieren photograph, (b) Schlieren photograph, (c) Schlieren interferometric photograph.

Figure 9-24a illustrates one type of initiation, Figs. 9-24b and 9-24c the other. In both cases the initiation starts as a simple wave phenomenon traveling away from the end wall behind the reflected shock. This wave overtakes the reflected shock some distance from the end wall to produce a detonation in the incident gas flow. The nature of this strictly one-dimensional initiation process has been rather extensively analyzed by assuming that the development of a detonation under these conditions is primarily controlled by the presence of a "reaction wave" triggered by reflected shock passage. This nonsteady "reaction wave" is fundamentally different from the "flames" discussed in Chapter 7 because its velocity is dictated by (1) the reflected shock velocity, and (2) a subtle interaction of local gas dynamics and reaction kinetics.

Two separate theoretical approaches have been used to model flows of the type illustrated in Fig. 9-24. In the first a constant-gamma gas containing a thermally neutral chain-branching reaction with an activation energy and oxygen concentration dependence similar to that of the high-temperature hydrogen–oxygen induction zone reaction was assumed to control the length of the induction zone along each particle path. This induction region was assumed to be followed by a region of heat addition in which the effect of the recombination reactions could be modeled by the equation

$$\frac{Q}{a_{rs}^2} = \frac{Q_{\max}}{a_{rs}^2}\,[1 - \exp(-t/\tau_{\mathrm{rec}})]$$

where τ_{rec} is a characteristic time and t is measured from the end of the induction zone for each fluid particle. This model is a reasonable representation of the recombination kinetics for the hydrogen–oxygen system. This model was programmed for an (x,t) method of characteristics calculation in a constant-gamma gas with the boundary conditions $u = 0$ at the end wall and $u = 0$ immediately downstream of the reflected shock. In this model, each fluid element of the quiescent gas downstream of the reflected shock initially starts "exploding" at a constant time after shock passage. Thus the system first produces a reaction wave which is traveling away from the end wall at the reflected shock velocity. The initial velocity of the reaction wave (i.e., the locus of the beginning of heat addition in the x,t plane) will therefore be highly subsonic relative to the gas. However, an adiabatic explosion occurring in the quiescent region at the end wall results in a local pressure rise and causes a compression wave to propagate ahead of the reaction wave. This process has been found to be very strongly self-accelerating because the compression process both heats and raises the density of a gas element which has not yet completed its induction reaction. Thus the induction period of an element of gas which is located some distance from the end wall is decreased by this process. This causes heat addition to occur earlier, and the compression wave is accelerated. This model has been very favorably compared to observations of initiation in the hydrogen–oxygen system. These results are summarized in Fig. 9-25. We note from Fig. 9-25a that the calculated reaction-wave–shock-wave pattern overlays the experimental reaction-wave–shock-wave pattern exactly (the reaction wave and reaction shock wave are unresolvable on the photograph). We also see that the calculated particle paths asymptotically approach the experimental contact surface motion in the figure. In Fig. 9-25b we have plotted the calculated (P, V) behavior of various individual particle paths which start at the condition rs, at different distances from the end wall of the tube. These show that an element of gas at the end wall undergoes a simple adiabatic explosion (in a vessel whose volume increases slightly during the process), while all other particles first undergo a shock transition (shown either as a dotted line or as an absence of a line) followed by a further isentropic compression process until heat addition occurs. At this point there is a discontinuous change of slope to a slope corresponding to the heat-addition region of the flow along that particle path. We note that for particles near the end wall the pressure still rises continuously during heat addition but that farther from the end wall the particles undergo an expansion process during heat addition. As is

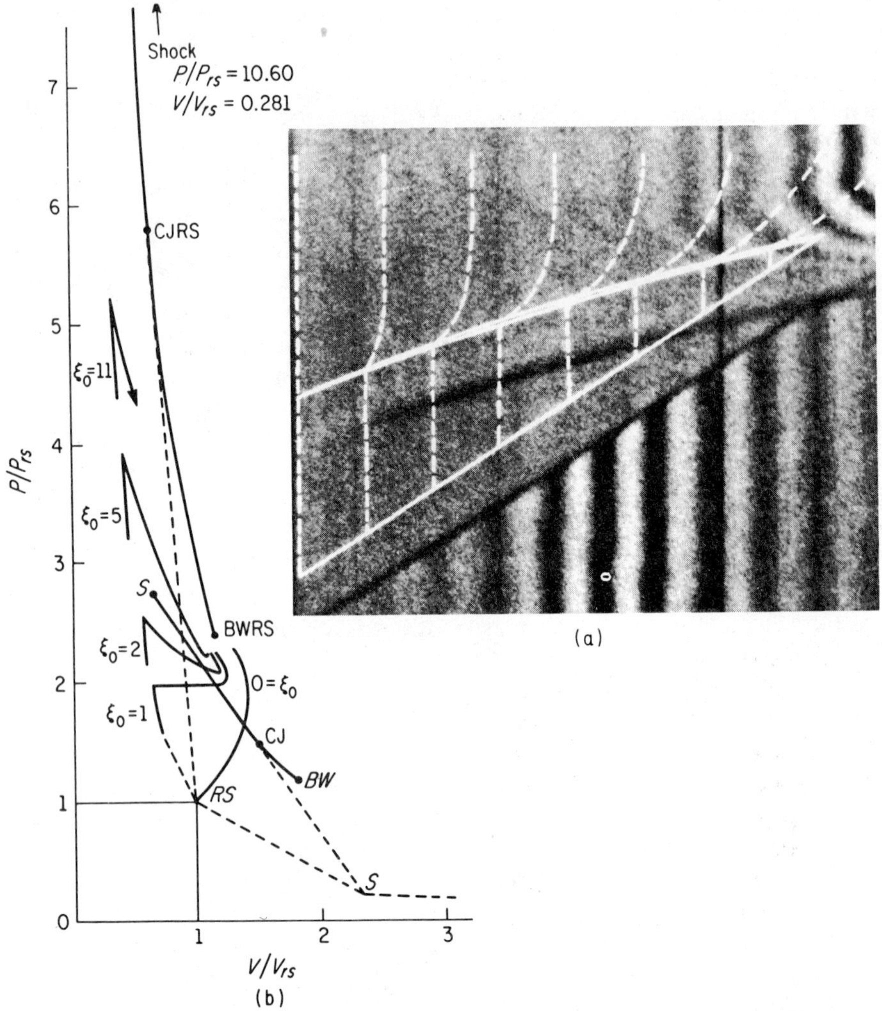

Fig. 9-25. (a) Comparison of an (x, t) experimental photograph with a theoretical wave pattern calculated for the initial conditions. Theoretical pattern is displaced upward about one-third of the delay time at the back wall on the left. (b) A (P, V) diagram for the process shown in (a). Each ξ_0 line is the (P, V) behavior of a particle path. ξ_0 is the dimensionless distance of the particle from the back wall at the moment it is overwritten by the reflected shock.

shown, this process could only be followed out to the dimensionless distance $\xi_0 = 11$ in the calculation. The CJ detonation behavior for the region behind the reflected shock is plotted on Fig. 9-25b for comparison to the behavior of these nonsteady path process lines. We note (without any real proof) that the process appears to be systematically approaching the heat-addition behavior dictated by steady CJ conditions behind the reflected shock. The point BWRS in this diagram has been calculated for the isentropic expansion of the gases leaving the traveling detonation at the point CJRS to the end-wall velocity through a rightward-facing expansion fan. We note further that the final pressure along each streamline in the nonsteady calculation appears to be approaching this pressure but that the final volume of each point will be slightly different for each path process. This results from the fact that entropy addition along the different streamlines is different due to processing by a variable-strength shock and to the assumption that Q_{max} = const. for each particle (which leads to a different chemical entropy change along each particle path). The different nature of the two processes shown in Fig. 9-24 may be explained by noting that the line S-CJ-BW in Fig. 9-25b is the locus of a CJ detonation occurring behind the incident shock in the flow. Thus if the acceleration process produces a local pressure above that of point S, the interaction of the reflected shock, and the reaction shock will initially produce an overdriven detonation wave in the incident gases as shown in Figs. 9-24b and 9-24c and in Fig. 9-25a. However, if the acceleration process has not progressed for a sufficient time to yield a pressure above the pressure at point S at the time of intersection, the compression wave will simply cause the reflected shock to accelerate to a CJ detonation wave as shown in Fig. 9-24a.

A few more comments about this model and its agreement with experiment may be made. First we note that the model implicitly assumes that transport phenomena are absent or not important. Thus the "reaction wave" of this model has no similarity to the "flame" of Chapter 7 and 8. Second, we have assumed that the local kinetics at any point in the flow may be modeled by assuming that the kinetics of an ordinary static system may be applied at each point in the flow. At present the validity of this assumption is justified only by the results. There is no *a priori* reason for its use except that it is the simplest assumption that one can make to model these nonsteady flows. Third, the model predicts that the reaction shock wave will always start as a sound wave at the end wall and that its acceleration starts at zero and goes through a maximum before again returning to zero. This behavior is primarily the result of the assumption that the rate of heat addition is finite along each particle path, and that heat addition starts along a subsonic reaction wave. Fourth, it is sometimes observed experimentally that a weak compression wave precedes the shock wave for some distance. The model discussed here does not predict this behavior because it assumes that there is a discontinuous change in the rate of heat addition at the start of the heat-addition process. If one were to remove this restriction the model would undoubtedly yield a relatively broad compression region ahead of the reaction shock in the neighborhood of the end wall. Lastly, we note that sub-CJ nonsteady shock waves are both observed experimentally and predicted theoretically for this case of homogeneous one-dimensional initiation. Thus we see that sub-CJ shock waves followed by an exothermic reaction are physically realizable as nonsteady processes.

In another theoretical study of reflected shock initiation Shear[21] has assumed that the reaction has a simple first-order thermal explosion mechanism and he has followed the flow pattern using Lax's[22] calculation technique. The (x,t) pattern that he obtained for a typical initiation calculation in this type of system is shown in Fig. 9-26. We note that this system exhibits the same type of self-acceleration behavior that a

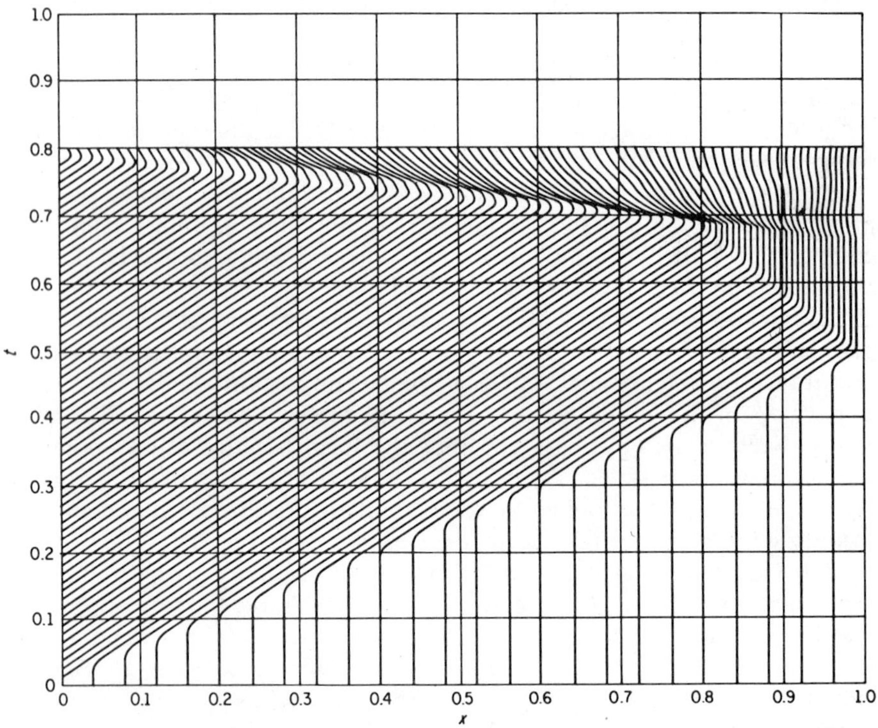

Fig. 9-26. (x, t) diagram of reflected shock initiation for an exothermic reaction. A first-order homogeneous reaction with an activation energy of 28.6 keal is assumed. The technique of Lax is used in the calculation. Particle paths are shown. (Courtesy R. E. Shear; originally appeared in *BRL MR* 1743, May 1966, Ballistics Research Laboratory, Aberdeen Proving Ground, Md.)

[21] R. E. Shear, "Reflected Shock Initiation of a Chemical Reaction," BRL Memo Report MR 1743, May 1966 (Aberdeen Proving Ground, Md.).

[22] P. D. Lax, *Comm. Pure and Appl. Math*, *7*, 159 (1954).

chain-branching mechanism exhibits and that once again one obtains a homogeneous and one-dimensional initiation process. This behavior has not yet been verified experimentally because of the difficulty of finding a chemical system which exhibits a purely thermal-explosion behavior at these high temperatures.

The fourth type of reflected shock-initiation behavior that has been observed in the hydrogen–oxygen system is illustrated in Fig. 9-27. In

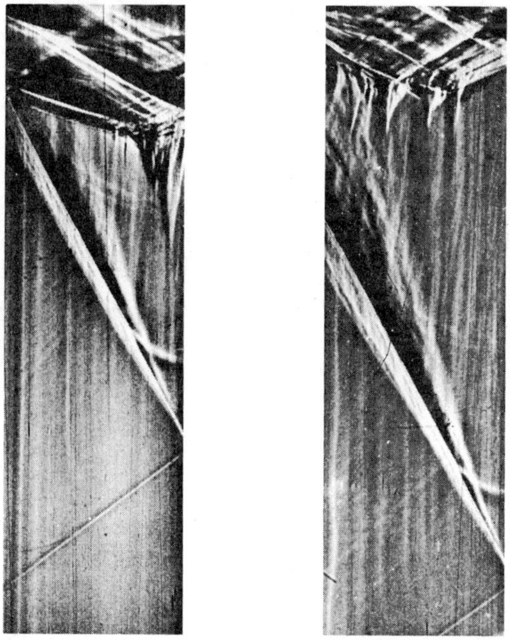

(a) (b)

Fig. 9-27. Reflected shock-initiation photographs similar to those shown in Fig. 9-24, illustrating the occurrence of hot-spot initiation at low temperatures. Hydrogen–oxygen system, time increasing upward, back wall to the right. (Courtesy R. I. Soloukhin, Novosibirsk State University, Novosibirsk, U.S.S.R.)

this case, even though the system is apparently homogeneous, hot spots of weak ignition occur and these each propagate spherically growing flames. When either these flames or their pressure waves coalesce they produce a local detonation and the initiation process is no longer one-dimensional. This type of initiation has been observed to occur in hydrogen–oxygen systems at high pressure and low temperature where the reaction

$$M + \mathrm{H} + \mathrm{O_2} \rightarrow \mathrm{HO_2} + M$$

becomes important during the chain-branching portion of the reaction. Currently no satisfactory reason may be given for this behavior except that it is observed to occur under these conditions.

An extension of the reaction theory of initiation to the process of initiation behind an accelerating incident shock has led to the observation that the application of these same principles may be used to predict the locus of the explosion line behind an accelerating-shock wave. The accelerating shock has been produced by passing an incident-step shock wave into a slowly converging channel so that the phenomenon may be approximately modeled as a one-dimensional process. In this case the locus of the explosion line may be determined by finding the fraction of explosion delay time, ψ, consumed at any point along a particle path by integrating the equation

$$\psi = \int_{t_s}^{t} \frac{dt}{\tau}$$

where τ is the local value of the explosion delay time and t_s the shock passage time along that particle path. A typical calculation of the flow properties and explosion delay locus ($\psi = 1$) is shown in Fig. 9-28. In

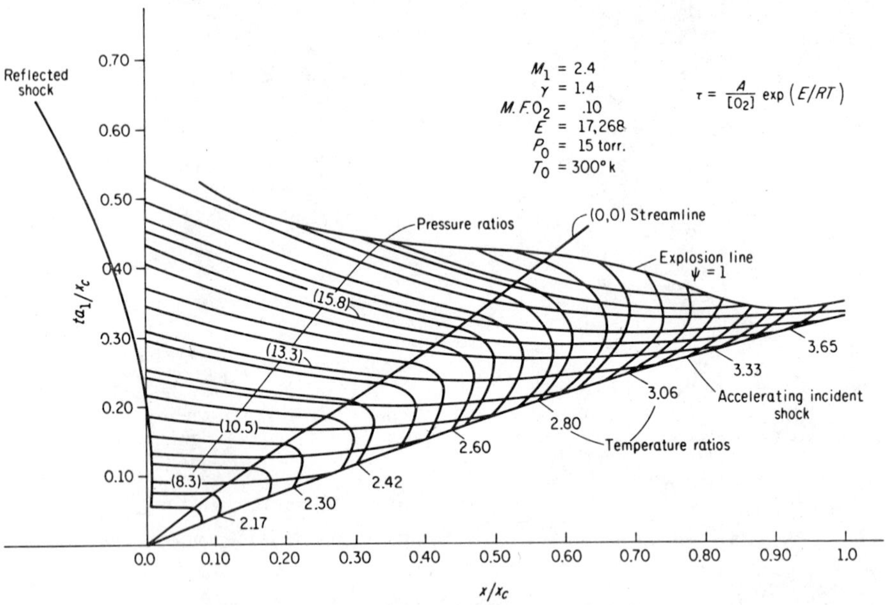

Fig. 9-28. Isopressure and isotemperature lines behind an accelerating shock wave produced by passing a step shock into a slowly convergent channel. The locus of the explosion line for high-temperature hydrogen–oxygen induction-delay time is shown. [Adapted from R. A. Strehlow, A. J. Cooker, and R. E. Cusey, *Comb. and Flame*, *11*, 339 (1967).]

this calculation heat was not added at the explosion line. Therefore its minimum in true time is the only explosion detail which may be calculated by this technique. Note that the isopressure and isotemperature lines in this plot are parallel in the region to the left of the streamline running through the point (0,0). This region is an isentropic flow region (because the reflected shock is weak), and the local pressure history therefore conforms to the local temperature history. However, to the right of the entrance streamline the temperature rises markedly because of the large entropy gradient produced by shock acceleration. Thus, as shown in this figure the locus of the explosion line obtained by integrating Eq. 9-27 along a particle path from the shock wave to the point where $\psi = 1$ exhibits a dip in the (x,t) plane, even though the total delay time along neighboring particle paths is decreasing monotonically as the shock strength increases. This behavior has been verified experimentally. A typical set of initiation photographs taken in schlieren and visible light is shown in Fig. 9-29. We note from Fig. 9-29 that hot-spot initiation occurs in the systems because initiation always occurs at rather low temperatures in a slowly converging channel. The hot-spot initiation that is observed has no relation to tube geometry from run to run and the overall initiation behavior is undoubtedly due to the nonsteady quasi-one-dimensional shock acceleration effects illustrated in Fig. 9-28.

The other techniques of initiation, from flames or sparks, have not been analyzed as extensively as the above techniques. However, direct-spark initiation is known to occur only under conditions where the temperature and pressure behind the initiating spherical shock wave are maintained at a high value for a sufficiently long time to cause a rapid exothermic reaction in the gas mixture. Thus direct initiation in this case is again seen to be an entirely different process than the spark ignition of flames. In the flame case the shock is weak and unimportant, and heat transfer from the hot kernel of spark-heated gas produces ignition. However, as in the other cases of detonation initiation discussed above, direct-spark initiation is found to be explainable by invoking a shock-generated reaction wave, without transport.

Flame initiation in a closed tube is a much more complex type of initiation then any of the above. In some cases the detonation is observed to appear well ahead of the flame and then propagate both a backward-facing retonation and a forward-facing detonation wave. However, in other cases the first apparent breakout of "detonation" appears to occur in a region which already contains flame and the process leading to initiation is clouded. It should be pointed out that in all cases of flame initiation the flow processes are three-dimensional and complex and the only valid general statement that can be made is that detonation always

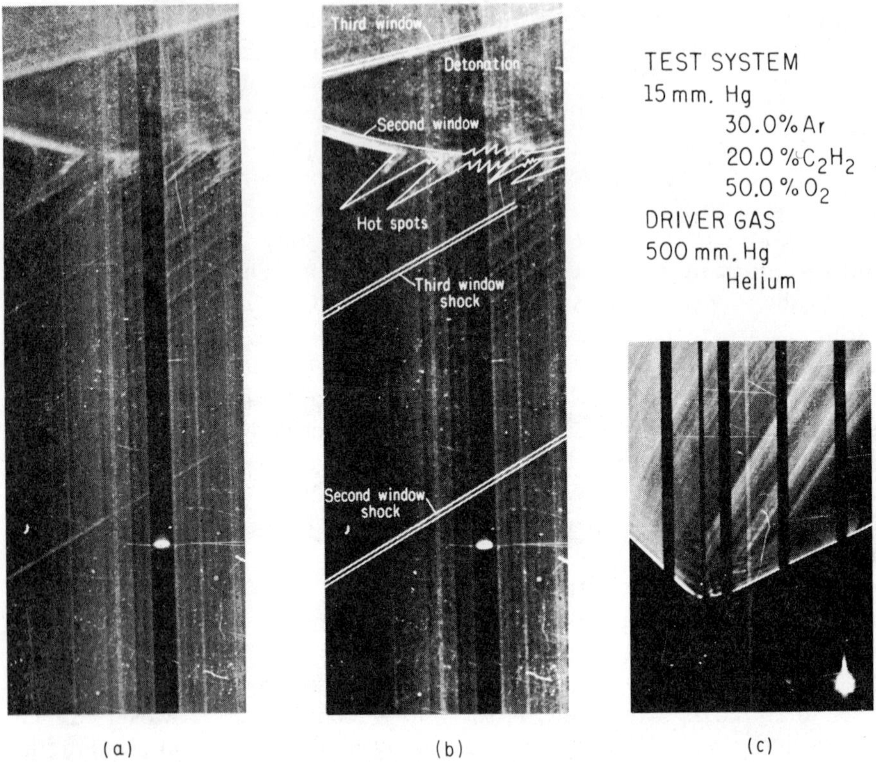

Fig. 9-29. An experimental initiation photograph taken through the last portion of a 20-ft-long convergence section with an overall area ratio of 14 to 1. The windows are 10 in. long. The end of the converging section occurred in the middle of the fifth window. See Fig. 9-28 for reference.

appears in a region which has been recently traversed by a strong shock wave.[23]

One interesting observation that has been made during initiation experiments is that transverse waves appear spontaneously even during the most nicely controlled one-dimensional initiations. Figure 9-30 is a smoked-foil record taken on the bottom wall of the shock tube during the initiation experiment of Fig. 9-24c. Note that the structure appears some distance from the point of initiation at the end wall, and that the transverse waves are weak at their appearance point and then grow in amplitude as they propagate away from the end wall. This is evident from the refraction of the waves at their intersections. We also note that the transverse waves can appear before the reaction shock intersects

[23]See, for example, P. A. Urtiew and A. K. Oppenheim, *Proc. Royal Soc. A295*, 13 (1966).

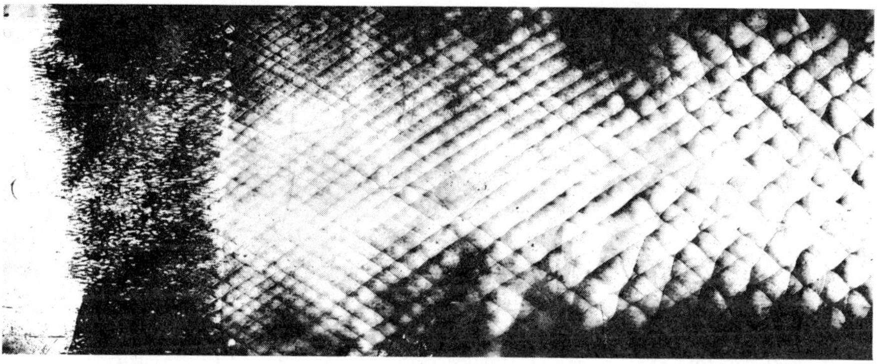

Fig. 9-30. A smoked-foil record taken during the initiation experiment of Fig. 9-24c. Back wall is to left. Discontinuity in writing pattern about $3\frac{1}{2}$ in. from back wall is point at which the reflected shock and reaction shock merged to produce a detonation wave. The original smoked-foil is 4 in. wide.

the reflected shock. This means that the occurrence of transverse waves is not limited to CJ detonation, but that they can appear spontaneously on a shock which is simply followed by an exothermic reaction. Also the smoke-foil records taken during initiation have always shown that the transverse waves appear with a regular spacing which has no relation to the tube's geometry. This supports the contention that acoustic-wave convolutions can be the source of the regular spacing of transverse phenomena. However, the large ratio that exists between the spacing of finite amplitude transverse waves, and the acoustic theory spacing shows that an acoustic mechanism is not operative during self-sustained propagation.

9-6. SUPERSONIC COMBUSTION

The term *supersonic combustion* describes a process which is relatively new technologically. In a supersonic-combustion device the purpose is to utilize a combustion process to produce thrust without allowing the flow to become subsonic at any point in the engine. We choose to discuss the principles of this phenomenon in this chapter because of its similarity to the "reaction wave" combustion of a one-dimensional detonation.

Let us again consider the Hugoniot and Rayleigh equations for one-dimensional flow with heat addition, Eqs. 9-1 and 9-2. As we showed earlier, these equations may be solved by eliminating the pressure ratio Y to yield Eq. 9-5 for the density ratio as a function of the extent of reaction and incident Mach number. Equations 9-1 and 9-2 may also be solved for Y by eliminating ϵ to yield an analogous equation for the pressure ratio between station 1 and 2:

$$Y = \frac{1 + \gamma M^2 \pm \gamma(M^2 - 1)\sqrt{1 - \phi}}{(\gamma + 1)}$$

where M is the Mach number of the approach flow. Once again we note that for $\phi = 0$ the solution obtained by using the negative sign is the trivial solution $Y = 1$, while that obtained by using the positive sign is the pressure ratio for a simple shock transition from state 1 to state 2. We now wish to examine the flow properties of the solution which starts at the trivial point $Y = 1$ and as heat is being added continuously approaches the Hugoniot along the lower Rayleigh path in a P, V plane. This corresponds to the classical "weak" detonation solution and represents a shockless transition in which the flow remains supersonic throughout.

We therefore see that strictly one-dimensional supersonic combustion would be obtained if we could trigger an exothermic reaction in a supersonic flow without the presence of a normal shock wave (which would simultaneously render the flow subsonic). One further requirement is that the initial Mach number be sufficiently high so that the full heat addition (by the chemical reaction) will not thermally choke the flow. In general, it has been observed that choking causes a shock to propagate upstream through the reaction zone and produce a detonation.

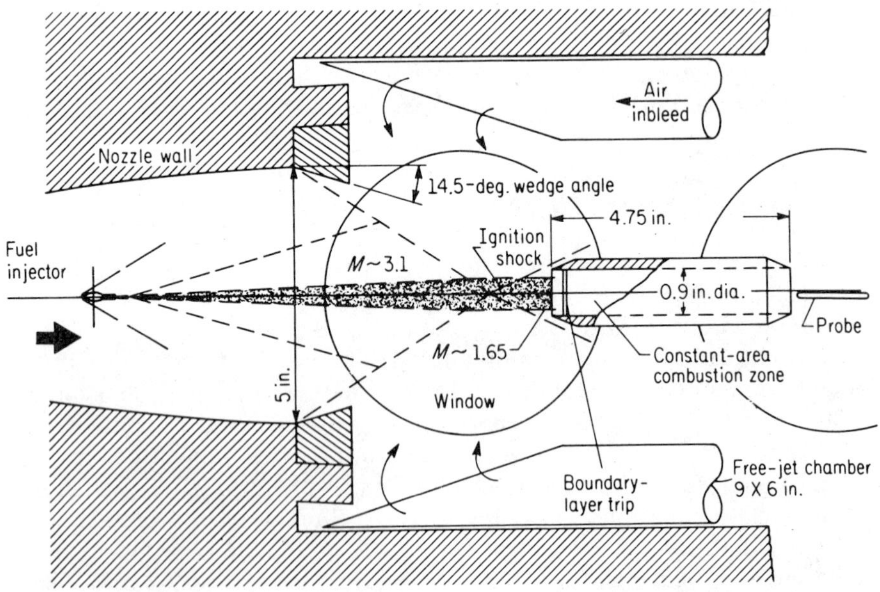

Fig. 9-31. Apparatus to produce supersonic combustion in a constant-area duct. Note the oblique ignition shock waves. (USAF, Arnold Engineering Development Center, Tullahoma, Tenn.

The real difficulty in producing supersonic combustion in a channel is that of providing the proper triggering mechanism. Two primary mechanisms of triggering have been found to be available, (1) an oblique shock, and (2) injection of fuel into a flowing stream whose static temperature is already well above the ignition temperature. The work of Rubins and co-workers[24] represents a good example of oblique-shock triggering in a one-dimensional steady flow. A drawing of their apparatus is shown in Fig. 9-31. In this experiment hydrogen is injected into the preheated air at a point where the static temperature is below the ignition temperature, and a pair of opposing oblique shocks are used to raise the temperature to the ignition temperature. To maintain one-dimensional flow the central portion of the flow (which contains a rather uniform hydrogen concentration) is intercepted by a tube which prevents lateral expansion. Pressure records for variety of hydrogen additions are shown in Fig. 9-32,

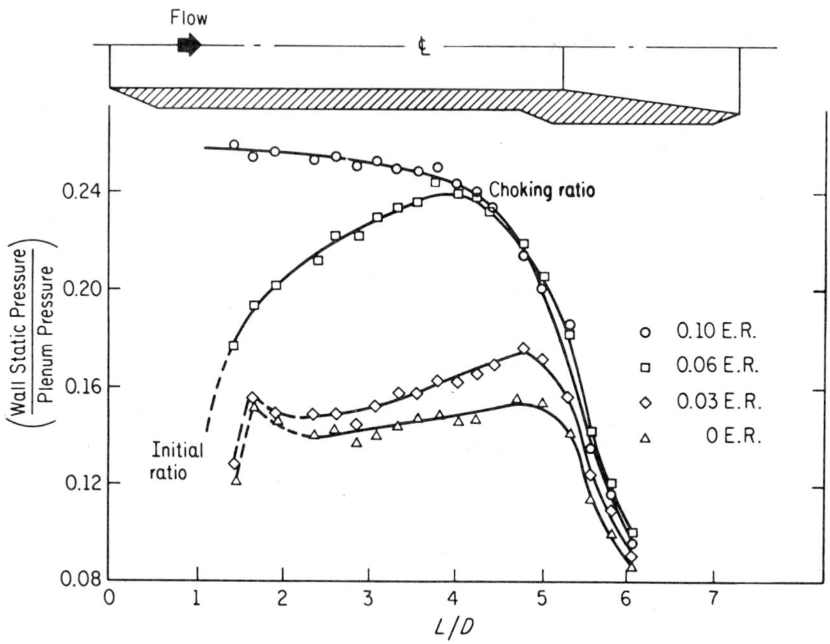

Fig. 9-32. Pressure records obtained in the apparatus shown in Fig. 9-31. Note the existence of a choking-equivalence ratio. Also note that for equivalence ratios below this critical value the pressure rise corresponds to the weak detonation solution and is a shockless transition to the final state, while above this critical value the pressure rise corresponds to the strong detonation solution and to an overdriven detonation. (USAF, Arnold Engineering Development Center, Tullahoma, Tenn.)

[24] P. M. Rubins and R. P. Rhodes, *AIAA J.*, *1*, 2778 (1963); P. M. Rubins and J. C. Cunningham, "Shock-Induced Supersonic Combustion in a Constant-Area Duct," AEDC-TDR-64-266 (January 1965), AEDC Arnold Air Force Station, Tullahoma, Tenn.

and two photographs showing unchoked and choked flow in the tube are shown in Fig. 9-33.

The other process, that of local fuel injection in a high-temperature supersonic stream is probably more desirable in an actual supersonic-

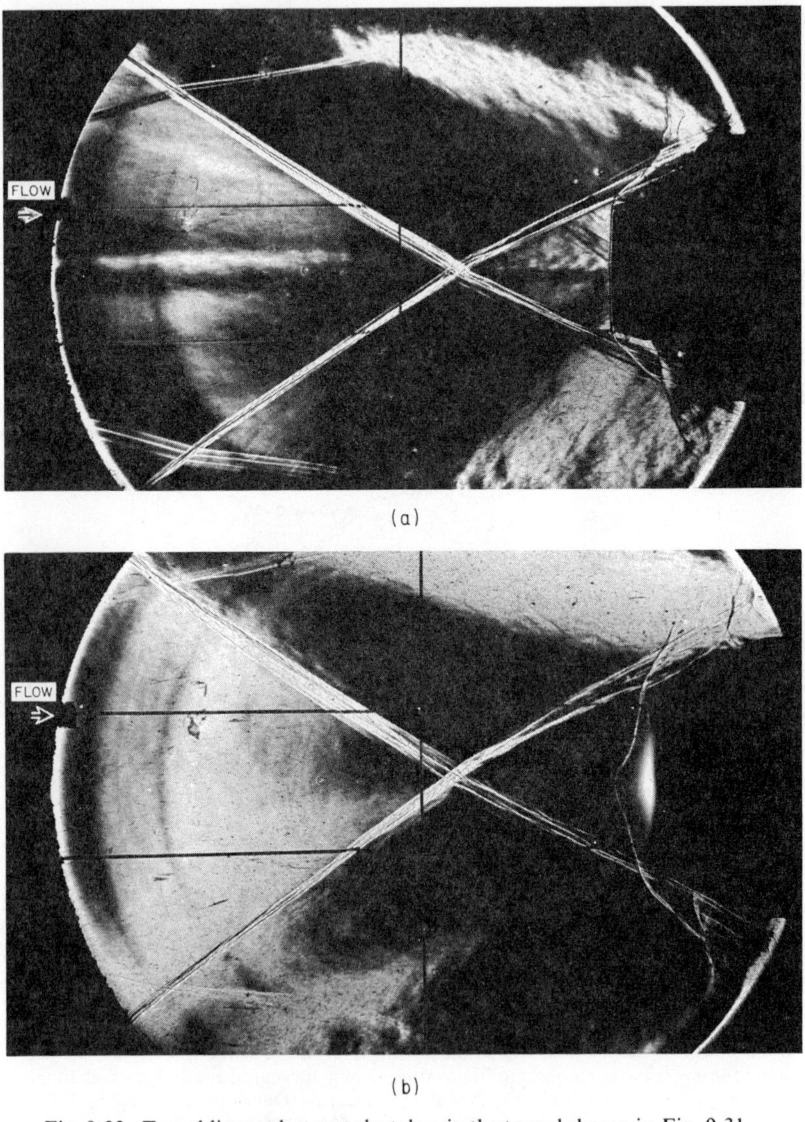

Fig. 9-33. Two shlieren photographs taken in the tunnel shown in Fig. 9-31. (a) Test section operating supersonic throughout. (b) Subsonic flow in test section. Notice the expelled normal shock wave. (Courtesy P. Rubins, ARO Inc., Arnold Engineering Development Center, Tullahoma, Tenn.)

combustion ramjet engine. In this case fuel is ejected at side ports and usually burns while mixing. An oblique shock is produced in this case and the flow is locally neither completely supersonic nor one-dimensional. However, as the flow remains primarily supersonic the pressure rise in the combustion chamber is kept to a minimum during heat addition.

In addition to the observation of true supersonic combustion, research in this area has led to the construction of "standing" detonation waves in supersonic streams. In this case the reaction is triggered by a normal shock wave in a flow containing fuel which has been injected upstream at a point where the static temperature is low. Three techniques for generating these normal shocks have been used. These are illustrated in Fig. 9-34. Both the normal shock produced in an axisym-

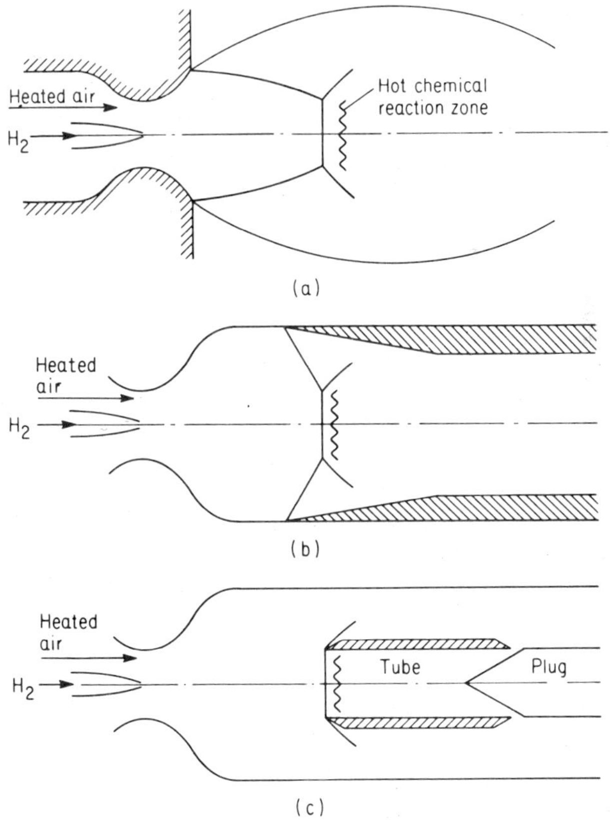

Fig. 9-34. Standing detonation experiments, (a) behind underexpanded shock bottle, (b) on Mach stem of two oblique shocks, (c) behind normal shock held on lip of tube by a choking plug.

metric shock bottle on an underexpanded jet (Fig. 9-34a) and the normal Mach-stem shock produced in a two-dimensional tunnel (Fig. 9-34b) have the disadvantage that lateral expansion of the flow is allowable downstream of the normal shock. Thus it is quite difficult to analyze these processes to determine or confirm kinetic data. However, Fig. 9-34c illustrates a technique which allows one to reasonably maintain one-dimensional flow and simultaneously measure pressure at a series of stations in the reaction zone. A one-dimensional flow with no lateral expansion is obtained in this case by advancing a control plug sufficiently to just produce choked flow so that the shock "stands" on the lip of the test chamber. If properly designed this apparatus can maintain a steady detonation for long periods of time and therefore be used to conveniently determine pressure, temperature, and concentration profiles throughout the reaction zone.

In all the "standing detonation" studies illustrated in Fig. 9-34 there has never been any evidence of transverse phenomena occurring on the normal shock. All of these "detonations" are maintained by external means, however, and therefore represent overdriven waves. This fact, of course, simply negates the requirement for transverse waves but does not eliminate their possibility. It is quite probable, however, that their occurrence is suppressed because in all cases studied to date the active reaction zone has always had a considerably greater length than the width of the test apparatus. At present the "one-dimensional" nature of these standing waves needs further investigation.

9-7. SOLID- AND LIQUID-PHASE DETONATION

There are many solids and liquids, both pure compounds and mixtures of compounds, which are capable of sustaining a detonation wave. In all cases these substances support highly exothermic decomposition reactions which produce copious quantities of gas. However, not all such highly exothermic compounds or mixtures detonate easily or even detonate at all. Therefore, three classes of exothermic compounds are usually defined relative to their detonatability:

1. The primary explosives. Primary explosives are the most sensitive explosive substances in commercial use. Indications are that they propagate a CJ (or high-order) detonation very soon after being initiated by friction. In this group are the metal styphnates, fulminates, and azides. In general these compounds detonate when heated, before melting.

2. Secondary explosives. These substances must usually be traversed by a strong shock wave before detonation appears. In general, they melt before detonating and in most cases of initiation by friction during con-

finement they burn before detonating. This group contains the substances trinitrotouluene (TNT), ammonium nitrate and pentaerythritol tetranitrate (PETN).

3. Propellants. These substances, even though highly exothermic, have not exhibited a detonative decomposition even when struck by a strong shock wave. Commercial solid-rocket propellants usually fall in this group. The combustion behavior of this group will be discussed in the next chapter.

It should be emphasized that the boundaries between these three groups are not, in general, clearly defined. For example, there are many extremely sensitive substances such as nitrogen trichloride or mixtures such as potassium perchlorate plus yellow phosphorus which are so sensitive that they have very little practical value. Consequently the detonation behavior of these extremely sensitive formulations has not been studied extensively. There is also a gray area between the primary and secondary explosive categories because there is no unambiquous definition of their differences. Also, there are many solid propellants which have not yet been *observed* to detonate, but which may be found to exhibit this property in the future.

In the following paragraphs we will discuss a few of the more important properties of detonation propagation in solid or liquid explosives in the same order that these properties were discussed in the previous sections on gaseous detonations.

The one-dimensional theory for the propagation of detonation in solids and liquids is based on the same equations of steady motion developed for gas-phase detonation, i.e., Eqs. 5-17 through 5-19. However, the extremely high initial density of solid and liquid explosives leads to detonation pressures which lie in the range of 70 to 700 kiloatm. These extremely high pressures cause two major complications in the application of the elementary theory. In the first place, the product gases no longer obey the ideal gas law even at the high temperatures encountered; second, the type of confinement required to produce one-dimensional flow is impossible to obtain except in the central area of a wave front whose breadth is very much larger than the reaction-zone thickness. The first complication means that calculated detonation velocities will not exhibit as good an agreement with experimental values as for the gas-phase case. Note that the only way to produce the temperatures and pressures encountered in condensed-phase detonations is to have a condensed-phase detonation. Thus, there is no experimental method available for independently measuring equation-of-state parameters to allow a truly independent calculation of detonation velocity. Furthermore, theoretical estimates are not sufficiently accurate to choose an unambiguous mathematical form for the equation of state. Therefore, detonation-propagation velocities themselves are usually used to confirm a particular form of the equation of state and the constants in that equation.

A rather simple but useful equation of state for these calculations is the Clausius equation of state:

$$P(V - b) = \mathrm{R}T \tag{9-27}$$

where b, the covolume, may be considered to be the effective volume of the molecules. Note that at these high pressures Eq. 9-27 is equivalent to Van der Waal's equation of state. Equation 9-27 has been used by Abel with b equal to a constant, by Cook with $b = b(V)$, and by Jones with $b = b(P)$. Another equation of state which has been used rather extensively is the Kistiakowsky and Wilson equation of state,

$$PV = n\mathrm{R}T[1 + x\exp(x)]$$

where $x = K/VT^{1/2}$ and K is an adjustable constant.

It should be pointed out that pressure cannot be measured directly in a detonation, but can only be inferred from the detonation velocity and an assumed equation of state. The procedure is to measure the detonation velocity of a relatively large diameter charge as a function of composition and initial density and compare this value to velocities calculated with an assumed equation of state. Pressures predicted by the equation of state which yields the best fit to the experimental data are then listed as the detonation pressures for that system. In the above we made no mention of the effect of composition on the equation of state. However, the equilibrium composition should have an effect, and there has been some attempt to write an equation of state which is a function of the equilibrium composition of the detonation products. This attempt has not been too successful because it is extremely difficult and yields answers that agree with experiment about as well as those obtained from empirical fits to equations which do not contain composition-dependent terms.

The other problem, that of confinement, arises because the CJ detonation pressure of most condensed explosives is well above the yield strength of all known substances. Therefore, even heavy metal walls can impose only an "inertial" confinement and true one-dimensionality can only be obtained in the central region of large-diameter charges. Figure 9-35 is a photograph of the shape of a detonation front for weak confinement in a cylindrical charge. Inadequate confinement causes the detonation velocity to drop appreciably and eventually can lead to extinguishment. It has been found experimentally that for unconfined charges there is a critical minimum-charge diameter and that charges with a diameter less than this critical value will not spontaneously propagate detonation.

Most commercial secondary explosives have critical diameters in the range of $\frac{1}{8}$ to 1 in. However, ammonium nitrate has a minimum charge

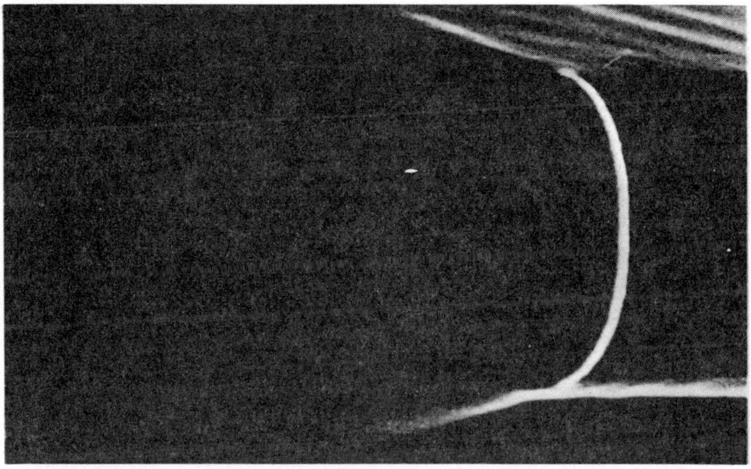

Fig. 9-35. Shape of a detonation front in a small-diameter unconfined charge. Propagation is from left to right. (With permission from J. Berger and J. Viard, *Physique des Explosifs Solides*, Dunod, Paris, 1962.)

diameter of approximately 9 in., and the nitrate-sulphate $2NH_4NO_3 = (NH_4)_2SO_4$ which was used commercially as a fertilizer after World War I has an even larger minimum-charge diameter. This fact resulted in an unfortunate accident soon after World War I. At that time it was thought that the nitrate–sulphate would be incapable of sustaining detonation (since the usual small sample impact tests showed it to be insensitive) and the material was handled in bulk like any other industrial chemical. This substance cakes rather easily, especially in humid atmospheres, and in order to recover the salt from storage a congealed pile of 9 million lb was dynamited on September 21, 1921, in the industrial town of Oppau, Germany (pop. 7,500). Instead of simply fracturing, the pile detonated as a unit. The disaster was the largest man-made explosion prior to the advent of the atom bomb. The explosion killed an estimated 1,100 people, left a hole 190 ft deep by 400 ft in diameter, and caused severe damage to a radius of 4 miles. It is best to assume that all exothermic compounds are dangerous.

The occurrence of transverse structure and its role in the propagation of condensed-phase detonation has not been settled as yet. Transverse waves have been observed in nitromethane—acetone mixtures and the theory for their growth in the gas phase certainly does not preclude their occurrence during the propagation of detonation in any homogeneous substance. However, in most high explosives the reaction zone is so thin that the investigation of transverse structure would be extremely difficult. Also, most commercial explosives are granular substances (if pressed) or fine-grain crystalline substances (if cast), and it is known that this type of inhomogeneity introduces a structural roughness in the detonation front which would overshadow any inherent transverse structure. In short, we know that condensed-phase detonation propagates as a self-sustaining phenomenon under the proper conditions, and we also know that the one-dimensional Chapman-Jouguet theory does not justify the observed stability of the wave front. Therefore we suspect that three-dimensional structure is important for propagation, but at the present time we do not have sufficient information to corroborate these suppositions.

Initiation of detonation in a condensed medium may be accomplished by friction in the more sensitive substances; Bowden[25] has shown that the friction of grit particles will ignite a particular explosive only if the melting point of the grit material is above a certain temperature. Since the temperature rise associated with frictional heating cannot exceed the melting temperature of the substance involved, this implies that there is a minimum hot-spot temperature required for initiation of primary explosives. In this connection it is interesting to note that most primary explosives detonate *before* melting.

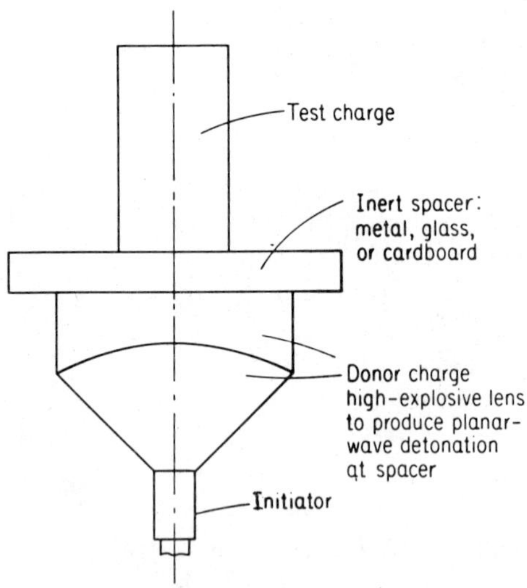

Fig. 9-36. Typical setup to study detonation initiation in a test charge of explosive.

[25] F. P. Bowden and A. Yoffe, *Initiation and Growth of Explosion*, Cambridge U. Press, London, 1952.

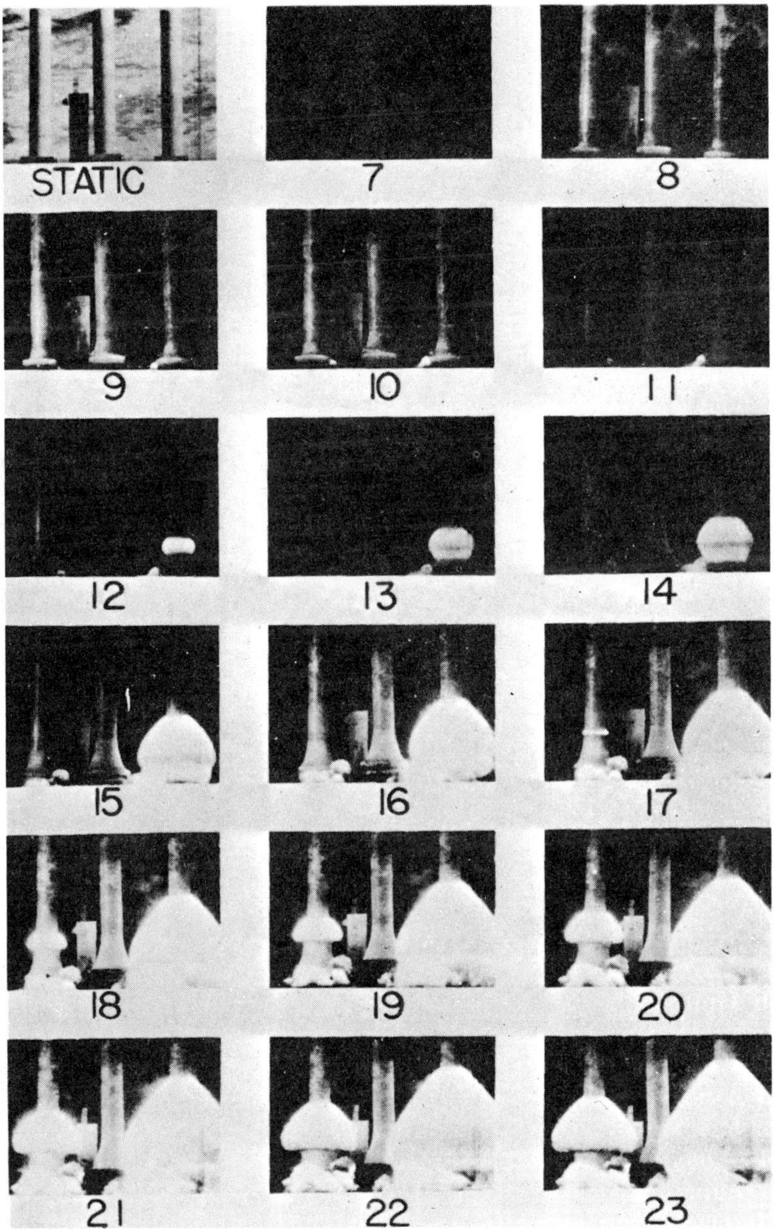

Fig. 9-37. Initiation of three sticks of high explosive by a planar-wave initiator similar to that shown in Fig. 9-36. Note that (1) the right-hand grain ignites first and detonates both forward and back towards the initiation plate, (2) the left-hand grain ignites later and propagates a detonation in the forward direction only, (3) the center charge does not ignite, and (4) all charges are extensively crushed by the initiating shock before initiation occurs. (With permission from M. A. Cook, *The Science of High Explosives*, *ACS Monograph No. 39*, New York: Reinhold, 1958.

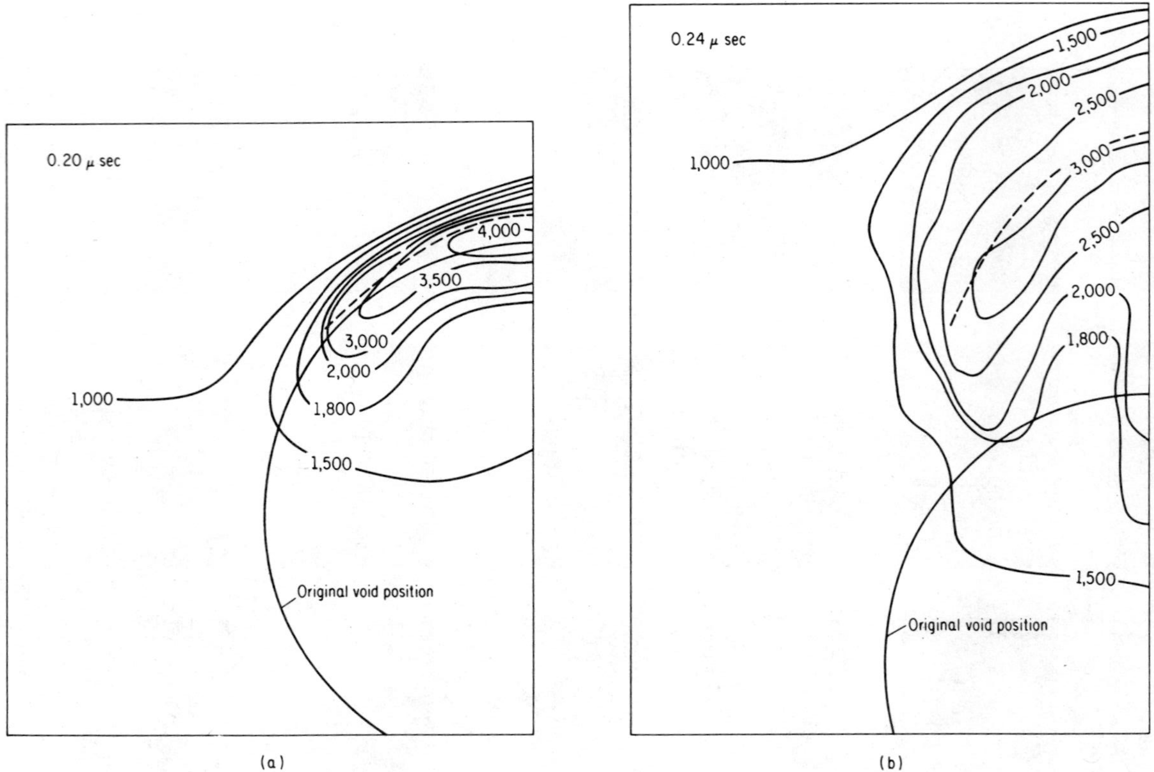

Fig. 9-38. Isotherms during collapse of a spherical bubble in nitromethane. The shock is an 85-kg bar shock which is normally insufficient to ignite nitromethane in the absence of voids. Temperatures above 1400°K may be assumed to cause instantaneous reaction in nitromethane. (Adapted from C. L. Mader, *Phys. Fluids*, *8*, 1811 (1965).)

Most secondary explosives cannot be ignited by friction alone and require the presence of a strong shock wave before ignition occurs. In practice, the strong shock is usually supplied by the detonation of a donor charge, and for testing purposes this shock is attenuated by passage through an appropriate thickness of inert material. A typical experimental setup is shown in Fig. 9-36 and a set of high-speed photographs of the initiation process are shown in Fig. 9-37. As can be seen from Fig. 9-37 the initiation process is not strictly one-dimensional and there is a considerable spreading of the high-explosive charge by the entering shock wave before initiation occurs. Experiments using different geometries have shown that initiation usually occurs first at the center of the charge some distance behind the entering shock and then propagates to overtake the entering shock wave and produce a high-order detonation in the undisturbed charge. Note that the process is quite similar hydrodynamically to reflected shock initiation in the gas phase.

An interesting two-dimensional study of shock initiation in a liquid high explosive has been reported by Mader.[26] He solved the axisymmetric nonsteady flow which occurs when a shock wave passes over a cavity in a liquid. His calculations were for a number of cavity shapes, but his spherical cavity calculation is most interesting because it may be easily duplicated experimentally by placing a small gas bubble in a liquid explosive. Mader has verified the observation that a single gas bubble of sufficient size greatly lowers the shock sensitivity of a liquid explosive because of the local production of high temperatures. Typical snapshots of the temperature distribution late in the bubble-collapse process are shown in Fig. 9-38. By using a realistic rate law for the decomposition of nitromethane, Mader's model quite adequately predicts the initiation behavior which is observed experimentally. This represents another case where initiation has been explained fully in terms of the interaction of reaction kinetics and nonsteady fluid dynamics with realistic initial and boundary conditions.

The reader is referred to the bibliography for more information on the behavior of high explosives.

BIBLIOGRAPHY

1. B. Lewis and G. von Elbe, *Combustion, Flames and Explosions in Gases*, 2d ed. (New York: Academic, 1961.)

2. A. Ferri, *Fundamental Data Obtained from Shock Tube Experiments*, Agardograph No. 41. (New York: Pergamon, 1961.)

3. B. P. Mullins and S. S. Penner, *Explosives, Detonations, Flammability and Ignition*, Agardograph No. 31. (New York: Pergamon, 1959.)

[26]C. L. Mader, *Phys. Fluids*, *8*, 1811 (1965).

4. J. V. Foa, *Elements of Flight Propulsion*. (New York: Wiley, 1960.)

5. S. S. Penner and F. A. Williams, *Detonation and Two-Phase Flow Progress in Astronautics and Rocketry*, Vol. 6. (New York: Academic, 1962.)

6. F. A. Williams, *Combustion Theory*. (Reading, Mass.: Addison-Wesley, 1965.)

7. Y. B. Zel'dovich and A. S. Kompaneets, *Theory of Detonation*. (New York: Academic, 1960.)

8. K. I. Shchelkin and Y. K. Troshin, *Gas Dynamics of Combustion*. (Baltimore: Mono Book, 1965.)

9. R. I. Soloukhin, *Shock Waves and Detonations in Gases*. (Baltimore: Mono Book, 1966.)

10. B. V. Voitsekhovskii, V. V. Mitrofanov, and M. E. Topchian, "Structure of the Detonation Front in Gases," *Izd-vo Sibirsk. Otdel. Akad Nauk. SSSR*, 1 Novosibirsk (1963).

11. E. S. Shchetinkov, *Physics of the Combustion of Gases*. (Moscow, 1965.)

12. F. P. Bowden and A. Yoffe, *Initiation and Growth of Explosion*. (London: Cambridge, 1952.)

13. J. Berger and J. Viard, *Physique des Explosifs Solides*. (Paris: Dunod, 1962.)

14. M. A. Cook, *The Science of High Explosives*, ACS Monograph No. 139. (New York: Reinhold, 1958.)

PROBLEMS

9-1. Calculate the Hugoniot curve for the detonative decomposition of nitric oxide (NO) at 300°K. Assume that only nitrogen (N_2), oxygen (O_2) and nitric oxide (NO) are at equilibrium along the Hugoniot. Calculate the CJ detonation velocity for this process by determining the tangency condition for the Rayleigh line and Hugoniot curves. Note that this process is pressure-independent because the equilibrium composition is independent of pressure. (Use enthalpies and equilibrium constants from tables in Appendix F; see also Tables 4-3 and 4-4.)

9-2. As the Mach number of a CJ detonation decreases the $\phi_{\|}$ plane moves towards the shock wave. Calculate the Mach number for $\phi_{\|} = 0.0$ as a function of gamma for a constant-gamma heat-addition detonation.

9-3. In the constant-gamma heat-addition model the weak detonation solution corresponds to a shockless flow transition during heat addition, while the strong solution corresponds to a relatively strong shock wave followed by heat addition. Show, for a slight overdrive ($\phi = 1 - \xi$ where $\xi \ll 1$), that the entropy rise is the same near the CJ point irrespective of the path (i.e., weak or strong) and that the first and second derivatives of the entropy with volume ratio ϵ are continuous through the CJ point, even though the physical nature of the process changes in a discontinuous manner.

10

Solid- and Liquid-Propellant Combustion in Rockets

10-1. INTRODUCTION

Both solid- and liquid-propellant rockets have played an important role in modern propulsion technology. While the steady-state theory of the rocket as a thrust producer has been adequately discussed in a number of texts, the combustion process which leads to this thrust production is usually not discussed in detail. We will therefore circumvent the external aspects of the rocket by looking in detail only at the processes that occur in the combustion chamber. Thus we will not discuss the effect of either recombination or of nonaxial flow in the nozzle. These can be handled adequately by techniques based on the reactive-flow equations introduced in Chapter 5 and the assumption of steady ($M = 1$) flow through the throat (see Sec. 5-6 for an example).

We will also not concern ourselves with the techniques of calculating the specific impulse (pounds of thrust per pound of propellant) for a particular propellant system. This calculation is ordinarily based on the assumption of full chemical equilibrium in the flow through the throat, and the techniques for this type of calculation have been discussed adequately in other texts and introduced in Chapters 4 and 5 of this text. We will therefore assume that for any system under consideration the equilibrium composition of the combustion products, the isobaric flame temperature, and the specific impulse are available as a function of the chamber pressure for steady operation. With this information available, the rocket designer must choose an operating pressure, throat area, and nozzle shape to yield the desired thrust. The choice of these variables in turn fixes the total mass flow through the nozzle for steady operation and therefore establishes the steady rate of production of combustion products in the chamber. In general, the size and shape of the chamber are then dictated by the requirement that combustion be complete in the plane of the nozzle's throat. For the case of liquid propellants, the design problem becomes primarily one of properly metering the fluid and oxidizer into the chamber and ensuring that they are dispersed and mixed so as to produce complete combustion and steady operation.

In the case of solid fuels, the rate of production of combustion products is primarily determined by the exposed surface area of the solid-propellant grain. This is because at a fixed pressure the regression rate of the surface of a solid-propellant grain is everywhere constant in a direction normal to the surface. Furthermore, the pressure-dependence of this regression rate for steady operation may usually be approximated by an equation of the form

$$r = B_p P^n \tag{10-1}$$

where B_p and n are constants and $n < 1$. Thus if a surface area s is exposed in the rocket chamber, the mass rate of production of product gases is given by the equation

$$\dot{m}_p = s r \rho_s = s \rho_s B_p P^n \tag{10-2}$$

where s is usually a slowly varying function of time. However, the equilibrium mass flow rate through a nozzle is, to a first approximation, simply proportional to the pressure

$$\dot{m}_e = B_e A P \tag{10-3}$$

where B_e is a constant dependent on the nozzle efficiency and the equilibrium combustion properties in the gas phase and A is the throat area. Steady operation is obtained when $\dot{m}_p = \dot{m}_e$. Equations 10-2 and 10-3 may therefore be combined to yield the equation

$$s = A\mathcal{K}(P) \tag{10-4}$$

where

$$\mathcal{K}(P) = B_e P^{(1-n)}/B_p \rho_s$$

Here $\mathcal{K}(P)$ is the ratio of the propellant surface area to nozzle throat area that will just produce the equilibrium combustion-chamber pressure P during steady operation. Note that Eq. 10-4 predicts a stable operating condition only if $n < 1$, since if $n > 1$ an incremental increase in pressure ΔP will cause $\Delta \dot{m}_e > \Delta \dot{m}_p$ and the pressure will continue to increase spontaneously with time, whereas if $n < 1$ an incremental increase in ΔP will cause $\Delta \dot{m}_e < \Delta \dot{m}_p$ and the pressure will spontaneously seek an equilibrium level. Most solid propellants have values of n in the range 0.4–0.85. There is no known combustible substance with a value of n greater than unity.

Equations 10-1 to 10-4 indicate that the relationship between throat area, exposed propellant surface area, and chamber pressure will be a dynamic relationship. In fact, in a solid-propellant rocket the initial surface shape and the value of $\mathcal{K}(P)$ determines the $s(t)$ relationship and therefore the specific $P(t)$ relationship for the rocket's operation.

However, the exact power-law dependence of Eq. 10-1 (with n = const.) is not, in general, required to allow the prediction of a $P(t)$ relationship for any specific initial surface shape and throat area. All that is needed is an experimentally determined plot of $\mathcal{K}(P)$ versus the chamber pressure. Numerical integration may then be used to determine $s(t)$ for a specific initial shape. In general, three types of instantaneous behavior of the grain surface area may be defined:

$$\frac{\partial s}{\partial t} > 0; \quad \text{progressive}$$

$$\frac{\partial s}{\partial t} = 0; \quad \text{neutral}$$

$$\frac{\partial s}{\partial t} < 0; \quad \text{regressive}$$

Typical examples of surfaces which exhibit these three behaviors are shown in Fig. 10-1.

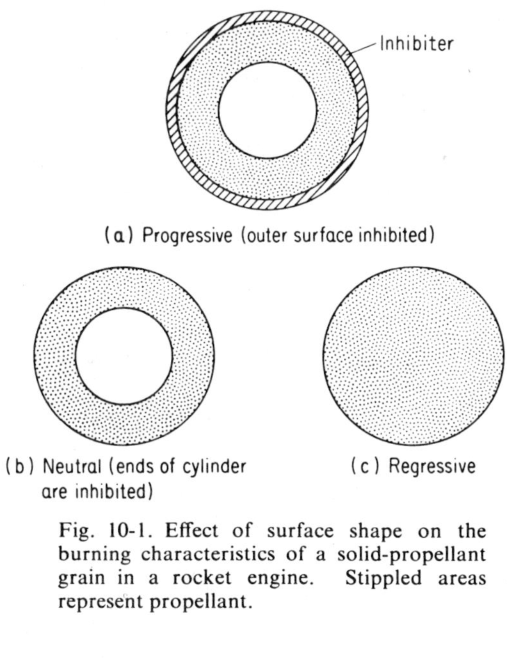

Fig. 10-1. Effect of surface shape on the burning characteristics of a solid-propellant grain in a rocket engine. Stippled areas represent propellant.

10-2. SOLID-PROPELLANT COMBUSTION

There are two major classifications of solid propellants in use today—*composite* and *double-base*. Composite propellants contain a crystalline inorganic oxidizer dispersed as a powder through a matrix of an organic polymer which ordinarily acts primarily as a fuel. Typical oxidizers are ammonium perchlorate, potassium perchlorate, or ammonium nitrate. Typical polymers are polyurethane, asphalt, or rubber type compounds. These propellants are formulated by mixing the oxidizer with the uncured polymer, casting the mix in a mold of the desired shape, and then curing the polymer either with thermal or chemical techniques. Double-base rocket propellants are related to the nitroglycerine–nitrocellulose formulations used in the manufacture of smokeless powder for guns. This material may be formed into grains either by extrusion or by casting. Extrusion of either a heated mix or an acetone-gelled mix is used to produce the smaller grains. Grains extruded with the acetone process may not be larger than about $\frac{1}{2}$ in. in diameter because they are cured by allowing the acetone to diffuse to the surface and evaporate. The size of hot extruded grains is limited by the size of available presses.

The casting process is used to produce the larger double-based grains. These castings are manufactured by first extruding small ($<\frac{1}{16}$ in. diameter) pellets of double-base propellant of low nitroglycerin content by the usual acetone or hot-extrusion process. The pellets (casting powder) are then placed in a mold and the mold evacuated prior to immersion of the entire pellet bed in liquid nitroglycerin. The filled mold is then cured at a high temperature. During this curing process the individual grains of casting powder absorb the interstitial nitroglycerin and swell to contact each other and produce a relatively uniform single grain of double-base propellant.

In many applications these basic formulations are modified for specific purposes. For example, powdered metals may be used as additives, or an exothermic binder may be used in a composite-propellant formulation. Modifications in composition are continually being made to alter the properties of existing propellants in an attempt to increase structural integrity or specific impulse, modify the burning rate, and decrease the propellant's tendency to burn unstably in a rocket engine.

Once a sample propellant is formulated, its burning rate must be determined experimentally. This is usually done in either a strand burner, window bomb, or an actual small experimental rocket engine. In the strand burner a stick of propellant is coated on its sides with a relatively inert material (inhibiter) so that the combustion wave will propagate as a one-dimensional wave throughout the entire length of the stick. Thin fuse wires are then placed in small holes drilled across the stick at suitable intervals and the prepared sample is placed in a high-pressure bomb of relatively large volume. The fuse wires are connected to an external timing device and the bomb pressurized to the desired pressure level. The sample is ignited at one end (by a heated wire, for example), and the arrival of the combustion wave at each fuse wire is timed by noting the instant at which the circuit resistance increases markedly.

In the window bomb, inhibitors may not be used because they tend to obscure the burning surface. In this case a flow of inert gas counter to the direction of regression has been used to retain a one-dimensional combustion front. A choked nozzle is usually used to maintain constant pressure in a window bomb.

Window-bomb experiments are primarily used for observing the details of the flame structure during combustion. A small experimental rocket engine may also be used to determine propellant burning rate if one knows the nozzle's throat area, the propellant surface area as a function of the fraction burned, and has available an experimental pressure-time curve. This method is usually less accurate than the strand or window-bomb techniques because it is difficult to estimate heat-transfer losses and the effects of nonideal flow in a small motor. However, the rocket motor technique is a reasonable good technique for determining the pressure exponent n because for most propellants $1 - n \ll 1$. Thus, a small change in the surface area of the propellant will cause a large change in the equilibrium operating pressure of the rocket engine. Regression rates for most solid propellants lie in the range of 0.2–3 in./sec.

The burning zone of a solid propellant is somewhat more complex than the equivalent region in a premixed gas flame. Furthermore, even though it is customary to discuss a one-dimensional solid-propellant com-

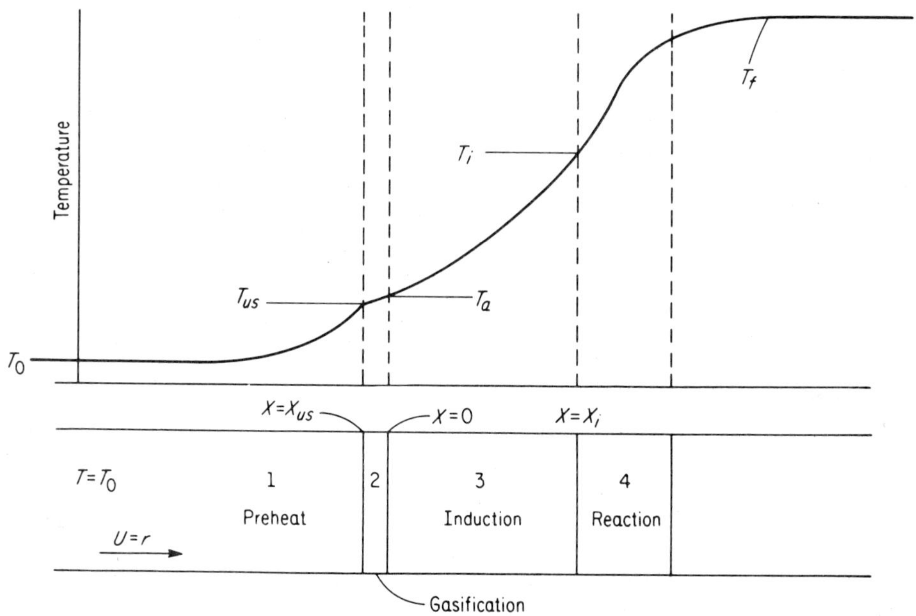

Fig. 10-2. The structure of the combustion zone of a typical solid propellant.

bustion wave, it is doubtful if one can ever hope to observe strictly one-dimensional combustion during the burning of a solid-propellant grain. The inherent three-dimensional structure that appears in the burning zone of a solid propellant does not arise from the type of transverse instability which leads to non-one-dimensional detonation fronts but mainly appears because of inhomogeneities in the structure of the propellant grain itself. It is observed that this burning roughness or lack of one-dimensionality becomes more severe as one changes from a simple extruded double-base to a composite propellant. It has also been observed that the presence of metal additives decreases the one-dimensionality of the burning process. We will, however, follow convention and discuss the structure of a solid-propellant combustion wave as though it were a strictly one-dimensional process. Figure 10-2 is a schematic of the structure in a coordinate system that is steady relative to the burning surface. There are four rather distinct and different regions involved in the combustion process. These are:

1. The Solid Phase Heat-Transfer Region

In this region the propellant is being heated by conduction from the hot surface but is not reacting. Thus the temperature gradient in this region is governed by the usual steady heat-conduction equation with no source terms.

$$\frac{\partial}{\partial x}\left(\kappa_s \frac{\partial T}{\partial x} - m_0 c_s T\right) = 0 \qquad (10\text{-}5)$$

where m_0 = mass flux, c_s = heat capacity, and κ_s is the heat conductivity of the solid. The boundary conditions for this region are $T = T_0$ at $x = -\infty$ and $T = T_{us}$ at $x = x_{us}$.

2. The Gasification Layer

In this region vaporization of the propellant takes place with some reaction. The exact nature of this region is at present unknown and in the case of composites is most unsatisfactorily treated as a one-dimensional flow region. The region must be included in the model, however, because gasification cannot be assumed to be either a thermally neutral or an infinitely rapid process, and because the rate of gasification primarily controls the rate of surface regression. Due to the lack of detailed information, one usually assumes that the rate of gasification in this layer is related to the average temperature of the layer through an Arrhenius temperature-dependence. The activation energy for gasification then becomes an important constant in the model. The relationship between T_a at $x = 0$ and T_{us} must be obtained by assuming an exothermicity or endothermicity for the gasification reaction and by knowing a surface-regression rate. More experimental data is needed before a truly adequate model of this zone may be constructed.

3. The Gas-Phase Induction Zone Region

The products of the gasification process next traverse a gas-phase induction zone without undergoing extensive reaction. For the pure double-base propellants this zone has been shown to contain a large quantity of nitric oxide and have a temperature equal to about one-half the flame temperature of the propellant. In composite propellants this zone is far from being one-dimensional because it is the region where the fuel and oxidizer mix through diffusional and turbulent mixing processes. Thus, the length of this induction zone is governed by either a thermal or chain-branching delay mechanism or by a mixing-delay mechanism. It is a region in which heat transfer is of primarily importance. A steady flow model for this region yields an equation similar to 10-5 with the inclusion of a heat source term.

4. The Hot-Reaction Zone Region

In this region the final chemical reactions occur and the temperature of the propellant gases approach the adiabatic flame temperature. The boundary between this zone and the induction zone (at x_i in Fig. 10-2) is never sharp but may be located for convenience by requiring that the reaction rate at the location x_i be a specified fraction of the maximum reaction rate of region 4. This is analogous to the definition used in Chapter 7 for a one-dimensional gas flame.

We thus see that the structure of a solid propellant's combustion zone differs from the structure of a gas-phase combustion zone mainly by the addition of a heat-transfer zone in the solid phase and of a solid-phase gasification zone. These differences are significant and they play an important role in the transient response of the burning rate to pressure fluctuations. This will be discussed in Sec. 10-4. The model as described has never been used to predict a burning rate for a solid propellant. In fact, the process is so complex that it is doubtful whether a burning rate will ever be calculated from first principles.

Solid propellants will usually support combustion in an inert atmosphere down to pressures of about 6 psia. In the case of double-base propellants, combustion at these low pressures is incomplete, the flame zone is absent and the propellant is said to "fizz-burn." Under these conditions the nitric oxide decomposition reactions do not go to completion, and only about half of the available energy appears as thermal energy in the product gases.

10-3. LIQUID-PROPELLANT AND HYBRID-PROPELLANT COMBUSTION

In a liquid-propellant rocket engine the fuel and oxidizer are usually introduced into the chamber independently. It is simply too dangerous to premix the two or use a single exothermic compound as a propellant. In general, there are two types of fuel–oxidizer combinations in use today: the hypergolic, which burn on contact or the nonhypergolic, which form relatively inert mixtures. Nonhypergolic systems require an independent ignition system and, in some cases, continuous ignition.

The design of efficient and reliable liquid-propellant rocket engines is mainly dependent on the existence of a proper arrangement for spray mixing in the chamber. The determination of this arrangement is primarily experimental at the present time and it appears that it will remain so in the future. There has been a considerable amount of work on spray-nozzle operation with and without combustion, but a discussion of this facet of combustor operation is outside the scope of this text.

Hybrid engines use a combination of solid and liquid propellants in their operation. One of the difficulties with the use of a composite solid propellant in an engine is that up to 80% of inorganic oxidizer must usually be included in the formulation to obtain a maximum specific impulse. However, such high concentrations of crystalline additives adversely affect the physical properties of the grain. One advantage of hybrid operation is that a fuel-rich grain with good physical properties may be burned with the injection of liquid oxidizer from the head end of the rocket. This increases the specific impulse of the entire system and has the added advantage of allowing more freedom in thrust programming for the rocket designer.

10-4. SOLID-PROPELLANT COMBUSTION INSTABILITY

A rocket engine is very susceptible to unstable operation. It is, after all, an enclosure in which heat is being released at a very rapid rate, and it therefore conforms to the general requirements for self-excited instability discussed in Sec. 8-6. The differences in behavior and the relative progress in understanding solid- and liquid-propellant rocket instabilities are sufficiently unique so that each rocket type will be discussed in a separate section.

Solid-propellant rocket motors have been observed to suffer from three relatively distinct types of instabilities: (1) sporadic instabilities, (2) low-pressure instabilities (chuffing), and (3) acoustic instabilities. Sporadic instabilities are not inherent to the system. They are caused by various types of structural failure of the rocket grain such as the opening of a fissure, the uncovery of an included bubble, or the loss of a large piece of propellant. These processes all cause the effective area of the burning surface to change rapidly and therefore cause the equilibrium operating pressure of the rocket to seek a new level rapidly. In addition, the loss of a sizable piece of propellant may cause a momentary nozzle blockage and lead to a destructive overpressure in the rocket engine. All of these difficulties may be remedied by choosing a solid propellant with the proper physical properties and by imposing adequate quality control during the manufacture of the grain.

Chuffing or low-pressure instability was at one time thought to be caused primarily by poor ignition; it is, in the sense that poor ignition of the grain will lead to low-pressure operation because the entire surface is not burning, which in turn will lead to the instability. It is now known that this type of instability is caused by the coupling of the response time of the solid-phase heat-conduction zone (region 1 of Fig. 10-2) and the characteristic exhaust time of the rocket engine. The solid-phase heat-conduction zone responds most slowly to a change in the operating pressure in the engine. Furthermore, a mathematical analysis of the effect of pressure change on regression rate indicates that when the propellant is subjected to a step pressure change, the burning rate does not respond instantaneously, and when it does respond it overshoots the new equilibrium burning rate momentarily. This behavior is shown schematically in Fig. 10-3. The time delay to the maximum overshoot has been found to be directly related to the thickness of the preheat zone in the propellant prior to the transient behavior. On the other hand, all rocket engines have a characteristic exhaust time that is proportional to the value of $L^* = V/A$ for the chamber, where V is the instantaneous free volume of the chamber and A the throat area. L^* is related to the characteristic exhaust time through the velocity of sound of the propellant gases at the throat. Since L^* is essentially independent of the chamber pressure, the rocket's characteristic response time due to nozzle outflow is itself independent of the chamber pressure. The time delay to overshoot, τ, shown in Fig. 10-3 is, however, proportional to the preheat zone thickness which in turn is a function of the regression rate. Thus this overshoot time increases markedly as the operating pressure of the rocket decreases. At the pressure where the overshoot time and the exhaust time are equal, the two will couple to produce an instability in which the entire chamber pressure is oscillating as a unit. A typical photograph of the growth of the pressure oscillation just prior to extinguishment is shown in Fig. 10-4a. It has been observed that this oscillation always leads to extinguishment of the grain. A correlation of extinction pressure for a number of grain shapes burning in a small rocket engine is shown in Fig. 10-4b. The slope of the limit line for the case of the unaluminized propellant

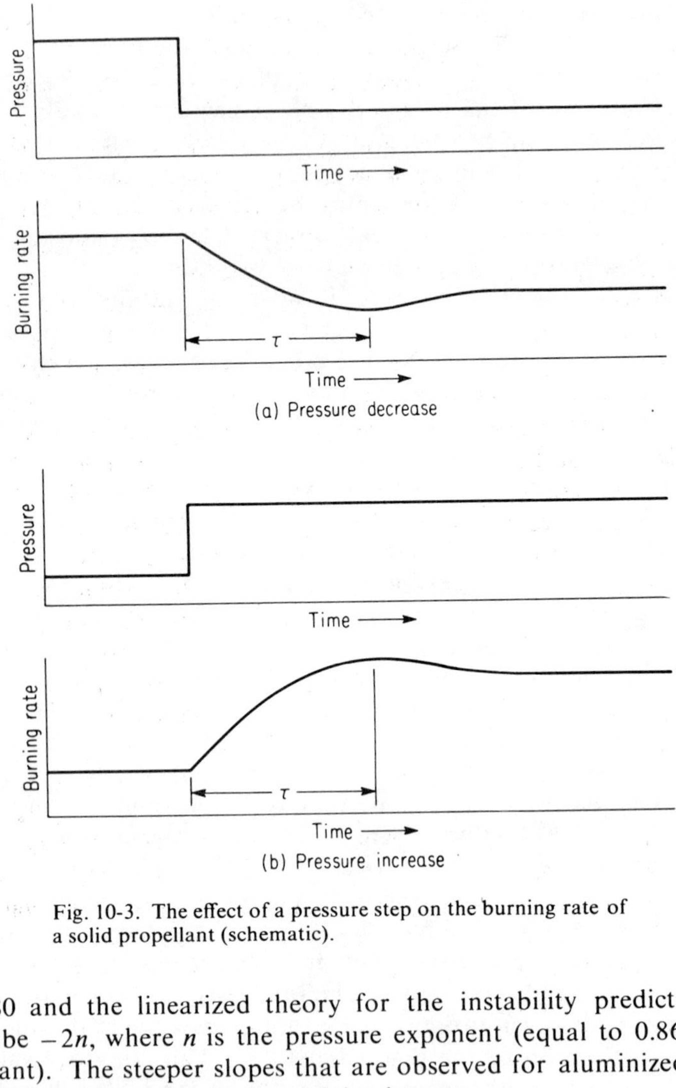

Fig. 10-3. The effect of a pressure step on the burning rate of a solid propellant (schematic).

is -1.80 and the linearized theory for the instability predicts that it should be $-2n$, where n is the pressure exponent (equal to 0.86 for this propellant). The steeper slopes that are observed for aluminized propellants have not been adequately explained as yet.

In all the firings shown in Fig. 10-4b the grain was burning regressively and stabily for at least 1 sec before extinguishment. L^* was measured by removing the unburned portion of the grain after the firing and calculating the residual chamber volume at extinguishment. The grain could be recovered in these cases only because the rockets were fired into a large vacuum chamber, and once the grain was extinguished by the instability it could not reignite because the chamber pressure itself fell below the minimum combustion pressure of the propellant. However,

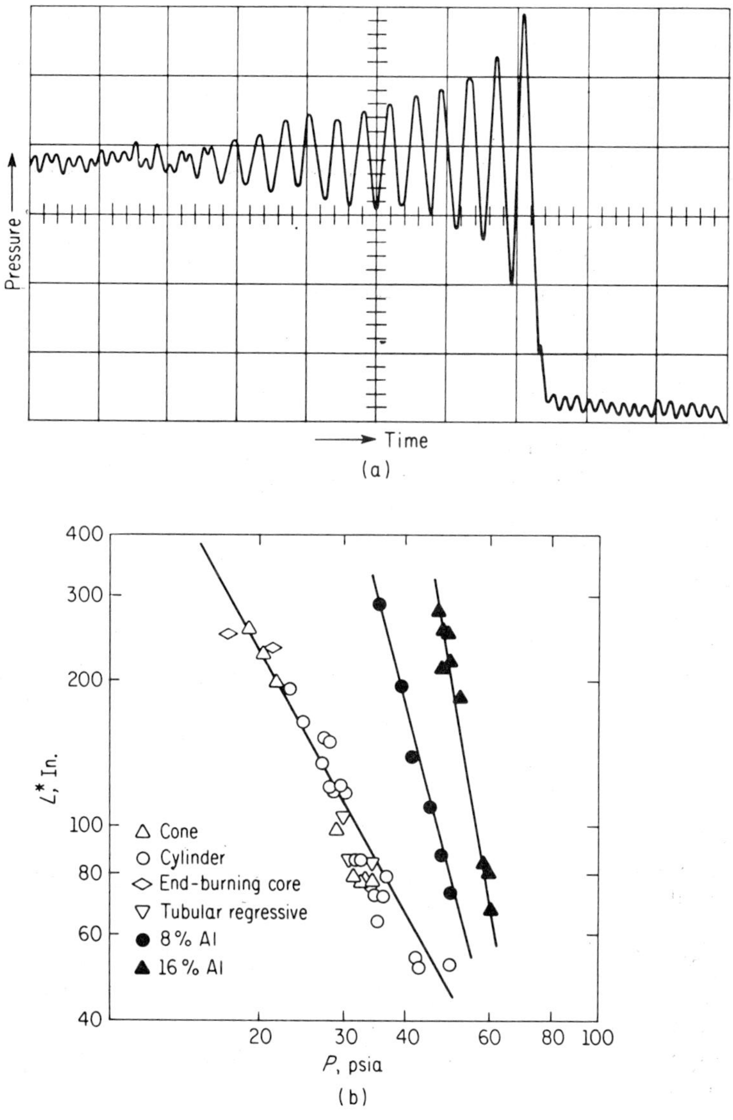

Fig. 10-4. (a) The low-frequency oscillation preceding L* extinguishment. 10 psi/major division on the ordinate and 0.25 sec/major division on the abscissa. [With permission from N. W. Ryan, *Tenth Symposium (International) on Combustion*, The Combustion Institute, Pittsburgh, Pa., 1965, p. 1082.] (b) The pressure-L* dependence of extinguishment for a number of regressive burning propellant shapes and for two different aluminum additions. Open symbols are for 0% Al additions. [Adapted from F. A. Anderson, R. A. Strehlow, and L. D. Strand, *AIAA Journal, 1*, 2269 (1963).]

it has been observed that when the chamber is operated in an external environment whose pressure is above the propellant's minimum burning pressure, the presence of hot product gases in the chamber and the existence of a residual preheat zone in the propellant can cause the extinguished grain to reignite and burn. This behavior is called *chuffing*. Chuffing has been observed to occur repeatedly at low chamber pressures until either all the propellant is consumed or, if the grain is progressive, until the equilibrium burning pressure exceeds the L^* limit for the propellant and stable combustion occurs. The theory for the reignition phase of the chuffing phenomenon is not complete, but it appears that reignition occurs by rapid decomposition of the solid phase in a thin zone, and that gas evolution during this process can be sufficiently rapid to momentarily pressurize the chamber to pressure levels above its equilibrium operating pressure.

This problem may be eliminated by designing the rocket engine to operate well above the L^* limit of that particular propellant, and by ensuring that the entire burning surface is ignited rapidly during the ignition phase. This problem is usually handled adequately by placing the ignitor at the head end of the grain so that the ignitor gases and the initial propellant gases sweep over the entire unignited surface during the early phases of the ignition process.

Acoustic instability is so labeled because in its initial stages the growth mechanism is a linearly coupled mechanism involving an acoustic mode of the chamber. Acoustic instability is important because it can (and usually does) lead to very large amplitude pressure fluctuations in the rocket engine. It is not uncommon to find a ± 400-psi fluctuation in a chamber whose average steady operating pressure is 600 psi. This large a fluctuating pressure can cause the average chamber pressure to change markedly, thereby changing the thrust level. It also may lead to a rapid temperature rise in the propellant grain, since the propellant is a viscoelastic material and will absorb energy if compressed at high amplitudes and high frequencies. In addition, these internal motor oscillations adversely effect any instrument package on the rocket and cause a weight penalty if one must design the chamber to contain the overpressures.

Acoustic instability occurs in a solid-propellant rocket engine when any of the possible acoustic modes of the chamber have been excited. The available modes and their frequencies may be found by solving a wave equation similar to Eq. 8-8 with $k = 0$. In this case one looks for a solution that is separable in space coordinates and time and is harmonic in time. For example, a rocket engine whose chamber is a right circular cylinder could possibly excite any of the frequencies of the chamber's longitudinal organ-pipe modes with a fundamental of wavelength $4L$, where L is the chamber length, and in addition, could excite the chamber's transverse, radial, and rotating modes. Cross-modes and coupled modes are also possible in this simple geometry.

The fundamental and higher modes of the more complex internal geometries such as that of an internal burning star-shaped grain are more difficult to determine by purely mathematical techniques. Nevertheless, in general, one must assume that any mode over a broad frequency range is available for excitation. This behavior differs somewhat from the behavior of the Rijke tube as discussed in Chapter 8. In the Rijke tube discussion we noted that instability of a particular standing mode of the tube (only longitudinal modes were considered) was dependent through a particular coupling mechanism on the heater location in the tube and on the direction of the drift velocity. In the case of a solid-propellant rocket, however, the region of active heat addition is always relatively close to the surface of the grain and, therefore, to the boundary of the "acoustic" chamber. This is particularly true for high-pressure operation in a large motor.

As McClure, Hart, and Bird[1] first pointed out, this localization of the possible source of instability allows one to greatly simplify the problem because the bulk of the chamber volume may be treated as a resonant cavity which is filled with inert gas, and the response of the regressing surface to an acoustic level fluctuation may be examined independently as the source of acoustic energy. In general, a particular acoustic mode of the chamber will produce a local oscillation behavior of the gas near the surface which can contain both a fluctuating pressure component (usually thought of as acting normal to the surface) and a fluctuating velocity component (usually assumed to be acting along the surface). Figure 10-5 illustrates this behavior schematically for a specific longitudinal mode in a cylindrical chamber. Note that an acoustic-level signal at a point on the surface at $x = L/3$ consists entirely of a pure velocity perturbation, while at the points $x = 0$ and $x = 2L/3$ it contains only a fluctuating pressure component. Points between these two locations are affected by a coupled transverse velocity–pressure perturbation whose nature changes with the location. In the framework of a linearized coupling theory one may assume, as a first approximation, that a purely pressure-coupled mode is acting independently of the velocity perturbations, even though both may be present.

The theory which discusses the amplification of a purely one-dimensional pressure perturbation by the burning surface is now highly developed. In short, the coupled equations for the one-dimensional model shown in Fig. 10-2 have been linearized to determine the reflection coefficient of the surface as a function of the frequency for an acoustic-

[1] F. T. McClure, R. W. Hart, and J. F. Bird, in M. Summerfield (ed.), *Solid Propellant Rocket Research*, Academic, New York, 1960.

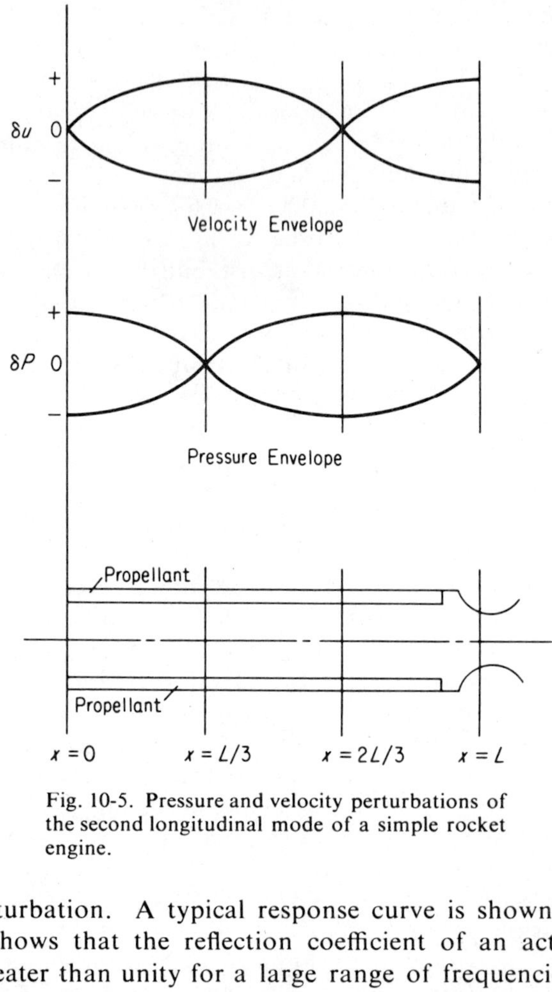

Fig. 10-5. Pressure and velocity perturbations of the second longitudinal mode of a simple rocket engine.

pressure perturbation. A typical response curve is shown in Fig. 10-6. This curve shows that the reflection coefficient of an actively burning surface is greater than unity for a large range of frequencies, and therefore that this propellant surface could be expected to spontaneously amplify incident acoustic-level signals in a broad frequency range. The important result here is not the quantitative shape of the curve shown in Fig. 10-6, since this is dependent in a very complex way on the numerical value of a number of constants which are very difficult to determine. It is instead, the qualitative proof that a regressing propellant surface can linearly amplify an incident acoustic signal over an extremely broad frequency range. In fact, because of the difficulty involved in determining the proper constants to use in the model, at the present time the reflection coefficient is determined by using an experimental technique. Figure 10-7 shows a drawing of a "T" burner which may be used for this determination. This chamber is operated with a port which may be either a

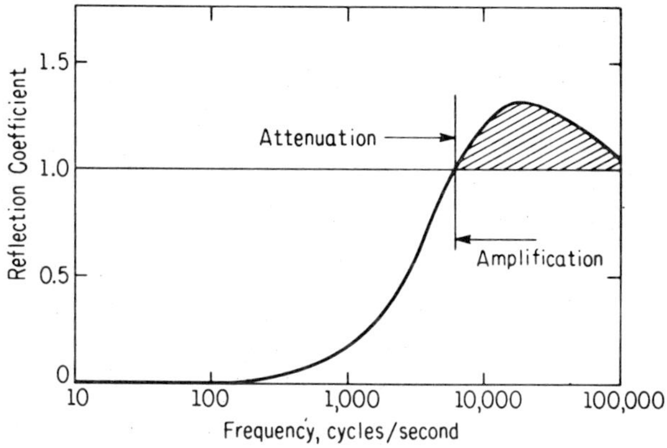

Fig. 10-6. The reflection of a burning solid-propellant surface as a function of the imposed frequency. This curve was calculated by McClure et al using a linearized theory. Note the broad frequency range in which the amplification per cycle is positive (reflection coefficient > 1). [Adapted with permission from F. T. McClure, R. W. Hart, and J. F. Bird, *Solid Propellant Rocket Research* (M. Summerfield, ed.), Academic, New York, 1960.]

sonic nozzle or a larger-apperture nozzle connected to a large volume, high-pressure chamber. Active propellant surfaces are placed at either end or both ends of the burner and the record of the high-frequency transducer is used to monitor the rate of growth of the first longitudinal mode of the chamber (wavelength = $2L$). Frequency is changed by changing the chamber length L from run to run. A typical pressure record and its analysis is shown in Fig. 10-8. Note that the fluctuating pressure amplitude is growing exponentially at the beginning of oscillation. This is typical for a linearly self-excited instability. The experi-

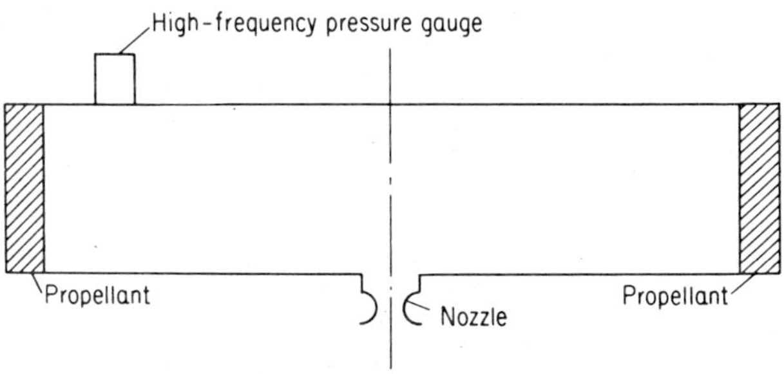

Fig. 10-7. The "T" burner used to elevate the instability behavior of a propellant to a pure pressure perturbation.

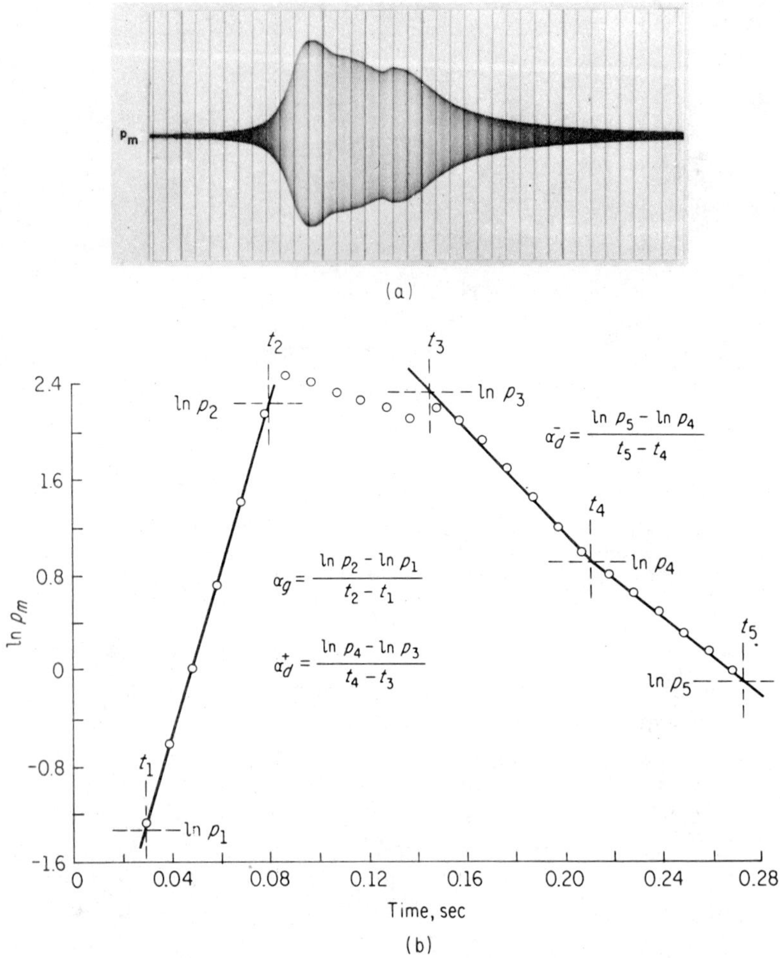

Fig. 10-8. Typical test record showing the oscillating pressure envelope and an analysis of its amplitude behavior. (Adapted from M. D. Horton, *AIAA Journal, 2,* 1112 (1964).)

mental technique is somewhat complicated by the fact that the chamber has losses which must be determined independently to obtain a true value for the reflection coefficient of the active propellant surface. However, the technique has verified the theoretical result and may be used quite effectively for studying the relative response of different propellants to an acoustic perturbation.

The behavior of a velocity-coupled perturbation has not been examined as fully as the purely pressure-coupled amplification behavior discussed above. However, there is ample evidence that velocity-coupled modes do occur in solid-propellant rocket engines. In this case, the presence of the velocity fluctuation must preferentially excite a pressure fluctuation to cause self-amplification, (Rayleigh's criterion, Chapter 8). Therefore, the mode cannot actively couple with the chamber oscillation as a linear phenomenon at the location of the pressure mode for that particular mode (i.e., at $x = L/3$ in Fig. 10-5). It must instead either couple in regions where both fluctuations are present or must couple by an essentially nonlinear mechanism. Such a mechanism is shown schematically in Fig. 10-9. This figure illustrates a weak shock wave which is

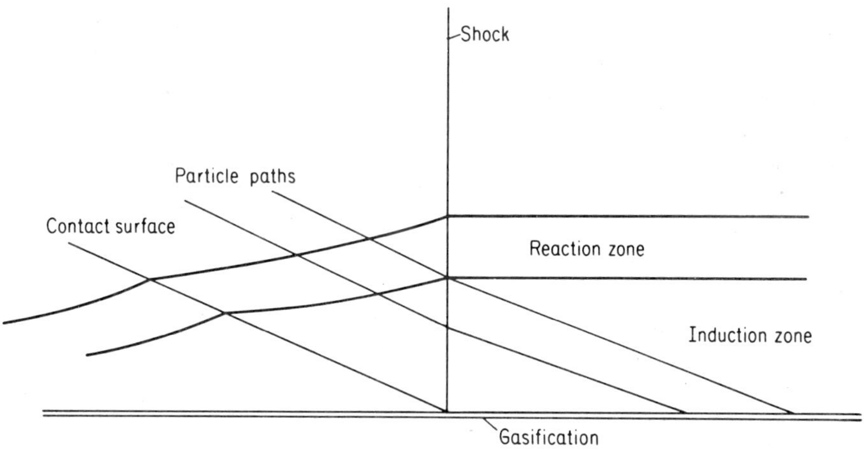

Fig. 10-9. Shock uncovery of the burning zone of a solid propellant in a shock-wave–burning-surface-oriented system (steady for a step shock wave). Note the movement of the reaction zone, the curvature of the particle paths at the shock and the contact surface. The shock is not flat in the actual case because a gradient of sound velocity exists in the neighborhood of the surface. (Schematic.)

uncovering the burning surface. Because of the shock compression and the extreme temperature sensitivity of the induction zone, the rate of heat release behind the shock wave will be enhanced, and this could lead to an active though nonlinear coupling mechanism. Even though we are still discussing the excitation of a standing mode of the chamber, in this case local excitation is occurring by means of a traveling wave, and the usual acoustic theory of coupling is no longer a valid approximation.

Experimental observations of this type of coupling may be made in a modified "T" burner in which the nozzle is located in a position so as to preferentially excite the second mode (L = wavelength). The propellant sample under study is placed in the wall of the chamber at a location well removed from the nozzle location. The nozzle's location is important to the mode type because a nozzle will greatly attenuate local pressure perturbations. It therefore favors an acoustic mode which has a pressure node at its location. A complete understanding of this type of coupling is not yet available.

It has been observed that the presence of fine, high-melting-point particles in the grain, particularly combustible particles, usually enhances the stability behavior of the propellant. This is particularly true for aluminum powder, and therefore a number of practical propellants have been formulated with high concentrations of aluminum powder to suppress unstable burning. The reason why aluminum is effective in suppressing instability is not adequately understood at present. However, two possibilities exist. In the first place, immediately above the propellant surface the aluminum concentration on a mass basis may be considerably above its average concentration because it appears in the gas phase at the solid-phase density and velocity. This is caused by the fact that it takes time for the propellant gases to accelerate the particles. This effect is shown schematically in Fig. 10-10 for a specific steady regression rate.

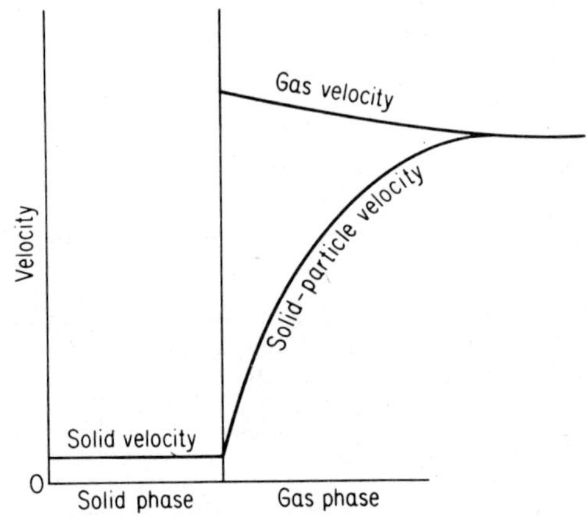

Fig. 10-10. Velocity of the gas and solid particles above a regressing surface. Surface velocity = 0.

Secondly, there is evidence that combustion of the aluminum in the gas phase above the surface has a stabilizing effect on the induction-zone thickness which in turn reduces the acoustic-reflection coefficient sufficiently to produce stable operation. The presence of a fine powder above the surface would also alter the response of the flow to the weak transverse shock wave illustrated in Fig. 10-9. There is considerable literature available on this subject. The interested reader is referred to the bibliography.

10-5. LIQUID-ROCKET ENGINE INSTABILITIES

The reader will note that the title of this section is different from that of the previous section. This difference arises because the instability behavior of solid-propellant engines has been quite definitely related to the behavior of the thin gas-phase region in the neighborhood of the burning surface, which constitutes the active heat-release region for the solid propellant. In the case of liquid rockets, however, the propellant is *pumped* into the chamber in an unmixed state and the active heat-release zone is somewhere in the interior of the chamber. Thus the coupling mechanisms which cause instability are much more strongly related to the transient behavior of the overall system in the case of a liquid engine.

As with solid-propellant engines, there are three major types of instabilities in liquid-propellant engines. However, two of these are fundamentally different from any of the solid-propellant rocket instabilities discussed above. The three major types are (1) pump-line coupling, (2) entropy instabilities (low frequency), and (3) acoustic instabilities.

Pump-line coupling problems can lead to a low-frequency instability in the liquid-propellant rocket engine. This type of instability is reminiscent of the instability first observed by Higgins and, in a liquid-propellant engine it is caused by a coupling of the characteristic relaxation time of the fuel feed rate and the exhaust time of the chamber. High-frequency excitation is seldom observed with this type of interaction because it is primarily controlled by the relatively slow inertial response of a column of liquid in the feed lines to a pressure fluctuation in the chamber. This instability may be remedied by operating the injectors at higher pressures, by changing the length of the injector lines, or by using positive-displacement pumps in the feed lines. It is not really a serious problem at the present time.

Entropy instability is a low-frequency instability which is unique to the liquid-propellant rocket engine. It is primarily a nonlinear phenomena and as such it appears to require a finite-amplitude pressure perturbation before it is excited. This instability travels as a longitudinal wave in the motor—from the nozzle to the injector plate at the velocity of sound and from the injector plate to the nozzle at the flow velocity. The instability occurs when a finite pressure pulse from the nozzle so alters the mixing and burning process as it uncovers the injector region that the local mixture burns incompletely and then passes through the nozzle as a non-equilibrium mixture. The oscillation is maintained if the conditions are such that the passage of this incompletely burned sample through the nozzle generates a new pressure pulse of the proper shape to again cause incomplete combustion at the injectors. This instability is called an *entropy instability* because the fluid motion must carry an entropy inhomogeneity through the nozzle to maintain the instability. It has sometimes been incorrectly called an "entropy-wave" instability, even though entropy cannot be propagated as a wave phenomenon. It has been found that entropy instabilities disappear if the fuel and oxidizer are injected in the proper manner. However, the problem has not been solved at the fundamental level, since the "proper" manner of injection must still be arrived at by trial-and-error techniques, and the design of injectors and injector configurations is still an empirical art.

Acoustic instabilities in liquid-propellant rockets, as in solid-propellant rockets, occur because the enclosure can support a variety of standing-wave modes, each with a characteristic acoustic frequency, and because coupling mechanisms are available which lead to slight amplification of the signal level during each acoustic cycle. However, it is not always obvious in the case of liquid-propellant rockets that the coupling mechanisms are linear mechanisms, and fundamental studies of liquid-propellant rocket instabilities have still not produced even a realistic categorization of all the possible coupling modes that may be available to the system. The major difficulty is that a large fraction of the volume of the engine always contains unburned and partially unmixed fuel and oxidizer, and the system is thus susceptible to the occurrence of catastrophic and self-energized pressure waves. Local detonation waves have been observed in certain circumstances and other high-amplitude oscillations are observed repeatedly. Even though a great body of empirical information on different stable designs has accumulated in the past 20 years, there is still always the possibility of observing unstable operation on the first firing of a newly designed engine or on the first firing of an engine with a new fuel–oxidizer system. The reader is referred to the bibliography for more information on this subject.

BIBLIOGRAPHY

1. *Progress in Aeronautics and Rocketry, Solid Propellant Rocket Research,* Vol. 1 (New York: Academic, 1960.)

2. *Progress in Aeronautics and Rocketry, Liquid Rockets and Propellants,* Vol. 2 (New York: Academic, 1960.)

3. *Progress in Aeronautics and Astronautics, Heterogeneous Combustion,* Vol. 15 (New York: Academic, 1964.)

4. G. P. Sutton, *Rocket Propulsion Elements,* 3d ed. (New York: Wiley, 1963.)

5. R. L. Wilkins, *Theoretical Evaluation of Chemical Propellants.* (Englewood Cliffs, N.J.: Prentice-Hall, 1963.)

6. B. Siegel and L. Schieler, *Energetics of Propellant Chemistry.* (New York: Wiley, 1964.)

7. F. A. Warren, *Rocket Propellants.* (New York: Reinhold, 1958.)

8. S. S. Penner, *Chemical Rocket Propulsion and Combustion Research.* (New York: Gordon and Breach, 1962.)

9. N. I. Melek-Pashayev, *Liquid Propellant Engines.* (New York: Pergamon, 1962.)

10. R. E. Wiech, Jr., and R. F. Strauss, *Fundamentals of Rocket Propulsion.* (New York: Reinhold, 1960.)

11. Chemical Propulsion Information Agency, Second Combustion Conference, Vol. 1 (May, 1966.)

12. B. A. Nikolaev, *Thermodynamic Assessment of Rocket Engines.* (New York: Pergamon, 1963.)

13. S. L. Bragg, *Rocket Engines.* (London: Newnes, 1962.)

14. B. Kit and D. S. Evered, *Rocket Propellant Handbook.* (New York: Macmillan, 1960.)

15. M. Barrére, *Rocket Propulsion.* (New York: Elsevier, 1960.)

16. R. N. Wimpress, *Internal Ballistics of Solid Fuel Rockets.* (New York: McGraw-Hill, 1950.)

PROBLEMS

10-1. Suppose a propellant of density 1.6 gm/cc has a value of the pressure exponent $n = 0.5$ in Eq. 10-1 and a regression rate of 1 in./sec at 1,000 psia. Assume the flame temperature is 2500°K, $\gamma = 1.2$, and the molecular weight of the product gases is constant = 24 gm/mole. Calculate the surface area required to maintain pressures of 200, 500, 1,000, and 5,000 psi if the nozzle has a ½-in. diameter throat.

10-2. For the propellant of Prob. 10-1 calculate a theoretical pressure time curve for the progressive and regressive cylinders shown in Fig. 10-1. Assume that the ends of the grains are inhibited.

10-3. Set up the acoustic equation for cylindrical coordinates and calculate the available longitudinal, transverse, and radial frequencies for a cylindrical chamber of size $L/D = 10$. Express your answers as ratios to the fundamental longitudinal mode. Assume that this is a classic organ-pipe mode of wavelength $4L$.

Appendices

APPENDIX A

Bridgman's Table of Thermodynamic Relationships*

First derivatives of thermodynamic functions may be evaluated by substitution in the table as follows:

$$\left(\frac{\partial A}{\partial B}\right)_c = \frac{(\partial A)_c}{(\partial B)_c}$$

(See Sec. 2-10 for details.)

*With permission from P. W. Bridgman, *Condensed Collection of Thermodynamic Formulas*, Harvard U. Press, Cambridge, 1925.

$$(\partial T)_p = -(\partial p)_T = 1$$

$$(\partial V)_p = -(\partial p)_V = \left(\frac{\partial V}{\partial T}\right)_p$$

$$(\partial S)_p = -(\partial p)_S = \frac{C_p}{T}$$

$$(\partial E)_p = -(\partial p)_E = C_p - p\left(\frac{\partial V}{\partial T}\right)_p$$

$$(\partial H)_p = -(\partial p)_H = C_p$$

$$(\partial F)_p = -(\partial p)_F = -S$$

$$(\partial A)_p = -(\partial p)_A = -\left[S + p\left(\frac{\partial V}{\partial T}\right)_p\right]$$

$$(\partial V)_T = -(\partial T)_V = -\left(\frac{\partial V}{\partial p}\right)_T$$

$$(\partial S)_T = -(\partial T)_S = \left(\frac{\partial V}{\partial T}\right)_p$$

$$(\partial E)_T = -(\partial T)_E = T\left(\frac{\partial V}{\partial T}\right)_p + p\left(\frac{\partial V}{\partial p}\right)_T$$

$$(\partial H)_T = -(\partial T)_H = -V + T\left(\frac{\partial V}{\partial T}\right)_p$$

$$(\partial F)_T = -(\partial T)_F = -V$$

$$(\partial A)_T = -(\partial T)_A = p\left(\frac{\partial V}{\partial p}\right)_T$$

$$(\partial S)_V = -(\partial V)_S = \frac{1}{T}\left[C_p\left(\frac{\partial V}{\partial p}\right)_T + T\left(\frac{\partial V}{\partial T}\right)_p^2\right]$$

$$(\partial E)_V = -(\partial V)_E = C_p\left(\frac{\partial V}{\partial p}\right)_T + T\left(\frac{\partial V}{\partial T}\right)_p^2$$

$$(\partial H)_V = -(\partial V)_H = C_p\left(\frac{\partial V}{\partial p}\right)_T + T\left(\frac{\partial V}{\partial T}\right)_p^2 - V\left(\frac{\partial V}{\partial T}\right)_p$$

$$(\partial F)_V = -(\partial V)_F = -\left[V\left(\frac{\partial V}{\partial T}\right)_p + S\left(\frac{\partial V}{\partial p}\right)_T\right]$$

$$(\partial A)_V = -(\partial V)_A = -S\left(\frac{\partial V}{\partial p}\right)_T$$

$$(\partial E)_S = -(\partial S)_E = \frac{p}{T}\left[C_p\left(\frac{\partial V}{\partial p}\right)_T + T\left(\frac{\partial V}{\partial T}\right)_p^2\right]$$

$$(\partial H)_S = -(\partial S)_H = -\frac{VC_p}{T}$$

$$(\partial F)_S = -(\partial S)_F = -\frac{1}{T}\left[VC_p - ST\left(\frac{\partial V}{\partial T}\right)_p\right]$$

$$(\partial A)_S = -(\partial S)_A = \frac{1}{T}\left\{p\left[C_p\left(\frac{\partial V}{\partial p}\right)_T + T\left(\frac{\partial V}{\partial T}\right)_p^2\right] + ST\left(\frac{\partial V}{\partial T}\right)_p\right\}$$

$$(\partial H)_E = -(\partial E)_H = -V\left[C_p - p\left(\frac{\partial V}{\partial T}\right)_p\right] - p\left[C_p\left(\frac{\partial V}{\partial p}\right)_T + T\left(\frac{\partial V}{\partial T}\right)_p^2\right]$$

$$(\partial F)_E = -(\partial E)_F = -V\left[C_p - p\left(\frac{\partial V}{\partial T}\right)_p\right] + S\left[T\left(\frac{\partial V}{\partial T}\right)_p + p\left(\frac{\partial V}{\partial p}\right)_T\right]$$

$$(\partial A)_E = -(\partial E)_A = p\left[C_p\left(\frac{\partial V}{\partial p}\right)_T + T\left(\frac{\partial V}{\partial T}\right)_p^2\right] + S\left[T\left(\frac{\partial V}{\partial T}\right)_p + p\left(\frac{\partial V}{\partial p}\right)_T\right]$$

$$(\partial F)_H = -(\partial H)_F = -V(C_p + S) + TS\left(\frac{\partial V}{\partial T}\right)_p$$

$$(\partial A)_H = -(\partial H)_A = -\left[S + p\left(\frac{\partial V}{\partial T}\right)_p\right]\left[V - T\left(\frac{\partial V}{\partial T}\right)_p\right] + pC_p\left(\frac{\partial V}{\partial p}\right)_T$$

$$(\partial A)_F = -(\partial F)_A = -S\left[V + p\left(\frac{\partial V}{\partial p}\right)_T\right] - pV\left(\frac{\partial V}{\partial T}\right)_p$$

APPENDIX B

Conversion of Energy Units*

The following are all used as energy units: erg, the fundamental unit of the cgs system; joule, 10^7 ergs; calorie, the heat required to raise 1 gm of water 1°C; kilocalorie, 10^3 calories; frequency in sec^{-1}, the energy of a photon of that frequency; wave number, in cm^{-1}, the energy of a photon of that wave number; volt, the energy of an electron accelerated through that number of volts; degree, the energy divided by k, the Boltzmann constant.

	ergs/molecule	sec^{-1}	cm^{-1}	ev	joules/mole	kcal/mole	°K
1 erg per molecule equals	1	1.5092×10^{26}	5.034×10^{15}	6.242×10^{11}	6.023×10^{16}	1.4391×10^{13}	7.245×10^{15}
Unit frequency in sec^{-1} equals	6.6258×10^{-27}	1	3.3356×10^{-11}	4.136×10^{-15}	3.9907×10^{-10}	9.536×10^{-14}	4.800×10^{-11}
Unit wave number in cm^{-1} equals	1.9864×10^{-16}	2.99796×10^{10}	1	1.240×10^{-4}	11.964	2.859×10^{-3}	1.4390
1 abs electron volt equals	1.6021×10^{-12}	2.418×10^{14}	8.065×10^{3}	1	9.6494×10^{4}	23.055	1.1606×10^{4}
1 abs joule per mole equals	1.6603×10^{-17}	2.506×10^{9}	8.358×10^{-2}	1.0369×10^{-5}	1	2.3894×10^{-4}	0.12028
1 kilocalorie per mole equals	6.949×10^{-14}	1.049×10^{13}	3.498×10^{2}	4.3375×10^{-2}	4.1852×10^{3}	1	5.034×10^{2}
1 degree Kelvin equals	1.3804×10^{-16}	2.083×10^{10}	0.6949	8.616×10^{-5}	8.3136	1.9864×10^{-3}	1

*Adapted from J. E. Mayer and M. G. Mayer, *Statistical Mechanics*, Wiley, New York, 1940.

APPENDIX C-1

MOLECULAR PARAMETERS FOR DIATOMIC MOLECULES*

Specie	ω_e Vibrational Constant, cm^{-1}	B_e Rotational Constant, cm^{-1}	r_e Equilibrium Radius, cm × 10^{+8}	D_0 Heat of Dissociation, cm^{-1}
H_2	4,405.3	60.848	0.741	36,150
N_2	2,357.6	1.998	1.097	78,700
O_2	1,580.2	1.445	1.207	41,300
CO	2,169.5	1.930	1.128	89,200
NO	1,903.6	1.704	1.151	52,300
CN	2,068.6	1.899	—	65,000
NH	3,315.0	16.668	1.038	29,000
CH	2,861.3	14.454	1.120	—
HCl	2,989.6	10.588	1.275	35,700
HBr	2,649.6	8.466	—	29,000
Cl_2	561.1	0.2408	1.988	20,000
Br_2	323.2	0.08092	—	15,900

*Data adapted from JANAF tables of thermochemical data, Dow Chemical Corporation, Midland, Mich.

APPENDIX C-2

STRUCTURE OF TRIATOMIC MOLECULES*

Specie and Structure	Bond Angle and Distances, cm × 10^8	Vibrational Constants and Degeneracies, cm^{-1}	Moments of Inertia Linear $I_A = I_B$, gm cm^2 Nonlinear $I_A I_B I_C$, gm^{3}cm^6 or B, cm^{-1}
CO_2 O=C=O	1.926 Linear 1.926	1342.86(1) 667.30(2) 2349.30(1)	$7.152 \times 10^{-39} I_A = I_B$
H_2O H–O–H	0.9584 104.45° 0.9584	3657.05(1) 1594.59(1) 3755.79(1)	$5.766 \times 10^{-120} I_A I_B I_C$
O_3 O=O=O	— — —	1110(1) 705(1) 1043(1)	3.5538 cm^{-1} B_A 0.4453 cm^{-1} B_B 0.3948 cm^{-1} B_C
N_2O N≡N=O	1.128 Linear 1.184	1276.5(1) 589.2(2) 2223.7(1)	0.4211 cm^{-1} B

Specie and Structure	Bond Angle and Distances, cm × 10^8	Vibrational Constants and Degeneracies, cm^{-1}	Moments of Inertia Linear $I_A = I_B$, gm cm^2 Nonlinear $I_A I_B I_C$, $gm^3 cm^6$ or B, cm^{-1}
NO_2 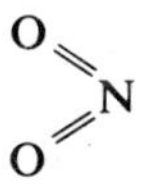	1.197 134°15′ 1.197	1357.8(1) 756.8(1) 1665.5(1)	$15.398 \times 10^{-117} I_A I_B I_C$
NH_2 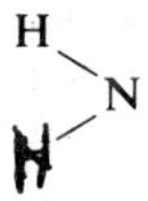	1.025 103.0° 1.025	3400(1) 1550(1) 3450(1)	$8.5697 \times 10^{-120} I_A I_B I_C$
HCN 	— Linear —	2100.24(1) 711.90(2) 3315.03(1)	$18.846 \times 10^{-39} I_A = I_B$
HCO 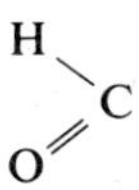	1.08 119°30′ 1.198	3000(1) 1860(1) 1083(1)	$4.73 \times 10^{-118} I_A I_B I_C$
CH_2 	1.03 Linear 1.03	2960(1) 1114(2) 3200(1)	$3.55 \times 10^{-40} I_A = I_B$

*Data adapted from JANAF tables of thermochemical data, Dow Chemical Corporation, Midland, Mich.

APPENDIX C-3

CHARACTERISTIC WAVE NUMBERS OF CHEMICAL BONDS*

Bond	Stretching Vibration, cm^{-1}	Bending Vibration, cm^{-1}
C—C (aliphatic)	910	650
C—C (aromatic)	1,500	600
C=C	1,200	910
C≡C	2,080	375
C—H	3,000	1,050
C—O	1,030	1,120
C=O	1,740	780
C—N	1,000	450
C≡N	2,220	240
C—Cl	650	260
C—F	1,050	1,200
C—Br	610	950
C—I	530	880
H—O	3,500	1,350

APPENDIX C-3 (*continued*)

Bond	Stretching Vibration, cm^{-1}	Bending Vibration, cm^{-1}
H—S	2,570	860
S=O	1,250	520
S—C	690	280
N—N bond in N_2O	2,220	590
N—N bond in N_2O_4	280	500
N—N	1,000	900
N—H	3,300	1,200
N—O	1,270	660
N=O	1,470	650
Si—H	2,180	930
Si—O	1,050	400
B—H	2,550	1,150
B—F	890	690

*R. V. Meghreblian, *J. Am. Roc. Soc.*, *21*, 127 (1951).

APPENDIX D

Van der Waals and Critical Constants of Gases*

a = (atm)(cubic centimeters per gram-mole)^2
b = cubic centimeters per gram-mole
T_c = critical temperature, degrees Kelvin
P_c = critical pressure, atm
ρ_c = critical vapor density, grams per cubic centimeter
V_c = critical molar volume, cubic centimeters per gram-mole

Gases	a	b	T_c	P_c	ρ_c	V_c
Argon	$(1.353)10^6$	32.31	151.2	48.0	0.533	75
Acetylene	$(4.417)10^6$	51.54	309.5	61.6	0.2304	113
Ammonia	$(4.197)10^6$	37.37	405.5	111.3	0.235	72.5
Carbon dioxide	$(3.606)10^6$	42.80	304.2	72.9	0.468	94
Carbon monoxide	$(1.456)10^6$	39.54	133.0	34.5	0.3012	93
Chlorine	$(6.491)10^6$	56.21	417.0	76.1	0.572	124
Ethylene	$(4.508)10^6$	57.50	283.1	50.5	0.226	124
Hydrogen chloride	$(3.672)10^6$	40.85	324.6	81.5	0.419	87
Hydrogen	$(0.2461)10^6$	26.68	33.3	12.8	0.03102	65
Methane	$(2.256)10^6$	42.71	190.7	45.8	0.162	99
Methyl chloride	$(7.470)10^6$	64.80	416.3	65.9	0.353	143
Nitrogen	$(1.351)10^6$	38.64	126.2	33.5	0.311	90
Oxygen	$(1.362)10^6$	31.87	154.4	49.7	0.432	74
Sulfur dioxide	$(6.773)10^6$	56.78	430.7	77.8	0.525	122
Water	$(5.454)10^6$	30.42	647.4	218.3	0.322	56

*With permission from O. A. Hougen, K. M. Watson, and R. A. Ragatz, *Chemical Process Principles*, Part II, *Thermodynamics*, 2d ed., Wiley, New York, 1959.

APPENDIX E-1

FUNCTIONS FOR PREDICTION OF TRANSPORT PROPERTIES OF GASES AT LOW DENSITIES*
(Collision Integrals)

kT/ϵ or $(kT/\epsilon)_{AB}$	Ω_μ or Ω_K (For viscosity and thermal conductivity)	Ω_D (For mass diffusivity)	B*	C*
0.30	2.785	2.662	−27.880581	
0.35	2.628	2.476	−18.754895	
0.40	2.492	2.318	−13.798835	
0.45	2.368	2.184	−10.754975	
0.50	2.257	2.066	−8.720205	
0.55	2.156	1.966	−7.2740858	
0.60	2.065	1.877	−6.1979708	
0.65	1.982	1.798	−5.3681918	
0.70	1.908	1.729	−4.7100370	−3.37664
0.75	1.841	1.667	−4.1759283	−1.79197
0.80	1.780	1.612	−3.7342254	−0.84953
0.85	1.725	1.562	−3.3631193	−0.27657
0.90	1.675	1.517	−3.0471143	0.07650
0.95	1.629	1.476	−2.7749102	0.29509
1.00	1.587	1.439	−2.5380814	0.42966
1.05	1.549	1.406	−2.3302208	0.51080
1.10	1.514	1.375	−2.1463742	0.55762
1.15	1.482	1.346	−1.9826492	0.58223
1.20	1.452	1.320	−1.8359492	0.59240
1.25	1.424	1.296	−1.7037784	0.59326
1.30	1.399	1.273	−1.5841047	0.58815
1.35	1.375	1.253	−1.4752571	0.57933
1.40	1.353	1.233	−1.3758479	0.56831
1.45	1.333	1.215	−1.2847160	0.55611
1.50	1.314	1.198	−1.2008832	0.54339
1.55	1.296	1.182	−1.1235183	0.53059
1.60	1.279	1.167	−1.0519115	0.51803
1.65	1.264	1.153	−0.98545337	0.50587
1.70	1.248	1.140	−0.92361639	0.49425
1.75	1.234	1.128	−0.86594279	0.48320
1.80	1.221	1.116	−0.81203328	0.47277
1.85	1.209	1.105	−0.76153734	0.46296
1.90	1.197	1.094	−0.71414733	0.45376
1.95	1.186	1.084	−0.66959030	0.44515

Appendix E-1 (*continued*)

kT/ϵ or $(kT/\epsilon)_{AB}$	Ω_μ or Ω_K (For viscosity and thermal conductivity)	Ω_D (For mass diffusivity)	B*	C*
2.00	1.175	1.075	−0.62762535	0.43710
2.10	1.156	1.057	−0.55063308	0.42260
2.20	1.138	1.041	−0.48170997	0.40999
2.30	1.122	1.026	−0.41967761	0.39900
2.40	1.107	1.012	−0.36357566	0.38943
2.50	1.093	0.9996	−0.31261340	0.38108
2.60	1.081	0.9878	−0.26613345	0.37378
2.70	1.069	0.9770	−0.22358626	0.36737
2.80	1.058	0.9672	−0.18450728	0.36173
2.90	1.048	0.9576	−0.14850215	0.35675
3.00	1.039	0.9490	−0.11523390	0.35234
3.10	1.030	0.9406	−0.08441245	0.34842
3.20	1.022	0.9328	−0.05578696	0.34491
3.30	1.014	0.9256	−0.02913997	0.34177
3.40	1.007	0.9186	−0.00428086	0.33894
3.50	0.9999	0.9120	0.01895684	0.33638
3.60	0.9932	0.9058	0.04072012	0.33407
3.70	0.9870	0.8998	0.06113882	0.33196
3.80	0.9811	0.8942	0.08032793	0.33002
3.90	0.9755	0.8888	0.09839014	0.32825
4.00	0.9700	0.8836	0.11541691	0.32662
4.10	0.9649	0.8788	0.13149021	0.32510
4.20	0.9600	0.8740	0.14668372	0.32369
4.30	0.9553	0.8694	0.16106381	0.32238
4.40	0.9507	0.8652	0.17469039	0.32115
4.50	0.9464	0.8610	0.18761774	0.32000
4.60	0.9422	0.8568	0.19989511	0.31891
4.70	0.9382	0.8530	0.21156728	0.31788
4.80	0.9343	0.8492	0.22267507	0.31690
4.90	0.9305	0.8456	0.23325577	0.31596
5.0	0.9269	0.8422	0.24334351	0.31508
6.0	0.8963	0.8124	0.32290437	0.30771
7.0	0.8727	0.7896	0.37608846	0.30166
8.0	0.8538	0.7712	0.41343396	0.29618
9.0	0.8379	0.7556	0.44059784	0.29103
10.0	0.8242	0.7424	0.46087529	0.28610
20.0	0.7432	0.6640	0.52537420	0.24643
30.0	0.7005	0.6232	0.52692546	0.21954
40.0	0.6718	0.5960	0.51857502	0.20012
50.0	0.6504	0.5756	0.50836143	0.18529

kT/ϵ or $(kT/\epsilon)_{AB}$	Ω_μ or Ω_K (For viscosity and thermal conductivity)	Ω_D (For mass diffusivity)	B*	C*
60.0	0.6335	0.5596	0.49821261	0.17347
70.0	0.6194	0.5464	0.48865069	0.16376
80.0	0.6076	0.5352	0.47979009	0.15560
90.0	0.5973	0.5256	0.47161504	0.14860
100.0	0.5882	0.5130	0.46406948	0.14251
200.0	0.5320	0.4644	0.41143168	0.10679
300.0	0.5016	0.4360	0.38012787	0.08943
400.0	0.4811	0.4170	0.35835117	0.07862

*With permission from J. O. Hirschfelder, C. F. Curtiss, and R. B. Bird, *Molecular Theory of Gases and Liquids*, Wiley, New York, 1954.

.APPENDIX E-2

LENNARD-JONES PARAMETERS[a]

Substance	Molecular Weight, M	Lennard-Jones Parameters: From Viscosity Data ϵ/k °K	σ_0 Å	From Second Virial Coefficients* ϵ/k °K	σ_0 Å	$b_0 = \frac{2}{3}\pi\mathfrak{N}\sigma_0^3$ (cc/mole)
Light elements:						
H_2	2.016	38.0	2.915	29.2	2.87	29.76
D_2	4.032	39.0	2.948	31.1	2.87	29.76
Noble gases:						
He	4.003	10.22	2.576	10.22	2.556	21.07
Ne	20.183	27.5	2.858	35.60	2.749	26.21
Ar	39.948	116	3.465	119.8	3.405	49.80
Kr	83.80	190	3.61	171	3.60	58.86
Xe	131.30	229	4.055	221	4.10	86.94
Simple polyatomic substances:						
Air	28.97[b]	84.0	3.689	99.2	3.522	55.11
N_2	28.02	79.8	3.749	95.05	3.698	63.78
O_2	32.00	88.0	3.541	117.5	3.58	57.75
CO	28.01	88.0	3.706	100.2	3.763	67.22
CO_2	44.05	213.0	3.897	189	4.486	113.9
NO	30.01	91.0	3.599	131	3.17	40.0
N_2O	44.02	237	3.816	189	4.59	122
C_2N_2*	52.05	339	4.38	—	—	—
HCN*	27.03	569	3.63	—	—	—
NH_3*	17.03	558	2.90	—	—	—
COS*	60.08	335	4.13	—	—	—
CS_2*	76.14	488	4.44	—	—	—
SO_2*	64.07	252	4.29	—	—	—
H_2S*	34.08	301	3.62	—	—	—
H_2O*	18.02	809	2.64	—	—	—
H_2O_2*	34.02	289	4.20	—	—	—

APPENDIX E-2 (*continued*)

Substance	Molecular Weight, M	Lennard-Jones Parameters: From Viscosity Data, ϵ/k °K	From Viscosity Data, σ_0 Å	From Second Virial Coefficients*, ϵ/k °K	From Second Virial Coefficients*, σ_0 Å	$b_0 = \frac{2}{3}\pi \mathfrak{N} \sigma_0^3$ (cc/mole)
Halogen compounds:						
F_2*	38.00	112	3.65	—	—	—
Cl_2	70.91	257	4.40	—	—	—
Br_2*	159.83	520	4.27	—	—	—
I_2*	253.81	550	4.98	—	—	—
HF*	20.01	330	3.15	—	—	—
HCl*	36.47	360	3.30	—	—	—
HBr*	80.92	449	3.35	—	—	—
HI*	127.91	324	4.12	—	—	—
Hydrocarbons:						
CH_4	16.04	144	3.796	148.2	3.817	70.16
C_2H_2	26.04	185	4.221	—	—	—
C_2H_4	28.05	205	4.232	199.2	4.523	116.7
C_2H_6	30.07	230	4.418	243	3.954	78
C_3H_8	44.09	254	5.016	242	5.637	226
n—C_4H_{10}	58.12	410	4.997	297	4.971	155
i—C_4H_{10}	58.12	313	5.341	—	—	—
n—C_5H_{12}*	72.15	345	5.77	—	—	—
C_6H_6	78.11	440	5.27	—	—	—
C_6H_{12}*	84.16	297	6.18	—	—	—
n—C_6H_{14}*	86.17	399	5.95	—	—	—

[a] Values of σ_0 and ϵ/k other than those marked with asterisk (*) are taken with permission from J. O. Hirschfelder, C. F. Curtiss, and R. B. Bird, *Molecular Theory of Gases and Liquids*, Wiley, New York, 1954, pp. 1110–1112; also addenda and corrigenda, p. 11. The above values computed from viscosity data are applicable for temperatures above 300°K. It is suggested that force constants derived from viscosity be used to calculate transport coefficients and those derived from equations of state data be used exclusively for equation of state work.

[b] For air, the molecular weight M has been calculated from the average composition of dry air, as given in *International Critical Tables*, Vol. I, McGraw-Hill, New York, 1926, p. 393.

*Values of σ_0 and ϵ/k from R. M. Fristrom and A. A. Westenberg, *Flame Structure*, McGraw-Hill, New York, 1965.

APPENDIX F

Thermodynamic Properties of Selected Compounds*

List of Contents

*The horizontal dashed lines in some of the tables denote the limits of temperature range over which either the solid, liquid, or gaseous state is the preferred equilibrium state of that substance at one atmosphere pressure. The equilibrium constant of formation, *Kp*, tabulated as a logarithm in the last column always contains units of atmospheres (to the appropriate exponent). Data in these tables adapted from JANAF tables of thermodynamic data, Dow Chemical Company, Midland, Michigan.

The complete JANNAF tables have been published as: Stull, D. R. and Prophet, H., JANNAF Thermochemical Tables, second edition, NSRDS-NBS 37, National Bureau of Standards, June 1971. They may be purchased from the Superintendent of Documents, US Government Printing Office, Washington, D.C. 20402 (Cat. No. C 13.48:37) for $9.75. Supplement tables containing correction tables and new species tables appear at irregular intervals in the Journal of Physical and Chemical Reference Data, published by the American Chemical Society and the American Institute of Physics for the National Bureau of Standards.

OXYGEN, DIATOMIC (O_2)

(Reference State—Ideal Gas)(mol. wt. = 32.00)Sept. 30, 1965

	cal mole^{-1} deg^{-1}			kcal mole^{-1}			
T, °K.	C_p°	S°	$-(F^\circ - H^\circ_{298})/T$	$H^\circ - H^\circ_{298}$	ΔH_f°	ΔF_f°	$Log_{10}K_P$
0	.000	.000	INFINITE	− 2.075	.000	.000	.000
100	6.958	41.395	55.205	− 1.381	.000	.000	.000
200	6.961	46.218	49.643	− .685	.000	.000	.000
298	7.020	49.004	49.004	.000	.000	.000	.000
300	7.023	49.047	49.004	.013	.000	.000	.000
400	7.196	51.091	49.282	.724	.000	.000	.000
500	7.431	52.722	49.812	1.455	.000	.000	.000
600	7.670	54.098	50.414	2.210	.000	.000	.000
700	7.883	55.297	51.028	2.988	.000	.000	.000
800	8.063	56.361	51.629	3.786	.000	.000	.000
900	8.212	57.320	52.209	4.600	.000	.000	.000
1000	8.336	58.192	52.765	5.427	.000	.000	.000
1100	8.439	58.991	53.295	6.266	.000	.000	.000
1200	8.527	59.729	53.801	7.114	.000	.000	.000
1300	8.604	60.415	54.283	7.971	.000	.000	.000
1400	8.674	61.055	54.744	8.835	.000	.000	.000
1500	8.738	61.656	55.185	9.706	.000	.000	.000

OXYGEN, DIATOMIC (O_2) (Continued)

	cal mole⁻¹ deg⁻¹			kcal mole⁻¹			
T, °K.	C_p°	S°	$-(F^\circ - H_{298}^\circ)/T$	$H^\circ - H_{298}^\circ$	ΔH_f°	ΔF_f°	$Log_{10}K_P$
1600	8.800	62.222	55.608	10.583	.000	.000	.000
1700	8.858	62.757	56.013	11.465	.000	.000	.000
1800	8.916	63.265	56.401	12.354	.000	.000	.000
1900	8.973	63.749	56.776	13.249	.000	.000	.000
2000	9.029	64.210	57.136	14.149	.000	.000	.000
2100	9.084	64.652	57.483	15.054	.000	.000	.000
2200	9.139	65.076	57.819	15.966	.000	.000	.000
2300	9.194	65.483	58.143	16.882	.000	.000	.000
2400	9.248	65.876	58.457	17.804	.000	.000	.000
2500	9.301	66.254	58.762	18.732	.000	.000	.000
2600	9.354	66.620	59.057	19.664	.000	.000	.000
2700	9.405	66.974	59.344	20.602	.000	.000	.000
2800	9.455	67.317	59.622	21.545	.000	.000	.000
2900	9.503	67.650	59.893	22.493	.000	.000	.000
3000	9.551	67.973	60.157	23.446	.000	.000	.000
3100	9.596	68.287	60.415	24.403	.000	.000	.000
3200	9.640	68.592	60.665	25.365	.000	.000	.000
3300	9.682	68.889	60.910	26.331	.000	.000	.000
3400	9.723	69.179	61.149	27.302	.000	.000	.000
3500	9.762	69.461	61.383	28.276	.000	.000	.000
3600	9.799	69.737	61.611	29.254	.000	.000	.000
3700	9.835	70.006	61.834	30.236	.000	.000	.000
3800	9.869	70.269	62.053	31.221	.000	.000	.000
3900	9.901	70.525	62.267	32.209	.000	.000	.000
4000	9.932	70.776	62.476	33.201	.000	.000	.000
4100	9.961	71.022	62.682	34.196	.000	.000	.000
4200	9.988	71.262	62.883	35.193	.000	.000	.000
4300	10.015	71.498	63.081	36.193	.000	.000	.000
4400	10.039	71.728	63.275	37.196	.000	.000	.000
4500	10.062	71.954	63.465	38.201	.000	.000	.000
4600	10.084	72.176	63.652	39.208	.000	.000	.000
4700	10.104	72.393	63.836	40.218	.000	.000	.000
4800	10.123	72.606	64.016	41.229	.000	.000	.000
4900	10.140	72.814	64.194	42.242	.000	.000	.000
5000	10.156	73.019	64.368	43.257	.000	.000	.000
5100	10.172	73.221	64.540	44.274	.000	.000	.000
5200	10.187	73.418	64.708	45.292	.000	.000	.000
5300	10.200	73.613	64.875	46.311	.000	.000	.000
5400	10.213	73.803	65.038	47.332	.000	.000	.000
5500	10.225	73.991	65.199	48.353	.000	.000	.000
5600	10.237	74.175	65.358	49.377	.000	.000	.000
5700	10.247	74.356	65.514	50.401	.000	.000	.000
5800	10.258	74.535	65.668	51.426	.000	.000	.000
5900	10.267	74.710	65.820	52.452	.000	.000	.000
6000	10.276	74.883	65.970	53.479	.000	.000	.000

Oxygen, Monatomic (O)

(Ideal Gas)(atomic wt. = 16.00) June 30, 1962

T, °K.	C_p°	S°	$-(F^\circ - H_{298}^\circ)/T$	$H^\circ - H_{298}^\circ$	ΔH_f°	ΔF_f°	$Log_{10}K_P$
	cal mole⁻¹ deg⁻¹			kcal mole⁻¹			
0	.000	.000	INFINITE	− 1.608	58.989	58.989	INFINITE
100	5.666	32.466	43.266	− 1.080	59.160	57.989	− 126.730
200	5.434	36.340	38.953	− .523	59.377	56.733	− 61.992
298	5.237	38.468	38.468	.000	59.559	55.395	− 40.604
300	5.235	38.501	38.468	.010	59.562	55.369	− 40.334
400	5.135	39.991	38.672	.528	59.725	53.946	− 29.473
500	5.081	41.131	39.055	1.038	59.870	52.485	− 22.940
600	5.049	42.054	39.480	1.544	59.998	50.995	− 18.574
700	5.029	42.831	39.905	2.048	60.113	49.486	− 15.449
800	5.015	43.501	40.313	2.550	60.216	47.960	− 13.101
900	5.006	44.092	40.701	3.052	60.311	46.422	− 11.272
1000	4.999	44.619	41.067	3.552	60.397	44.875	− 9.807
1100	4.994	45.095	41.412	4.051	60.477	43.318	− 8.606
1200	4.990	45.529	41.737	4.551	60.553	41.755	− 7.604
1300	4.987	45.928	42.044	5.049	60.623	40.186	− 6.755
1400	4.984	46.298	42.335	5.548	60.689	38.611	− 6.027
1500	4.982	46.642	42.611	6.046	60.752	37.032	− 5.395
1600	4.981	46.963	42.873	6.544	60.812	35.448	− 4.842
1700	4.979	47.265	43.123	7.042	60.869	33.862	− 4.353
1800	4.979	47.550	43.361	7.540	60.922	32.271	− 3.918
1900	4.978	47.819	43.588	8.038	60.973	30.678	− 3.529
2000	4.978	48.074	43.806	8.536	61.020	29.082	− 3.178
2100	4.978	48.317	44.015	9.034	61.066	27.484	− 2.860
2200	4.979	48.549	44.216	9.532	61.108	25.884	− 2.571
2300	4.980	48.770	44.409	10.029	61.147	24.282	− 2.307
2400	4.981	48.982	44.596	10.527	61.184	22.679	− 2.065
2500	4.984	49.185	44.775	11.026	61.219	21.073	− 1.842
2600	4.986	49.381	44.948	11.524	61.251	19.467	− 1.636
2700	4.990	49.569	45.116	12.023	61.281	17.859	− 1.446
2800	4.994	49.751	45.278	12.522	61.309	16.251	− 1.268
2900	4.999	49.926	45.436	13.022	61.334	14.642	− 1.103
3000	5.004	50.096	45.588	13.522	61.358	13.031	− .949
3100	5.010	50.260	45.736	14.023	61.380	11.420	− .805
3200	5.017	50.419	45.880	14.524	61.401	9.807	− .670
3300	5.025	50.573	46.020	15.026	61.420	8.194	− .543
3400	5.033	50.724	46.156	15.529	61.437	6.581	− .423
3500	5.041	50.870	46.289	16.033	61.454	4.967	− .310
3600	5.050	51.012	46.418	16.537	61.469	3.354	− .204
3700	5.060	51.150	46.544	17.043	61.484	1.739	− .103
3800	5.070	51.285	46.667	17.549	61.498	.125	− .007
3900	5.081	51.417	46.787	18.057	61.511	− 1.492	.084
4000	5.091	51.546	46.904	18.565	61.524	− 3.107	.170

Oxygen, Monatomic (O) (Continued)

T, °K.	C_p°	S°	$-(F^\circ - H^\circ_{298})/T$	$H^\circ - H^\circ_{298}$	ΔH_f°	ΔF_f°	$Log_{10}K_P$
	cal mole⁻¹ deg⁻¹			kcal mole⁻¹			
4100	5.103	51.672	47.019	19.075	61.536	− 4.723	.252
4200	5.114	51.795	47.131	19.586	61.549	− 6.339	.330
4300	5.126	51.915	47.241	20.098	61.561	− 7.955	.404
4400	5.138	52.033	47.349	20.611	61.572	− 9.573	.475
4500	5.150	52.149	47.454	21.126	61.584	− 11.189	.543
4600	5.162	52.262	47.558	21.641	61.596	− 12.805	.608
4700	5.174	52.373	47.659	22.158	61.608	− 14.423	.671
4800	5.186	52.482	47.758	22.676	61.620	− 16.041	.730
4900	5.198	52.589	47.856	23.195	61.633	− 17.661	.788
5000	5.210	52.695	47.951	23.715	61.646	− 19.279	.843
5100	5.222	52.798	48.045	24.237	61.659	− 20.896	.895
5200	5.234	52.899	48.138	24.760	61.673	− 22.517	.946
5300	5.246	52.999	48.229	25.284	61.687	− 24.134	.995
5400	5.258	53.097	48.318	25.809	61.702	− 25.755	1.042
5500	5.269	53.194	48.406	26.335	61.718	− 27.373	1.088
5600	5.280	53.289	48.492	26.863	61.733	− 28.995	1.132
5700	5.292	53.383	48.577	27.392	61.750	− 30.616	1.174
5800	5.302	53.475	48.661	27.921	61.767	− 32.234	1.215
5900	5.313	53.565	48.743	28.452	61.785	− 33.856	1.254
6000	5.323	53.655	48.824	28.984	61.803	− 35.476	1.292

Nitrogen, Diatomic (N_2)

(Reference State—Ideal Gas)(mol. wt. = 28.01) Sept. 30, 1965

T, °K.	C_p°	S°	$-(F^\circ - H^\circ_{298})/T$	$H^\circ - H^\circ_{298}$	ΔH_f°	ΔF_f°	$Log_{10}K_P$
	cal mole⁻¹ deg⁻¹			kcal mole⁻¹			
0	.000	.000	INFINITE	− 2.072	.000	.000	.000
100	6.956	38.170	51.957	− 1.379	.000	.000	.000
200	6.957	42.992	46.407	− .683	.000	.000	.000
298	6.961	45.770	45.770	.000	.000	.000	.000
300	6.961	45.813	45.770	.013	.000	.000	.000
400	6.990	47.818	46.043	.710	.000	.000	.000
500	7.069	49.386	46.561	1.413	.000	.000	.000
600	7.196	50.685	47.143	2.125	.000	.000	.000
700	7.350	51.806	47.731	2.853	.000	.000	.000
800	7.512	52.798	48.303	3.596	.000	.000	.000
900	7.670	53.692	48.853	4.355	.000	.000	.000
1000	7.815	54.507	49.378	5.129	.000	.000	.000
1100	7.945	55.258	49.879	5.917	.000	.000	.000
1200	8.061	55.955	50.357	6.718	.000	.000	.000
1300	8.162	56.604	50.813	7.529	.000	.000	.000
1400	8.252	57.212	51.248	8.350	.000	.000	.000
1500	8.330	57.784	51.665	9.179	.000	.000	.000
1600	8.398	58.324	52.065	10.015	.000	.000	.000
1700	8.458	58.835	52.448	10.858	.000	.000	.000
1800	8.512	59.320	52.816	11.707	.000	.000	.000
1900	8.559	59.782	53.171	12.560	.000	.000	.000
2000	8.601	60.222	53.513	13.418	.000	.000	.000

T, °K.	C_p°	S°	$-(F^\circ - H_{298}^\circ)/T$	$H^\circ - H_{298}^\circ$	ΔH_f°	ΔF_f°	$Log_{10}K_P$
	cal mole⁻¹ deg⁻¹			kcal mole⁻¹			
2100	8.638	60.642	53.842	14.280	.000	.000	.000
2200	8.672	61.045	54.160	15.146	.000	.000	.000
2300	8.703	61.431	54.468	16.015	.000	.000	.000
2400	8.731	61.802	54.766	16.886	.000	.000	.000
2500	8.756	62.159	55.055	17.761	.000	.000	.000
2600	8.779	62.503	55.335	18.638	.000	.000	.000
2700	8.800	62.835	55.606	19.517	.000	.000	.000
2800	8.820	63.155	55.870	20.398	.000	.000	.000
2900	8.838	63.465	56.127	21.280	.000	.000	.000
3000	8.855	63.765	56.376	22.165	.000	.000	.000
3100	8.871	64.055	56.619	23.051	.000	.000	.000
3200	8.886	64.337	56.856	23.939	.000	.000	.000
3300	8.900	64.611	57.087	24.829	.000	.000	.000
3400	8.914	64.877	57.312	25.719	.000	.000	.000
3500	8.927	65.135	57.532	26.611	.000	.000	.000
3600	8.939	65.387	57.747	27.505	.000	.000	.000
3700	8.950	65.632	57.957	28.399	.000	.000	.000
3800	8.962	65.871	58.162	29.295	.000	.000	.000
3900	8.972	66.104	58.362	30.191	.000	.000	.000
4000	8.983	66.331	58.559	31.089	.000	.000	.000
4100	8.993	66.553	58.751	31.988	.000	.000	.000
4200	9.002	66.770	58.940	32.888	.000	.000	.000
4300	8.012	66.982	59.124	33.788	.000	.000	.000
4400	9.021	67.189	59.305	34.690	.000	.000	.000
4500	9.030	67.392	59.482	35.593	.000	.000	.000
4600	9.039	67.591	59.657	36.496	.000	.000	.000
4700	9.048	67.785	59.827	37.400	.000	.000	.000
4800	9.057	67.976	59.995	38.306	.000	.000	.000
4900	9.066	68.162	60.160	39.212	.000	.000	.000
5000	9.074	68.346	60.322	40.119	.000	.000	.000
5100	9.083	68.525	60.481	41.027	.000	.000	.000
5200	9.091	68.702	60.637	41.935	.000	.000	.000
5300	9.100	68.875	60.791	42.845	.000	.000	.000
5400	9.109	69.045	60.942	43.755	.000	.000	.000
5500	9.118	69.213	61.091	44.667	.000	.000	.000
5600	9.127	69.377	61.238	45.579	.000	.000	.000
5700	9.136	69.539	61.382	46.492	.000	.000	.000
5800	9.145	69.698	61.524	47.406	.000	.000	.000
5900	9.155	69.854	61.664	48.321	.000	.000	.000
6000	9.165	70.008	61.802	49.237	.000	.000	.000

Nitrogen, Monatomic (N)

(Ideal Gas)(atomic wt. = 14.01)March 31, 1961

T, °K.	C_p°	S°	$-(F^\circ - H_{298}^\circ)/T$	$H^\circ - H_{298}^\circ$	ΔH_f°	ΔF_f°	$Log_{10}K_P$
	cal mole⁻¹ deg⁻¹			kcal mole⁻¹			
0	.000	.000	INFINITE	− 1.481	112.520	112.520	INFINITE
100	4.968	31.187	41.031	− .984	112.670	111.459	− 243.583
200	4.968	34.631	37.069	− .488	112.819	110.191	− 120.405
298	4.968	36.614	36.614	.000	112.965	108.870	− 79.800

NITROGEN, MONATOMIC (N) (Continued)

T, °K.	C_p° (cal mole⁻¹ deg⁻¹)	S° (cal mole⁻¹ deg⁻¹)	$-(F^\circ - H^\circ_{298})/T$ (cal mole⁻¹ deg⁻¹)	$H^\circ - H^\circ_{298}$ (kcal mole⁻¹)	ΔH°_f (kcal mole⁻¹)	ΔF°_f (kcal mole⁻¹)	$\text{Log}_{10}K_P$
300	4.968	36.645	36.614	.009	112.968	108.845	− 79.289
400	4.968	38.074	36.809	.506	113.116	107.448	− 58.704
500	4.968	39.183	37.177	1.003	113.261	106.014	− 46.336
600	4.968	40.089	37.589	1.500	113.402	104.551	− 38.081
700	4.968	40.855	38.002	1.996	113.535	103.066	− 32.177
800	4.968	41.518	38.401	2.493	113.660	101.562	− 27.744
900	4.968	42.103	38.781	2.990	113.778	100.042	− 24.292
1000	4.968	42.627	39.140	3.487	113.887	98.510	− 21.528
1100	4.968	43.100	39.479	3.984	113.990	96.967	− 19.265
1200	4.968	43.532	39.799	4.481	114.087	95.415	− 17.377
1300	4.968	43.930	40.101	4.977	114.178	93.856	− 15.778
1400	4.968	44.298	40.388	5.474	114.264	92.289	− 14.406
1500	4.968	44.641	40.660	5.971	114.346	90.716	− 13.217
1600	4.968	44.962	40.919	6.468	114.425	89.139	− 12.175
1700	4.968	45.263	41.166	6.965	114.501	87.556	− 11.256
1800	4.968	45.547	41.402	7.461	114.573	85.968	− 10.437
1900	4.969	45.815	41.627	7.958	114.644	84.378	− 9.705
2000	4.969	46.070	41.843	8.455	114.712	82.784	− 9.046
2100	4.970	46.313	42.050	8.952	114.778	81.185	− 8.449
2200	4.971	46.544	42.249	9.449	114.843	79.583	− 7.905
2300	4.972	46.765	42.440	9.946	114.906	77.979	− 7.409
2400	4.975	46.977	42.625	10.444	114.968	76.373	− 6.954
2500	4.978	47.180	42.803	10.941	115.028	74.763	− 6.535
2600	4.982	47.375	42.975	11.439	115.088	73.151	− 6.149
2700	4.987	47.563	43.142	11.938	115.148	71.537	− 5.790
2800	4.993	47.745	43.303	12.437	115.207	69.922	− 5.457
2900	5.001	47.920	43.459	12.936	115.265	68.303	− 5.147
3000	5.011	48.090	43.611	13.437	115.324	66.683	− 4.858
3100	5.022	48.254	43.758	13.939	115.383	65.060	− 4.587
3200	5.035	48.414	43.901	14.441	115.442	63.435	− 4.332
3300	5.050	48.569	44.040	14.946	115.503	61.810	− 4.093
3400	5.067	48.720	44.175	15.451	115.564	60.181	− 3.868
3500	5.086	48.867	44.307	15.959	115.626	58.551	− 3.656
3600	5.107	49.011	44.436	16.469	115.690	56.921	− 3.455
3700	5.130	49.151	44.562	16.980	115.755	55.287	− 3.265
3800	5.156	49.288	44.684	17.495	115.823	53.653	− 3.086
3900	5.183	49.422	44.804	18.012	115.892	52.015	− 2.915
4000	5.213	49.554	44.921	18.531	115.964	50.376	− 2.752
4100	5.245	49.683	45.036	19.054	116.038	48.735	− 2.598
4200	5.278	49.810	45.148	19.580	116.116	47.094	− 2.450
4300	5.314	49.934	45.258	20.110	116.196	45.449	− 2.310
4400	5.351	50.057	45.365	20.643	116.280	43.803	− 2.176
4500	5.390	50.178	45.471	21.180	116.367	42.154	− 2.047

T, °K.	C_p° (cal mole⁻¹ deg⁻¹)	S°	$-(F^\circ - H_{298}^\circ)/T$	$H^\circ - H_{298}^\circ$ (kcal mole⁻¹)	ΔH_f°	ΔF_f°	$Log_{10}K_P$
4600	5.431	50.297	45.575	21.721	116.458	40.505	− 1.924
4700	5.473	50.414	45.676	22.266	116.552	38.853	− 1.807
4800	5.517	50.530	45.776	22.816	116.651	37.199	− 1.694
4900	5.562	50.644	45.874	23.370	116.753	35.542	− 1.585
5000	5.608	50.757	45.971	23.928	116.860	33.883	− 1.481
5100	5.655	50.868	46.066	24.491	116.971	32.222	− 1.381
5200	5.703	50.978	46.159	25.059	117.087	30.560	− 1.284
5300	5.751	51.087	46.251	25.632	117.207	28.894	− 1.191
5400	5.800	51.195	46.342	26.210	117.332	27.228	− 1.102
5500	5.850	51.302	46.431	26.792	117.461	25.557	− 1.016
5600	5.900	51.408	46.519	27.380	117.595	23.884	− .932
5700	5.950	51.513	46.606	27.972	117.734	22.211	− .852
5800	6.000	51.617	46.691	28.570	117.878	20.534	− .774
5900	6.050	51.720	46.775	29.172	118.026	18.853	− .698
6000	6.100	51.822	46.859	29.780	118.179	17.172	− .625

Hydrogen, Diatomic (H_2)

(Reference State—Ideal Gas)(mol. wt. = 2.016)March 31, 1961

T, °K.	C_p° (cal mole⁻¹ deg⁻¹)	S°	$-(F^\circ - H_{298}^\circ)/T$	$H^\circ - H_{298}^\circ$ (kcal mole⁻¹)	ΔH_f°	ΔF_f°	$Log_{10}K_P$
0	.000	.000	INFINITE	− 2.024	.000	.000	.000
100	5.393	24.387	37.035	− 1.265	.000	.000	.000
200	6.518	28.520	31.831	− .662	.000	.000	.000
298	6.892	31.208	31.208	.000	.000	.000	.000
300	6.894	31.251	31.208	.013	.000	.000	.000
400	6.975	33.247	31.480	.707	.000	.000	.000
500	6.993	34.806	31.995	1.406	.000	.000	.000
600	7.009	36.082	32.573	2.106	.000	.000	.000
700	7.036	37.165	33.153	2.808	.000	.000	.000
800	7.087	38.i07	33.715	3.514	.000	.000	.000
900	7.148	38.946	34.250	4.226	.000	.000	.000
1000	7.219	39.702	34.758	4.944	.000	.000	.000
1100	7.300	40.394	35.240	5.670	.000	.000	.000
1200	7.390	41.033	35.696	6.404	.000	.000	.000
1300	7.490	41.628	36.130	7.148	.000	.000	.000
1400	7.600	42.187	36.543	7.902	.000	.000	.000
1500	7.720	42.716	36.937	8.668	.000	.000	.000
1600	7.823	43.217	37.314	9.446	.000	.000	.000
1700	7.921	43.695	37.675	10.233	.000	.000	.000
1800	8.016	44.150	38.022	11.030	.000	.000	.000
1900	8.108	44.586	38.356	11.836	.000	.000	.000
2000	8.195	45.004	38.678	12.651	.000	.000	.000
2100	8.279	45.406	38.989	13.475	.000	.000	.000
2200	8.358	45.793	39.290	14.307	.000	.000	.000
2300	8.434	46.166	39.581	15.146	.000	.000	.000
2400	8.506	46.527	39.863	15.993	.000	.000	.000
2500	8.575	46.875	40.136	16.848	.000	.000	.000

Hydrogen, Diatomic (H_2) (Continued)

T, °K.	C_p° (cal mole⁻¹ deg⁻¹)	S°	$-(F^\circ - H^\circ_{298})/T$	$H^\circ - H^\circ_{298}$ (kcal mole⁻¹)	ΔH_f°	ΔF_f°	$Log_{10} K_P$
2600	8.639	47.213	40.402	17.708	.000	.000	.000
2700	8.700	47.540	40.660	18.575	.000	.000	.000
2800	8.757	47.857	40.912	19.448	.000	.000	.000
2900	8.810	48.166	41.157	20.326	.000	.000	.000
3000	8.859	48.465	41.395	21.210	.000	.000	.000
3100	8.911	48.756	41.628	22.098	.000	.000	.000
3200	8.962	49.040	41.855	22.992	.000	.000	.000
3300	9.012	49.317	42.077	23.891	.000	.000	.000
3400	9.061	49.586	42.294	24.794	.000	.000	.000
3500	9.110	49.850	42.506	25.703	.000	.000	.000
3600	9.158	50.107	42.714	26.616	.000	.000	.000
3700	9.205	50.359	42.917	27.535	.000	.000	.000
3800	9.252	50.605	43.116	28.457	.000	.000	.000
3900	9.297	50.846	43.311	29.385	.000	.000	.000
4000	9.342	51.082	43.502	30.317	.000	.000	.000
4100	9.386	51.313	43.690	31.253	.000	.000	.000
4200	9.429	51.540	43.874	32.194	.000	.000	.000
4300	9.472	51.762	44.055	33.139	.000	.000	.000
4400	9.514	51.980	44.233	34.088	.000	.000	.000
4500	9.555	52.194	44.407	35.042	.000	.000	.000
4600	9.595	52.405	44.579	35.999	.000	.000	.000
4700	9.634	52.612	44.748	36.961	.000	.000	.000
4800	9.673	52.815	44.914	37.926	.000	.000	.000
4900	9.711	53.015	45.077	38.895	.000	.000	.000
5000	9.748	53.211	45.238	39.868	.000	.000	.000
5100	9.785	53.405	45.396	40.845	.000	.000	.000
5200	9.822	53.595	45.552	41.825	.000	.000	.000
5300	9.859	53.783	45.705	42.809	.000	.000	.000
5400	9.895	53.967	45.857	43.797	.000	.000	.000
5500	9.930	54.149	46.006	44.788	.000	.000	.000
5600	9.965	54.328	46.153	45.783	.000	.000	.000
5700	10.000	54.505	46.298	46.781	.000	.000	.000
5800	10.034	54.679	46.441	47.783	.000	.000	.000
5900	10.067	54.851	46.582	48.788	.000	.000	.000
6000	10.100	55.020	46.721	49.796	.000	.000	.000

Hydrogen, Monatomic (H)

(Ideal Gas)(atomic wt. = 1.008)Sept. 30, 1965

T, °K.	C_p° (cal mole⁻¹ deg⁻¹)	S°	$-(F^\circ - H^\circ_{298})/T$	$H^\circ - H^\circ_{298}$ (kcal mole⁻¹)	ΔH_f°	ΔF_f°	$Log_{10} K_P$
0	.000	.000	INFINITE	− 1.481	51.631	51.631	INFINITE
100	4.968	21.965	31.809	− .984	51.748	50.771	− 110.954
200	4.968	25.408	27.847	− .488	51.943	49.714	− 54.322
298	4.968	27.392	27.392	.000	52.100	48.585	− 35.612
300	4.968	27.423	27.392	.009	52.103	48.563	− 35.377
400	4.968	28.852	27.587	.506	52.253	47.361	− 25.876
500	4.968	29.961	27.955	1.003	52.400	46.121	− 20.158

T, °K.	cal mole⁻¹ deg⁻¹			kcal mole⁻¹			
	C_p°	S°	$-(F^\circ - H_{298}^\circ)/T$	$H^\circ - H_{298}^\circ$	ΔH_f°	ΔF_f°	$\log_{10} K_P$
600	4.968	30.867	28.367	1.500	52.547	44.851	−16.336
700	4.968	31.632	28.780	1.996	52.692	43.558	−13.599
800	4.968	32.296	29.179	2.493	52.836	42.242	−11.539
900	4.968	32.881	29.559	2.990	52.977	40.910	−9.934
1000	4.968	33.404	29.917	3.487	53.115	39.562	−8.646
1100	4.968	33.878	30.256	3.984	53.249	38.200	−7.589
1200	4.968	34.310	30.576	4.481	53.379	36.826	−6.707
1300	4.968	34.708	30.879	4.977	53.503	35.441	−5.958
1400	4.968	35.076	31.166	5.474	53.623	34.048	−5.315
1500	4.968	35.419	31.438	5.971	53.737	32.645	−4.756
1600	4.968	35.739	31.697	6.468	53.845	31.236	−4.266
1700	4.968	36.041	31.944	6.965	53.948	29.819	−3.833
1800	4.968	36.325	32.179	7.461	54.046	28.396	−3.448
1900	4.968	36.593	32.405	7.958	54.140	26.970	−3.102
2000	4.968	36.848	32.621	8.455	54.230	25.538	−2.790
2100	4.968	37.090	32.828	8.952	54.314	24.102	−2.508
2200	4.968	37.322	33.027	9.449	54.395	22.659	−2.251
2300	4.968	37.542	33.218	9.945	54.472	21.217	−2.016
2400	4.968	37.754	33.403	10.442	54.546	19.769	−1.800
2500	4.968	37.957	33.581	10.939	54.615	18.316	−1.601
2600	4.968	38.152	33.753	11.436	54.682	16.864	−1.417
2700	4.968	38.339	33.919	11.933	54.745	15.409	−1.247
2800	4.968	38.520	34.081	12.430	54.806	13.949	−1.089
2900	4.968	38.694	34.237	12.926	54.863	12.491	−.941
3000	4.968	38.862	34.388	13.423	54.918	11.030	−.803
3100	4.968	39.025	34.535	13.920	54.971	9.565	−.674
3200	4.968	39.183	34.678	14.417	55.021	8.099	−.553
3300	4.968	39.336	34.817	14.914	55.068	6.632	−.439
3400	4.968	39.484	34.952	15.410	55.113	5.164	−.332
3500	4.968	39.628	35.083	15.907	55.156	3.695	−.231
3600	4.968	39.768	35.212	16.404	55.196	2.224	−.135
3700	4.968	39.904	35.337	16.901	55.233	.753	−.044
3800	4.968	40.037	35.459	17.398	55.269	−.722	.042
3900	4.968	40.166	35.578	17.895	55.302	−2.196	.123
4000	4.968	40.292	35.694	18.391	55.333	−3.671	.201
4100	4.968	40.414	35.808	18.888	55.362	−5.144	.274
4200	4.968	40.534	35.919	19.385	55.388	−6.621	.345
4300	4.968	40.651	36.027	19.882	55.412	−8.099	.412
4400	4.968	40.765	36.134	20.379	55.435	−9.575	.476
4500	4.968	40.877	36.238	20.875	55.454	−11.056	.537
4600	4.968	40.986	36.340	21.372	55.473	−12.531	.595
4700	4.968	41.093	36.440	21.669	55.489	−14.010	.651
4800	4.968	41.198	36.538	22.366	55.503	−15.492	.705
4900	4.968	41.300	36.634	22.863	55.515	−16.968	.757
5000	4.968	41.400	36.728	23.359	55.525	−18.447	.806

HYDROGEN, MONATOMIC (H) (Continued)

	cal mole⁻¹ deg⁻¹			kcal mole⁻¹			
T, °K.	C_p°	S°	$-(F^\circ - H^\circ_{298})/T$	$H^\circ - H^\circ_{298}$	ΔH_f°	ΔF_f°	$Log_{10}K_P$
5100	4.968	41.499	36.821	23.856	55.534	− 19.928	.854
5200	4.968	41.595	36.912	24.353	55.541	− 21.406	.900
5300	4.968	41.690	37.001	24.850	55.545	− 22.887	.944
5400	4.968	41.783	37.089	25.347	55.548	− 24.369	.986
5500	4.968	41.874	37.175	25.844	55.550	− 25.848	1.027
5600	4.968	41.963	37.260	26.340	55.549	− 27.326	1.066
5700	4.968	42.051	37.343	26.837	55.547	− 28.805	1.104
5800	4.968	42.138	37.425	27.334	55.543	− 30.289	1.141
5900	4.968	42.223	37.506	27.831	55.537	− 31.768	1.177
6000	4.968	42.306	37.585	28.328	55.530	− 33.246	1.211

CARBON (C)

(Reference State—Graphite)(atomic wt. = 12.01)March 31, 1961

	cal mole⁻¹ deg⁻¹			kcal mole⁻¹			
T, °K.	C_p°	S°	$-(F^\circ - H^\circ_{298})/T$	$H^\circ - H^\circ_{298}$	ΔH_f°	ΔF_f°	$Log_{10}K_P$
0	.000	.000	INFINITE	− .252	.000	.000	.000
100	.395	.210	2.591	− .238	.000	.000	.000
200	1.202	.720	1.520	− .160	.000	.000	.000
298	2.038	1.359	1.359	.000	.000	.000	.000
300	2.054	1.372	1.359	.004	.000	.000	.000
400	2.851	2.075	1.450	.250	.000	.000	.000
500	3.496	2.784	1.646	.569	.000	.000	.000
600	4.038	3.471	1.893	.947	.000	.000	.000
700	4.440	4.126	2.166	1.372	.000	.000	.000
800	4.740	4.739	2.449	1.831	.000	.000	.000
900	4.970	5.311	2.736	2.318	.000	.000	.000
1000	5.149	5.844	3.020	2.824	.000	.000	.000
1100	5.304	6.342	3.300	3.347	.000	.000	.000
1200	5.430	6.809	3.573	3.883	.000	.000	.000
1300	5.527	7.248	3.839	4.432	.000	.000	.000
1400	5.605	7.661	4.098	4.988	.000	.000	.000
1500	5.669	8.050	4.348	5.552	.000	.000	.000
1600	5.721	8.417	4.591	6.122	.000	.000	.000
1700	5.765	8.765	4.827	6.696	.000	.000	.000
1800	5.803	9.096	5.055	7.275	.000	.000	.000
1900	5.836	9.411	5.276	7.857	.000	.000	.000
2000	5.865	9.711	5.490	8.442	.000	.000	.000
2100	5.891	9.998	5.698	9.029	.000	.000	.000
2200	5.914	10.272	5.900	9.620	.000	.000	.000
2300	5.936	10.536	6.095	10.212	.000	.000	.000
2400	5.956	10.789	6.286	10.807	.000	.000	.000
2500	5.974	11.032	6.471	11.403	.000	.000	.000
2600	5.992	11.267	6.651	12.002	.000	.000	.000
2700	6.009	11.493	6.826	12.602	.000	.000	.000
2800	6.026	11.712	6.997	13.203	.000	.000	.000
2900	6.042	11.924	7.163	13.807	.000	.000	.000
3000	6.057	12.129	7.325	14.412	.000	.000	.000

T, °K.	cal mole⁻¹ deg⁻¹ C_p°	S°	$-(F^\circ - H^\circ_{298})/T$	kcal mole⁻¹ $H^\circ - H^\circ_{298}$	ΔH_f°	ΔF_f°	$Log_{10}K_P$
3100	6.073	12.328	7.483	15.018	.000	.000	.000
3200	6.088	12.521	7.638	15.626	.000	.000	.000
3300	6.103	12.708	7.788	16.236	.000	.000	.000
3400	6.119	12.891	7.936	16.847	.000	.000	.000
3500	6.134	13.068	8.080	17.460	.000	.000	.000
3600	6.150	13.241	8.221	18.074	.000	.000	.000
3700	6.165	13.410	8.359	18.690	.000	.000	.000
3800	6.181	13.575	8.494	19.307	.000	.000	.000
3900	6.197	13.736	8.626	19.926	.000	.000	.000
4000	6.213	13.893	8.756	20.546	.000	.000	.000
4100	6.230	14.046	8.883	21.168	.000	.000	.000
4200	6.247	14.197	9.008	21.792	.000	.000	.000
4300	6.264	14.344	9.130	22.418	.000	.000	.000
4400	6.281	14.488	9.250	23.045	.000	.000	.000
4500	6.299	14.629	9.368	23.674	.000	.000	.000
4600	6.317	14.768	9.484	24.305	.000	.000	.000
4700	6.335	14.904	9.598	24.937	.000	.000	.000
4800	6.354	15.038	9.710	25.572	.000	.000	.000
4900	6.373	15.169	9.820	26.208	.000	.000	.000
5000	6.392	15.298	9.928	26.846	.000	.000	.000
5100	6.412	15.424	10.035	27.487	.000	.000	.000
5200	6.432	15.549	10.140	28.129	.000	.000	.000
5300	6.452	15.672	10.243	28.773	.000	.000	.000
5400	6.473	15.793	10.345	29.419	.000	.000	.000
5500	6.494	15.912	10.445	30.068	.000	.000	.000
5600	6.516	16.029	10.543	30.718	.000	.000	.000
5700	6.538	16.144	10.641	31.371	.000	.000	.000
5800	6.560	16.258	10.737	32.026	.000	.000	.000
5900	6.583	16.371	10.831	32.683	.000	.000	.000
6000	6.606	16.481	10.924	33.342	.000	.000	.000

Carbon, Monatomic (C)

(Ideal Gas)(atomic wt. = 12.01)March 31, 1961

T, °K.	cal mole⁻¹ deg⁻¹ C_p°	S°	$-(F^\circ - H^\circ_{298})/T$	kcal mole⁻¹ $H^\circ - H^\circ_{298}$	ΔH_f°	ΔF_f°	$Log_{10}K_P$
0	.000	.000	INFINITE	−1.562	169.576	169.576	INFINITE
100	5.085	32.283	42.203	− .992	170.132	166.925	−364.796
200	4.997	35.770	38.217	− .489	170.557	163.547	−178.707
298	4.981	37.761	37.761	.000	170.886	160.033	−117.302
300	4.981	37.792	37.761	.009	170.891	159.965	−116.529
400	4.975	39.224	37.956	.507	171.143	156.283	− 85.385
500	4.973	40.334	38.325	1.004	171.321	152.546	− 66.675
600	4.971	41.240	38.738	1.502	171.441	148.779	− 54.190
700	4.970	42.006	39.151	1.999	171.513	144.996	− 45.268
800	4.970	42.670	39.551	2.496	171.551	141.206	− 38.574
900	4.970	43.255	39.930	2.993	171.561	137.411	− 33.366
1000	4.969	43.779	40.290	3.490	171.552	133.616	− 29.200

CARBON, MONATOMIC (C) (Continued)

T, °K.	C_p° (cal mole^{-1} deg^{-1})	S°	$-(F^\circ - H^\circ_{298})/T$	$H^\circ - H^\circ_{298}$ (kcal mole^{-1})	ΔH_f°	ΔF_f°	$\text{Log}_{10}K_P$
1100	4.969	44.253	40.629	3.986	171.525	129.824	− 25.792
1200	4.970	44.685	40.949	4.483	171.486	126.035	− 22.953
1300	4.971	45.083	41.252	4.980	171.434	122.249	− 20.551
1400	4.972	45.451	41.539	5.478	171.376	118.469	− 18.493
1500	4.975	45.794	41.811	5.975	171.309	114.692	− 16.710
1600	4.978	46.116	42.070	6.473	171.237	110.919	− 15.150
1700	4.984	46.418	42.317	6.971	171.161	107.151	− 13.775
1800	4.990	46.703	42.553	7.469	171.080	103.388	− 12.552
1900	4.998	46.973	42.779	7.969	170.998	99.631	− 11.460
2000	5.008	47.229	42.995	8.469	170.913	95.876	− 10.476
2100	5.019	47.474	43.202	8.970	170.827	92.128	− 9.587
2200	5.032	47.708	43.402	9.473	170.739	88.380	− 8.779
2300	5.046	47.932	43.594	9.977	170.651	84.641	− 8.042
2400	5.061	48.147	43.779	10.482	170.561	80.903	− 7.367
2500	5.077	48.354	43.958	10.989	170.472	77.168	− 6.746
2600	5.094	48.553	44.131	11.497	170.381	73.438	− 6.173
2700	5.112	48.746	44.298	12.008	170.292	69.710	− 5.642
2800	5.130	48.932	44.460	12.520	170.203	65.987	− 5.150
2900	5.149	49.112	44.618	13.034	170.113	62.267	− 4.692
3000	5.168	49.287	44.771	13.550	170.024	58.549	− 4.265
3100	5.187	49.457	44.919	14.067	169.935	54.836	− 3.866
3200	5.206	49.622	45.063	14.587	169.847	51.124	− 3.491
3300	5.224	49.782	45.204	15.108	169.758	47.413	− 3.140
3400	5.243	49.938	45.341	15.632	169.671	43.709	− 2.809
3500	5.261	50.091	45.474	16.157	169.583	40.003	− 2.498
3600	5.279	50.239	45.605	16.684	169.496	36.303	− 2.204
3700	5.296	50.384	45.732	17.213	169.409	32.605	− 1.926
3800	5.313	50.526	45.856	17.743	169.322	28.910	− 1.663
3900	5.329	50.664	45.978	18.275	169.235	25.217	− 1.413
4000	5.345	50.799	46.097	18.809	169.149	21.526	− 1.176
4100	5.360	50.931	46.213	19.344	169.062	17.834	− .951
4200	5.375	51.060	46.327	19.881	168.975	14.149	− .736
4300	5.388	51.187	46.438	20.419	168.887	10.462	− .532
4400	5.402	51.311	46.548	20.959	168.800	6.778	− .337
4500	5.414	51.433	46.655	21.499	168.711	3.096	− .150
4600	5.426	51.552	46.760	22.041	168.622	− .582	.028
4700	5.437	51.668	46.863	22.585	168.534	− 4.259	.198
4800	5.448	51.783	46.965	23.129	168.443	− 7.933	.361
4900	5.459	51.896	47.064	23.674	168.352	− 11.608	.518
5000	5.468	52.006	47.162	24.221	168.261	− 15.279	.668
5100	5.477	52.114	47.258	24.768	168.167	− 18.954	.812
5200	5.486	52.221	47.352	25.316	168.073	− 22.620	.951
5300	5.494	52.325	47.445	25.865	167.978	− 26.284	1.084
5400	5.502	52.428	47.536	26.415	167.882	− 29.947	1.212
5500	5.509	52.529	47.626	26.966	167.784	− 33.611	1.335

T, °K.	cal mole⁻¹ deg⁻¹ C_p°	S°	$-(F^\circ - H^\circ_{298})/T$	kcal mole⁻¹ $H^\circ - H^\circ_{298}$	ΔH_f°	ΔF_f°	$Log_{10}K_P$
5600	5.516	52.628	47.715	27.517	167.685	– 37.272	1.455
5700	5.523	52.726	47.802	28.069	167.584	– 40.934	1.569
5800	5.529	52.822	47.888	28.621	167.481	– 44.591	1.680
5900	5.535	52.917	47.972	29.175	167.378	– 48.243	1.787
6000	5.541	53.010	48.055	29.728	167.272	– 51.901	1.890

CARBON, DIATOMIC (C_2)

(Ideal Gas)(mol. wt. = 24.02)Sept. 30, 1961

T, °K.	cal mole⁻¹ deg⁻¹ C_p°	S°	$-(F^\circ - H^\circ_{298})/T$	kcal mole⁻¹ $H^\circ - H^\circ_{298}$	ΔH_f°	ΔF_f°	$Log_{10}K_P$
0	.000	.000	INFINITE	– 2.530	197.000	197.000	INFINITE
100	7.114	37.919	56.249	– 1.833	197.669	193.919	– 423.790
200	9.670	43.549	48.574	– 1.005	198.341	189.919	– 207.524
298	10.317	47.628	47.628	.000	199.026	185.636	– 136.068
300	10.300	47.692	47.628	.019	199.037	185.553	– 135.169
400	9.472	50.536	48.022	1.006	199.532	180.977	– 98.877
500	8.868	52.581	48.739	1.921	199.809	176.303	– 77.058
600	8.593	54.170	49.516	2.792	199.924	171.587	– 62.498
700	8.501	55.486	50.278	3.646	199.928	166.864	– 52.095
800	8.496	56.620	51.001	4.495	199.859	162.145	– 44.294
900	8.530	57.622	51.682	5.346	199.736	157.436	– 38.229
1000	8.582	58.524	52.322	6.202	199.580	152.744	– 33.381
1100	8.642	59.345	52.924	7.063	199.395	148.068	– 29.417
1200	8.706	60.099	53.491	7.930	199.190	143.413	– 26.118
1300	8.772	60.799	54.026	8.804	198.966	138.773	– 23.329
1400	8.839	61.451	54.534	9.685	198.735	134.154	– 20.941
1500	8.907	62.063	55.015	10.572	198.494	129.549	– 18.874
1600	8.975	62.640	55.474	11.466	198.248	124.958	– 17.068
1700	9.042	63.186	55.912	12.367	198.001	120.385	– 15.476
1800	9.108	63.705	56.331	13.274	197.750	115.827	– 14.063
1900	9.172	64.199	56.732	14.188	197.500	111.283	– 12.800
2000	9.255	64.672	57.117	15.110	197.252	106.752	– 11.665
2100	9.334	65.125	57.488	16.039	197.007	102.236	– 10.639
2200	9.407	65.561	57.845	16.976	196.762	97.724	– 9.708
2300	9.475	65.981	58.189	17.921	196.523	93.232	– 8.859
2400	9.538	66.386	58.523	18.871	196.283	88.745	– 8.081
2500	9.600	66.776	58.845	19.828	196.048	84.268	– 7.366
2600	9.660	67.154	59.157	20.791	195.813	79.801	– 6.708
2700	9.717	67.520	59.460	21.760	195.582	75.341	– 6.098
2800	9.772	67.874	59.754	22.735	195.355	70.895	– 5.533
2900	9.825	68.218	60.040	23.714	195.126	66.454	– 5.008
3000	9.875	68.552	60.319	24.699	194.901	62.020	– 4.518
3100	9.922	68.876	60.589	25.689	194.679	57.596	– 4.060
3200	9.967	69.192	60.853	26.684	194.458	53.178	– 3.632
3300	10.009	69.499	61.111	27.683	194.237	48.761	– 3.229
3400	10.049	69.799	61.362	28.686	194.018	44.360	– 2.851
3500	10.087	70.091	61.607	29.692	193.798	39.957	– 2.495

CARBON, DIATOMIC (C_2) (Continued)

T, °K.	C_p° (cal mole⁻¹ deg⁻¹)	S°	$-(F^\circ - H^\circ_{298})/T$	$H^\circ - H^\circ_{298}$ (kcal mole⁻¹)	ΔH_f°	ΔF_f°	$Log_{10}K_P$
3600	10.123	70.375	61.847	30.703	193.581	35.565	– 2.159
3700	10.157	70.653	62.081	31.717	193.363	31.180	– 1.842
3800	10.188	70.924	62.310	32.734	193.146	26.803	– 1.541
3900	10.218	71.189	62.534	33.754	192.928	22.430	– 1.257
4000	10.245	71.448	62.754	34.778	192.712	18.062	– .987
4100	10.271	71.702	62.969	35.803	192.493	13.693	– .730
4200	10.295	71.950	63.180	36.832	192.274	9.340	– .486
4300	10.318	72.192	63.387	37.862	192.052	4.985	– .253
4400	10.339	72.430	63.590	38.895	191.831	.636	– .032
4500	10.359	72.662	63.789	39.930	191.608	– 3.710	.180
4600	10.378	72.890	63.984	40.967	191.383	– 8.045	.382
4700	10.396	73.113	64.176	42.006	191.158	– 12.378	.576
4800	10.412	73.332	64.364	43.046	190.928	– 16.703	.760
4900	10.427	73.547	64.550	44.088	190.698	– 21.027	.938
5000	10.441	73.758	64.732	45.131	190.465	– 25.345	1.108
5100	10.454	73.965	64.911	46.176	190.228	– 29.668	1.271
5200	10.466	74.168	65.087	47.222	189.990	– 33.974	1.428
5300	10.478	74.368	65.260	48.269	189.749	– 38.275	1.578
5400	10.489	74.563	65.431	49.318	189.506	– 42.573	1.723
5500	10.499	74.756	65.598	50.367	189.257	– 46.869	1.862
5600	10.509	74.945	65.764	51.418	189.008	– 51.161	1.997
5700	10.518	75.131	65.926	52.469	188.753	– 55.454	2.126
5800	10.526	75.314	66.087	53.521	188.495	– 59.735	2.251
5900	10.534	75.494	66.245	54.574	188.234	– 64.005	2.371
6000	10.542	75.671	66.400	55.628	187.970	– 68.287	2.487

CARBON, TRIATOMIC (C_3)

(Ideal Gas)(mol. wt. = 36.03)Dec. 31, 1960.

T, °K.	C_p° (cal mole⁻¹ deg⁻¹)	S°	$-(F^\circ - H^\circ_{298})/T$	$H^\circ - H^\circ_{298}$ (kcal mole⁻¹)	ΔH_f°	ΔF_f°	$Log_{10}K_P$
0	.000	.000	INFINITE	– 2.319	188.104	188.104	INFINITE
100	7.047	41.986	58.209	– 1.622	188.759	184.626	– 403.480
200	8.207	47.180	51.511	– .866	189.281	180.282	– 196.993
298	9.388	50.688	50.688	.000	189.670	175.777	– 128.842
300	9.407	50.746	50.688	.017	189.675	175.690	– 127.984
400	10.311	53.581	51.068	1.005	189.925	170.987	– 93.419
500	11.029	55.962	51.815	2.073	190.036	166.237	– 72.659
600	11.622	58.027	52.682	3.207	190.036	161.475	– 58.814
700	12.115	59.857	53.579	4.395	189.949	156.720	– 48.928
800	12.524	61.502	54.468	5.627	189.804	151.986	– 41.519
900	12.862	62.997	55.334	6.897	189.616	147.269	– 35.760
1000	13.141	64.368	56.170	8.198	189.396	142.572	– 31.158
1100	13.373	65.631	56.973	9.524	189.156	137.903	– 27.398
1200	13.565	66.803	57.744	10.871	188.895	133.254	– 24.268
1300	13.727	67.896	58.484	12.236	188.619	128.629	– 21.624
1400	13.862	68.918	59.193	13.615	188.330	124.026	– 19.360
1500	13.977	69.878	59.873	15.007	188.030	119.442	– 17.402

	cal mole⁻¹ deg⁻¹			kcal mole⁻¹			
T, °K.	C_p°	S°	$-(F^\circ - H_{298}^\circ)/T$	$H^\circ - H_{298}^\circ$	ΔH_f°	ΔF_f°	$Log_{10} K_P$
1600	14.074	70.784	60.527	16.410	187.720	114.878	-15.691
1700	14.158	71.639	61.156	17.822	187.407	110.337	-14.184
1800	14.230	72.451	61.761	19.241	187.092	105.815	-12.847
1900	14.292	73.222	62.344	20.667	186.772	101.305	-11.652
2000	14.347	73.956	62.907	22.099	186.452	96.812	-10.579
2100	14.394	74.658	63.450	23.537	186.129	92.341	-9.610
2200	14.436	75.328	63.974	24.978	185.797	87.884	-8.730
2300	14.473	75.971	64.482	26.424	185.461	83.440	-7.928
2400	14.506	76.587	64.974	27.873	185.119	79.011	-7.195
2500	14.536	77.180	65.450	29.325	184.774	74.601	-6.521
2600	14.562	77.751	65.912	30.780	184.426	70.195	-5.900
2700	14.586	78.301	66.361	32.237	184.074	65.812	-5.327
2800	14.607	78.832	66.797	33.697	183.719	61.438	-4.795
2900	14.626	79.345	67.221	35.158	183.362	57.080	-4.301
3000	14.644	79.841	67.633	36.622	183.002	52.731	-3.841
3100	14.660	80.321	68.035	38.087	182.637	48.394	-3.412
3200	14.674	80.787	68.426	39.554	182.271	44.070	-3.010
3300	14.687	81.239	68.808	41.022	181.900	39.751	-2.632
3400	14.700	81.677	69.180	42.491	181.524	35.453	-2.279
3500	14.711	82.103	69.543	43.962	181.144	31.164	-1.946
3600	14.721	82.518	69.898	45.433	180.758	26.880	-1.632
3700	14.731	82.922	70.244	46.906	180.368	22.609	-1.335
3800	14.739	83.314	70.583	48.380	179.973	18.349	-1.055
3900	14.748	83.697	70.914	49.854	179.572	14.109	-.791
4000	14.755	84.071	71.239	51.329	179.166	9.862	-.539
4100	14.762	84.435	71.556	52.805	178.755	5.643	-.301
4200	14.769	84.791	71.867	54.282	178.339	1.425	-.074
4300	14.775	85.139	72.172	55.759	177.917	-2.781	.141
4400	14.781	85.479	72.470	57.236	177.489	-6.978	.347
4500	14.786	85.811	72.763	58.715	177.057	-11.168	.542
4600	14.791	86.136	73.050	60.194	176.619	-15.352	.729
4700	14.796	86.454	73.332	61.673	176.175	-19.518	.908
4800	14.800	86.765	73.609	63.153	175.726	-23.668	1.078
4900	14.804	87.071	73.880	64.633	175.271	-27.818	1.241
5000	14.808	87.370	74.147	66.114	174.811	-31.968	1.397
5100	14.812	87.663	74.409	67.595	174.345	-36.092	1.547
5200	14.815	87.951	74.667	69.076	173.873	-40.220	1.690
5300	14.819	88.233	74.920	70.558	173.396	-44.323	1.828
5400	14.822	88.510	75.169	72.040	172.913	-48.433	1.960
5500	14.825	88.782	75.414	73.522	172.424	-52.520	2.087
5600	14.827	89.049	75.655	75.005	171.930	-56.618	2.210
5700	14.830	89.312	75.893	76.487	171.429	-60.695	2.327
5800	14.833	89.570	76.126	77.971	170.924	-64.751	2.440
5900	14.835	89.823	76.356	79.454	170.412	-68.822	2.549
6000	14.837	90.072	76.583	80.938	169.895	-72.874	2.654

CARBON, TETRATOMIC (C_4)
(Ideal Gas)(mol. wt. = 48.04)Dec. 31, 1960.

	cal mole⁻¹ deg⁻¹			kcal mole⁻¹			
T, °K.	C_p°	S°	$-(F^\circ - H^\circ_{298})/T$	$H^\circ - H^\circ_{298}$	ΔH_f°	ΔF_f°	$\text{Log}_{10} K_p$
0	.000	.000	INFINITE	− 2.825	240.500	240.500	INFINITE
100	8.062	47.001	65.601	− 1.860	241.409	236.797	− 517.494
200	10.720	53.335	59.140	− 1.161	241.796	231.711	− 253.190
298	12.803	58.083	58.083	.000	242.321	226.629	− 166.116
300	12.833	58.162	58.083	.024	242.329	226.531	− 165.020
400	14.250	62.057	58.605	1.381	242.702	221.206	− 120.855
500	15.297	65.356	59.633	2.861	242.906	215.804	− 94.323
600	16.148	68.223	60.831	4.435	242.968	210.374	− 76.625
700	16.854	70.767	62.072	6.086	242.919	204.944	− 63.983
800	17.440	73.057	63.304	7.802	242.799	199.531	− 54.507
900	17.925	75.140	64.506	9.571	242.624	194.132	− 47.139
1000	18.326	77.050	65.666	11.384	242.409	188.751	− 41.250
1100	18.660	78.812	66.782	13.234	242.171	183.399	− 36.436
1200	18.937	80.448	67.853	15.114	241.907	178.067	− 32.429
1300	19.170	81.974	68.882	17.020	241.625	172.759	− 29.042
1400	19.365	83.402	69.868	18.947	241.328	167.473	− 26.142
1500	19.530	84.743	70.816	20.892	241.017	162.208	− 23.632
1600	19.670	86.008	71.726	22.852	240.693	156.961	− 21.439
1700	19.791	87.205	72.602	24.825	240.366	151.741	− 19.507
1800	19.895	88.339	73.445	26.809	240.038	146.541	− 17.792
1900	19.984	89.417	74.257	28.804	239.705	141.351	− 16.258
2000	20.064	90.444	75.041	30.806	239.371	136.179	− 14.880
2100	20.132	91.425	75.798	32.816	239.033	131.033	− 13.636
2200	20.192	92.363	76.530	34.832	238.685	125.899	− 12.506
2300	20.246	93.261	77.238	36.854	238.331	120.779	− 11.476
2400	20.293	94.124	77.924	38.881	237.970	115.676	− 10.533
2500	20.336	94.953	78.588	40.913	237.606	110.592	− 9.667
2600	20.373	95.752	79.233	42.948	237.237	105.511	− 8.869
2700	20.408	96.521	79.859	44.987	236.864	100.457	− 8.131
2800	20.438	97.264	80.468	47.029	236.486	95.411	− 7.447
2900	20.466	97.982	81.059	49.075	236.108	90.384	− 6.811
3000	20.491	98.676	81.635	51.123	235.724	85.364	− 6.218
3100	20.514	99.348	82.196	53.173	235.334	80.358	− 5.665
3200	20.535	100.000	82.742	55.225	234.942	75.365	− 5.147
3300	20.554	100.632	83.275	57.280	234.545	70.376	− 4.661
3400	20.572	101.246	83.794	59.336	234.141	65.413	− 4.205
3500	20.588	101.842	84.301	61.394	233.731	60.459	− 3.775
3600	20.603	102.423	84.797	63.454	233.315	55.508	− 3.370
3700	20.617	102.987	85.281	65.515	232.892	50.573	− 2.987
3800	20.629	103.537	85.754	67.577	232.462	45.649	− 2.625
3900	20.641	104.073	86.217	69.640	232.025	40.749	− 2.283
4000	20.652	104.596	86.670	71.705	231.582	35.838	− 1.958

T, °K.	cal mole^{-1} deg^{-1}			kcal mole^{-1}			
	C_p°	S°	$-(F^\circ - H_{298}^\circ)/T$	$H^\circ - H_{298}^\circ$	ΔH_f°	ΔF_f°	$Log_{10}K_P$
4100	20.662	105.106	87.113	73.771	231.132	30.961	-1.650
4200	20.671	105.604	87.548	75.837	230.674	26.084	-1.357
4300	20.680	106.091	87.973	77.905	230.210	21.219	-1.078
4400	20.689	106.566	88.390	79.973	229.738	16.365	-.813
4500	20.696	107.031	88.799	82.043	229.260	11.518	-.559
4600	20.703	107.486	89.201	84.113	228.774	6.676	-.317
4700	20.710	107.931	89.594	86.183	228.280	1.856	-.086
4800	20.716	108.367	89.981	88.255	227.780	-2.944	.134
4900	20.722	108.795	90.361	90.327	227.272	-7.745	.345
5000	20.728	109.213	90.733	92.399	226.756	-12.550	.549
5100	20.733	109.624	91.100	94.472	226.233	-17.321	.742
5200	20.738	110.026	91.460	96.546	225.703	-22.100	.929
5300	20.743	110.421	91.814	98.620	225.165	-26.848	1.107
5400	20.748	110.809	92.162	100.694	224.619	-31.607	1.279
5500	20.752	111.190	92.505	102.769	224.066	-36.337	1.444
5600	20.755	111.564	92.842	104.845	223.506	-41.083	1.603
5700	20.759	111.931	93.173	106.920	222.937	-45.802	1.756
5800	20.763	112.292	93.500	108.996	222.361	-50.496	1.903
5900	20.766	112.647	93.821	111.073	221.778	-55.211	2.045
6000	20.770	112.996	94.138	113.150	221.187	-59.904	2.182

CARBON, PENTATOMIC (C_5)

(Ideal Gas)(mol. wt. = 60.06)Dec. 31, 1960.

T, °K.	cal mole^{-1} deg^{-1}			kcal mole^{-1}			
	C_p°	S°	$-(F^\circ - H_{298}^\circ)/T$	$H^\circ - H_{298}^\circ$	ΔH_f°	ΔF_f°	$Log_{10}K_P$
0	.000	.000	INFINITE	-3.331	240.298	240.298	INFINITE
100	9.077	46.146	67.126	-2.098	241.461	236.956	-517.842
200	13.233	53.620	60.900	-1.456	241.713	231.717	-253.196
298	16.218	59.608	59.608	.000	242.374	226.634	-166.119
300	16.259	59.708	59.608	.030	242.384	226.536	-165.023
400	18.189	64.663	60.271	1.757	242.881	221.174	-120.838
500	19.565	68.880	61.582	3.649	243.178	215.708	-94.281
600	20.674	72.548	63.110	5.663	243.302	210.198	-76.561
700	21.593	75.807	64.695	7.778	243.292	204.679	-63.900
800	22.356	78.741	66.271	9.976	243.195	199.174	-54.409
900	22.988	81.412	67.807	12.245	243.034	193.680	-47.030
1000	23.511	83.862	69.292	14.570	242.824	188.202	-41.130
1100	23.947	86.124	70.720	16.944	242.588	182.754	-36.308
1200	24.309	88.224	72.093	19.357	242.321	177.325	-32.294
1300	24.613	90.182	73.410	21.804	242.033	171.921	-28.901
1400	24.868	92.015	74.674	24.278	241.727	166.540	-25.997
1500	25.083	93.739	75.888	26.776	241.405	161.180	-23.483
1600	25.266	95.363	77.055	29.294	241.068	155.838	-21.285
1700	25.424	96.900	78.177	31.828	240.727	150.525	-19.350
1800	25.560	98.357	79.258	34.378	240.387	145.235	-17.633
1900	25.676	99.742	80.300	36.940	240.039	139.952	-16.097
2000	25.781	101.062	81.306	39.513	239.692	134.688	-14.717

Carbon, Pentatomic (C_5) (Continued)

	cal mole⁻¹ deg⁻¹			kcal mole⁻¹			
T, °K.	C_p°	S°	$-(F^\circ - H^\circ_{298})/T$	$H^\circ - H^\circ_{298}$	ΔH_f°	ΔF_f°	$Log_{10} K_P$
2100	25.870	102.322	82.277	42.095	239.339	129.452	−13.472
2200	25.948	103.527	83.215	44.686	238.975	124.229	−12.340
2300	26.019	104.682	84.124	47.285	238.604	119.021	−11.309
2400	26.080	105.791	85.004	49.890	238.224	113.829	−10.365
2500	26.136	106.857	85.857	52.501	237.840	108.660	−9.499
2600	26.184	107.883	86.684	55.117	237.451	103.491	−8.699
2700	26.230	108.872	87.488	57.737	237.056	98.352	−7.961
2800	26.269	109.827	88.269	60.362	236.656	93.222	−7.276
2900	26.306	110.749	89.028	62.991	236.255	88.111	−6.640
3000	26.338	111.641	89.767	65.623	235.847	83.008	−6.047
3100	26.368	112.505	90.487	68.259	235.433	77.920	−5.493
3200	26.396	113.343	91.188	70.897	235.016	72.846	−4.975
3300	26.421	114.156	91.872	73.538	234.592	67.774	−4.488
3400	26.444	114.945	92.539	76.181	234.160	62.733	−4.032
3500	26.465	115.712	93.190	78.826	233.720	57.700	−3.603
3600	26.485	116.457	93.826	81.474	233.273	52.670	−3.197
3700	26.503	117.183	94.447	84.123	232.817	47.657	−2.815
3800	26.519	117.890	95.055	86.775	232.354	42.656	−2.453
3900	26.534	118.579	95.649	89.427	231.881	37.683	−2.112
4000	26.549	119.251	96.231	92.081	231.400	32.695	−1.786
4100	26.562	119.907	96.800	94.737	230.911	27.748	−1.479
4200	26.573	120.547	97.358	97.394	230.413	22.797	−1.186
4300	26.585	121.173	97.905	100.052	229.906	17.861	−.908
4400	26.597	121.784	98.441	102.711	229.390	12.936	−.643
4500	26.606	122.382	98.966	105.371	228.865	8.019	−.389
4600	26.615	122.967	99.481	108.032	228.331	3.107	−.148
4700	26.624	123.539	99.987	110.694	227.788	−1.780	0.83
4800	26.632	124.100	100.484	113.357	227.236	−6.643	.302
4900	26.640	124.649	100.971	116.020	226.674	−11.510	.513
5000	26.648	125.187	101.450	118.685	226.104	−16.382	.716
5100	26.654	125.715	101.921	121.350	225.524	−21.214	.909
5200	26.661	126.233	102.383	124.016	224.935	−26.057	1.095
5300	26.667	126.740	102.838	126.682	224.336	−30.862	1.273
5400	26.674	127.239	103.286	129.349	223.728	−35.683	1.444
5500	26.679	127.729	103.725	132.017	223.111	−40.469	1.608
5600	26.683	128.209	104.158	134.685	222.484	−45.276	1.767
5700	26.688	128.682	104.584	137.353	221.847	−50.052	1.919
5800	26.693	129.146	105.004	140.022	221.201	−54.796	2.065
5900	26.697	129.602	105.417	142.692	220.546	−59.569	2.206
6000	26.703	130.051	105.824	145.362	219.881	−64.314	2.343

CARBON MONOXIDE (CO)

(Ideal Gas)(mol. wt. = 28.01)Sept. 30, 1965.

	cal mole⁻¹ deg⁻¹			kcal mole⁻¹			
T, °K.	C_p°	S°	$-(F^\circ - H^\circ_{298})/T$	$H^\circ - H^\circ_{298}$	ΔH_f°	ΔF_f°	$\text{Log}_{10} K_P$
0	.000	.000	INFINITE	- 2.072	- 27.200	- 27.200	INFINITE
100	6.956	39.613	53.401	- 1.379	- 26.876	- 28.741	62.809
200	6.957	44.435	47.851	- .683	- 26.599	- 30.718	33.566
298	6.965	47.214	47.214	.000	- 26.417	- 32.783	24.029
300	6.965	47.257	47.214	.013	- 26.414	- 32.823	23.910
400	7.013	49.265	47.488	.711	- 26.318	- 34.975	19.109
500	7.121	50.841	48.006	1.417	- 26.296	- 37.144	16.235
600	7.276	52.152	48.591	2.137	- 26.332	- 39.311	14.318
700	7.450	53.287	49.182	2.873	- 26.409	- 41.468	12.946
800	7.624	54.293	49.759	3.627	- 26.514	- 43.612	11.914
900	7.786	55.200	50.314	4.397	- 26.637	- 45.744	11.108
1000	7.931	56.028	50.845	5.183	- 26.771	- 47.859	10.459
1100	8.057	56.790	51.351	5.983	- 26.914	- 49.962	9.926
1200	8.168	57.496	51.834	6.794	- 27.062	- 52.049	9.479
1300	8.263	58.154	52.295	7.616	- 27.218	- 54.126	9.099
1400	8.346	58.769	52.736	8.446	- 27.376	- 56.189	8.771
1500	8.417	59.348	53.158	9.285	- 27.537	- 58.241	8.485
1600	8.480	59.893	53.562	10.130	- 27.700	- 60.284	8.234
1700	8.535	60.409	53.950	10.980	- 27.865	- 62.315	8.011
1800	8.583	60.898	54.322	11.836	- 28.032	- 64.337	7.811
1900	8.626	61.363	54.681	12.697	- 28.201	- 66.349	7.631
2000	8.664	61.807	55.026	13.561	- 28.372	- 68.353	7.469
2100	8.698	62.230	55.359	14.430	- 28.543	- 70.346	7.321
2200	8.728	62.635	55.680	15.301	- 28.719	- 72.335	7.185
2300	8.756	63.024	55.991	16.175	- 28.894	- 74.311	7.061
2400	8.781	63.397	56.292	17.052	- 29.074	- 76.282	6.946
2500	8.804	63.756	56.584	17.931	- 29.254	- 78.247	6.840
2600	8.825	64.102	56.866	18.813	- 29.438	- 80.202	6.741
2700	8.844	64.435	57.140	19.696	- 29.623	- 82.153	6.649
2800	8.863	64.757	57.407	20.582	- 29.810	- 84.093	6.563
2900	8.879	65.069	57.666	21.469	- 30.001	- 86.028	6.483
3000	8.895	65.370	57.917	22.357	- 30.194	- 87.957	6.407
3100	8.910	65.662	58.163	23.248	- 30.388	- 89.878	6.336
3200	8.924	65.945	58.401	24.139	- 30.586	- 91.795	6.269
3300	8.937	66.220	58.634	25.032	- 30.786	- 93.707	6.206
3400	8.949	66.487	58.861	25.927	- 30.988	- 95.609	6.145
3500	8.961	66.746	59.083	26.822	- 31.192	- 97.509	6.088
3600	8.973	66.999	59.299	27.719	- 31.399	- 99.400	6.034
3700	8.984	67.245	59.511	28.617	- 31.608	- 101.286	5.982
3800	8.994	67.485	59.717	29.516	- 31.818	- 103.164	5.933
3900	9.004	67.718	59.919	30.416	- 32.031	- 105.039	5.886
4000	9.014	67.946	60.117	31.316	- 32.247	- 106.908	5.841

Carbon Monoxide (CO) (Continued)

T, °K.	C_p°	S°	$-(F^\circ - H_{298}^\circ)/T$	$H^\circ - H_{298}^\circ$	ΔH_f°	ΔF_f°	$Log_{10}K_P$
	cal mole^{-1} deg^{-1}			kcal mole^{-1}			
4100	9.024	68.169	60.311	32.218	−32.464	−108.774	5.798
4200	9.033	68.387	60.501	33.121	−32.684	−110.630	5.756
4300	9.042	68.599	60.687	34.025	−32.906	−112.483	5.717
4400	9.051	68.807	60.869	34.930	−33.130	−114.333	5.679
4500	9.059	69.011	61.047	35.835	−33.356	−116.177	5.642
4600	9.068	69.210	61.223	36.741	−33.584	−118.012	5.607
4700	9.076	69.405	61.395	37.649	−33.814	−119.845	5.573
4800	9.084	69.596	61.564	38.557	−34.046	−121.672	5.540
4900	9.092	69.784	61.729	39.465	−34.280	−123.497	5.508
5000	9.100	69.967	61.892	40.375	−34.516	−125.315	5.477
5100	9.107	70.148	62.052	41.285	−34.755	−127.132	5.448
5200	9.115	70.325	62.210	42.196	−34.995	−128.941	5.419
5300	9.123	70.498	62.365	43.108	−35.237	−130.741	5.391
5400	9.130	70.669	62.517	44.021	−35.480	−132.542	5.364
5500	9.138	70.836	62.667	44.934	−35.727	−134.336	5.338
5600	9.145	71.001	62.814	45.849	−35.974	−136.129	5.312
5700	9.153	71.163	62.959	46.763	−36.225	−137.919	5.288
5800	9.160	71.322	63.102	47.679	−36.476	−139.698	5.264
5900	9.167	71.479	63.242	48.595	−36.730	−141.473	5.240
6000	9.175	71.633	63.381	49.513	−36.985	−143.249	5.218

Carbon Dioxide (CO_2)

(Ideal Gas)(mol. wt. = 44.05)Sept. 30, 1965.

T, °K.	C_p°	S°	$-(F^\circ - H_{298}^\circ)/T$	$H^\circ - H_{298}^\circ$	ΔH_f°	ΔF_f°	$Log_{10}K_P$
	cal mole^{-1} deg^{-1}			kcal mole^{-1}			
0	.000	.000	INFINITE	−2.238	−93.965	−93.965	INFINITE
100	6.981	42.758	58.188	−1.543	−93.997	−94.100	205.645
200	7.734	47.769	51.849	−.816	−94.028	−94.191	102.922
298	8.874	51.072	51.072	.000	−94.054	−94.265	69.095
300	8.896	51.127	51.072	.016	−94.055	−94.267	68.670
400	9.877	53.830	51.434	.958	−94.070	−94.335	51.540
500	10.666	56.122	52.148	1.987	−94.091	−94.399	41.260
600	11.310	58.126	52.981	3.087	−94.124	−94.458	34.405
700	11.846	59.910	53.845	4.245	−94.169	−94.510	29.506
800	12.293	61.522	54.706	5.453	−94.218	−94.556	25.830
900	12.667	62.992	55.546	6.702	−94.270	−94.596	22.970
1000	12.980	64.344	56.359	7.984	−94.321	−94.628	20.680
1100	13.243	65.594	57.143	9.296	−94.371	−94.658	18.806
1200	13.466	66.756	57.896	10.632	−94.419	−94.681	17.243
1300	13.656	67.841	58.620	11.988	−94.469	−94.701	15.920
1400	13.815	68.859	59.315	13.362	−94.515	−94.716	14.785
1500	13.953	69.817	59.984	14.750	−94.562	−94.728	13.801
1600	14.074	70.722	60.627	16.152	−94.607	−94.739	12.940
1700	14.177	71.578	61.246	17.565	−94.650	−94.746	12.180
1800	14.269	72.391	61.843	18.987	−94.696	−94.750	11.504
1900	14.352	73.165	62.418	20.418	−94.742	−94.751	10.898
2000	14.424	73.903	62.974	21.857	−94.788	−94.752	10.353

T, °K.	cal mole⁻¹ deg⁻¹ C_p°	S°	$-(F^\circ - H^\circ_{298})/T$	kcal mole⁻¹ $H^\circ - H^\circ_{298}$	ΔH_f°	ΔF_f°	$Log_{10}K_P$
2100	14.489	74.608	63.512	23.303	− 94.834	− 94.746	9.860
2200	14.547	75.284	64.031	24.755	− 94.885	− 94.744	9.411
2300	14.600	75.931	64.535	26.212	− 94.936	− 94.735	9.001
2400	14.648	76.554	65.023	27.674	− 94.991	− 94.724	8.625
2500	14.692	77.153	65.496	29.141	− 95.048	− 94.714	8.280
2600	14.734	77.730	65.956	30.613	− 95.107	− 94.698	7.960
2700	14.771	78.286	66.402	32.088	− 95.170	− 94.683	7.664
2800	14.807	78.824	66.836	33.567	− 95.235	− 94.662	7.388
2900	14.841	79.344	67.259	35.049	− 95.305	− 94.639	7.132
3000	14.873	79.848	67.670	36.535	− 95.377	− 94.615	6.892
3100	14.902	80.336	68.071	38.024	− 95.451	− 94.587	6.668
3200	14.930	80.810	68.461	39.515	− 95.530	− 94.560	6.458
3300	14.956	81.270	68.843	41.010	− 95.611	− 94.531	6.260
3400	14.982	81.717	69.215	42.507	− 95.696	− 94.495	6.074
3500	15.006	82.151	69.578	44.006	− 95.784	− 94.462	5.898
3600	15.030	82.574	69.933	45.508	− 95.874	− 94.421	5.732
3700	15.053	82.986	70.280	47.012	− 95.968	− 94.379	5.574
3800	15.075	83.388	70.620	48.518	− 96.064	− 94.331	5.425
3900	15.097	83.780	70.953	50.027	− 96.162	− 94.286	5.283
4000	15.119	84.162	71.278	51.538	− 96.263	− 94.237	5.149
4100	15.139	84.536	71.597	53.051	− 96.367	− 94.186	5.020
4200	15.159	84.901	71.909	54.566	− 96.473	− 94.130	4.898
4300	15.179	85.258	72.216	56.082	− 96.583	− 94.072	4.781
4400	15.197	85.607	72.516	57.601	− 96.694	− 94.015	4.670
4500	15.216	85.949	72.811	59.122	− 96.807	− 93.954	4.563
4600	15.234	86.284	73.100	60.644	− 96.923	− 93.885	4.460
4700	15.254	86.611	73.384	62.169	− 97.040	− 93.818	4.362
4800	15.272	86.933	73.663	63.695	− 97.160	− 93.746	4.268
4900	15.290	87.248	73.937	65.223	− 97.281	− 93.678	4.178
5000	15.306	87.557	74.206	66.753	− 97.404	− 93.603	4.091
5100	15.327	87.860	74.471	68.285	− 97.530	− 93.528	4.008
5200	15.349	88.158	74.731	69.819	− 97.656	− 93.450	3.927
5300	15.371	88.451	74.988	71.355	− 97.783	− 93.361	3.850
5400	15.393	88.738	75.239	72.893	− 97.912	− 93.280	3.775
5500	15.415	89.021	75.488	74.433	− 98.042	− 93.190	3.703
5600	15.437	89.299	75.732	75.976	− 98.173	− 93.104	3.633
5700	15.459	89.572	75.972	77.521	− 98.305	− 93.017	3.566
5800	15.481	89.841	76.209	79.068	− 98.438	− 92.918	3.501
5900	15.503	90.106	76.442	80.617	− 98.572	− 92.820	3.438
6000	15.525	90.367	76.672	82.168	− 98.707	− 92.724	3.377

Carbon Suboxide (C_3O_2)
(Ideal Gas)(mol. wt. = 68.03)Sept. 30, 1965.

T, °K.	cal mole^{-1} deg^{-1}			kcal mole^{-1}			
	C_p°	S°	$-(F^\circ - H_{298}^\circ)/T$	$H^\circ - H_{298}^\circ$	ΔH_f°	ΔF_f°	$Log_{10}K_P$
0	.000	.000	INFINITE	−3.593	−24.142	−24.142	INFINITE
100	10.841	51.981	78.464	−2.648	−23.952	−24.935	54.493
200	13.399	60.186	67.435	−1.450	−23.668	−26.026	28.439
298	16.009	66.049	66.049	.000	−23.380	−27.246	19.971
300	16.050	66.148	66.049	.030	−23.375	−27.271	19.866
400	17.917	71.036	66.703	1.733	−23.121	−28.609	15.630
500	19.310	75.190	67.995	3.597	−22.945	−30.003	13.113
600	20.433	78.813	69.503	5.586	−22.845	−31.426	11.446
700	21.366	82.035	71.067	7.678	−22.806	−32.858	10.258
800	22.145	84.941	72.623	9.854	−22.805	−34.295	9.369
900	22.794	87.588	74.140	12.102	−22.832	−35.733	8.677
1000	23.335	90.018	75.608	14.410	−22.869	−37.163	8.122
1100	23.786	92.264	77.022	16.766	−22.921	−38.592	7.667
1200	24.163	94.350	78.380	19.164	−22.979	−40.011	7.287
1300	24.480	96.297	79.684	21.597	−23.050	−41.429	6.965
1400	24.747	98.121	80.937	24.059	−23.120	−42.836	6.687
1500	24.973	99.837	82.140	26.545	−23.197	−44.243	6.446
1600	25.167	101.455	83.297	29.052	−23.277	−45.648	6.235
1700	25.333	102.986	84.411	31.577	−23.356	−47.043	6.048
1800	25.476	104.438	85.483	34.118	−23.441	−48.434	5.880
1900	25.600	105.818	86.517	36.672	−23.528	−49.816	5.730
2000	25.709	107.134	87.516	39.238	−23.617	−51.199	5.595
2100	25.804	108.391	88.480	41.813	−23.708	−52.572	5.471
2200	25.888	109.594	89.413	44.398	−23.808	−53.952	5.359
2300	25.962	110.746	90.315	46.991	−23.907	−55.314	5.256
2400	26.028	111.852	91.190	49.590	−24.015	−56.676	5.161
2500	26.087	112.916	92.038	52.196	−24.125	−58.040	5.074
2600	26.140	113.940	92.860	54.807	−24.243	−59.392	4.992
2700	26.187	114.928	93.660	57.424	−24.364	−60.747	4.917
2800	26.230	115.881	94.436	60.045	−24.489	−62.088	4.846
2900	26.268	116.802	95.192	62.670	−24.624	−63.426	4.780
3000	26.304	117.693	95.927	65.298	−24.764	−64.763	4.718
3100	26.336	118.556	96.643	67.930	−24.907	−66.090	4.659
3200	26.365	119.393	97.341	70.565	−25.058	−67.419	4.604
3300	26.391	120.204	98.022	73.203	−25.216	−68.746	4.553
3400	26.416	120.993	98.686	75.843	−25.380	−70.059	4.503
3500	26.438	121.759	99.334	78.486	−25.550	−71.379	4.457
3600	26.459	122.504	99.967	81.131	−25.725	−72.683	4.412
3700	26.478	123.229	100.586	83.778	−25.908	−73.982	4.370
3800	26.496	123.935	101.191	86.427	−26.095	−75.271	4.329
3900	26.512	124.624	101.784	89.077	−26.290	−76.565	4.290
4000	26.527	125.295	102.363	91.729	−26.490	−77.850	4.253

T, °K.	C_p°	S°	$-(F^\circ - H^\circ_{298})/T$	$H^\circ - H^\circ_{298}$	ΔH_f°	ΔF_f°	$Log_{10}K_P$
	cal mole⁻¹ deg⁻¹			kcal mole⁻¹			
4100	26.541	125.950	102.930	94.382	− 26.698	− 79.137	4.218
4200	26.555	126.590	103.486	97.037	− 26.912	− 80.407	4.184
4300	26.567	127.215	104.031	99.693	− 27.134	− 81.679	4.151
4400	26.578	127.826	104.565	102.351	− 27.360	− 82.950	4.120
4500	26.589	128.423	105.088	105.009	− 27.594	− 84.213	4.090
4600	26.599	129.008	105.602	107.668	− 27.835	− 85.463	4.060
4700	26.609	129.580	106.106	110.329	− 28.080	− 86.713	4.032
4800	26.618	130.140	106.601	112.990	− 28.335	− 87.951	4.004
4900	26.626	130.689	107.087	115.652	− 28.594	− 89.197	3.978
5000	26.634	131.227	107.564	118.315	− 28.860	− 90.430	3.952
5100	26.641	131.755	108.033	120.979	− 29.136	− 91.672	3.928
5200	26.648	132.272	108.495	123.644	− 29.415	− 92.892	3.904
5300	26.655	132.780	108.948	126.309	− 29.701	− 94.102	3.880
5400	26.661	133.278	109.394	128.974	− 29.995	− 95.313	3.857
5500	26.667	133.767	109.833	131.641	− 30.296	− 96.516	3.835
5600	26.673	134.248	110.264	134.308	− 30.603	− 97.725	3.814
5700	26.678	134.720	110.689	136.975	− 30.919	− 98.931	3.793
5800	26.683	135.184	111.108	139.643	− 31.241	− 100.116	3.772
5900	26.688	135.640	111.520	142.312	− 31.569	− 101.289	3.752
6000	26.692	136.089	111.925	144.981	− 31.904	− 102.482	3.733

FORMYL (HCO)

(Ideal Gas)(mol. wt. = 29.02)March 31, 1961.

T, °K.	C_p°	S°	$-(F^\circ - H^\circ_{298})/T$	$H^\circ - H^\circ_{298}$	ΔH_f°	ΔF_f°	$Log_{10}K_P$
	cal mole⁻¹ deg⁻¹			kcal mole⁻¹			
0	.000	.000	INFINITE	− 2.387	− 2.986	− 2.986	INFINITE
100	7.949	44.930	60.855	− 1.593	− 2.941	− 4.118	8.998
200	7.999	50.447	54.429	− .796	− 2.864	− 5.334	5.829
298	8.264	53.683	53.683	.000	− 2.900	− 6.543	4.796
300	8.271	53.734	53.683	.015	− 2.902	− 6.566	4.783
400	8.703	56.171	54.012	.863	− 3.002	− 7.773	4.247
500	9.184	58.164	54.649	1.758	− 3.142	− 8.950	3.912
600	9.660	59.881	55.381	2.700	− 3.305	− 10.097	3.678
700	10.108	61.404	56.135	3.689	− 3.481	− 11.214	3.501
800	10.518	62.781	56.881	4.720	− 3.661	− 12.307	3.362
900	10.888	64.042	57.607	5.791	− 3.840	− 13.378	3.248
1000	11.216	65.206	58.310	6.896	− 4.013	− 14.428	3.153
1100	11.505	66.289	58.986	8.033	− 4.182	− 15.462	3.072
1200	11.758	67.301	59.638	9.196	− 4.346	− 16.479	3.001
1300	11.979	68.251	60.264	10.383	− 4.508	− 17.484	2.939
1400	12.172	69.146	60.867	11.591	− 4.665	− 18.475	2.884
1500	12.341	69.992	61.447	12.817	− 4.822	− 19.456	2.835
1600	12.489	70.793	62.007	14.059	− 4.978	− 20.429	2.790
1700	12.618	71.554	62.546	15.314	− 5.131	− 21.388	2.750
1800	12.732	72.279	63.067	16.582	− 5.285	− 22.341	2.712
1900	12.832	72.970	63.570	17.860	− 5.440	− 23.283	2.678
2000	12.921	73.630	64.057	19.148	− 5.594	− 24.219	2.646

Formyl (HCO) (Continued)

T, °K.	cal mole^{-1} deg^{-1}			kcal mole^{-1}			
	C_p°	S°	$-(F^\circ - H^\circ_{298})/T$	$H^\circ - H^\circ_{298}$	ΔH_f°	ΔF_f°	$\text{Log}_{10}K_P$
2100	13.000	74.263	64.528	20.444	− 5.750	− 25.145	2.617
2200	13.071	74.869	64.984	21.747	− 5.909	− 26.067	2.589
2300	13.133	75.452	65.427	23.058	− 6.068	− 26.978	2.563
2400	13.190	76.012	65.856	24.374	− 6.232	− 27.883	2.539
2500	13.240	76.551	66.273	25.695	− 6.398	− 28.784	2.516
2600	13.286	77.071	66.678	27.022	− 6.566	− 29.675	2.494
2700	13.327	77.574	67.073	28.353	− 6.738	− 30.562	2.474
2800	13.365	78.059	67.456	29.687	− 6.912	− 31.440	2.454
2900	13.399	78.529	67.830	31.025	− 7.091	− 32.311	2.435
3000	13.430	78.983	68.194	32.367	− 7.273	− 33.179	2.417
3100	13.458	79.424	68.550	33.711	− 7.457	− 34.039	2.400
3200	13.484	79.852	68.896	35.058	− 7.646	− 34.894	2.383
3300	13.508	80.267	69.234	36.408	− 7.839	− 35.744	2.367
3400	13.530	80.671	69.565	37.760	− 8.035	− 36.586	2.352
3500	13.550	81.063	69.888	39.114	− 8.235	− 37.425	2.337
3600	13.569	81.445	70.204	40.470	− 8.439	− 38.255	2.322
3700	13.586	81.817	70.513	41.828	− 8.648	− 39.080	2.308
3800	13.602	82.180	70.815	43.187	− 8.859	− 39.897	2.294
3900	13.617	82.533	71.111	44.548	− 9.075	− 40.711	2.281
4000	13.631	82.878	71.401	45.911	− 9.294	− 41.520	2.268
4100	13.644	83.215	71.685	47.275	− 9.518	− 42.325	2.256
4200	13.656	83.544	71.963	48.640	− 9.746	− 43.119	2.244
4300	13.667	83.866	72.236	50.006	− 9.978	− 43.912	2.232
4400	13.678	84.180	72.504	51.373	− 10.214	− 44.700	2.220
4500	13.688	84.487	72.767	52.741	− 10.454	− 45.484	2.209
4600	13.697	84.788	73.025	54.111	− 10.698	− 46.255	2.198
4700	13.706	85.083	73.279	55.481	− 10.946	− 47.025	2.187
4800	13.714	85.372	73.527	56.852	− 11.198	− 47.789	2.176
4900	13.722	85.654	73.772	58.224	− 11.453	− 48.551	2.165
5000	13.729	85.932	74.013	59.596	− 11.712	− 49.306	2.155
5100	13.736	86.204	74.249	60.969	− 11.977	− 50.057	2.145
5200	13.742	86.470	74.481	62.343	− 12.244	− 50.802	2.135
5300	13.749	86.732	74.710	63.718	− 12.515	− 51.535	2.125
5400	13.754	86.989	74.935	65.093	− 12.791	− 52.272	2.115
5500	13.760	87.242	75.157	66.469	− 13.070	− 52.999	2.106
5600	13.765	87.490	75.375	67.845	− 13.353	− 53.725	2.097
5700	13.770	87.733	75.589	69.222	− 13.640	− 54.446	2.087
5800	13.775	87.973	75.801	70.599	− 13.932	− 55.158	2.078
5900	13.779	88.208	76.009	71.977	− 14.226	− 55.862	2.069
6000	13.784	88.440	76.214	73.355	− 14.525	− 56.570	2.060

Formaldehyde (CH_2O)

(Ideal Gas)(mol. wt. = 30.03)March 31, 1961.

T, °K.	cal mole^{-1} deg^{-1}			kcal mole^{-1}			
	C_p°	S°	$-(F^\circ - H^\circ_{298})/T$	$H^\circ - H^\circ_{298}$	ΔH_f°	ΔF_f°	$\text{Log}_{10}K_P$
0	.000	.000	INFINITE	− 2.395	− 26.782	− 26.782	INFINITE
100	7.949	43.479	59.481	− 1.600	− 27.116	− 26.928	58.849
200	8.007	48.996	53.016	− .804	− 27.341	− 26.669	29.141
298	8.461	52.261	52.261	.000	− 27.700	− 26.266	19.253

	cal mole⁻¹ deg⁻¹			kcal mole⁻¹			
T, °K.	C_p°	S°	$-(F^\circ - H^\circ_{298})/T$	$H^\circ - H^\circ_{298}$	ΔH_f°	ΔF_f°	$\text{Log}_{10}K_P$
300	8.475	52.313	52.261	.016	− 27.708	− 26.258	19.128
400	9.385	54.869	52.603	.906	− 28.113	− 25.713	14.048
500	10.460	57.077	53.281	1.898	− 28.505	− 25.068	10.957
600	11.524	59.079	54.083	2.998	− 28.860	− 24.347	8.868
700	12.505	60.931	54.931	4.200	− 29.174	− 23.568	7.358
800	13.380	62.659	55.790	5.495	− 29.443	− 22.749	6.214
900	14.149	64.280	56.644	6.872	− 29.672	− 21.898	5.317
1000	14.817	65.806	57.485	8.322	− 29.860	− 21.024	4.595
1100	15.395	67.246	58.307	9.833	− 30.017	− 20.134	4.000
1200	15.893	68.608	59.110	11.398	− 30.146	− 19.228	3.502
1300	16.322	69.897	59.890	13.009	− 30.256	− 18.314	3.079
1400	16.693	71.121	60.649	14.660	− 30.347	− 17.391	2.715
1500	17.013	72.284	61.386	16.346	− 30.427	− 16.461	2.398
1600	17.291	73.391	62.102	18.062	− 30.498	− 15.531	2.121
1700	17.534	74.446	62.798	19.803	− 30.558	− 14.592	1.876
1800	17.746	75.455	63.473	21.567	− 30.615	− 13.652	1.658
1900	17.932	76.419	64.129	23.351	− 30.666	− 12.707	1.462
2000	18.095	77.343	64.767	25.153	− 30.715	− 11.761	1.285
2100	18.240	78.230	65.387	26.970	− 30.761	− 10.811	1.125
2200	18.369	79.081	65.990	28.800	− 30.810	− 9.862	.980
2300	18.483	79.900	66.577	30.643	− 30.856	− 8.907	.846
2400	18.585	80.689	67.149	32.497	− 30.905	− 7.950	.724
2500	18.677	81.450	67.706	34.360	− 30.957	− 6.997	.612
2600	18.760	82.184	68.249	32.232	− 31.010	− 6.035	.507
2700	18.834	82.893	68.778	38.112	− 31.066	− 5.075	.411
2800	18.902	83.580	69.295	39.998	− 31.125	− 4.111	.321
2900	18.963	84.244	69.799	41.892	− 31.188	− 3.142	.237
3000	19.019	84.888	70.291	43.791	− 31.254	− 2.176	.159
3100	19.070	85.512	70.772	45.695	− 31.322	− 1.205	.085
3200	19.116	86.119	71.242	47.605	− 31.396	− .233	.016
3300	19.159	86.707	71.702	49.518	− 31.474	.741	− .049
3400	19.198	87.280	72.152	51.436	− 31.556	1.718	− .110
3500	19.235	87.837	72.592	53.358	− 31.643	2.697	− .168
3600	19.268	88.379	73.023	55.283	− 31.734	3.680	− .223
3700	19.299	88.908	73.445	57.212	− 31.831	4.666	− .276
3800	19.327	89.423	73.859	59.143	− 31.932	5.657	− .325
3900	19.354	89.925	74.264	61.077	− 32.039	6.647	− .372
4000	19.379	90.415	74.662	63.014	− 32.150	7.640	− .417
4100	19.401	90.894	75.052	64.953	− 32.266	8.634	− .460
4200	19.423	91.362	75.435	66.894	− 32.389	9.636	− .501
4300	19.443	91.819	75.811	68.837	− 32.516	10.637	− .541
4400	19.462	92.267	76.180	70.782	− 32.649	11.639	− .578
4500	19.479	92.704	76.542	72.729	− 32.787	12.645	− .614
4600	19.496	93.132	76.898	74.678	− 32.930	13.662	− .649
4700	19.511	93.552	77.248	76.628	− 33.079	14.677	− .682
4800	19.526	93.963	77.592	78.580	− 33.232	15.695	− .715
4900	19.539	94.366	77.930	80.534	− 33.390	16.714	− .745
5000	19.552	94.760	78.263	82.488	− 33.554	17.736	− .775

FORMALDEHYDE (CH_2O) (Continued)

T, °K.	cal mole⁻¹ deg⁻¹			kcal mole⁻¹			
	C_p°	S°	$-(F^\circ - H^\circ_{298})/T$	$H^\circ - H^\circ_{298}$	ΔH_f°	ΔF_f°	$\text{Log}_{10}K_P$
5100	19.564	95.148	78.590	84.444	− 33.725	18.763	− .804
5200	19.576	95.528	78.912	86.401	− 33.899	19.793	− .832
5300	19.587	95.901	79.229	88.359	− 34.078	20.834	− .859
5400	19.597	96.267	79.541	90.318	− 34.264	21.867	− .885
5500	19.607	96.627	79.849	92.278	− 34.454	22.911	− .910
5600	19.616	96.980	80.151	94.240	− 34.650	23.952	− .935
5700	19.625	97.327	80.450	96.202	− 34.851	24.998	− .958
5800	19.633	97.669	80.744	98.164	− 35.058	26.051	− .982
5900	19.641	98.004	81.033	100.128	− 35.269	27.110	− 1.004
6000	19.648	98.334	81.319	102.093	− 35.485	28.164	− 1.026

HYDROXYL (OH)

(Ideal Gas)(mol. wt. = 17.01)Mar. 31, 1966.

T, °K.	cal mole⁻¹ deg⁻¹			kcal mole⁻¹			
	C_p°	S°	$-(F^\circ - H^\circ_{298})/T$	$H^\circ - H^\circ_{298}$	ΔH_f°	ΔF_f°	$\text{Log}_{10}K_P$
0	.000	.000	INFINITE	− 2.192	9.289	9.289	INFINITE
100	7.798	35.726	50.398	− 1.467	9.278	9.001	− 19.672
200	7.356	40.985	44.541	− .711	9.393	8.671	− 9.475
298	7.167	43.880	43.880	.000	9.432	8.307	− 6.089
300	7.165	43.925	43.881	.013	9.432	8.299	− 6.046
400	7.087	45.974	44.160	.725	9.442	7.920	− 4.327
500	7.055	47.551	44.687	1.432	9.434	7.540	− 3.296
600	7.057	48.837	45.275	2.137	9.411	7.163	− 2.609
700	7.090	49.927	45.863	2.845	9.379	6.791	− 2.120
800	7.150	50.877	46.432	3.556	9.338	6.424	− 1.755
900	7.233	51.724	46.974	4.275	9.294	6.062	− 1.472
1000	7.332	52.491	47.488	5.003	9.250	5.706	− 1.247
1100	7.439	53.195	47.975	5.742	9.206	5.353	− 1.064
1200	7.549	53.847	48.437	6.491	9.164	5.005	− .912
1300	7.659	54.455	48.877	7.252	9.124	4.661	− .784
1400	7.766	55.027	49.296	8.023	9.086	4.318	− .674
1500	7.867	55.566	49.696	8.805	9.050	3.980	− .580
1600	7.963	56.077	50.079	9.596	9.014	3.642	− .497
1700	8.053	56.563	50.447	10.397	8.980	3.307	− .425
1800	8.137	57.025	50.799	11.207	8.947	2.975	− .361
1900	8.214	57.467	51.139	12.024	8.914	2.645	− .304
2000	8.286	57.891	51.466	12.849	8.881	2.313	− .253
2100	8.353	58.296	51.782	13.681	8.849	1.988	− .207
2200	8.415	58.686	52.087	14.520	8.815	1.662	− .165
2300	8.472	59.062	52.382	15.364	8.782	1.336	− .127
2400	8.526	59.424	52.668	16.214	8.748	1.014	− .092
2500	8.576	59.773	52.945	17.069	8.711	.690	− .060
2600	8.622	60.110	53.214	17.929	8.675	.372	− .031
2700	8.665	60.436	53.476	18.794	8.637	.054	− .004
2800	8.706	60.752	53.730	19.662	8.598	− .264	.021
2900	8.744	61.058	53.977	20.535	8.557	− .578	.044
3000	8.780	61.355	54.218	21.411	8.515	− .893	.065

T, °K.	C_p°	S°	$-(F^\circ - H^\circ_{298})/T$	$H^\circ - H^\circ_{298}$	ΔH_f°	ΔF_f°	$Log_{10}K_P$
	cal mole⁻¹ deg⁻¹			kcal mole⁻¹			
3100	8.814	61.644	54.453	22.291	8.472	− 1.208	.085
3200	8.846	61.924	54.682	23.174	8.427	− 1.519	.104
3300	8.876	62.197	54.906	24.060	8.381	− 1.829	.121
3400	8.905	62.462	55.124	24.949	8.333	− 2.137	.137
3500	8.933	62.721	55.338	25.841	8.283	− 2.446	.153
3600	8.959	62.973	55.546	26.735	8.232	− 2.751	.167
3700	8.984	63.218	55.750	27.632	8.179	− 3.052	.180
3800	9.008	63.458	55.950	28.532	8.125	− 3.355	.193
3900	9.031	63.693	56.145	29.434	8.069	− 3.660	.205
4000	9.053	63.922	56.337	30.338	8.011	− 3.961	.216
4100	9.074	64.145	56.525	31.245	7.952	− 4.256	.227
4200	9.095	64.364	56.709	32.153	7.892	− 4.553	.237
4300	9.115	64.579	56.889	33.064	7.830	− 4.851	.247
4400	9.134	64.788	57.066	33.976	7.766	− 5.144	.255
4500	9.153	64.994	57.240	34.890	7.701	− 5.439	.264
4600	9.171	65.195	57.411	35.807	7.635	− 5.726	.272
4700	9.189	65.393	57.579	36.725	7.567	− 6.018	.280
4800	9.206	65.586	57.744	37.644	7.499	− 6.304	.287
4900	9.223	65.776	57.906	38.566	7.429	− 6.592	.294
5000	9.239	65.963	58.065	39.489	7.358	− 6.882	.301
5100	9.255	66.146	58.222	40.414	7.286	− 7.162	.307
5200	9.271	66.326	58.376	41.340	7.213	− 7.448	.313
5300	9.286	66.502	58.527	42.268	7.140	− 7.722	.318
5400	9.301	66.676	58.677	43.197	7.065	− 8.007	.324
5500	9.316	66.847	58.824	44.128	6.989	− 8.284	.329
5600	9.330	67.015	58.968	45.060	6.912	− 8.563	.334
5700	9.344	67.180	59.111	45.994	6.835	− 8.837	.339
5800	9.358	67.343	59.252	46.929	6.757	− 9.112	.343
5900	9.372	67.503	59.390	47.866	6.678	− 9.385	.348
6000	9.386	67.661	59.527	48.803	6.598	− 9.659	.352

HYDROPEROXYL (HO_2)

(Ideal Gas)(mol. wt. = 33.01)Mar. 31, 1964.

T, °K.	C_p°	S°	$-(F^\circ - H^\circ_{298})/T$	$H^\circ - H^\circ_{298}$	ΔH_f°	ΔF_f°	$Log_{10}K_P$
	cal mole⁻¹ deg⁻¹			kcal mole⁻¹			
0	.000	.000	INFINITE	− 2.390	5.697	5.697	INFINITE
100	7.949	45.618	61.574	− 1.596	5.399	6.209	− 13.568
200	8.003	51.135	55.132	− .799	5.214	7.085	− 7.742
298	8.338	54.383	54.383	.000	5.000	8.049	− 5.900
300	8.347	54.434	54.383	.015	4.996	8.067	− 5.877
400	8.907	56.910	54.717	.877	4.800	9.122	− 4.984
500	9.479	58.960	55.366	1.797	4.639	10.222	− 4.468
600	9.980	60.734	56.116	2.771	4.508	11.351	− 4.134
700	10.405	62.305	56.890	3.791	4.399	12.501	− 3.903
800	10.769	63.719	57.657	4.850	4.307	13.663	− 3.732
900	11.087	65.006	58.403	5.943	4.230	14.838	− 3.603
1000	11.365	66.189	59.123	7.066	4.167	16.021	− 3.501

HYDROPEROXYL (HO_2) (Continued)

	cal mole⁻¹ deg⁻¹			kcal mole⁻¹			
T, °K.	C_p°	S°	$-(F^\circ - H^\circ_{298})/T$	$H^\circ - H^\circ_{298}$	ΔH_f°	ΔF_f°	$Log_{10}K_P$
1100	11.612	67.284	59.816	8.215	4.114	17.209	− 3.419
1200	11.831	68.304	60.481	9.387	4.071	18.401	− 3.351
1300	12.025	69.258	61.120	10.580	4.035	19.597	− 3.294
1400	12.197	70.156	61.734	11.791	4.005	20.795	− 3.246
1500	12.350	71.003	62.323	13.019	3.979	21.996	− 3.205
1600	12.485	71.804	62.891	14.261	3.955	23.197	− 3.168
1700	12.606	72.565	63.438	15.515	3.934	24.401	− 3.137
1800	12.714	73.288	63.965	16.782	3.913	25.605	− 3.109
1900	12.810	73.979	64.474	18.058	3.891	26.812	− 3.084
2000	12.895	74.638	64.966	19.343	3.869	28.017	− 3.061
2100	12.972	75.269	65.442	20.637	3.845	29.226	− 3.041
2200	13.041	75.874	65.902	21.937	3.818	30.435	− 3.023
2300	13.104	76.455	66.349	23.245	3.790	31.645	− 3.007
2400	13.160	77.014	66.781	24.558	3.757	32.859	− 2.992
2500	13.210	77.552	67.202	25.876	3.720	34.069	− 2.978
2600	13.256	78.071	67.610	27.200	3.682	35.286	− 2.966
2700	13.298	78.572	68.007	28.528	3.638	36.502	− 2.954
2800	13.336	79.057	68.393	29.859	3.590	37.719	− 2.944
2900	13.371	79.525	68.768	31.195	3.539	38.941	− 2.935
3000	13.403	79.979	69.135	32.534	3.483	40.162	− 2.926
3100	13.432	80.419	69.492	33.875	3.423	41.386	− 2.918
3200	13.459	80.846	69.840	35.220	3.359	42.610	− 2.910
3300	13.484	81.261	70.180	36.567	3.291	43.838	− 2.903
3400	13.507	81.663	70.511	37.917	3.218	45.067	− 2.897
3500	13.528	82.055	70.836	39.268	3.141	46.299	− 2.891
3600	13.547	82.437	71.153	40.622	3.060	47.534	− 2.886
3700	13.565	82.808	71.463	41.978	2.974	48.771	− 2.881
3800	13.582	83.170	71.766	43.335	2.886	50.011	− 2.876
3900	13.598	83.523	72.063	44.694	2.793	51.250	− 2.872
4000	13.612	83.867	72.354	46.055	2.695	52.493	− 2.868
4100	13.626	84.204	72.639	47.417	2.594	53.741	− 2.864
4200	13.638	84.532	72.918	48.780	2.490	54.989	− 2.861
4300	13.650	84.853	73.192	50.144	2.382	56.242	− 2.858
4400	13.662	85.167	73.460	51.510	2.270	57.493	− 2.856
4500	13.672	85.474	73.724	52.876	2.154	58.749	− 2.853
4600	13.682	85.775	73.983	54.244	2.037	60.013	− 2.851
4700	13.691	86.069	74.237	55.613	1.914	61.274	− 2.849
4800	13.700	86.358	74.486	56.982	1.790	62.538	− 2.847
4900	13.708	86.640	74.731	58.353	1.663	63.802	− 2.846
5000	13.716	86.917	74.972	59.724	1.533	65.069	− 2.844
5100	13.723	87.189	75.209	61.096	1.399	66.346	− 2.843
5200	13.730	87.455	75.442	62.468	1.264	67.616	− 2.842
5300	13.736	87.717	75.671	63.842	1.126	68.900	− 2.841
5400	13.743	87.974	75.897	65.216	.985	70.174	− 2.840
5500	13.748	88.226	76.119	66.590	.843	71.460	− 2.839

T, °K.	C_p° (cal mole⁻¹ deg⁻¹)	S° (cal mole⁻¹ deg⁻¹)	$-(F^\circ - H_{298}^\circ)/T$ (cal mole⁻¹ deg⁻¹)	$H^\circ - H_{298}^\circ$ (kcal mole⁻¹)	ΔH_f° (kcal mole⁻¹)	ΔF_f° (kcal mole⁻¹)	$Log_{10}K_P$
5600	13.754	88.474	76.337	67.965	.697	72.742	− 2.839
5700	13.759	88.717	76.552	69.341	.550	74.029	− 2.838
5800	13.764	88.957	76.764	70.717	.400	75.323	− 2.838
5900	13.769	89.192	76.973	72.094	.248	76.615	− 2.838
6000	13.774	89.423	77.178	73.471	.094	77.911	− 2.838

WATER (H_2O)

(Ideal Gas)(mol. wt. = 18.02)Mar. 31, 1961.

T, °K.	C_p° (cal mole⁻¹ deg⁻¹)	S° (cal mole⁻¹ deg⁻¹)	$-(F^\circ - H_{298}^\circ)/T$ (cal mole⁻¹ deg⁻¹)	$H^\circ - H_{298}^\circ$ (kcal mole⁻¹)	ΔH_f° (kcal mole⁻¹)	ΔF_f° (kcal mole⁻¹)	$Log_{10}K_P$
0	.000	.000	INFINITE	− 2.367	− 57.103	− 57.103	INFINITE
100	7.961	36.396	52.202	− 1.581	− 57.433	− 56.557	123.600
200	7.969	41.916	45.837	− .784	− 57.579	− 55.635	60.792
298	8.025	45.106	45.106	.000	− 57.798	− 54.636	40.048
300	8.027	45.155	45.106	.015	− 57.803	− 54.617	39.786
400	8.186	47.484	45.422	.825	− 58.042	− 53.519	29.240
500	8.415	49.334	46.026	1.654	− 58.277	− 52.361	22.886
600	8.676	50.891	46.710	2.509	− 58.500	− 51.156	18.633
700	8.954	52.249	47.406	3.390	− 58.710	− 49.915	15.583
800	9.246	53.464	48.089	4.300	− 58.905	− 48.646	13.289
900	9.547	54.570	48.749	5.240	− 59.084	− 47.352	11.498
1000	9.851	55.592	49.382	6.209	− 59.246	− 46.040	10.062
1100	10.152	56.545	49.991	7.210	− 59.391	− 44.712	8.883
1200	10.444	57.441	50.575	8.240	− 59.519	− 43.371	7.899
1300	10.723	58.288	51.136	9.298	− 59.634	− 42.022	7.064
1400	10.987	59.092	51.675	10.384	− 59.734	− 40.663	6.347
1500	11.233	59.859	52.196	11.495	− 59.824	− 39.297	5.725
1600	11.462	60.591	52.698	12.630	− 59.906	− 37.927	5.180
1700	11.674	61.293	53.183	13.787	− 59.977	− 36.549	4.699
1800	11.869	61.965	53.652	14.964	− 60.041	− 35.170	4.270
1900	12.048	62.612	54.107	16.160	− 60.099	− 33.786	3.886
2000	12.214	63.234	54.548	17.373	− 60.150	− 32.401	3.540
2100	12.366	63.834	54.976	18.602	− 60.198	− 31.012	3.227
2200	12.505	64.412	55.392	19.846	− 60.242	− 29.621	2.942
2300	12.634	64.971	55.796	21.103	− 60.282	− 28.229	2.682
2400	12.753	65.511	56.190	22.372	− 60.321	− 26.832	2.443
2500	12.863	66.034	56.573	23.653	− 60.359	− 25.439	2.224
2600	12.965	66.541	56.947	24.945	− 60.393	− 24.040	2.021
2700	13.059	67.032	57.311	26.246	− 60.428	− 22.641	1.833
2800	13.146	67.508	57.667	27.556	− 60.462	− 21.242	1.658
2900	13.228	67.971	58.014	28.875	− 60.496	− 19.838	1.495
3000	13.304	68.421	58.354	30.201	− 60.530	− 18.438	1.343
3100	13.374	68.858	58.685	31.535	− 60.562	− 17.034	1.201
3200	13.441	69.284	59.010	32.876	− 60.596	− 15.630	1.067
3300	13.503	69.698	59.328	34.223	− 60.631	− 14.223	.942
3400	13.562	70.102	59.639	35.577	− 60.666	− 12.818	.824
3500	13.617	70.496	59.943	36.936	− 60.703	− 11.409	.712

Water (H_2O) (Continued)

T, °K.	cal mole⁻¹ deg⁻¹			kcal mole⁻¹			
	C_p°	S°	$-(F^\circ - H_{298}^\circ)/T$	$H^\circ - H_{298}^\circ$	ΔH_f°	ΔF_f°	$\text{Log}_{10}K_P$
3600	13.669	70.881	60.242	38.300	-60.741	-10.000	.607
3700	13.718	71.256	60.534	39.669	-60.782	-8.589	.507
3800	13.764	71.622	60.821	41.043	-60.822	-7.177	.413
3900	13.808	71.980	61.103	42.422	-60.865	-5.766	.323
4000	13.850	72.331	61.379	43.805	-60.910	-4.353	.238
4100	13.890	72.673	61.651	45.192	-60.957	-2.938	.157
4200	13.927	73.008	61.917	46.583	-61.006	-1.522	.079
4300	13.963	73.336	62.179	47.977	-61.056	-.105	.005
4400	13.997	73.658	62.436	49.375	-61.109	1.311	-.065
4500	14.030	73.973	62.689	50.777	-61.164	2.729	-.133
4600	14.061	74.281	62.938	52.181	-61.220	4.154	-.197
4700	14.091	74 584	63.182	53.589	-61.279	5.576	-.259
4800	14.120	74.881	63.423	55.000	-61.339	6.998	-.319
4900	14.148	75.172	63.660	56.413	-61.401	8.422	-.376
5000	14.174	75.459	63.893	57.829	-61.465	9.844	-.430
5100	14.201	75.740	64.122	59.248	-61.532	11.275	-.483
5200	14.228	76.016	64.348	60.669	-61.600	12.700	-.534
5300	14.254	76.287	64.571	62.093	-61.669	14.135	-.583
5400	14.279	76.553	64.791	63.520	-61.741	15.560	-.630
5500	14.303	76.816	65.007	64.949	-61.813	16.995	-.675
5600	14.328	77.074	65.220	66.381	-61.889	18.426	-.719
5700	14.351	77.327	65.430	67.815	-61.965	19.862	-.762
5800	14.375	77.577	65.637	69.251	-62.043	21.299	-.803
5900	14.398	77.823	65.842	70.690	-62.122	22.736	-.842
6000	14.422	78.065	66.044	72.131	-62.203	24.174	-.880

Nitrous Oxide (N_2O)

(Ideal Gas)(mol. wt. = 44.02)Dec. 31, 1964.

T, °K.	cal mole⁻¹ deg⁻¹			kcal mole⁻¹			
	C_p°	S°	$-(F^\circ - H_{298}^\circ)/T$	$H^\circ - H_{298}^\circ$	ΔH_f°	ΔF_f°	$\text{Log}_{10}K_P$
0	.000	.000	INFINITE	-2.290	20.430	20.430	INFINITE
100	7.015	43.988	59.924	-1.594	20.084	21.573	-47.145
200	8.033	49.108	53.353	-.849	19.786	23.185	-25.334
298	9.230	52.546	52.546	.000	19.610	24.896	-18.248
300	9.250	52.603	52.546	.017	19.608	24.928	-18.159
400	10.201	55.400	52.921	.992	19.530	26.716	-14.596
500	10.953	57.761	53.659	2.051	19.520	28.514	-12.463
600	11.565	59.813	54.517	3.178	19.558	30.311	-11.040
700	12.070	61.635	55.406	4.360	19.623	32.097	-10.021
800	12.486	63.275	56.289	5.589	19.710	33.873	-9.253
900	12.830	64.766	57.149	6.855	19.810	35.638	-8.654
1000	13.113	66.133	57.980	8.153	19.920	37.391	-8.171
1100	13.348	67.394	58.779	9.476	20.036	39.132	-7.774
1200	13.544	68.565	59.547	10.821	20.156	40.863	-7.442
1300	13.707	69.655	60.283	12.184	20.279	42.583	-7.158
1400	13.845	70.676	60.989	13.562	20.404	44.293	-6.914
1500	13.961	71.635	61.667	14.952	20.530	45.996	-6.701

T, °K.	C_p° (cal mole⁻¹ deg⁻¹)	S°	$-(F^\circ - H^\circ_{298})/T$	$H^\circ - H^\circ_{298}$ (kcal mole⁻¹)	ΔH_f°	ΔF_f°	$Log_{10}K_P$
1600	14.060	72.540	62.319	16.353	20.657	47.691	– 6.514
1700	14.145	73.395	62.945	17.764	20.783	49.377	– 6.347
1800	14.218	74.205	63.548	19.182	20.908	51.054	– 6.198
1900	14.282	74.976	64.130	20.607	21.032	52.728	– 6.065
2000	14.337	75.710	64.690	22.038	21.155	54.392	– 5.943
2100	14.385	76.411	65.232	23.474	21.277	56.049	– 5.833
2200	14.428	77.081	65.755	24.915	21.396	57.703	– 5.732
2300	14.466	77.723	66.262	26.359	21.513	59.350	– 5.639
2400	14.499	78.339	66.752	27.808	21.630	60.993	– 5.554
2500	14.529	78.932	67.228	29.259	21.742	62.630	– 5.475
2600	14.556	79.502	67.689	30.713	21.853	64.262	– 5.401
2700	14.580	80.052	68.137	32.170	21.962	65.894	– 5.333
2800	14.602	80.583	68.572	33.629	22.069	67.517	– 5.270
2900	14.621	81.095	68.995	35.090	22.174	69.140	– 5.210
3000	14.639	81,591	69.406	36.553	22.275	70.757	– 5.154
3100	14.655	82.072	69.807	38.018	22.376	72.371	– 5.102
3200	14.670	82.537	70.198	39.484	22.473	73.981	– 5.052
3300	14.684	82.989	70.579	40.952	22.568	75.591	– 5.006
3400	14.696	83.427	70.950	42.421	22.661	77.196	– 4.962
3500	14.707	83.853	71.313	43.891	22.752	78.797	– 4.920
3600	14.718	84.268	71.667	45.363	22.841	80.400	– 4.881
3700	14.727	84.671	72.013	46.835	22.928	81.995	– 4.843
3800	14.736	85.064	72.351	48.308	23.013	83.591	– 4.807
3900	14.745	85.447	72.682	49.782	23.097	85.183	– 4.773
4000	14.752	85.820	73.006	51.257	23.177	86.774	– 4.741
4100	14.760	86.185	73.323	52.733	23.257	88.365	– 4.710
4200	14.766	86.541	73.633	54.209	23.334	89.951	– 4.680
4300	14.772	86.888	73.937	55.686	23.411	91.537	– 4.652
4400	14.778	87.228	74.236	57.163	23.485	93.120	– 4.625
4500	14.784	87.560	74.528	58.641	23.558	94.703	– 4.599
4600	14.789	87.885	74.815	60.120	23.630	96.288	– 4.574
4700	14.794	88.203	75.096	61.599	23.700	97.860	– 4.550
4800	14.798	88.515	75.373	63.079	23.768	99.441	– 4.527
4900	14.802	88.820	75.644	64.559	23.836	101.011	– 4.505
5000	14.806	89.119	75.911	66.039	23.902	102.590	– 4.484
5100	14.810	89.412	76.172	67.520	23.966	104.156	– 4.463
5200	14.814	89.700	76.430	69.001	24.030	105.733	– 4.444
5300	14.817	89.982	76.683	70.483	24.092	107.305	– 4.425
5400	14.820	90.259	76.932	71.965	24.154	108.872	– 4.406
5500	14.823	90.531	77.176	73.447	24.213	110.446	– 4.388
5600	14.826	90.798	77.417	74.929	24.272	112.010	– 4.371
5700	14.829	91.060	77.654	76.412	24.329	113.575	– 4.354
5800	14.831	91.318	77.888	77.895	24.386	115.142	– 4.338
5900	14.834	91.572	78.117	79.378	24.441	116.706	– 4.323
6000	14.836	91.821	78.344	80.862	24.495	118.267	– 4.308

Nitric Oxide (NO)

(Ideal Gas)(mol. wt. = 30.01)June 30, 1963.

	cal mole⁻¹ deg⁻¹			kcal mole⁻¹			
T, °K.	C_p°	S°	$-(F^\circ - H^\circ_{298})/T$	$H^\circ - H^\circ_{298}$	ΔH_f°	ΔF_f°	$Log_{10} K_P$
0	.000	.000	INFINITE	− 2.197	·21.456	21.456	INFINITE
100	7.721	42.286	56.801	− 1.451	21.503	21.256	− 46.453
200	7.271	47.477	51.003	− .705	21.558	20.984	− 22.929
298	7.133	50.347	50.347	.000	21.580	20.697	− 15.171
300	7.132	50.392	50.348	.013	21.580	20.692	− 15.073
400	7.157	52.444	50.627	.727	21.590	20.394	− 11.142
500	7.287	54.053	51.157	1.448	21.594	20.095	− 8.783
600	7.466	55.397	51.755	2.186	21.598	19.795	− 7.210
700	7.655	56.562	52.360	2.942	21.601	19.494	− 6.086
800	7.832	57.596	52.951	3.716	21.605	19.192	− 5.243
900	7.988	58.528	53.520	4.507	21.610	18.890	− 4.587
1000	8.123	59.377	54.064	5.313	21.615	18.588	− 4.062
1100	8.238	60.157	54.583	6.131	21.620	18.285	− 3.633
1200	8.336	60.878	55.078	6.960	21.624	17.981	− 3.275
1300	8.419	61.548	55.550	7.798	21.628	17.678	− 2.972
1400	8.491	62.175	56.001	8.644	21.631	17.373	− 2.712
1500	8.552	62.763	56.432	9.496	21.633	17.069	− 2.487
1600	8.605	63.317	56.845	10.354	21.635	16.765	− 2.290
1700	8.651	63.840	57.242	11.217	21.635	16.461	− 2.116
1800	8.692	64.335	57.622	12.084	12.633	16.156	− 1.962
1900	8.727	64.806	57.988	12.955	21.630	15.853	− 1.823
2000	8.759	65.255	58.340	13.829	21.626	15.548	− 1.699
2100	8.788	65.683	58.680	14.706	21.619	15.244	− 1.586
2200	8.813	66.092	59.007	15.587	21.611	14.941	− 1.484
2300	8.837	66.484	59.324	16.469	21.601	14.637	− 1.391
2400	8.858	66.861	59.630	17.354	21.589	14.336	− 1.305
2500	8.877	67.223	59.927	18.241	21.574	14.033	− 1.227
2600	8.895	67.571	60.214	19.129	21.558	13.732	− 1.154
2700	8.912	67.908	60.493	20.020	21.540	13.432	− 1.087
2800	8.927	68.232	60.763	20.911	21.520	13.132	− 1.025
2900	8.941	68.545	61.026	21.805	21.498	12.834	− .967
3000	8.955	68.849	61.282	22.700	21.474	12.535	− .913
3100	8.968	69.143	61.531	23.596	21.449	12.237	− .863
3200	8.980	69.427	61.773	24.493	21.421	11.940	− .815
3300	8.991	69.704	62.010	25.392	21.392	11.644	− .771
3400	9.002	69.973	62.240	26.291	21.361	11.349	− .729
3500	9.012	70.234	62.464	27.192	21.329	11.054	− .690
3600	9.022	70.488	62.684	28.094	21.294	10.762	− .653
3700	9.032	70.735	62.898	28.997	21.259	10.470	− .618
3800	9.041	70.976	63.108	29.900	21.222	10.179	− .585
3900	9.050	71.211	63.312	30.805	21.185	9.889	− .554
4000	9.058	71.440	63.513	31.710	21.145	9.598	− .524

T, °K.	C_p° (cal mole⁻¹ deg⁻¹)	S°	$-(F^\circ - H^\circ_{298})/T$	$H^\circ - H^\circ_{298}$ (kcal mole⁻¹)	ΔH_f°	ΔF_f°	$Log_{10} K_P$
4100	9.066	71.664	63.709	32.616	21.104	9.311	− .496
4200	9.074	71.882	63.901	33.523	21.063	9.024	− .470
4300	9.082	72.096	64.089	34.431	21.021	8.739	− .444
4400	9.090	72.305	64.273	35.340	20.977	8.452	− .420
4500	9.097	72.509	64.454	36.249	20.932	8.169	− .397
4600	9.105	72.709	64.631	37.159	20.887	7.888	− .375
4700	9.112	72.905	64.805	38.070	20.841	7.605	− .354
4800	9.119	73.097	64.976	38.982	20.794	7.324	− .333
4900	9.125	73.285	65.144	39.894	20.747	7.040	− .314
5000	9.132	73.470	65.308	40.807	20.699	6.763	− .296
5100	9.139	73.651	65.470	41.720	20.650	6.484	− .278
5200	9.145	73.828	65.629	42.634	20.601	6.207	− .261
5300	9.152	74.002	65.786	43.549	20.551	5.932	− .245
5400	9.158	74.173	65.939	44.465	20.501	5.654	− .229
5500	9.164	74.342	66.091	45.381	20.451	5.383	− .214
5600	9.170	74.507	66.239	46.298	20.400	5.107	− .199
5700	9.176	74.669	66.386	47.215	20.348	4.835	− .185
5800	9.182	74.829	66.530	48.133	20.297	4.566	− .172
5900	9.188	74.986	66.672	49.051	20.245	4.292	− .159
6000	9.194	75.140	66.812	49.970	20.192	4.024	− .147

Nitrogen Dioxide (NO_2)

(Ideal Gas)(mol. wt. = 46.01)Sept. 30, 1964.

T, °K.	C_p° (cal mole⁻¹ deg⁻¹)	S°	$-(F^\circ - H^\circ_{298})/T$	$H^\circ - H^\circ_{298}$ (kcal mole⁻¹)	ΔH_f°	ΔF_f°	$Log_{10} K_P$
0	.000	.000	INFINITE	− 2.435	8.586	8.586	INFINITE
100	7.953	48.387	64.785	− 1.640	8.326	9.545	− 20.859
200	8.218	53.954	58.130	− .835	8.099	10.853	− 11.859
298	8.837	57.343	57.343	.000	7.910	12.247	− 8.977
300	8.850	57.398	57.344	.016	7.907	12.274	− 8.941
400	9.601	60.046	57.700	.939	7.770	13.751	− 7.513
500	10.327	62.268	58.397	1.936	7.684	15.258	− 6.669
600	10.955	64.208	59.207	3.001	7.638	16.778	− 6.111
700	11.469	65.937	60.047	4.123	7.618	18.302	− 5.714
800	11.881	67.496	60.882	5.291	7.617	19.828	− 5.417
900	12.208	68.915	61.697	6.496	7.629	21.355	− 5.185
1000	12.468	70.215	62.485	7.730	7.649	22.879	− 5.000
1100	12.677	71.414	63.243	8.988	7.674	24.400	− 4.848
1200	12.847	72.524	63.971	10.265	7.702	25.921	− 4.721
1300	12.985	73.558	64.669	11.556	7.731	27.438	− 4.612
1400	13.099	74.525	65.339	12.861	7.761	28.951	− 4.519
1500	13.193	75.432	65.982	14.176	7.790	30.464	− 4.438
1600	13.273	76.286	66.599	15.499	7.818	31.975	− 4.367
1700	13.340	77.093	67.193	16.830	7.846	33.484	− 4.304
1800	13.398	77.857	67.764	18.167	7.869	34.992	− 4.248
1900	13.447	78.583	68.315	19.509	7.890	36.498	− 4.198
2000	13.490	79.274	68.846	20.856	7.908	38.002	− 4.152

NITROGEN DIOXIDE (NO_2) (Continued)

T, °K.	C_p°	S°	$-(F^\circ - H^\circ_{298})/T$	$H^\circ - H^\circ_{298}$	ΔH_f°	ΔF_f°	$Log_{10} K_P$
	cal mole⁻¹ deg⁻¹			kcal mole⁻¹			
2100	13.527	79.933	69.358	22.207	7.923	39.507	– 4.111
2200	13.560	80.563	69.853	23.561	7.932	41.010	– 4.074
2300	13.588	81.166	70.332	24.919	7.939	42.514	– 4.040
2400	13.614	81.745	70.796	26.279	7.942	44.019	– 4.008
2500	13.636	82.301	71.245	27.641	7.939	45.520	– 3.979
2600	13.656	82.836	71.680	29.006	7.933	47.025	– 3.953
2700	13.674	83.352	72.103	30.373	7.922	48.529	– 3.928
2800	13.690	83.850	72.514	31.741	7.907	50.031	– 3.905
2900	13.705	84.330	72.913	33.111	7.888	51.540	– 3.884
3000	13.718	84.795	73.301	34.482	7.863	53.045	– 3.864
3100	13.730	85.245	73.679	35.854	7.836	54.551	– 3.846
3200	13.741	85.681	74.048	37.228	7.803	56.058	– 3.828
3300	13.751	86.104	74.407	38.602	7.767	57.565	– 3.812
3400	13.760	86.515	74.757	39.978	7.726	59.075	– 3.797
3500	13.768	86.914	75.099	41.354	7.683	60.583	– 3.783
3600	13.776	87.302	75.432	42.731	7.635	62.097	– 3.770
3700	13.783	87.679	75.758	44.109	7.584	63.613	– 3.757
3800	13.790	88.047	76.077	45.488	7.529	65.128	– 3.746
3900	13.796	88.405	76.388	46.867	7.473	66.643	– 3.734
4000	13.801	88.755	76.693	48.247	7.412	68.158	– 3.724
4100	13.806	89.096	76.991	49.627	7.347	69.678	– 3.714
4200	13.811	89.428	77.284	51.008	7.281	71.201	– 3.705
4300	13.816	89.753	77.570	52.390	7.213	72.727	– 3.696
4400	13.820	90.071	77.850	53.771	7.140	74.247	– 3.688
4500	13.824	90.382	78.125	55.154	7.066	75.772	– 3.680
4600	13.828	90.686	78.395	56.536	6.990	77.304	– 3.673
4700	13.831	90.983	78.660	57.919	6.911	78.833	– 3.666
4800	13.834	91.274	78.920	59.302	6.830	80.366	– 3.659
4900	13.837	91.559	79.175	60.686	6.748	81.894	– 3.652
5000	13.840	91.839	79.425	62.070	6.663	83.428	– 3.646
5100	13.843	92.113	79.671	63.454	6.577	84.966	– 3.641
5200	13.846	92.382	79.913	64.838	6.489	86.501	– 3.635
5300	13.848	92.646	80.151	66.223	6.400	88.044	– 3.630
5400	13.850	92.905	80.385	67.608	6.309	89.579	– 3.625
5500	13.852	93.159	80.615	68.993	6.217	91.128	– 3.621
5600	13.854	93.408	80.841	70.379	6.122	92.673	– 3.617
5700	13.856	93.654	81.063	71.764	6.027	94.215	– 3.612
5800	13.858	93.895	81.283	73.150	5.931	95.767	– 3.608
5900	13.860	94.132	81.498	74.536	5.833	97.313	– 3.605
6000	13.862	94.365	81.711	75.922	5.734	98.866	– 3.601

NITROXYL (HNO)
(Ideal Gas)(mol. wt. = 31.02)Mar. 31, 1963.

T, °K.	cal $mole^{-1}$ deg^{-1}			kcal $mole^{-1}$			
	C_p°	S°	$-(F^\circ - H^\circ_{298})/T$	$H^\circ - H^\circ_{298}$	ΔH_f°	ΔF_f°	$Log_{10} K_p$
0	.000	.000	INFINITE	-2.387	24.498	24.498	INFINITE
100	7.949	43.977	59.902	-1.592	24.215	25.018	-54.674
200	7.995	49.493	53.475	-.796	24.018	25.893	-28.293
298	8.279	52.729	52.729	.000	23.800	26.859	-19.688
300	8.287	52.781	52.730	.015	23.796	26.878	-19.580
400	8.788	55.230	53.060	.868	23.598	27.937	-15.263
500	9.342	57.251	53.702	1.775	23.438	29.041	-12.693
600	9.869	59.001	54.442	2.735	23.315	30.174	-10.990
700	10.345	60.559	55.207	3.747	23.222	31.325	-9.779
800	10.766	61.968	55.965	4.803	23.155	32.486	-8.874
900	11.134	63.258	56.705	5.898	23.107	33.656	-8.172
1000	11.454	64.448	57.421	7.028	23.078	34.830	-7.612
1100	11.731	65.553	58.110	8.187	23.061	36.006	-7.153
1200	11.970	66.584	58.774	9.373	23.055	37.183	-6.772
1300	12.177	67.551	59.412	10.580	23.056	38.361	-6.449
1400	12.356	68.460	60.026	11.807	23.064	39.537	-6.172
1500	12.511	69.318	60.618	13.051	23.074	40.714	-5.932
1600	12.646	70.130	61.187	14.309	23.087	41.889	-5.722
1700	12.764	70.900	61.736	15.579	23.101	43.065	-5.536
1800	12.867	71.633	62.265	16.861	23.116	44.238	-5.371
1900	12.958	72.331	62.777	18.152	23.130	45.412	-5.223
2000	13.038	72.998	63.271	19.452	23.143	46.584	-5.090
2100	13.108	73.635	63.750	20.760	23.155	47.756	-4.970
2200	13.171	74.247	64.213	22.074	23.164	48.927	-4.860
2300	13.227	74.834	64.662	23.394	23.172	50.097	-4.760
2400	13.277	75.398	65.098	24.719	23.177	51.269	-4.668
2500	13.322	75.940	65.521	26.049	23.178	52.437	-4.584
2600	13.363	76.464	65.932	27.383	23.178	53.609	-4.506
2700	13.399	76.969	66.331	28.721	23.174	54.780	-4.434
2800	13.432	77.457	66.720	30.063	23.167	55.949	-4.367
2900	13.463	77.929	67.098	31.408	23.158	57.123	-4.305
3000	13.490	78.385	67.467	32.755	23.145	58.293	-4.246
3100	13.515	78.828	67.826	34.106	23.130	59.464	-4.192
3200	13.538	79.258	68.177	35.458	23.110	60.636	-4.141
3300	13.559	79.675	68.519	36.813	23.088	61.810	-4.093
3400	13.578	80.080	68.853	38.170	23.063	62.983	-4.048
3500	13.596	80.473	69.180	39.529	23.034	64.157	-4.006
3600	13.612	80.857	69.499	40.889	23.002	65.333	-3.966
3700	13.628	81.230	69.811	42.251	22.966	66.510	-3.928
3800	13.642	81.594	70.116	43.615	22.928	67.688	-3.893
3900	13.655	81.948	70.415	44.980	22.887	68.866	-3.859
4000	13.667	82.294	70.707	46.346	22.842	70.045	-3.827

NITROXYL (HNO) (Continued)

T, °K.	C_p° (cal mole⁻¹ deg⁻¹)	S° (cal mole⁻¹ deg⁻¹)	$-(F^\circ - H_{298}^\circ)/T$ (cal mole⁻¹ deg⁻¹)	$H^\circ - H_{298}^\circ$ (kcal mole⁻¹)	ΔH_f° (kcal mole⁻¹)	ΔF_f° (kcal mole⁻¹)	$\text{Log}_{10} K_P$
4100	13.678	82.632	70.994	47.713	22.794	71.226	– 3.796
4200	13.689	82.961	71.275	49.081	22.744	72.408	– 3.768
4300	13.699	83.283	71.551	50.451	22.691	73.592	– 3.740
4400	13.708	83.599	71.821	51.821	22.634	74.774	– 3.714
4500	13.717	83.907	72.086	53.192	22.574	75.959	– 3.689
4600	13.725	84.208	72.346	54.564	22.513	77.151	– 3.665
4700	13.732	84.503	72.602	55.937	22.448	78.338	– 3.643
4800	13.739	84.793	72.853	57.311	22.380	79.528	– 3.621
4900	13.746	85.076	73.100	58.685	22.310	80.716	– 3.600
5000	13.752	85.354	73.342	60.060	22.238	81.909	– 3.580
5100	13.758	85.626	73.580	61.435	22.162	83.104	– 3.561
5200	13.764	85.893	73.814	62.812	22.086	84.299	– 3.543
5300	13.770	86.156	74.045	64.188	22.006	85.499	– 3.525
5400	13.775	86.413	74.271	65.565	21.923	86.693	– 3.509
5500	13.779	86.666	74.494	66.943	21.839	87.898	– 3.493
5600	13.784	86.914	74.714	68.321	21.752	89.096	– 3.477
5700	13.788	87.158	74.930	69.700	21.663	90.301	– 3.462
5800	13.792	87.398	75.143	71.079	21.571	91.508	– 3.448
5900	13.796	87.634	75.353	72.458	21.478	92.712	– 3.434
6000	13.800	87.866	75.559	73.838	21.382	93.921	– 3.421

IMIDOGEN (NH)
(Ideal Gas)(mol. wt. = 15.01)Dec. 31, 1965.

T, °K.	C_p° (cal mole⁻¹ deg⁻¹)	S° (cal mole⁻¹ deg⁻¹)	$-(F^\circ - H_{298}^\circ)/T$ (cal mole⁻¹ deg⁻¹)	$H^\circ - H_{298}^\circ$ (kcal mole⁻¹)	ΔH_f° (kcal mole⁻¹)	ΔF_f° (kcal mole⁻¹)	$\text{Log}_{10} K_P$
0	.000	.000	INFINITE	– 2.060	80.988	80.988	INFINITE
100	6.961	35.688	49.486	– 1.380	80.946	80.502	– 175.929
200	6.963	40.514	43.932	– .684	80.989	80.037	– 87.456
298	6.966	43.295	43.295	.000	81.000	79.567	– 58.321
300	6.966	43.338	43.295	.013	81.000	79.558	– 57.955
400	6.973	45.343	43.568	.710	81.001	79.077	– 43.204
500	6.994	46.901	44.085	1.408	80.998	78.596	– 34.353
600	7.041	48.179	44.664	2.110	80.994	78.117	– 28.453
700	7.119	49.270	45.246	2.817	80.987	77.638	– 24.238
800	7.223	50.228	45.810	3.534	80.979	77.159	– 21.078
900	7.343	51.085	46.349	4.262	80.972	76.682	– 18.620
1000	7.471	51.865	46.862	5.003	80.967	76.206	– 16.654
1100	7.601	52.583	47.350	5.757	80.963	75.730	– 15.046
1200	7.727	53.250	47.814	6.523	80.962	75.255	– 13.705
1300	7.847	53.874	48.257	7.302	80.963	74.778	– 12.571
1400	7.960	54.459	48.679	8.092	80.966	74.303	– 11.599
1500	8.064	55.012	49.083	8.894	80.970	73.827	– 10.756
1600	8.161	55.536	49.470	9.705	80.974	73.350	– 10.019
1700	8.250	56.033	49.842	10.526	80.980	72.874	– 9.368
1800	8.331	56.507	50.199	11.355	80.986	72.397	– 8.790
1900	8.406	56.959	50.543	12.192	80.994	71.921	– 8.272
2000	8.474	57.392	50.875	13.036	81.001	71.443	– 7.807

T, °K.	cal mole^{-1} deg^{-1} C_p°	S°	$-(F^\circ - H^\circ_{298})/T$	kcal mole^{-1} $H^\circ - H^\circ_{298}$	ΔH_f°	ΔF_f°	$Log_{10}K_P$
2100	8.538	57.807	51.195	13.886	81.009	70.964	- 7.385
2200	8.596	58.206	51.505	14.743	81.016	70.485	- 7.002
2300	8.650	58.589	51.804	15.605	81.025	70.007	- 6.652
2400	8.701	58.959	52.095	16.473	81.033	69.527	- 6.331
2500	8.747	59.315	52.377	17.345	81.041	69.046	- 6.036
2600	8.791	59.659	52.650	18.222	81.049	68.567	- 5.763
2700	8.832	59.991	52.916	19.103	81.057	68.088	- 5.511
2800	8.870	60.313	53.174	19.989	81.066	67.606	- 5.277
2900	8.907	60.625	53.426	20.878	81.075	67.127	- 5.059
3000	8.941	60.928	53.671	21.770	81.082	66.643	- 4.855
3100	8.973	61.221	53.910	22.666	81.091	66.163	- 4.664
3200	9.004	61.507	54.143	23.565	81.099	65.680	- 4.486
3300	9.034	61.784	54.370	24.466	81.106	65.200	- 4.318
3400	9.062	62.054	54.592	25.371	81.115	64.718	- 4.160
3500	9.089	62.317	54.809	26.279	81.122	64.236	- 4.011
3600	9.114	62.574	55.021	27.189	81.128	63.751	- 3.870
3700	9.139	62.824	55.229	28.102	81.135	63.269	- 3.737
3800	9.163	63.068	55.432	29.017	81.141	62.787	- 3.611
3900	9.186	63.306	55.631	29.934	81.146	62.305	- 3.491
4000	9.209	63.539	55.825	30.854	81.151	61.821	- 3.378
4100	9.230	63.767	56.016	31.776	81.155	61.336	- 3.269
4200	9.251	63.989	56.204	32.700	81.159	60.856	- 3.167
4300	9.272	64.207	56.387	33.626	81.163	60.372	- 3.068
4400	9.292	64.421	56.567	34.554	81.165	59.885	- 2.974
4500	9.311	64.630	56.744	35.484	81.167	59.400	- 2.885
4600	9.330	64.835	56.918	36.417	81.169	58.919	- 2.799
4700	9.349	65.035	57.088	37.351	81.170	58.438	- 2.717
4800	9.367	65.232	57.256	38.286	81.170	57.955	- 2.639
4900	9.385	65.426	57.421	39.224	81.170	57.467	- 2.563
5000	9.403	65.616	57.583	40.163	81.170	56.982	- 2.491
5100	9.420	65.802	57.742	41.104	81.168	56.500	- 2.421
5200	9.437	65.985	57.899	42.047	81.167	56.018	- 2.354
5300	9.454	66.165	58.053	42.992	81.165	55.534	- 2.290
5400	9.470	66.342	58.205	43.938	81.162	55.048	- 2.228
5500	9.486	66.516	58.355	44.886	81.158	54.566	- 2.168
5600	9.502	66.687	58.502	45.835	81.154	54.081	- 2.111
5700	9.518	66.855	58.647	46.786	81.150	53.602	- 2.055
5800	9.534	67.021	58.790	47.739	81.144	53.116	- 2.001
5900	9.549	67.184	58.931	48.693	81.139	52.633	- 1.950
6000	9.565	67.344	59.070	49.649	81.132	52.152	- 1.900

AMIDOGEN (NH_2)

(Ideal Gas)(mol. wt. = 16.02)Dec. 31, 1965.

T, °K.	cal mole^{-1} deg^{-1} C_p°	S°	$-(F^\circ - H^\circ_{298})/T$	kcal mole^{-1} $H^\circ - H^\circ_{298}$	ΔH_f°	ΔF_f°	$Log_{10}K_P$
0	.000	.000	INFINITE	- 2.373	40.757	40.757	INFINITE
100	7.949	37.792	53.574	- 1.578	40.450	41.015	- 89.635
200	7.954	43.303	47.219	- .783	40.291	41.633	- 45.492
298	8.024	46.488	46.488	.000	40.070	42.337	- 31.033

AMIDOGEN (NH_2) (Continued)

T, °K.	cal mole^{-1} deg^{-1}			kcal mole^{-1}			
	C_p°	S°	$-(F^\circ - H^\circ_{298})/T$	$H^\circ - H^\circ_{298}$	ΔH_f°	ΔF_f°	$\text{Log}_{10}K_p$
300	8.027	46.538	46.488	.015	40.065	42.351	– 30.851
400	8.220	48.871	46.805	.826	39.834	43.148	– 23.574
500	8.492	50.733	47.410	1.661	39.619	44.002	– 19.232
600	8.804	52.308	48.099	2.526	39.427	44.897	– 16.353
700	9.142	53.691	48.801	3.423	39.259	45.823	– 14.306
800	9.491	54.934	49.491	4.355	39.113	46.770	– 12.776
900	9.841	56.072	50.160	5.321	38.988	47.736	– 11.591
1000	10.180	57.127	50.804	6.323	38.884	48.713	– 10.646
1100	10.501	58.112	51.424	7.357	38.798	49.700	– 9.874
1200	10.799	59.039	52.021	8.422	38.729	50.695	– 9.232
1300	11.071	59.914	52.595	9.516	38.673	51.694	– 8.690
1400	11.319	60.744	53.147	10.635	38.628	52.697	– 8.226
1500	11.542	61.533	53.680	11.779	38.591	53.704	– 7.824
1600	11.743	62.284	54.195	12.943	38.559	54.711	– 7.473
1700	11.923	63.001	54.692	14.126	38.534	55.724	– 7.163
1800	12.084	63.688	55.173	15.327	38.513	56.733	– 6.888
1900	12.229	64.345	55.638	16.543	38.497	57.748	– 6.642
2000	12.359	64.975	56.089	17.772	38.482	58.762	– 6.421
2100	12.475	65.581	56.527	19.014	38.469	59.776	– 6.221
2200	12.580	66.164	56.952	20.267	38.457	60.790	– 6.039
2300	12.675	66.725	57.365	21.530	38.446	61.806	– 5.873
2400	12.760	67.267	57.766	22.802	38.436	62.822	– 5.720
2500	12.838	67.789	58.157	24.082	38.423	63.837	– 5.580
2600	12.908	68.294	58.537	25.369	38.412	64.855	– 5.451
2700	12.972	68.783	58.907	26.663	38.400	65.871	– 5.332
2800	13.031	69.255	59.269	27.963	38.386	66.889	– 5.221
2900	13.084	69.714	59.621	29.269	38.373	67.908	– 5.117
3000	13.133	70.158	59.965	30.580	38.357	68.926	– 5.021
3100	13.177	70.589	60.301	31.895	38.342	69.945	– 4.931
3200	13.218	71.008	60.629	33.215	38.324	70.965	– 4.846
3300	13.256	71.416	60.949	34.539	38.303	71.985	– 4.767
3400	13.291	71.812	61.263	35.866	38.283	73.005	– 4.693
3500	13.323	72.198	61.570	37.197	38.259	74.027	– 4.622
3600	13.353	72.573	61.870	38.531	38.232	75.051	– 4.556
3700	13.381	72.940	62.165	39.868	38.203	76.073	– 4.493
3800	13.407	73.297	62.453	41.207	38.172	77.098	– 4.434
3900	13.431	73.645	62.735	42.549	38.138	78.125	– 4.378
4000	13.453	73.986	63.013	43.893	38.102	79.148	– 4.324
4100	13.474	74.318	63.284	45.239	38.062	80.176	– 4.274
4200	13.493	74.643	63.551	46.588	38.020	81.204	– 4.225
4300	13.511	74.961	63.812	47.938	37.975	82.231	– 4.179
4400	13.529	75.272	64.069	49.290	37.927	83.258	– 4.135
4500	13.545	75.576	64.322	50.644	37.875	84.288	– 4.093

T, °K.	C_p°	S°	$-(F^\circ - H_{298}^\circ)/T$	$H^\circ - H_{298}^\circ$	ΔH_f°	ΔF_f°	$Log_{10}K_P$
	cal mole⁻¹ deg⁻¹			kcal mole⁻¹			
4600	13.560	75.874	64.570	51.999	37.822	85.324	− 4.054
4700	13.574	76.166	64.813	53.356	37.765	86.356	− 4.015
4800	13.587	76.451	65.053	54.714	37.705	87.394	− 3.979
4900	13.600	76.732	65.288	56.073	37.642	88.426	− 3.944
5000	13.612	77.007	65.520	57.434	37.576	89.461	− 3.910
5100	13.623	77.276	65.748	58.795	37.507	90.503	− 3.878
5200	13.633	77.541	65.972	60.158	37.436	91.542	− 3.847
5300	13.643	77.801	66.193	61.522	37.360	92.584	− 3.818
5400	13.653	78.056	66.410	62.887	37.282	93.623	− 3.789
5500	13.662	78.306	66.624	64.252	37.201	94.673	− 3.762
5600	13.670	78.553	66.835	65.619	37.117	95.712	− 3.735
5700	13.678	78.795	67.043	66.986	37.029	96.763	− 3.710
5800	13.686	79.033	67.247	68.355	36.939	97.810	− 3.685
5900	13.694	79.267	67.449	69.724	36.845	98.860	− 3.662
6000	13.701	79.497	67.648	71.093	36.749	99.911	− 3.639

AMMONIA (NH_3)

(Ideal Gas)(mol. wt. = 17.03)Sept. 30, 1965.

T, °K.	C_p°	S°	$-(F^\circ - H_{298}^\circ)/T$	$H^\circ - H_{298}^\circ$	ΔH_f°	ΔF_f°	$Log_{10}K_P$
	cal mole⁻¹ deg⁻¹			kcal mole⁻¹			
0	.000	.000	INFINITE	− 2.404	− 9.302	− 9.302	INFINITE
100	7.950	37.210	53.303	− 1.609	− 9.988	− 8.146	17.801
200	8.064	42.740	46.794	− .811	− 10.446	− 6.139	6.708
298	8.515	46.033	46.033	.000	− 10.970	− 3.915	2.869
300	8.526	46.085	46.033	.016	− 10.980	− 3.871	2.820
400	9.241	48.633	46.375	.903	− 11.482	− 1.424	.778
500	10.036	50.780	47.047	1.867	− 11.919	1.142	− .499
600	10.808	52.679	47.830	2.909	− 12.282	3.790	− 1.380
700	11.538	54.400	48.647	4.027	− 12.582	6.494	− 2.027
800	12.225	55.986	49.467	5.215	− 12.824	9.235	− 2.523
900	12.868	57.464	50.274	6.470	− 13.016	12.005	− 2.915
1000	13.467	58.851	51.063	7.787	− 13.163	14.792	− 3.233
1100	14.030	60.161	51.831	9.163	− 13.271	17.594	− 3.495
1200	14.550	61.404	52.578	10.592	− 13.343	20.405	− 3.716
1300	15.030	62.588	53.303	12.071	− 13.385	23.218	− 3.903
1400	15.460	63.718	54.007	13.596	− 13.402	26.034	− 4.064
1500	15.850	64.798	54.690	15.162	− 13.399	28.853	− 4.204
1600	16.205	65.833	55.355	16.765	− 13.381	31.666	− 4.325
1700	16.520	66.825	56.000	18.402	− 13.347	34.483	− 4.433
1800	16.762	67.776	56.628	20.066	− 13.303	37.294	− 4.528
1900	16.995	68.689	57.239	21.754	− 13.250	40.104	− 4.613
2000	17.220	69.566	57.834	23.465	− 13.191	42.911	− 4.689
2100	17.429	70.411	58.413	25.197	− 13.125	45.715	− 4.757
2200	17.630	71.227	58.977	26.950	− 13.053	48.514	− 4.819
2300	17.825	72.015	59.527	28.723	− 12.973	51.310	− 4.875
2400	18.014	72.777	60.063	30.515	− 12.887	54.107	− 4.927
2500	18.195	73.516	60.586	32.326	− 12.797	56.893	− 4.973

Ammonia (NH_3) (Continued)

T, °K.	cal mole⁻¹ deg⁻¹			kcal mole⁻¹			
	C_p°	S°	$-(F^\circ - H_{298}^\circ)/T$	$H^\circ - H_{298}^\circ$	ΔH_f°	ΔF_f°	$\text{Log}_{10}K_p$
2600	18.370	74.234	61.097	34.154	− 12.697	59.679	− 5.016
2700	18.537	74.930	61.597	35.999	− 12.592	62.461	− 5.056
2800	18.698	75.607	62.085	37.861	− 12.480	65.237	− 5.092
2900	18.853	76.266	62.563	39.739	− 12.360	68.015	− 5.125
3000	19.000	76.908	63.030	41.631	− 12.236	70.780	− 5.156
3100	19.115	77.532	63.488	43.537	− 12.105	73.546	− 5.185
3200	19.228	78.141	63.937	45.454	− 11.973	76.307	− 5.211
3300	19.341	78.734	64.376	47.383	− 11.838	79.067	− 5.236
3400	19.452	79.314	64.807	49.322	− 11.698	81.814	− 5.259
3500	19.563	79.879	65.230	51.273	− 11.557	84.565	− 5.280
3600	19.672	80.432	65.644	53.235	− 11.412	87.307	− 5.300
3700	19.781	80.972	66.051	55.207	− 11.265	90.051	− 5.319
3800	19.888	81.501	66.451	57.191	− 11.112	92.787	− 5.336
3900	19.994	82.019	66.843	59.185	− 10.958	95.520	− 5.353
4000	20.100	82.527	67.229	61.190	− 10.800	98.246	− 5.368
4100	20.205	83.024	67.608	63.205	− 10.639	100.972	− 5.382
4200	20.308	83.512	67.981	65.231	− 10.474	103.694	− 5.396
4300	20.411	83.991	68.348	67.267	− 10.306	106.409	− 5.408
4400	20.512	84.462	68.709	69.313	− 10.134	109.117	− 5.420
4500	20.613	84.924	69.064	71.369	− 9.961	111.823	− 5.431
4600	20.712	85.378	69.414	73.435	− 9.781	114.534	− 5.441
4700	20.811	85.824	69.758	75.511	− 9.600	117.236	− 5.451
4800	20.908	86.264	70.098	77.597	− 9.415	119.928	− 5.460
4900	21.005	86.696	70.432	79.693	− 9.226	122.621	− 5.469
5000	21.100	87.121	70.761	81.798	− 9.033	125.309	− 5.477
5100	21.195	87.540	71.086	83.913	− 8.838	127.995	− 5.485
5200	21.288	87.952	71.407	86.037	− 8.638	130.678	− 5.492
5300	21.381	88.359	71.723	88.170	− 8.436	133.355	− 5.499
5400	21.472	88.759	72.035	90.313	− 8.230	136.026	− 5.505
5500	21.563	89.154	72.342	92.465	− 8.021	138.697	− 5.511
5600	21.652	89.543	72.646	94.625	− 7.809	141.361	− 5.517
5700	21.740	89.927	72.946	96.795	− 7.592	144.028	− 5.522
5800	21.828	90.306	73.242	98.974	− 7.374	146.683	− 5.527
5900	21.914	90.680	73.534	101.161	− 7.152	149.337	− 5.532
6000	22.000	91.049	73.823	103.356	− 6.926	151.984	− 5.536

Hydrogen Cyanide (HCN)

(Ideal Gas)(mol. wt. = 27.03)Mar. 31, 1961.

T, °K.	cal mole⁻¹ deg⁻¹			kcal mole⁻¹			
	C_p°	S°	$-(F^\circ - H_{298}^\circ)/T$	$H^\circ - H_{298}^\circ$	ΔH_f°	ΔF_f°	$\text{Log}_{10}K_p$
0	.000	.000	INFINITE	− 2.209	31.291	31.291	INFINITE
100	6.971	40.025	55.175	− 1.515	31.245	30.391	− 66.416
200	7.589	44.992	48.967	− .795	31.238	29.533	− 32.271
298	8.594	48.213	48.213	.000	31.200	28.704	− 21.040
300	8.611	48.266	48.213	.016	31.199	28.689	− 20.899
400	9.421	50.860	48.561	.920	31.161	27.858	− 15.220
500	10.042	53.032	49.244	1.894	31.115	27.037	− 11.817

	cal mole⁻¹ deg⁻¹			kcal mole⁻¹			
T, °K.	C_p°	S°	$-(F^\circ - H_{298}^\circ)/T$	$H^\circ - H_{298}^\circ$	ΔH_f°	ΔF_f°	$\text{Log}_{10}K_P$
600	10.562	54.910	50.035	2.925	31.062	26.226	– 9.552
700	11.025	56.573	50.853	4.004	31.002	25.426	– 7.938
800	11.447	58.073	51.663	5.128	30.942	24.633	– 6.729
900	11.832	59.444	52.452	6.293	30.884	23.847	– 5.791
1000	12.182	60.709	53.216	7.494	30.833	23.068	– 5.041
1100	12.499	61.885	53.951	8.728	30.787	22.293	– 4.429
1200	12.785	62.986	54.659	9.992	30.748	21.524	– 3.920
1300	13.042	64.019	55.339	11.284	30.713	20.756	– 3.489
1400	13.273	64.994	55.994	12.600	30.686	19.992	– 3.121
1500	13.479	65.917	56.625	13.938	30.662	19.230	– 2.802
1600	13.665	66.793	57.234	15.295	30.643	18.466	– 2.522
1700	13.833	67.627	57.821	16.670	30.629	17.707	– 2.276
1800	13.984	68.422	58.388	18.061	30.618	16.946	– 2.057
1900	14.122	69.182	58.936	19.467	30.612	16.188	– 1.862
2000	14.246	69.909	59.467	20.885	30.610	15.429	– 1.686
2100	14.361	70.607	59.981	22.315	30.610	14.671	– 1.527
2200	14.465	71.278	60.479	23.757	30.612	13.909	– 1.382
2300	14.561	71.923	60.963	25.208	30.618	13.152	– 1.250
2400	14.650	72.544	61.432	26.669	30.624	12.393	– 1.129
2500	14.733	73.144	61.889	28.138	30.633	11.630	– 1.017
2600	14.809	73.723	62.333	29.615	30.643	10.872	– .914
2700	14.881	74.284	62.765	31.100	30.655	10.109	– .818
2800	14.948	74.826	63.186	32.591	30.669	9.350	– .730
2900	15.011	75.352	63.597	34.089	30.683	8.589	– .647
3000	15.071	75.862	63.997	35.593	30.698	7.826	– .570
3100	15.127	76.357	64.388	37.103	30.716	7.064	– .498
3200	15.181	76.838	64.770	38.619	30.733	6.300	– .430
3300	15.232	77.306	65.142	40.139	30.750	5.535	– .367
3400	15.280	77.761	65.507	41.665	30.768	4.771	– .307
3500	15.326	78.205	65.863	43.195	30.786	4.004	– .250
3600	15.371	78.637	66.212	44.730	30.804	3.240	– .197
3700	15.414	79.059	66.554	46.269	30.822	2.475	– .146
3800	15.455	79.471	66.888	47.813	30.840	1.711	– .098
3900	15.495	79.873	67.216	49.360	30.857	.944	– .053
4000	15.533	80.265	67.537	50.912	30.875	.178	– .010
4100	15.570	80.649	67.852	52.467	30.891	– .594	.032
4200	15.606	81.025	68.162	54.026	30.907	– 1.357	.071
4300	15.641	81.393	68.465	55.588	30.922	– 2.128	.108
4400	15.675	81.753	68.763	57.154	30.936	– 2.897	.144
4500	15.708	82.105	69.056	58.723	30.950	– 3.670	.178
4600	15.740	82.451	69.343	60.296	30.963	– 4.436	.211
4700	15.772	82.790	69.626	61.871	30.975	– 5.204	.242
4800	15.803	83.122	69.903	63.450	30.985	– 5.973	.272
4900	15.833	83.448	70.176	65.032	30.995	– 6.743	.301
5000	15.862	83.768	70.445	66.616	31.003	– 7.516	.328

Hydrogen Cyanide (HCN) (Continued)

T, °K.	C_p° (cal mole⁻¹ deg⁻¹)	S°	$-(F^\circ - H^\circ_{298})/T$	$H^\circ - H^\circ_{298}$ (kcal mole⁻¹)	ΔH_f°	ΔF_f°	$Log_{10}K_P$
5100	15.891	84.083	70.709	68.204	31.010	− 8.289	.355
5200	15.919	84.392	70.969	69.795	31.016	− 9.058	.381
5300	15.947	84.695	71.226	71.388	31.020	− 9.827	.405
5400	15.974	84.993	71.478	72.984	31.023	− 10.596	.429
5500	16.001	85.287	71.726	74.583	31.025	− 11.368	.452
5600	16.027	85.575	71.971	76.184	31.025	− 12.141	.474
5700	16.053	85.859	72.212	77.788	31.024	− 12.913	.495
5800	16.078	86.139	72.450	79.395	31.020	− 13.684	.516
5900	16.102	86.414	72.684	81.004	31.015	− 14.451	.535
6000	16.126	86.684	72.915	82.615	31.009	− 15.227	.555

Cyano (CN)

(Ideal Gas)(mol. wt. = 26.02)Dec. 31, 1962.

T, °K.	C_p° (cal mole⁻¹ deg⁻¹)	S°	$-(F^\circ - H^\circ_{298})/T$	$H^\circ - H^\circ_{298}$ (kcal mole⁻¹)	ΔH_f°	ΔF_f°	$Log_{10}K_P$
0	.000	.000	INFINITE	− 2.073	108.215	108.215	INFINITE
100	6.956	40.804	54.593	− 1.379	108.553	106.399	− 232.523
200	6.957	45.626	49.043	− .683	108.819	104.136	− 113.789
298	6.969	48.406	48.406	.000	109.000	101.796	− 74.615
300	6.969	48.449	48.406	.013	109.002	101.751	− 74.122
400	7.029	50.460	48.680	.712	109.107	99.317	− 54.262
500	7.155	52.041	49.199	1.421	109.145	96.863	− 42.337
600	7.326	53.360	49.786	2.145	109.135	94.407	− 34.386
700	7.510	54.503	50.380	2.887	109.088	91.956	− 28.709
800	7.689	55.518	50.960	3.647	109.018	89.514	− 24.453
900	7.852	56.433	51.518	4.424	108.928	87.080	− 21.145
1000	7.996	57.268	52.052	5.216	108.828	84.657	− 18.501
1100	8.123	58.036	52.561	6.022	108.717	82.245	− 16.340
1200	8.236	58.748	53.048	6.840	108.598	79.845	− 14.541
1300	8.337	59.411	53.512	7.669	108.473	77.453	− 13.020
1400	8.432	60.033	53.956	8.508	108.345	75.073	− 11.719
1500	8.522	60.618	54.381	9.355	108.214	72.701	− 10.592
1600	8.612	61.170	54.788	10.212	108.083	70.336	− 9.607
1700	8.702	61.695	55.179	11.078	107.953	67.981	− 8.739
1800	8.794	62.195	55.555	11.953	107.824	65.634	− 7.969
1900	8.890	62.673	55.917	12.837	107.700	63.294	− 7.280
2000	8.989	63.132	56.266	13.731	107.580	60.960	− 6.661
2100	9.091	63.573	56.604	14.635	107.466	58.633	− 6.102
2200	9.196	63.998	56.930	15.549	107.356	56.308	− 5.593
2300	9.302	64.409	57.247	16.474	107.254	53.992	− 5.130
2400	9.411	64.807	57.553	17.410	107.160	51.678	− 4.706
2500	9.520	65.194	57.851	18.356	107.073	49.367	− 4.315
2600	9.628	65.569	58.141	19.314	106.993	47.061	− 3.956
2700	9.735	65.935	58.423	20.282	106.921	44.756	− 3.623
2800	9.840	66.291	58.698	21.260	106.858	42.455	− 3.314
2900	9.942	66.638	58.965	22.250	106.803	40.157	− 3.026
3000	10.040	66.976	59.227	23.249	106.754	37.850	− 2.758

T, °K.	C_p° (cal mole⁻¹ deg⁻¹)	S° (cal mole⁻¹ deg⁻¹)	$-(F^\circ - H^\circ_{298})/T$ (cal mole⁻¹ deg⁻¹)	$H^\circ - H^\circ_{298}$ (kcal mole⁻¹)	ΔH_f° (kcal mole⁻¹)	ΔF_f° (kcal mole⁻¹)	$Log_{10}K_P$
3100	10.135	67.307	59.482	24.257	106.714	35.564	-2.507
3200	10.225	67.630	59.732	25.276	106.680	33.269	-2.272
3300	10.310	67.946	59.976	26.302	106.652	30.974	-2.051
3400	10.390	68.255	60.215	27.337	106.631	28.683	-1.844
3500	10.464	68.558	60.449	28.380	106.615	26.387	-1.648
3600	10.534	68.853	60.678	29.430	106.604	24.096	-1.463
3700	10.598	69.143	60.903	30.487	106.597	21.805	-1.288
3800	10.657	69.426	61.124	31.550	106.595	19.515	-1.122
3900	10.710	69.704	61.340	32.618	106.596	17.225	-.965
4000	10.759	69.976	61.553	33.691	106.601	14.933	-.816
4100	10.802	70.242	61.761	34.770	106.608	12.639	-.674
4200	10.841	70.503	61.966	35.852	106.616	10.349	-.539
4300	10.876	70.758	62.168	36.938	106.626	8.056	-.409
4400	10.906	71.008	62.366	38.027	106.637	5.763	-.286
4500	10.933	71.254	62.561	39.119	106.648	3.468	-.168
4600	10.956	71.494	62.752	40.213	106.660	1.178	-.056
4700	10.975	71.730	62.941	41.310	106.673	-1.116	.052
4800	10.992	71.961	63.126	42.408	106.683	-3.407	.155
4900	11.005	72.188	63.309	43.508	106.694	-5.703	.254
5000	11.016	72.411	63.489	44.609	106.704	-7.995	.349
5100	11.024	72.629	63.666	45.711	106.711	-10.296	.441
5200	11.031	72.843	63.840	46.814	106.717	-12.586	.529
5300	11.035	73.053	64.012	47.917	106.722	-14.880	.614
5400	11.038	73.260	64.182	49.021	106.724	-17.173	.695
5500	11.039	73.462	64.348	50.125	106.723	-19.466	.773
5600	11.039	73.661	64.513	51.229	106.721	-21.762	.849
5700	11.037	73.856	64.675	52.332	106.715	-24.059	.922
5800	11.035	74.048	64.835	53.436	106.707	-26.352	.993
5900	11.031	74.237	64.993	54.539	106.696	-28.644	1.061
6000	11.027	74.422	65.149	55.642	106.682	-30.942	1.127

CYANOGEN (C_2N_2)

(Ideal Gas)(mol. wt. = 52.04)Mar. 31, 1961.

T, °K.	C_p° (cal mole⁻¹ deg⁻¹)	S° (cal mole⁻¹ deg⁻¹)	$-(F^\circ - H^\circ_{298})/T$ (cal mole⁻¹ deg⁻¹)	$H^\circ - H^\circ_{298}$ (kcal mole⁻¹)	ΔH_f° (kcal mole⁻¹)	ΔF_f° (kcal mole⁻¹)	$Log_{10}K_P$
0	.000	.000	INFINITE	-3.018	73.428	73.428	INFINITE
100	8.700	45.654	68.408	-2.275	73.450	72.742	-158.970
200	11.690	52.663	58.898	-1.247	73.626	71.978	-78.650
298	13.563	57.711	57.711	.000	73.870	71.117	-52.128
300	13.590	57.795	57.711	.025	73.874	71.100	-51.794
400	14.771	61.878	58.260	1.447	74.107	70.140	-38.321
500	15.615	65.269	59.332	2.968	74.287	69.125	-30.213
600	16.305	68.178	60.569	4.565	74.415	68.079	-24.797
700	16.899	70.737	61.843	6.226	74.499	67.017	-20.923
800	17.415	73.028	63.100	7.942	74.554	65.945	-18.015
900	17.858	75.106	64.321	9.707	74.586	64.865	-15.751
1000	18.237	77.007	65.496	11.512	74.604	63.784	-13.939

Cyanogen (C_2N_2) (Continued)

T, °K.	C_p°	S°	$-(F^\circ - H^\circ_{298})/T$	$H^\circ - H^\circ_{298}$	ΔH_f°	ΔF_f°	$\text{Log}_{10}K_p$
	cal mole⁻¹ deg⁻¹			kcal mole⁻¹			
1100	18.559	78.761	66.623	13.352	74.610	62.700	– 12.457
1200	18.833	80.388	67.703	15.222	74.608	61.619	– 11.222
1300	19.066	81.905	68.738	17.117	74.594	60.536	– 10.177
1400	19.264	83.325	69.729	19.034	74.578	59.458	– 9.281
1500	19.434	84.660	70.681	20.969	74.556	58.378	– 8.505
1600	19.579	85.919	71.594	22.920	74.531	57.299	– 7.826
1700	19.705	87.110	72.472	24.885	74.505	56.222	– 7.228
1800	19.815	88.240	73.317	26.861	74.475	55.147	– 6.695
1900	19.910	89.314	74.131	28.847	74.444	54.077	– 6.220
2000	19.993	90.337	74.916	30.842	74.412	53.006	– 5.792
2100	20.066	91.314	75.674	32.845	74.379	51.938	– 5.405
2200	20.131	92.249	76.406	34.855	74.342	50.865	– 5.053
2300	20.188	93.145	77.115	36.871	74.306	49.803	– 4.732
2400	20.239	94.006	77.801	38.893	74.267	48.738	– 4.438
2500	20.285	94.833	78.465	40.919	74.227	47.672	– 4.167
2600	20.326	95.629	79.110	42.949	74.183	46.612	– 3.918
2700	20.363	96.397	79.736	44.984	74.140	45.549	– 3.687
2800	20.396	97.138	80.345	47.022	74.096	44.496	– 3.473
2900	20.426	97.855	80.936	49.063	74.047	43.439	– 3.273
3000	20.454	98.548	81.512	51.107	73.997	42.385	– 3.088
3100	20.479	99.219	82.072	53.154	73.947	41.333	– 2.914
3200	20.502	99.869	82.618	55.203	73.893	40.279	– 2.751
3300	20.522	100.500	83.151	57.254	73.836	39.228	– 2.598
3400	20.542	101.113	83.670	59.307	73.778	38.183	– 2.454
3500	20.559	101.709	84.177	61.362	73.716	37.131	– 2.318
3600	20.575	102.288	84.672	63.419	73.653	36.089	– 2.191
3700	20.590	102.852	85.156	65.477	73.587	35.047	– 2.070
3800	20.604	103.402	85.629	67.537	73.519	34.012	– 1.956
3900	20.617	103.937	86.091	69.598	73.447	32.973	– 1.848
4000	20.629	104.459	86.544	71.660	73.373	31.937	– 1.745
4100	20.640	104.969	86.987	73.724	73.296	30.895	– 1.647
4200	20.651	105.466	87.421	75.788	73.215	29.871	– 1.554
4300	20.660	105.952	87.847	77.854	73.130	28.835	– 1.465
4400	20.669	106.427	88.264	79.920	73.043	27.806	– 1.381
4500	20.678	106.892	88.672	81.988	72.953	26.774	– 1.300
4600	20.686	107.346	89.073	84.056	72.859	25.753	– 1.223
4700	20.693	107.791	89.467	86.125	72.763	24.732	– 1.150
4800	20.700	108.227	89.853	88.195	72.661	23.714	– 1.080
4900	20.707	108.654	90.233	90.265	72.556	22.693	– 1.012
5000	20.713	109.072	90.605	92.336	72.448	21.676	– .947
5100	20.719	109.483	90.971	94.407	72.333	20.652	– .885
5200	20.724	109.885	91.331	96.480	72.217	19.645	– .826
5300	20.730	110.280	91.685	98.552	72.096	18.636	– .768
5400	20.734	110.667	92.033	100.626	71.972	17.635	– .714
5500	20.739	111.048	92.375	102.699	71.841	16.628	– .661

T, °K.	C_p°	S°	$-(F^\circ - H^\circ_{298})/T$	$H^\circ - H^\circ_{298}$	ΔH_f°	ΔF_f°	$Log_{10} K_P$
	cal mole⁻¹ deg⁻¹			kcal mole⁻¹			
5600	20.744	111.422	92.712	104.773	71.708	15.621	− .610
5700	20.748	111.789	93.044	106.848	71.570	14.617	− .560
5800	20.752	112.150	93.370	108.923	71.427	13.620	− .513
5900	20.756	112.504	93.691	110.998	71.279	12.631	− .468
6000	20.759	112.853	94.008	113.074	71.128	11.630	− .424

METHYLIDYNE (CH)

(Ideal Gas)(mol. wt. = 13.02)Mar. 31, 1961.

T, °K.	C_p°	S°	$-(F^\circ - H^\circ_{298})/T$	$H^\circ - H^\circ_{298}$	ΔH_f°	ΔF_f°	$Log_{10} K_P$
	cal mole⁻¹ deg⁻¹			kcal mole⁻¹			
0	.000	.000	INFINITE	− 2.086	141.183	141.183	INFINITE
100	6.994	36.101	49.922	− 1.382	141.494	139.124	− 304.042
200	6.972	40.939	44.360	− .684	141.813	136.621	− 149.285
298	6.972	43.722	43.722	.000	142.006	134.028	− 98.240
300	6.972	43.765	43.722	.013	142.008	133.978	− 97.598
400	6.985	45.772	43.996	.711	142.113	131.283	− 71.726
500	7.026	47.335	44.513	1.411	142.145	128.571	− 56.196
600	7.108	48.622	45.094	2.117	142.123	125.857	− 45.841
700	7.224	49.726	45.679	2.833	142.063	123.151	− 38.448
800	7.361	50.700	46.247	3.563	141.980	120.455	− 32.905
900	7.508	51.575	46.791	4.306	141.881	117.769	− 28.597
1000	7.654	52.374	47.310	5.064	141.774	115.095	− 25.153
1100	7.794	53.110	47.804	5.837	141.660	112.432	− 22.337
1200	7.924	53.794	48.275	6.623	141.544	109.781	− 19.993
1300	8.044	54.433	48.724	7.421	141.421	107.139	− 18.011
1400	8.153	55.033	49.154	8.231	141.298	104.508	− 16.314
1500	8.252	55.599	49.565	9.052	141.171	101.885	− 14.844
1600	8.342	56.135	49.959	9.881	141.042	99.268	− 13.559
1700	8.423	56.643	50.337	10.720	140.913	96.661	− 12.426
1800	8.496	57.126	50.701	11.566	140.781	94.062	− 11.420
1900	8.563	57.588	51.051	12.419	140.649	91.471	− 10.521
2000	8.624	58.028	51.389	13.278	140.516	88.886	− 9.713
2100	8.680	58.451	51.716	14.143	140.383	86.309	− 8.982
2200	8.731	58.856	52.031	15.014	140.246	83.735	− 8.318
2300	8.779	59.245	52.336	15.889	140.110	81.171	− 7.713
2400	8.823	59.619	52.632	16.770	139.972	78.612	− 7.158
2500	8.864	59.980	52.919	17.654	139.833	76.056	− 6.648
2600	8.902	60.329	53.197	18.542	139.692	73.509	− 6.179
2700	8.938	60.665	53.467	19.434	139.550	70.964	− 5.744
2800	8.971	60.991	53.730	20.330	139.408	68.427	− 5.341
2900	9.003	61.306	53.986	21.228	139.264	65.896	− 4.966
3000	9.033	61.612	54.235	22.130	139.119	63.367	− 4.616
3100	9.062	61.909	54.478	23.035	138.974	60.845	− 4.289
3200	9.089	62.197	54.715	23.943	138.826	58.328	− 3.983
3300	9.115	62.477	54.946	24.853	138.677	55.813	− 3.696
3400	9.140	62.749	55.171	25.766	138.527	53.305	− 3.426
3500	9.164	63.015	55.392	26.681	138.375	50.799	− 3.172

Methylidyne (CH) (Continued)

T, °K.	C_p°	S°	$-(F^\circ - H^\circ_{298})/T$	$H^\circ - H^\circ_{298}$	ΔH_f°	ΔF_f°	$Log_{10}K_P$
	cal mole⁻¹ deg⁻¹			kcal mole⁻¹			
3600	9.187	63.273	55.607	27.598	138.222	48.299	− 2.932
3700	9.210	63.525	55.818	28.518	138.066	45.804	− 2.705
3800	9.231	63.771	56.024	29.440	137.911	43.315	− 2.491
3900	9.252	64.011	56.225	30.364	137.752	40.828	− 2.288
4000	9.273	64.246	56.423	31.291	137.592	38.345	− 2.095
4100	9.293	64.475	56.617	32.219	137.430	35.863	− 1.912
4200	9.312	64.699	56.806	33.149	137.266	33.391	− 1.737
4300	9.331	64.918	56.993	34.081	137.100	30.918	− 1.571
4400	9.350	65.133	57.175	35.015	136.932	28.450	− 1.413
4500	9.368	65.343	57.354	35.951	136.762	25.984	− 1.262
4600	9.385	65.550	57.530	36.889	136.590	23.527	− 1.118
4700	9.403	65.752	57.703	37.828	136.417	21.071	− .980
4800	9.420	65.950	57.873	38.769	136.240	18.620	− .848
4900	9.437	66.144	58.040	39.712	136.063	16.171	− .721
5000	9.453	66.335	58.204	40.657	135.883	13.725	− .600
5100	9.470	66.522	58.365	41.603	135.699	11.281	− .483
5200	9.486	66.706	58.524	42.551	135.515	8.844	− .372
5300	9.502	66.887	58.680	43.500	135.329	6.413	− .264
5400	9.518	67.065	58.833	44.451	135.140	3.982	− .161
5500	9.533	67.240	58.985	45.404	134.948	1.555	− .062
5600	9.549	67.412	59.133	46.358	134.754	− .870	.034
5700	9.564	67.581	59.280	47.314	134.558	− 3.293	.126
5800	9.579	67.747	59.425	48.271	134.359	− 5.710	.215
5900	9.594	67.911	59.567	49.229	134.158	− 8.118	.301
6000	9.609	68.073	59.708	50.190	133.955	− 10.534	.384

Methylene (CH_2)

(Ideal Gas)(mol. wt. = 14.03)Dec. 31, 1962.

T, °K.	C_p°	S°	$-(F^\circ - H^\circ_{298})/T$	$H^\circ - H^\circ_{298}$	ΔH_f°	ΔF_f°	$Log_{10}K_P$
	cal mole⁻¹ deg⁻¹			kcal mole⁻¹			
0	.000	.000	INFINITE	− 2.096	95.180	95.180	INFINITE
100	6.956	35.555	49.633	− 1.408	95.095	93.999	− 205.426
200	7.040	40.388	43.939	− .710	95.112	92.882	− 101.492
298	7.492	43.271	43.271	.000	95.000	91.809	− 67.294
300	7.504	43.317	43.271	.014	94.997	91.789	− 66.865
400	8.167	45.566	43.573	.797	94.840	90.743	− 49.577
500	8.799	47.458	44.165	1.646	94.671	89.737	− 39.222
600	9.357	49.112	44.855	2.554	94.501	88.766	− 32.331
700	9.857	50.593	45.571	3.515	94.335	87.824	− 27.419
800	10.317	51.939	46.284	4.524	94.179	86.905	− 23.740
900	10.742	53.179	46.982	5.578	94.034	86.004	− 20.884
1000	11.133	54.332	47.660	6.672	93.904	85.118	− 18.602
1100	11.489	55.410	48.316	7.803	93.786	84.245	− 16.737
1200	11.810	56.423	48.950	8.968	93.681	83.384	− 15.186
1300	12.098	57.380	49.562	10.164	93.584	82.528	− 13.874
1400	12.356	58.286	50.153	11.387	93.497	81.683	− 12.751
1500	12.585	59.147	50.724	12.634	93.414	80.843	− 11.778

T, °K.	C_p° (cal mole⁻¹ deg⁻¹)	S° (cal mole⁻¹ deg⁻¹)	$-(F^\circ - H^\circ_{298})/T$ (cal mole⁻¹ deg⁻¹)	$H^\circ - H^\circ_{298}$ (kcal mole⁻¹)	ΔH_f° (kcal mole⁻¹)	ΔF_f° (kcal mole⁻¹)	$Log_{10}K_P$
1600	12.789	59.966	51.276	13.903	93.335	80.004	− 10.928
1700	12.971	60.747	51.811	15.191	93.262	79.175	− 10.178
1800	13.132	61.493	52.328	16.497	93.192	78.348	− 9.512
1900	13.276	62.207	52.829	17.817	93.124	77.526	− 8.917
2000	13.405	62.891	53.315	19.151	93.058	76.706	− 8.382
2100	13.520	63.548	53.787	20.498	92.994	75.892	− 7.898
2200	13.623	64.179	54.245	21.855	92.928	75.077	− 7.458
2300	13.716	64.787	54.690	23.222	92.864	74.269	− 7.057
2400	13.799	65.372	55.123	24.598	92.798	73.462	− 6.689
2500	13.875	65.937	55.545	25.982	92.731	72.655	− 6.351
2600	13.943	66.483	55.955	27.373	92.663	71.855	− 6.040
2700	14.005	67.010	56.355	28.770	92.593	71.054	− 5.751
2800	14.062	67.521	56.744	30.174	92.523	70.258	− 5.484
2900	14.113	68.015	57.125	31.582	92.449	69.467	− 5.235
3000	14.160	68.494	57.496	32.996	92.374	68.673	− 5.003
3100	14.204	68.959	57.858	34.414	92.298	67.885	− 4.786
3200	14.243	69.411	58.212	35.837	92.219	67.099	− 4.582
3300	14.280	69.850	58.558	37.263	92.136	66.314	− 4.392
3400	14.313	70.277	58.896	38.693	92.052	65.533	− 4.212
3500	14.344	70.692	59.228	40.125	91.962	64.754	− 4.043
3600	14.373	71.096	59.552	41.561	91.871	63.977	− 3.884
3700	14.400	71.491	59.869	43.000	91.775	63.205	− 3.733
3800	14.424	71.875	60.180	44.441	91.677	62.436	− 3.591
3900	14.447	72.250	60.485	45.885	91.574	61.669	− 3.456
4000	14.469	72.616	60.783	47.331	91.468	60.904	− 3.327
4100	14.488	72.973	61.076	48.778	91.357	60.138	− 3.205
4200	14.507	73.323	61.364	50.228	91.242	59.382	− 3.090
4300	14.524	73.664	61.646	51.680	91.123	58.622	− 2.979
4400	14.541	73.998	61.923	53.133	91.000	57.866	− 2.874
4500	14.556	74.325	62.195	54.588	90.872	57.111	− 2.774
4600	14.570	74.646	62.462	56.044	90.740	56.367	− 2.678
4700	14.584	74.959	62.725	57.502	90.604	55.622	− 2.586
4800	14.597	75.266	62.983	58.961	90.463	54.880	− 2.499
4900	14.609	75.567	63.236	60.421	90.318	54.140	− 2.415
5000	14.620	75.863	63.486	61.883	90.169	53.401	− 2.334
5100	14.631	76.152	63.732	63.345	90.013	52.665	− 2.257
5200	14.641	76.436	63.973	64.809	89.855	51.934	− 2.183
5300	14.650	76.715	64.211	66.273	89.691	51.212	− 2.112
5400	14.659	76.989	64.445	67.739	89.523	50.485	− 2.043
5500	14.668	77.258	64.676	69.205	89.349	49.764	− 1.977
5600	14.676	77.523	64.903	70.672	89.171	49.043	− 1.914
5700	14.684	77.783	65.126	72.140	88.988	48.327	− 1.853
5800	14.691	78.038	65.347	73.609	88.800	47.615	− 1.794
5900	14.698	78.289	65.564	75.079	88.608	46.911	− 1.738
6000	14.705	78.536	65.778	76.549	88.411	46.199	− 1.683

METHYL (CH_3)

(Ideal Gas)(mol. wt. = 15.04)Dec. 31, 1962.

	cal mole⁻¹ deg⁻¹			kcal mole⁻¹			
T,°K.	C_p°	S°	$-(F^\circ - H^\circ_{298})/T$	$H^\circ - H^\circ_{298}$	ΔH_f°	ΔF_f°	$Log_{10}K_p$
0	.000	.000	INFINITE	− 2.423	32.805	32.805	INFINITE
100	7.951	37.235	53.517	− 1.628	32.447	32.403	− 70.813
200	8.142	42.781	46.916	− .827	32.266	32.410	− 35.414
298	8.773	46.137	46.137	.000	31.940	32.546	− 23.856
300	8.788	46.191	46.137	.016	31.933	32.550	− 23.711
400	9.669	48.838	46.492	.938	31.568	32.811	− 17.926
500	10.595	51.095	47.192	1.951	31.213	33.163	− 14.495
600	11.499	53.107	48.013	3.056	30.890	33.583	− 12.232
700	12.359	54.945	48.874	4.250	30.606	34.056	− 10.632
800	13.157	56.648	49.740	5.526	30.364	34.565	− 9.442
900	13.884	58.240	50.597	6.879	30.162	35.103	− 8.524
1000	14.535	59.737	51.437	8.300	30.000	35.660	− 7.793
1100	15.110	61.150	52.256	9.783	29.871	36.232	− 7.198
1200	15.616	62.487	53.054	11.320	29.771	36.817	− 6.705
1300	16.057	63.755	53.829	12.904	29.690	37.406	− 6.288
1400	16.443	64.959	54.581	14.530	29.629	38.004	− 5.932
1500	16.779	66.106	55.311	16.191	29.577	38.605	− 5.624
1600	17.073	67.198	56.020	17.884	29.533	39.204	− 5.355
1700	17.331	68.241	56.709	19.605	29.499	39.812	− 5.118
1800	17.557	69.238	57.377	21.349	29.469	40.418	− 4.907
1900	17.756	70.193	58.027	23.115	29.444	41.029	− 4.719
2000	17.932	71.108	58.658	24.900	29.421	41.639	− 4.550
2100	18.088	71.987	59.272	26.701	29.399	42.251	− 4.397
2200	18.227	72.832	59.870	28.517	29.376	42.862	− 4.258
2300	18.351	73.645	60.451	30.346	29.355	43.478	− 4.131
2400	18.462	74.428	61.017	32.187	29.330	44.093	− 4.015
2500	18.562	75.184	61.569	34.038	29.303	44.705	− 3.908
2600	18.652	75.914	62.106	35.899	29.275	45.324	− 3.810
2700	18.733	76.619	62.631	37.768	29.243	45.940	− 3.718
2800	18.807	77.302	63.143	39.645	29.210	46.558	− 3.634
2900	18.874	77.963	63.642	41.529	29.173	47.182	− 3.556
3000	18.934	78.604	64.131	43.419	29.132	47.801	− 3.482
3100	18.990	79.226	64.608	45.316	29.091	48.424	− 3.414
3200	19.041	79.829	65.074	47.217	29.043	49.049	− 3.350
3300	19.088	80.416	65.530	49.124	28.991	49.674	− 3.290
3400	19.131	80.986	65.976	51.035	28.937	50.301	− 3.233
3500	19.170	81.541	66.413	52.950	28.875	50.931	− 3.180
3600	19.207	82.082	66.841	54.869	28.811	51.561	− 3.130
3700	19.241	82.609	67.260	56.791	28.739	52.196	− 3.083
3800	19.272	83.122	67.671	58.717	28.664	52.833	− 3.038
3900	19.301	83.623	68.073	60.646	28.582	53.471	− 2.996
4000	19.328	84.112	68.468	62.577	28.496	54.110	− 2.956

T, °K.	C_p° (cal mole⁻¹ deg⁻¹)	S°	$-(F^\circ - H^\circ_{298})/T$	$H^\circ - H^\circ_{298}$ (kcal mole⁻¹)	ΔH_f°	ΔF_f°	$Log_{10}K_P$
4100	19.354	84.590	68.855	64.511	28.404	54.749	− 2.918
4200	19.377	85.057	69.236	66.448	28.305	55.397	− 2.882
4300	19.399	85.513	69.609	68.387	28.200	56.039	− 2.848
4400	19.420	85.959	69.975	70.327	28.090	56.686	− 2.815
4500	19.439	86.396	70.336	72.270	27.973	57.333	− 2.784
4600	19.457	86.823	70.689	74.215	27.852	57.993	− 2.755
4700	19.474	87.242	71.037	76.162	27.723	58.651	− 2.727
4800	19.490	87.652	71.379	78.110	27.589	59.311	− 2.700
4900	19.505	88.054	71.715	80.060	27.449	59.974	− 2.675
5000	19.519	88.448	72.046	82.011	27.303	60.635	− 2.650
5100	19.533	88.835	72.371	83.964	27.149	61.303	− 2.627
5200	19.545	89.214	72.692	85.917	26.991	61.973	− 2.605
5300	19.557	89.587	73.007	87.873	26.826	62.654	− 2.583
5400	19.569	89.952	73.317	89.829	26.654	63.327	− 2.563
5500	19.579	90.311	73.623	91.786	26.476	64.009	− 2.543
5600	19.589	90.664	73.924	93.745	26.292	64.690	− 2.525
5700	19.599	91.011	74.221	95.704	26.102	65.377	− 2.507
5800	19.608	91.352	74.513	97.664	25.904	66.066	− 2.489
5900	19.617	91.687	74.802	99.626	25.701	66.766	− 2.473
6000	19.625	92.017	75.086	101.588	25.492	67.455	− 2.457

METHANE (CH_4)

(Ideal Gas)(mol. wt. = 16.04)Mar. 31, 1961.

T, °K.	C_p° (cal mole⁻¹ deg⁻¹)	S°	$-(F^\circ - H^\circ_{298})/T$	$H^\circ - H^\circ_{298}$ (kcal mole⁻¹)	ΔH_f°	ΔF_f°	$Log_{10}K_P$
0	.000	.000	INFINITE	− 2.396	− 15.991	− 15.991	INFINITE
100	7.949	35.706	51.716	− 1.601	− 16.728	− 15.400	33.656
200	8.001	41.222	45.247	− .805	− 17.216	− 13.909	15.198
298	8.518	44.490	44.490	.000	− 17.895	− 12.145	8.902
300	8.535	44.543	44.490	.016	− 17.909	− 12.110	8.822
400	9.680	47.144	44.837	.923	− 18.636	− 10.066	5.500
500	11.076	49.453	45.533	1.960	− 19.316	− 7.845	3.429
600	12.483	51.597	46.367	3.138	− 19.916	− 5.493	2.001
700	13.813	53.622	47.260	4.454	− 20.429	− 3.046	.951
800	15.041	55.548	48.176	5.897	− 20.857	− .533	.146
900	16.157	57.385	49.098	7.458	− 21.207	2.029	− .493
1000	17.160	59.141	50.016	9.125	− 21.482	4.625	− 1.011
1100	18.052	60.819	50.922	10.887	− 21.696	7.247	− 1.440
1200	18.842	62.424	51.814	12.732	− 21.854	9.887	− 1.801
1300	19.538	63.960	52.690	14.652	− 21.971	12.535	− 2.107
1400	20.150	65.431	53.548	16.637	− 22.050	15.195	− 2.372
1500	20.688	66.840	54.387	18.679	− 22.104	17.859	− 2.602
1600	21.161	68.191	55.208	20.772	− 22.137	20.520	− 2.803
1700	21.579	69.486	56.010	22.910	− 22.148	23.189	− 2.981
1800	21.947	70.730	56.794	25.086	− 22.144	25.854	− 3.139
1900	22.273	71.926	57.559	27.298	− 22.127	28.522	− 3.281
2000	22.562	73.076	58.306	29.540	− 22.099	31.187	− 3.408

ACETYLENE (C_2H_2) (Continued)

T, °K.	C_p° (cal mole⁻¹ deg⁻¹)	S°	$-(F^\circ - H^\circ_{298})/T$	$H^\circ - H^\circ_{298}$ (kcal mole⁻¹)	ΔH°_f	ΔF°_f	$\text{Log}_{10} K_P$
5100	22.129	96.374	78.038	93.510	51.881	− 9.935	.426
5200	22.174	96.804	78.395	95.725	51.832	− 11.144	.468
5300	22.219	97.227	78.746	97.945	51.780	− 12.348	.509
5400	22.263	97.642	79.093	100.169	51.724	− 13.559	.549
5500	22.309	98.051	79.434	102.397	51.663	− 14.767	.587
5600	22.349	98.454	79.770	104.630	51.601	− 15.977	.624
5700	22.393	98.850	80.101	106.867	51.534	− 17.188	.659
5800	22.433	99.239	80.428	109.108	51.463	− 18.394	.693
5900	22.474	99.623	80.750	111.354	51.390	− 19.588	.726
6000	22.521	100.001	81.067	113.603	51.313	− 20.802	.758

ETHYLENE (C_2H_4)
(Ideal Gas)(mol. wt. = 28.05)Sept. 30, 1965.

T, °K.	C_p° (cal mole⁻¹ deg⁻¹)	S°	$-(F^\circ - H^\circ_{298})/T$	$H^\circ - H^\circ_{298}$ (kcal mole⁻¹)	ΔH°_f	ΔF°_f	$\text{Log}_{10} K_P$
0	.000	.000	INFINITE	− 2.514	14.578	14.578	INFINITE
100	7.952	43.125	60.316	− 1.719	13.827	14.434	− 31.544
200	8.451	48.721	53.267	− .909	13.275	15.227	− 16.638
298	10.250	52.396	52.396	.000	12.540	16.338	− 11.975
300	10.292	52.459	52.396	.019	12.525	16.361	− 11.918
400	12.679	55.745	52.828	1.167	11.793	17.752	− 9.699
500	14.933	58.821	53.722	2.550	11.140	19.319	− 8.444
600	16.889	61.721	54.816	4.143	10.577	21.008	− 7.652
700	18.574	64.454	55.999	5.918	10.098	22.788	− 7.114
800	20.039	67.033	57.219	7.851	9.701	24.628	− 6.728
900	21.320	69.468	58.446	9.920	9.372	26.514	− 6.438
1000	22.443	71.774	59.665	12.109	9.113	28.431	− 6.213
1100	23.427	73.960	60.866	14.404	8.910	30.373	− 6.034
1200	24.290	76.036	62.044	16.791	8.757	32.334	− 5.889
1300	25.044	78.011	63.197	19.258	8.638	34.302	− 5.766
1400	25.706	79.892	64.323	21.797	8.557	36.282	− 5.664
1500	26.285	81.686	65.421	24.397	8.497	38.266	− 5.575
1600	26.794	83.399	66.492	27.051	8.455	40.246	− 5.497
1700	27.242	85.037	67.535	29.753	8.435	42.237	− 5.430
1800	27.636	86.605	68.551	32.498	8.428	44.224	− 5.369
1900	27.986	88.109	69.541	35.279	8.433	46.215	− 5.316
2000	28.296	89.552	70.506	38.094	8.448	48.204	− 5.267
2100	28.571	90.940	71.446	40.937	8.469	50.192	− 5.223
2200	28.818	92.275	72.362	43.807	8.493	52.174	− 5.183
2300	29.038	93.561	73.256	46.700	8.524	54.163	− 5.146
2400	29.236	94.801	74.128	49.614	8.554	56.148	− 5.113
2500	29.414	95.998	74.979	52.546	8.584	58.124	− 5.081
2600	29.575	97.155	75.810	55.496	8.616	60.109	− 5.052
2700	29.721	98.274	76.621	58.461	8.647	62.085	− 5.025
2800	29.853	99.357	77.414	61.440	8.678	64.064	− 5.000
2900	29.973	100.407	78.189	64.431	8.705	66.047	− 4.977
3000	30.083	101.425	78.947	67.434	8.730	68.019	− 4.955

T, °K.	C_p° (cal mole⁻¹ deg⁻¹)	S°	$-(F^\circ - H_{298}^\circ)/T$	$H^\circ - H_{298}^\circ$ (kcal mole⁻¹)	ΔH_f°	ΔF_f°	$\mathrm{Log}_{10} K_P$
3100	30.184	102.413	79.688	70.447	8.755	69.996	− 4.934
3200	30.276	103.373	80.413	73.470	8.774	71.971	− 4.915
3300	30.360	104.305	81.123	76.502	8.788	73.947	− 4.897
3400	30.438	105.213	81.818	79.542	8.800	75.920	− 4.880
3500	30.510	106.096	82.499	82.590	8.804	77.894	− 4.864
3600	30.577	106.957	83.167	85.644	8.804	79.864	− 4.848
3700	30.638	107.795	83.821	88.705	8.795	81.844	− 4.834
3800	30.695	108.613	84.463	91.772	8.784	83.822	− 4.821
3900	30.748	109.411	85.092	94.844	8.762	85.798	− 4.808
4000	30.797	110.190	85.710	97.921	8.735	87.775	− 4.796
4100	30.843	110.951	86.317	101.003	8.701	89.746	− 4.784
4200	30.886	111.695	86.912	104.090	8.658	91.729	− 4.773
4300	30.926	112.422	87.497	107.180	8.606	93.703	− 4.762
4400	30.964	113.134	88.071	110.275	8.549	95.678	− 4.752
4500	30.999	113.830	88.636	113.373	8.481	97.653	− 4.742
4600	31.032	114.512	89.191	116.475	8.407	99.643	− 4.734
4700	31.063	115.180	89.737	119.579	8.323	101.628	− 4.725
4800	31.093	115.834	90.274	122.687	8.231	103.617	− 4.718
4900	31.120	116.475	90.802	125.798	8.132	105.608	− 4.710
5000	31.146	117.104	91.322	128.911	8.023	107.593	− 4.703
5100	31.171	117.721	91.834	132.027	7.903	109.582	− 4.696
5200	31.194	118.327	92.337	135.145	7.777	111.574	− 4.689
5300	31.216	118.921	92.833	138.266	7.642	113.583	− 4.683
5400	31.236	119.505	93.322	141.388	7.496	115.577	− 4.677
5500	31.256	120.078	93.803	144.513	7.341	117.583	− 4.672
5600	31.275	120.641	94.277	147.639	7.177	119.586	− 4.667
5700	31.292	121.195	94.745	150.768	7.004	121.591	− 4.662
5800	31.309	121.740	95.205	153.898	6.820	123.597	− 4.657
5900	31.325	122.275	95.660	157.030	6.628	125.625	− 4.653
6000	31.340	122.802	96.108	160.163	6.427	127.627	− 4.649

CHLORINE, DIATOMIC (Cl_2)

(Reference State—Ideal Gas)(mol. wt. = 70.91)Sept. 30, 1965

T, °K.	C_p° (cal mole⁻¹ deg⁻¹)	S°	$-(F^\circ - H_{298}^\circ)/T$	$H^\circ - H_{298}^\circ$ (kcal mole⁻¹)	ΔH_f°	ΔF_f°	$\mathrm{Log}_{10} K_P$
0	.000	.000	INFINITE	− 2.194	.000	.000	.000
100	7.001	45.150	60.130	− 1.498	.000	.000	.000
200	7.576	50.156	54.016	− .772	.000	.000	.000
298	8.111	53.289	53.289	.000	.000	.000	.000
300	8.119	53.339	53.289	.015	.000	.000	.000
400	8.437	55.724	53.612	.845	.000	.000	.000
500	8.624	57.628	54.231	1.698	.000	.000	.000
600	8.741	59.212	54.933	2.567	.000	.000	.000
700	8.821	60.565	55.643	3.445	.000	.000	.000
800	8.878	61.747	56.334	4.331	.000	.000	.000
900	8.922	62.796	56.995	5.221	.000	.000	.000
1000	8.956	63.737	57.623	6.115	.000	.000	.000

CHLORINE, DIATOMIC (Cl_2) (Continued)

T, °K.	cal mole⁻¹ deg⁻¹ C_p°	S°	$-(F^\circ - H^\circ_{298})/T$	kcal mole⁻¹ $H^\circ - H^\circ_{298}$	ΔH_f°	ΔF_f°	$Log_{10} K_P$
1100	8.985	64.592	58.218	7.012	.000	.000	.000
1200	9.010	65.375	58.782	7.912	.000	.000	.000
1300	9.032	66.097	59.318	8.814	.000	.000	.000
1400	9.051	66.767	59.826	9.718	.000	.000	.000
1500	9.069	67.392	60.310	10.624	.000	.000	.000
1600	9.086	67.978	60.771	11.532	.000	.000	.000
1700	9.102	68.530	61.211	12.441	.000	.000	.000
1800	9.117	69.050	61.633	13.352	.000	.000	.000
1900	9.133	69.544	62.036	14.264	.000	.000	.000
2000	9.149	70.013	62.423	15.179	.000	.000	.000
2100	9.166	70.459	62.795	16.094	.000	.000	.000
2200	9.184	70.886	63.154	17.012	.000	.000	.000
2300	9.203	71.295	63.499	17.931	.000	.000	.000
2400	9.223	71.687	63.832	18.852	.000	.000	.000
2500	9.245	72.064	64.154	19.776	.000	.000	.000
2600	9.268	72.427	64.465	20.701	.000	.000	.000
2700	9.293	72.777	64.766	21.629	.000	.000	.000
2800	9.319	73.116	65.058	22.560	.000	.000	.000
2900	9.346	73.443	65.342	23.493	.000	.000	.000
3000	9.374	73.760	65.617	24.429	.000	.000	.000
3100	9.403	74.068	65.885	25.368	.000	.000	.000
3200	9.432	74.367	66.145	26.310	.000	.000	.000
3300	9.461	74.658	66.399	27.255	.000	.000	.000
3400	9.490	74.941	66.646	28.202	.000	.000	.000
3500	9.518	75.216	66.887	29.153	.000	.000	.000
3600	9.546	75.485	67.122	30.106	.000	.000	.000
3700	9.573	75.747	67.352	31.062	.000	.000	.000
3800	9.598	76.002	67.576	32.020	.000	.000	.000
3900	9.622	76.252	67.795	32.981	.000	.000	.000
4000	9.645	76.496	68.010	33.945	.000	.000	.000
4100	9.666	76.734	68.220	34.910	.000	.000	.000
4200	9.685	76.967	68.425	35.878	.000	.000	.000
4300	9.702	77.195	68.626	36.847	.000	.000	.000
4400	9.718	77.419	68.824	37.818	.000	.000	.000
4500	9.732	77.637	69.017	38.791	.000	.000	.000
4600	9.743	77.851	69.207	39.764	.000	.000	.000
4700	9.754	78.061	69.393	40.739	.000	.000	.000
4800	9.762	78.266	69.576	41.715	.000	.000	.000
4900	9.768	78.468	69.755	42.692	.000	.000	.000
5000	9.774	78.665	69.931	43.669	.000	.000	.000
5100	9.778	78.859	70.105	44.646	.000	.000	.000
5200	9.783	79.049	70.275	45.624	.000	.000	.000
5300	9.787	79.235	70.442	46.603	.000	.000	.000
5400	9.790	79.418	70.607	47.582	.000	.000	.000
5500	9.792	79.598	70.768	48.561	.000	.000	.000

T, °K.	C_p° (cal mole⁻¹ deg⁻¹)	S°	$-(F^\circ - H_{298}^\circ)/T$	$H^\circ - H_{298}^\circ$ (kcal mole⁻¹)	ΔH_f°	ΔF_f°	$Log_{10}K_P$
5600	9.794	79.774	70.928	49.540	.000	.000	.000
5700	9.795	79.947	71.084	50.520	.000	.000	.000
5800	9.797	80.118	71.239	51.499	.000	.000	.000
5900	9.797	80.285	71.391	52.479	.000	.000	.000
6000	9.798	80.450	71.540	53.459	.000	.000	.000

Chlorine, Monatomic (Cl)

(Ideal Gas)(at. wt. = 35.46)Mar. 31, 1961.

T, °K.	C_p° (cal mole⁻¹ deg⁻¹)	S°	$-(F^\circ - H_{298}^\circ)/T$	$H^\circ - H_{298}^\circ$ (kcal mole⁻¹)	ΔH_f°	ΔF_f°	$Log_{10}K_P$
0	.000	.000	INFINITE	− 1.499	28.520	28.520	INFINITE
100	4.969	33.955	43.977	− 1.002	28.669	27.531	− 60.166
200	5.038	37.412	39.927	− .503	28.805	26.338	− 28.780
298	5.219	39.456	39.456	.000	28.922	25.102	− 18.400
300	5.223	39.488	39.456	.010	28.924	25.079	− 18.269
400	5.370	41.013	39.663	.540	29.040	23.779	− 12.992
500	5.436	42.220	40.058	1.081	29.154	22.451	− 9.813
600	5.445	43.212	40.503	1.625	29.264	21.100	− 7.685
700	5.424	44.050	40.952	2.169	29.368	19.731	− 6.160
800	5.389	44.772	41.385	2.710	29.466	18.347	− 5.012
900	5.351	45.405	41.797	3.247	29.558	16.952	− 4.116
1000	5.314	45.967	42.187	3.780	29.644	15.547	− 3.398
1100	5.280	46.472	42.554	4.310	29.725	14.133	− 2.808
1200	5.249	46.930	42.900	4.836	29.801	12.713	− 2.315
1300	5.221	47.349	43.226	5.359	29.873	11.286	− 1.897
1400	5.197	47.735	43.534	5.880	29.941	9.853	− 1.538
1500	5.175	48.092	43.827	6.399	30.006	8.416	− 1.226
1600	5.156	48.426	44.104	6.915	30.068	6.975	− .953
1700	5.140	48.738	44.367	7.430	30.128	5.530	− .711
1800	5.125	49.031	44.618	7.943	30.184	4.080	− .495
1900	5.112	49.308	44.858	8.455	30.239	2.629	− .302
2000	5.101	49.570	45.087	8.966	30.292	1.174	− .128
2100	5.091	49.819	45.306	9.476	30.343	− .283	.029
2200	5.081	50.055	45.517	9.984	30.393	− 1.742	.173
2300	5.073	50.281	45.719	10.492	30.440	− 3.204	.304
2400	5.066	50.497	45.914	10.999	30.487	− 4.668	.425
2500	5.059	50.703	46.101	11.505	30.532	− 6.133	.536
2600	5.053	50.902	46.282	12.011	30.575	− 7.601	.639
2700	5.048	51.092	46.457	12.516	30.617	− 9.069	.734
2800	5.043	51.276	46.626	13.020	30.658	− 10.540	.823
2900	5.038	51.453	46.789	13.524	30.698	− 12.013	.905
3000	5.034	51.623	46.947	14.028	30.737	− 13.487	.982
3100	5.031	51.788	47.101	14.531	30.775	− 14.960	1.055
3200	5.027	51.948	47.250	15.034	30.812	− 16.436	1.122
3300	5.024	52.103	47.395	15.537	30.848	− 17.914	1.186
3400	5.021	52.253	47.535	16.039	30.883	− 19.392	1.246
3500	5.018	52.398	47.672	16.541	30.917	− 20.871	1.303

CHLORINE, MONATOMIC (Cl) (Continued)

T, °K.	C_p° (cal mole⁻¹ deg⁻¹)	S°	$-(F^\circ - H^\circ_{298})/T$	$H^\circ - H^\circ_{298}$ (kcal mole⁻¹)	ΔH_f°	ΔF_f°	$Log_{10}K_P$
3600	5.016	52.539	47.805	17.043	30.950	− 22.352	1.357
3700	5.013	52.677	47.935	17.544	30.982	− 23.833	1.408
3800	5.011	52.810	48.062	18.045	31.013	− 25.314	1.456
3900	5.009	52.941	48.185	18.546	31.044	− 26.798	1.502
4000	5.007	53.067	48.306	19.047	31.073	− 28.281	1.545
4100	5.006	53.191	48.423	19.548	31.102	− 29.765	1.587
4200	5.004	53.312	48.538	20.048	31.130	− 31.249	1.626
4300	5.002	53.429	48.651	20.548	31.157	− 32.736	1.664
4400	5.001	53.544	48.761	21.049	31.184	− 34.222	1.700
4500	5.000	53.657	48.868	21.549	31.209	− 35.709	1.734
4600	4.998	53.767	48.973	22.049	31.234	− 37.196	1.767
4700	4.997	53.874	49.077	22.548	31.258	− 38.683	1.799
4800	4.996	53.979	49.178	23.048	31.281	− 40.173	1.829
4900	4.995	54.082	49.277	23.548	31.305	− 41.660	1.858
5000	4.994	54.183	49.374	24.047	31.326	− 43.149	1.886
5100	4.993	54.282	49.469	24.546	31.347	− 44.641	1.913
5200	4.992	54.379	49.563	25.046	31.368	− 46.130	1.939
5300	4.991	54.474	49.654	25.545	31.388	− 47.620	1.964
5400	4.990	54.567	49.744	26.044	31.407	− 49.112	1.988
5500	4.990	54.659	49.833	26.543	31.425	− 50.601	2.011
5600	4.989	54.749	49.920	27.042	31.443	− 52.094	2.033
5700	4.988	54.837	50.005	27.541	31.460	− 53.585	2.054
5800	4.988	54.924	50.090	28.039	31.476	− 55.077	2.075
5900	4.987	55.009	50.172	28.538	31.492	− 56.570	2.095
6000	4.986	55.093	50.254	29.037	31.507	− 58.064	2.115

HYDROGEN CHLORIDE (HCl)

(Ideal Gas)(mol. wt. = 36.47)Mar. 31, 1961.

T, °K.	C_p° (cal mole⁻¹ deg⁻¹)	S°	$-(F^\circ - H^\circ_{298})/T$	$H^\circ - H^\circ_{298}$ (kcal mole⁻¹)	ΔH_f°	ΔF_f°	$Log_{10}K_P$
0	.000	.000	INFINITE	− 2.065	− 21.926	− 21.926	INFINITE
100	6.959	37.041	50.835	− 1.379	− 21.968	− 22.195	48.505
200	6.961	41.866	45.282	− .683	− 21.936	− 22.442	24.522
298	6.964	44.645	44.645	.000	− 21.970	− 22.685	16.627
300	6.964	44.688	44.646	.013	− 21.971	− 21.689	16.528
400	6.973	46.693	44.919	.710	− 22.036	− 22.919	12.522
500	7.004	48.252	45.435	1.408	− 22.114	− 23.131	10.110
600	7.069	49.534	46.014	2.112	− 22.195	− 23.327	8.496
700	7.167	50.630	46.597	2.823	− 22.274	− 23.509	7.340
800	7.289	51.595	47.163	3.546	− 22.347	− 23.681	6.469
900	7.423	52.461	47.704	4.281	− 22.413	− 23.844	5.790
1000	7.559	53.250	48.220	5.030	− 22.470	− 23.999	5.245
1100	7.693	53.977	48.711	5.793	− 22.519	− 24.150	4.798
1200	7.819	54.652	49.178	6.569	− 22.560	− 24.296	4.425
1300	7.936	55.283	49.624	7.356	− 22.596	− 24.439	4.108
1400	8.043	55.875	50.049	8.156	− 22.627	− 24.580	3.837
1500	8.141	56.433	50.456	8.965	− 22.654	− 24.718	3.601

T, °K.	cal mole⁻¹ deg⁻¹			kcal mole⁻¹			
	C_p°	S°	$-(F^\circ - H_{298}^\circ)/T$	$H^\circ - H_{298}^\circ$	ΔH_f°	ΔF_f°	$\text{Log}_{10} K_P$
1600	8.229	56.961	50.847	9.783	−22.679	−24.856	3.395
1700	8.310	57.463	51.221	10.610	−22.701	−24.990	3.213
1800	8.383	57.940	51.581	11.445	−22.721	−25.125	3.050
1900	8.449	58.395	51.928	12.287	−22.739	−25.257	2.905
2000	8.509	58.830	52.262	13.135	−22.756	−25.390	2.774
2100	8.564	59.246	52.585	13.988	−22.774	−25.521	2.656
2200	8.614	59.646	52.897	14.847	−22.790	−25.651	2.548
2300	8.660	60.030	53.199	15.711	−22.806	−25.781	2.450
2400	8.703	60.399	53.491	16.579	−22.821	−25.909	2.359
2500	8.742	60.755	53.775	17.451	−22.838	−26.039	2.276
2600	8.778	61.099	54.050	18.327	−22.854	−26.166	2.199
2700	8.812	61.431	54.317	19.207	−22.871	−26.292	2.128
2800	8.844	61.752	54.577	20.090	−22.888	−26.420	2.062
2900	8.874	62.063	54.830	20.976	−22.905	−26.545	2.000
3000	8.902	62.364	55.076	21.864	−22.923	−26.671	1.943
3100	8.928	62.656	55.316	22.756	−22.941	−26.795	1.889
3200	8.953	62.940	55.549	23.650	−22.960	−26.919	1.838
3300	8.977	63.216	55.778	24.547	−22.980	−27.042	1.791
3400	8.999	63.484	56.000	25.445	−23.000	−27.166	1.746
3500	9.021	63.745	56.218	26.346	−23.021	−27.288	1.704
3600	9.041	64.000	56.431	27.249	−23.044	−27.410	1.664
3700	9.061	64.248	56.639	28.154	−23.067	−27.531	1.626
3800	9.080	64.490	56.842	29.062	−23.091	−27.650	1.590
3900	9.098	64.726	57.041	29.970	−23.117	−27.771	1.556
4000	9.115	64.956	57.236	30.881	−23.143	−27.889	1.524
4100	9.132	65.182	57.427	31.793	−23.171	−28.007	1.493
4200	9.149	65.402	57.614	32.707	−23.200	−28.124	1.463
4300	9.165	65.617	57.798	33.623	−23.230	−28.243	1.435
4400	9.180	65.828	57.978	34.540	−23.261	−28.359	1.409
4500	9.195	66.035	58.155	35.459	−23.293	−28.476	1.383
4600	9.210	66.237	58.328	36.379	−23.327	−28.589	1.358
4700	9.224	66.435	58.499	37.301	−23.361	−28.702	1.335
4800	9.238	66.630	58.666	38.224	−23.397	−28.817	1.312
4900	9.252	66.820	58.831	39.149	−23.434	−28.928	1.290
5000	9.265	67.007	58.992	40.075	−23.472	−29.041	1.269
5100	9.278	67.191	59.151	41.002	−23.512	−29.152	1.249
5200	9.291	67.371	59.308	41.930	−23.552	−29.262	1.230
5300	9.304	67.548	59.461	42.860	−23.594	−29.369	1.211
5400	9.317	67.722	59.613	43.791	−23.637	−29.481	1.193
5500	9.329	67.893	59.762	44.723	−23.680	−29.586	1.176
5600	9.341	68.062	59.909	45.657	−23.726	−29.695	1.159
5700	9.353	68.227	60.053	46.591	−23.772	−29.800	1.143
5800	9.365	68.390	60.195	47.527	−23.820	−29.905	1.127
5900	9.376	68.550	60.336	48.464	−23.868	−30.009	1.112
6000	9.388	68.708	60.474	49.403	−23.917	−30.116	1.097

BROMINE, DIATOMIC (Br_2)

(Reference State)(mol. wt. = 159.83)Sept. 30, 1961.

T, °K.	C_p° (cal mole⁻¹ deg⁻¹)	S° (cal mole⁻¹ deg⁻¹)	$-(F^\circ - H_{298}^\circ)/T$ (cal mole⁻¹ deg⁻¹)	$H^\circ - H_{298}^\circ$ (kcal mole⁻¹)	ΔH_f° (kcal mole⁻¹)	ΔF_f° (kcal mole⁻¹)	$Log_{10} K_P$
0	.000	.000	INFINITE	− 5.859	.000	.000	.000
100	10.419	12.870	64.780	− 5.191	.000	.000	.000
200	12.851	20.898	40.998	− 4.020	.000	.000	.000
298	18.090	36.384	36.384	.000	.000	.000	.000
300	18.077	36.496	36.384	.033	.000	.000	.000
400	8.775	61.203	40.520	8.273	.000	.000	.000
500	8.857	63.172	44.860	9.156	.000	.000	.000
600	8.908	64.791	48.051	10.044	.000	.000	.000
700	8.944	66.167	50.543	10.937	.000	.000	.000
800	8.970	67.363	52.573	11.833	.000	.000	.000
900	8.992	68.421	54.276	12.731	.000	.000	.000
1000	9.511	69.370	55.739	13.631	.000	.000	.000
1100	9.027	70.229	57.018	14.533	.000	.000	.000
1200	9.042	71.015	58.152	15.436	.000	.000	.000
1300	9.056	71.740	59.169	16.341	.000	.000	.000
1400	9.069	72.411	60.092	17.247	.000	.000	.000
1500	9.082	73.037	60.934	18.155	.000	.000	.000
1600	9.094	73.624	61.709	19.064	.000	.000	.000
1700	9.106	74.176	62.426	19.974	.000	.000	.000
1800	9.118	74.696	63.094	20.885	.000	.000	.000
1900	9.128	75.190	63.717	21.797	.000	.000	.000
2000	9.141	75.658	64.303	22.711	.000	.000	.000
2100	9.152	76.104	64.854	23.625	.000	.000	.000
2200	9.163	76.530	65.375	24.541	.000	.000	.000
2300	9.174	76.938	65.869	25.458	.000	.000	.000
2400	9.185	77.329	66.339	26.376	.000	.000	.000
2500	9.196	77.704	66.786	27.295	.000	.000	.000
2600	9.206	78.065	67.213	28.215	.000	.000	.000
2700	9.217	78.412	67.621	29.136	.000	.000	.000
2800	9.228	78.748	68.013	30.059	.000	.000	.000
2900	9.238	79.072	68.388	30.982	.000	.000	.000
3000	9.249	79.385	68.750	31.906	.000	.000	.000
3100	9.260	79.689	69.098	32.832	.000	.000	.000
3200	9.270	79.983	69.433	33.758	.000	.000	.000
3300	9.281	80.268	69.757	34.686	.000	.000	.000
3400	9.291	80.545	70.071	35.614	.000	.000	.000
3500	9.302	80.815	70.374	36.544	.000	.000	.000
3600	9.312	81.077	70.667	37.475	.000	.000	.000
3700	9.323	81.332	70.952	38.406	.000	.000	.000
3800	9.333	81.581	71.229	39.339	.000	.000	.000
3900	9.344	81.824	71.497	40.273	.000	.000	.000
4000	9.354	82.060	71.758	41.208	.000	.000	.000

	cal mole⁻¹ deg⁻¹			kcal mole⁻¹			
T, °K.	C_p°	S°	$-(F^\circ - H^\circ_{298})/T$	$H^\circ - H^\circ_{298}$	ΔH_f°	ΔF_f°	$\log_{10} K_P$
4100	9.365	82.291	72.012	42.144	.000	.000	.000
4200	9.375	82.517	72.260	43.081	.000	.000	.000
4300	9.386	82.738	72.501	44.019	.000	.000	.000
4400	9.396	82.954	72.736	44.958	.000	.000	.000
4500	9.406	83.165	72.965	45.898	.000	.000	.000
4600	9.417	83.372	73.189	46.839	.000	.000	.000
4700	9.427	83.575	73.408	47.781	.000	.000	.000
4800	9.438	83.773	73.622	48.725	.000	.000	.000
4900	9.448	83.968	73.831	49.669	.000	.000	.000
5000	9.458	84.159	74.036	50.614	.000	.000	.000
5100	9.469	84.346	74.236	51.561	.000	.000	.000
5200	9.479	84.530	74.432	52.508	.000	.000	.000
5300	9.490	84.711	74.625	53.456	.000	.000	.000
5400	9.500	84.888	74.813	54.406	.000	.000	.000
5500	9.510	85.063	74.998	55.356	.000	.000	.000
5600	9.521	85.234	75.179	56.308	.000	.000	.000
5700	9.531	85.403	75.357	57.261	.000	.000	.000
5800	9.542	85.569	75.532	58.214	.000	.000	.000
5900	9.552	85.732	75.703	59.169	.000	.000	.000
6000	9.562	85.892	75.872	60.125	.000	.000	.000

BROMINE (Br_2)

(Liquid)(mol. wt. = 159.83)Sept. 30, 1961.

	cal mole⁻¹ deg⁻¹			kcal mole⁻¹			
T, °K.	C_p°	S°	$-(F^\circ - H^\circ_{298})/T$	$H^\circ - H^\circ_{298}$	ΔH_f°	ΔF_f°	$\log_{10} K_P$
0							
100							
200							
298	18.090	36.384	36.384	.000	.000	.000	.000
300	18.077	36.496	36.384	.033	.000	.000	.000
400	18.000	41.676	37.091	1.834	– 6.439	1.372	– .749
500	18.000	45.692	38.425	3.634	– 5.522	3.218	– 1.406
600	18.000	48.974	39.918	5.434	– 4.610	4.880	– 1.777
700	18.000	51.749	41.415	7.234	– 3.703	6.390	– 1.995
800	18.000	54.153	42.860	9.034	– 2.799	7.769	– 2.122
900	18.000	56.273	44.235	10.834	– 1.897	9.036	– 2.194
1000	18.000	58.169	45.535	12.634	– .997	10.204	– 2.230

BROMINE, DIATOMIC (Br_2)

(Ideal Gas)(mol. wt. = 159.83)Sept. 30, 1961.

	cal mole⁻¹ deg⁻¹			kcal mole⁻¹			
T, °K.	C_p°	S°	$-(F^\circ - H^\circ_{298})/T$	$H^\circ - H^\circ_{298}$	ΔH_f°	ΔF_f°	$\log_{10} K_P$
0	.000	.000	INFINITE	– 2.324	10.922	10.922	INFINITE
100	7.385	49.841	66.037	– 1.620	10.958	7.261	15.869
200	8.268	55.270	59.426	– .831	10.576	3.701	4.045
298	8.616	58.647	58.647	.000	7.387	.749	.549
300	8.620	58.700	58.647	.016	7.370	.709	.516
400	8.775	61.204	58.987	.887	.000	.000	.000
500	8.857	63.171	59.634	1.769	.000	.000	.000

Bromine, Diatomic (Br_2) (Continued)

T, °K.	cal mole⁻¹ deg⁻¹ C_p°	S°	$-(F^\circ - H^\circ_{298})/T$	kcal mole⁻¹ $H^\circ - H^\circ_{298}$	ΔH_f°	ΔF_f°	$Log_{10}K_P$
600	8.908	64.791	60.363	2.657	.000	.000	.000
700	8.944	66.167	61.096	3.550	.000	.000	.000
800	8.970	67.363	61.806	4.445	.000	.000	.000
900	8.992	68.421	62.484	5.344	.000	.000	.000
1000	9.011	69.369	63.126	6.244	.000	.000	.000
1100	9.027	70.229	63.733	7.146	.000	.000	.000
1200	9.042	71.015	64.307	8.049	.000	.000	.000
1300	9.056	71.739	64.852	8.954	.000	.000	.000
1400	9.069	72.411	65.368	9.860	.000	.000	.000
1500	9.082	73.037	65.859	10.768	.000	.000	.000
1600	9.094	73.624	66.326	11.677	.000	.000	.000
1700	9.106	74.175	66.771	12.587	.000	.000	.000
1800	9.118	74.696	67.197	13.498	.000	.000	.000
1900	9.129	75.190	67.605	14.410	.000	.000	.000
2000	9.141	75.658	67.996	15.324	.000	.000	.000
2100	9.152	76.104	68.372	16.238	.000	.000	.000
2200	9.163	76.530	68.733	17.154	.000	.000	.000
2300	9.174	76.938	69.081	18.071	.000	.000	.000
2400	9.185	77.329	69.416	18.989	.000	.000	.000
2500	9.196	77.704	69.740	19.908	.000	.000	.000
2600	9.206	78.065	70.054	20.828	.000	.000	.000
2700	9.217	78.412	70.357	21.749	.000	.000	.000
2800	9.228	78.748	70.651	22.672	.000	.000	000
2900	9.238	79.072	70.935	23.595	.000	.000	.000
3000	9.249	79.385	71.212	24.519	.000	.000	.000
3100	9.260	79.688	71.480	25.445	.000	.000	.000
3200	9.270	79.983	71.742	26.371	.000	.000	.000
3300	9.281	80.268	71.996	27.299	.000	.000	.000
3400	9.291	80.545	72.243	28.227	.000	.000	.000
3500	9.302	80.815	72.484	29.157	.000	.000	.000
3600	9.312	81.077	72.719	30.088	.000	.000	.000
3700	9.323	81.332	72.949	31.019	.000	.000	.000
3800	9.333	81.581	73.172	31.952	.000	.000	.000
3900	9.344	81.823	73.391	32.886	.000	.000	.000
4000	9.354	82.060	73.605	33.821	.000	.000	.000
4100	9.365	82.291	73.814	34.757	.000	.000	.000
4200	9.375	82.517	74.019	35.694	.000	.000	.000
4300	9.386	82.738	74.219	36.632	.000	.000	.000
4400	9.396	82.954	74.415	37.571	.000	.000	.000
4500	9.406	83.165	74.607	38.511	.000	.000	.000
4600	9.417	83.372	74.795	39.452	.000	.000	.000
4700	9.427	83.574	74.980	40.394	.000	.000	.000
4800	9.438	83.773	75.161	41.338	.000	.000	.000
4900	9.448	83.968	75.339	42.282	.000	.000	.000
5000	9.458	84.159	75.513	43.227	.000	.000	.000

	cal mole⁻¹ deg⁻¹			kcal mole⁻¹			
T, °K.	C_p°	S°	$-(F^\circ - H_{298}^\circ)/T$	$H^\circ - H_{298}^\circ$	ΔH_f°	ΔF_f°	$Log_{10} K_P$
5100	9.469	84.346	75.685	44.174	.000	.000	.000
5200	9.479	84.530	75.853	45.121	.000	.000	.000
5300	9.490	84.711	76.018	46.070	.000	.000	.000
5400	9.500	84.888	76.181	47.019	.000	.000	.000
5500	9.510	85.063	76.341	47.970	.000	.000	.000
5600	9.521	85.234	76.498	48.921	.000	.000	.000
5700	9.531	85.403	76.653	49.874	.000	.000	.000
5800	9.542	85.569	76.805	50.827	.000	.000	.000
5900	9.552	85.732	76.955	51.782	.000	.000	.000
6000	9.562	85.892	77.103	52.738	.000	.000	.000

Bromine, Monatomic (Br)

(Ideal Gas)(atomic wt. = 79.92)Sept. 30, 1961.

	cal mole⁻¹ deg⁻¹			kcal mole⁻¹			
T, °K.	C_p°	S°	$-(F^\circ - H_{298}^\circ)/T$	$H^\circ - H_{298}^\circ$	ΔH_f°	ΔF_f°	$Log_{10} K_P$
0	.000	.000	INFINITE	- 1.481	28.188	28.188	INFINITE
100	4.968	36.378	46.222	- .984	28.351	25.357	- 55.415
200	4.968	39.821	42.260	- .488	28.262	22.388	- 24.463
298	4.968	41.805	41.805	.000	26.740	19.700	- 14.440
300	4.968	41.836	41.805	.009	26.733	19.656	- 14.319
400	4.968	43.265	42.000	.506	23.110	18.044	- 9.858
500	4.971	44.374	42.368	1.003	23.165	16.771	- 7.330
600	4.979	45.281	42.780	1.500	23.218	15.487	- 5.641
700	4.997	46.050	43.194	1.999	23.271	14.194	- 4.431
800	5.026	46.719	43.593	2.500	23.324	12.894	- 3.522
900	5.063	47.313	43.974	3.005	23.379	11.587	- 2.814
1000	5.107	47.848	44.335	3.513	23.438	10.274	- 2.245
1100	5.153	48.337	44.677	4.026	23.500	8.954	- 1.779
1200	5.199	48.788	45.001	4.544	23.566	7.629	- 1.389
1300	5.243	49.206	45.309	5.066	23.635	6.299	- 1.059
1400	5.284	49.596	45.601	5.592	23.709	4.962	- .775
1500	5.320	49.961	45.880	6.122	23.785	3.620	- .527
1600	5.351	50.306	46.146	6.656	23.864	2.274	- .311
1700	5.377	50.631	46.400	7.192	23.945	.922	- .119
1800	5.398	50.939	46.644	7.731	24.029	- .435	.053
1900	5.415	51.231	46.878	8.272	24.113	- 1.796	.207
2000	5.428	51.509	47.102	8.814	24.199	- 3.162	.346
2100	5.437	51.775	47.319	9.357	24.285	- 4.532	.472
2200	5.443	52.028	47.527	9.901	24.371	- 5.907	.587
2300	5.446	52.270	47.728	10.446	24.457	- 7.284	.692
2400	5.447	52.501	47.922	10.991	24.543	- 8.666	.789
2500	5.445	52.724	48.110	11.535	24.628	- 10.052	.879
2600	5.442	52.937	48.291	12.080	24.712	- 11.440	.962
2700	5.438	53.143	48.467	12.624	24.796	- 12.833	1.039
2800	5.432	53.340	48.638	13.167	24.878	- 14.228	1.110
2900	5.425	53.531	48.803	13.710	24.959	- 15.626	1.178
3000	5.418	53.715	48.964	14.252	25.039	- 17.027	1.240

Bromine, Monatomic (Br) (Continued)

T, °K.	cal mole⁻¹ deg⁻¹ C_p°	S°	$-(F^\circ - H^\circ_{298})/T$	kcal mole⁻¹ $H^\circ - H^\circ_{298}$	ΔH_f°	ΔF_f°	$\text{Log}_{10} K_P$
3100	5.410	53.892	49.120	14.794	25.118	−18.430	1.299
3200	5.402	54.064	49.272	15.334	25.195	−19.836	1.355
3300	5.393	54.230	49.420	15.874	25.271	−21.245	1.407
3400	5.384	54.391	49.563	16.413	25.346	−22.656	1.456
3500	5.375	54.547	49.704	16.951	25.419	−24.068	1.503
3600	5.366	54.698	49.840	17.488	25.490	−25.484	1.547
3700	5.357	54.845	49.973	18.024	25.561	−26.901	1.589
3800	5.348	54.988	50.104	18.559	25.630	−28.319	1.629
3900	5.339	55.126	50.231	19.093	25.697	−29.739	1.666
4000	5.330	55.261	50.355	19.627	25.763	−31.163	1.703
4100	5.321	55.393	50.476	20.159	25.827	−32.587	1.737
4200	5.312	55.521	50.595	20.691	25.890	−34.012	1.770
4300	5.304	55.646	50.711	21.222	25.952	−35.438	1.801
4400	5.295	55.768	50.824	21.752	26.013	−36.866	1.831
4500	5.287	55.887	50.935	22.281	26.072	−38.297	1.860
4600	5.279	56.003	51.044	22.809	26.130	−39.727	1.887
4700	5.271	56.116	51.151	23.337	26.186	−41.159	1.914
4800	5.264	56.227	51.256	23.863	26.241	−42.594	1.939
4900	5.256	56.336	51.358	24.389	26.295	−44.027	1.964
5000	5.249	56.442	51.459	24.915	26.348	−45.463	1.987
5100	5.242	56.546	51.557	25.439	26.399	−46.901	2.010
5200	5.236	56.647	51.654	25.963	26.449	−48.338	2.032
5300	5.229	56.747	51.749	26.486	26.498	−49.776	2.052
5400	5.223	56.845	51.843	27.009	26.546	−51.217	2.073
5500	5.217	56.940	51.935	27.531	26.593	−52.656	2.092
5600	5.211	57.034	52.025	28.052	26.638	−54.099	2.111
5700	5.205	57.126	52.114	28.573	26.683	−55.540	2.129
5800	5.199	57.217	52.201	29.093	26.726	−56.982	2.147
5900	5.194	57.306	52.287	29.613	26.769	−58.426	2.164
6000	5.189	57.393	52.371	30.132	26.810	−59.873	2.181

Hydrogen Bromide (HBr)

(Ideal Gas)(mol. wt. = 80.92)Sept. 30, 1961.

T, °K.	cal mole⁻¹ deg⁻¹ C_p°	S°	$-(F^\circ - H^\circ_{298})/T$	kcal mole⁻¹ $H^\circ - H^\circ_{298}$	ΔH_f°	ΔF_f°	$\text{Log}_{10} K_P$
0	.000	.000	INFINITE	−2.067	−6.785	−6.785	INFINITE
100	6.959	39.833	53.627	−1.379	−6.811	−8.932	19.520
200	6.961	44.657	48.074	−.683	−7.002	−10.992	12.011
298	6.964	47.437	47.437	.000	−8.660	−12.727	9.329
300	6.965	47.480	47.438	.013	−8.670	−12.752	9.290
400	6.980	49.486	47.711	.710	−12.440	−13.344	7.291
500	7.027	51.047	48.228	1.410	−12.531	−13.560	5.927
600	7.116	52.336	48.808	2.117	−12.618	−13.758	5.011
700	7.238	53.442	49.393	2.834	−12.698	−13.941	4.352
800	7.381	54.417	49.961	3.565	−12.768	−14.114	3.856
900	7.529	55.295	50.506	4.311	−12.828	−14.278	3.467
1000	7.675	56.096	51.025	5.071	−12.877	−14.437	3.155

	cal mole^{-1} deg^{-1}			kcal mole^{-1}			
T, °K.	C_p°	S°	$-(F^\circ - H_{298}^\circ)/T$	$H^\circ - H_{298}^\circ$	ΔH_f°	ΔF_f°	$\text{Log}_{10}K_P$
1100	7.812	56.834	51.520	5.845	− 12.916	− 14.591	2.899
1200	7.939	57.519	51.992	6.633	− 12.947	− 14.741	2.685
1300	8.054	58.159	52.442	7.433	− 12.972	− 14.890	2.503
1400	8.158	58.760	52.872	8.243	− 12.991	− 15.037	2.347
1500	8.251	59.326	53.284	9.064	− 13.008	− 15.182	2.212
1600	8.335	59.861	53.678	9.893	− 13.022	− 15.327	2.094
1700	8.410	60.369	54.057	10.731	− 13.033	− 15.470	1.989
1800	8.478	60.852	54.421	11.575	− 13.042	− 15.614	1.896
1900	8.539	61.312	54.772	12.426	− 13.050	− 15.756	1.812
2000	8.595	61.751	55.110	13.283	− 13.058	− 15.899	1.737
2100	8.646	62.172	55.436	14.145	− 13.065	− 16.041	1.669
2200	8.692	62.575	55.752	15.012	− 13.072	− 16.182	1.607
2300	8.734	62.962	56.057	15.883	− 13.079	− 16.323	1.551
2400	8.773	63.335	56.352	16.758	− 13.086	− 16.463	1.499
2500	8.809	63.694	56.639	17.638	− 13.094	− 16.605	1.452
2600	8.843	64.040	56.917	18.520	− 13.101	− 16.744	1.407
2700	8.874	64.374	57.187	19.406	− 13.109	− 16.885	1.367
2800	8.903	64.698	57.449	20.295	− 13.119	− 17.025	1.329
2900	8.930	65.011	57.705	21.187	− 13.127	− 17.163	1.293
3000	8.956	65.314	57.953	22.081	− 13.137	− 17.303	1.260
3100	8.981	65.608	58.196	22.978	− 13.147	− 17.442	1.230
3200	9.004	65.893	58.432	23.877	− 13.158	− 17.580	1.201
3300	9.026	66.171	58.662	24.778	− 13.170	− 17.718	1.173
3400	9.047	66.440	58.887	25.682	− 13.182	− 17.857	1.148
3500	9.067	66.703	59.106	26.588	− 13.196	− 17.992	1.123
3600	9.086	66.959	59.321	27.495	− 13.210	− 18.130	1.101
3700	9.104	67.208	59.531	28.405	− 13.226	− 18.226	1.079
3800	9.122	67.451	59.736	29.316	− 13.242	− 18.402	1.058
3900	9.139	67.688	59.937	30.229	− 13.260	− 18.537	1.039
4000	9.156	67.920	60.134	31.144	− 13.278	− 18.673	1.020
4100	9.172	68.146	60.326	32.060	− 13.298	− 18.808	1.003
4200	9.188	68.367	60.515	32.978	− 13.319	− 18.941	.986
4300	9.203	68.584	60.700	33.898	− 13.341	− 19.075	.969
4400	9.218	68.795	60.882	34.819	− 13.364	− 19.208	.954
4500	9.232	69.003	61.060	35.742	− 13.388	− 19.342	.939
4600	9.247	69.206	61.235	36.666	− 13.413	− 19.472	.925
4700	9.260	69.405	61.407	37.591	− 13.440	− 19.602	.911
4800	9.274	69.600	61.575	38.518	− 13.468	− 19.735	.899
4900	9.287	69.791	61.741	39.446	− 13.496	− 19.864	.886
5000	9.301	69.979	61.904	40.375	− 13.526	− 19.995	.874
5100	9.313	70.163	62.064	41.306	− 13.557	− 20.124	.862
5200	9.326	70.344	62.222	42.238	− 13.589	− 20.253	.851
5300	9.339	70.522	62.376	43.171	− 13.621	− 20.379	.840
5400	9.351	70.697	62.529	44.106	− 13.656	− 20.509	.830
5500	9.363	70.868	62.679	45.041	− 13.691	− 20.633	.820

HYDROGEN BROMIDE (HBr) (Continued)

T, °K.	C_p°	S°	$-(F^\circ - H^\circ_{298})/T$	$H^\circ - H^\circ_{298}$	ΔH_f°	ΔF_f°	$Log_{10}K_P$
	cal mole⁻¹ deg⁻¹			kcal mole⁻¹			
5600	9.375	71.037	62.827	45.978	− 13.727	− 20.762	.810
5700	9.387	71.203	62.972	46.916	− 13.765	− 20.885	.801
5800	9.399	71.366	63.116	47.855	− 13.803	− 21.010	.792
5900	9.410	71.527	63.257	48.796	− 13.843	− 21.134	.783
6000	9.422	71.686	63.396	49.737	− 13.883	− 21.260	.774

FLUORINE, DIATOMIC (F_2)
(Reference State—Ideal Gas)(mol. wt. = 38.00)Dec. 31, 1960.

T, °K.	C_p°	S°	$-(F^\circ - H^\circ_{298})/T$	$H^\circ - H^\circ_{298}$	ΔH_f°	ΔF_f°	$Log_{10}K_P$
	cal mole⁻¹ deg⁻¹			kcal mole⁻¹			
0	.000	.000	INFINITE	−2.110	.000	.000	.000
100	6.959	40.695	54.846	−1.415	.000	.000	.000
200	7.098	45.544	49.118	− .715	.000	.000	.000
298	7.490	48.447	48.447	.000	.000	.000	.000
300	7.498	48.493	48.447	.014	.000	.000	.000
400	7.900	50.707	48.746	.784	.000	.000	.000
500	8.207	52.505	49.324	1.591	.000	.000	.000
600	8.429	54.022	49.983	2.423	.000	.000	.000
700	8.591	55.334	50.656	3.274	.000	.000	.000
800	8.713	56.489	51.314	4.140	.000	.000	.000
900	8.808	57.521	51.948	5.016	.000	.000	.000
1000	8.885	58.453	52.553	5.901	.000	.000	.000
1100	8.948	59.303	53.128	6.793	.000	.000	.000
1200	9.003	60.084	53.676	7.690	.000	.000	.000
1300	9.050	60.807	54.197	8.593	.000	.000	.000
1400	9.092	61.479	54.693	9.500	.000	.000	.000
1500	9.130	62.108	55.167	10.411	.000	.000	.000
1600	9.166	62.698	55.619	11.326	.000	.000	.000
1700	9.198	63.255	56.052	12.244	.000	.000	.000
1800	9.229	63.781	56.467	13.166	.000	.000	.000
1900	9.259	64.281	56.865	14.090	.000	.000	.000
2000	9.287	64.757	57.248	15.017	.000	.000	.000
2100	9.314	65.210	57.616	15.947	.000	.000	.000
2200	9.341	65.644	57.972	16.880	.000	.000	.000
2300	9.366	66.060	58.314	17.815	.000	.000	.000
2400	9.391	66.459	58.645	18.753	.000	.000	.000
2500	9.416	66.843	58.966	19.694	.000	.000	.000
2600	9.440	67.213	59.276	20.637	.000	.000	.000
2700	9.464	67.570	59.576	21.582	.000	.000	.000
2800	9.488	67.914	59.868	22.529	.000	.000	.000
2900	9.511	68.248	60.151	23.479	.000	.000	.000
3000	9.534	68.570	60.427	24.432	.000	.000	.000
3100	9.557	68.883	60.694	25.386	.000	.000	.000
3200	9.579	69.187	60.955	26.343	.000	.000	.000
3300	9.602	69.482	61.209	27.302	.000	.000	.000
3400	9.624	69.769	61.457	28.263	.000	.000	.000
3500	9.646	70.049	61.698	29.227	.000	.000	.000

T, °K.	cal mole⁻¹ deg⁻¹ C_p°	S°	$-(F^\circ - H^\circ_{298})/T$	kcal mole⁻¹ $H^\circ - H^\circ_{298}$	ΔH_f°	ΔF_f°	$Log_{10}K_P$
3600	9.668	70.321	61.934	30.192	.000	.000	.000
3700	9.690	70.586	62.164	31.160	.000	.000	.000
3800	9.712	70.845	62.389	32.130	.000	.000	.000
3900	9.734	71.097	62.609	33.103	.000	.000	.000
4000	9.755	71.344	62.825	34.077	.000	.000	.000
4100	9.777	71.585	63.035	35.054	.000	.000	.000
4200	9.799	71.821	63.242	36.033	.000	.000	.000
4300	9.820	72.052	63.444	37.013	.000	.000	.000
4400	9.842	72.278	63.642	37.997	.000	.000	.000
4500	9.863	72.499	63.836	38.982	.000	.000	.000
4600	9.884	72.716	64.027	39.969	.000	.000	.000
4700	9.906	72.929	64.214	40.959	.000	.000	.000
4800	9.927	73.138	64.398	41.950	.000	.000	.000
4900	9.948	73.343	64.578	42.944	.000	.000	.000
5000	9.970	73.544	64.756	43.940	.000	.000	.000
5100	9.991	73.741	64.930	44.938	.000	.000	.000
5200	10.012	73.936	65.101	45.938	.000	.000	.000
5300	10.033	74.127	65.270	46.940	.000	.000	.000
5400	10.054	74.314	65.436	47.945	.000	.000	.000
5500	10.075	74.499	65.599	48.951	.000	.000	.000
5600	10.097	74.681	65.759	49.960	.000	.000	.000
5700	10.118	74.860	65.917	50.971	.000	.000	.000
5800	10.139	75.036	66.073	51.983	.000	.000	.000
5900	10.160	75.209	66.226	52.998	.000	.000	.000
6000	10.181	75.380	66.378	54.015	.000	.000	.000

Fluorine, Monatomic (F)

(Ideal Gas)(mol. wt. = 19.00)Sept. 30, 1965.

T, °K.	cal mole⁻¹ deg⁻¹ C_p°	S°	$-(F^\circ - H^\circ_{298})/T$	kcal mole⁻¹ $H^\circ - H^\circ_{298}$	ΔH_f°	ΔF_f°	$Log_{10}K_P$
0	.000	.000	INFINITE	- 1.558	18.357	18.357	INFINITE
100	5.068	32.116	42.710	- 1.059	18.508	17.331	- 37.875
200	5.403	35.746	38.415	- .534	18.684	16.089	- 17.580
298	5.437	37.917	37.917	.000	18.860	14.777	- 10.832
300	5.436	37.951	37.917	.010	18.863	14.752	- 10.746
400	5.361	39.505	38.129	.550	19.018	13.358	- 7.298
500	5.282	40.693	38.528	1.082	19.147	11.927	- 5.213
600	5.218	41.650	38.971	1.607	19.256	10.472	- 3.814
700	5.169	42.450	39.412	2.126	19.349	9.001	- 2.810
800	5.133	43.138	39.836	2.641	19.431	7.517	- 2.053
900	5.105	43.741	40.237	3.153	19.505	6.023	- 1.462
1000	5.083	44.277	40.615	3.663	19.572	4.522	- .988
1100	5.066	44.761	40.970	4.170	19.633	3.013	- .599
1200	5.052	45.201	41.305	4.676	19.691	1.500	- .273
1300	5.041	45.605	41.620	5.180	19.744	- .018	.003
1400	5.032	45.978	41.918	5.684	19.794	- 1.540	.240
1500	5.025	46.325	42.201	6.187	19.841	- 3.065	.447

Fluorine, Monatomic (F) (Continued)

T, °K.	C_p° (cal mole⁻¹ deg⁻¹)	S° (cal mole⁻¹ deg⁻¹)	$-(F^\circ - H^\circ_{298})/T$ (cal mole⁻¹ deg⁻¹)	$H^\circ - H^\circ_{298}$ (kcal mole⁻¹)	ΔH_f° (kcal mole⁻¹)	ΔF_f° (kcal mole⁻¹)	$\mathrm{Log}_{10} K_P$
1600	5.018	46.649	42.469	6.689	19.886	− 4.594	.627
1700	5.013	46.954	42.724	7.191	19.929	− 6.126	.788
1800	5.009	47.240	42.967	7.692	19.969	− 7.660	.930
1900	5.005	47.511	43.199	8.192	20.007	− 9.197	1.058
2000	5.001	47.767	43.421	8.693	20.044	− 10.733	1.173
2100	4.999	48.011	43.634	9.193	20.079	− 12.273	1.277
2200	4.996	48.244	43.838	9.692	20.112	− 13.816	1.372
2300	4.994	48.466	44.034	10.192	20.144	− 15.358	1.459
2400	4.992	48.678	44.224	10.691	20.175	− 16.902	1.539
2500	4.990	48.882	44.406	11.190	20.203	− 18.448	1.613
2600	4.988	49.078	44.582	11.689	20.231	− 19.995	1.681
2700	4.987	49.266	44.752	12.188	20.257	− 21.542	1.744
2800	4.986	49.447	44.916	12.687	20.282	− 23.090	1.802
2900	4.985	49.622	45.076	13.185	20.306	− 24.639	1.857
3000	4.984	49.791	45.230	13.683	20.327	− 26.191	1.908
3100	4.983	49.954	45.380	14.182	20.349	− 27.740	1.956
3200	4.982	50.113	45.525	14.680	20.369	− 29.294	2.001
3300	4.981	50.266	45.666	15.178	20.387	− 30.845	2.043
3400	4.980	50.415	45.804	15.676	20.405	− 32.399	2.082
3500	4.980	50.559	45.938	16.174	20.421	− 33.950	2.120
3600	4.979	50.699	46.068	16.672	20.436	− 35.502	2.155
3700	4.978	50.836	46.195	17.170	20.450	− 37.059	2.189
3800	4.978	50.968	46.319	17.668	20.463	− 38.610	2.220
3900	4.977	51.098	46.440	18.166	20.474	− 40.169	2.251
4000	4.977	51.224	46.558	18.663	20.485	− 41.723	2.280
4100	4.977	51.347	46.673	19.161	20.494	− 43.279	2.307
4200	4.976	51.467	46.786	19.659	20.502	− 44.835	2.333
4300	4.976	51.584	46.896	20.156	20.510	− 46.390	2.358
4400	4.975	51.698	47.004	20.654	20.515	− 47.944	2.381
4500	4.975	51.810	47.110	21.151	20.520	− 49.502	2.404
4600	4.975	51.919	47.213	21.649	20.524	− 51.056	2.426
4700	4.975	52.026	47.314	22.146	20.527	− 52.612	2.446
4800	4.974	52.131	47.413	22.644	20.529	− 54.169	2.466
4900	4.974	52.233	47.511	23.141	20.529	− 55.722	2.485
5000	4.974	52.334	47.606	23.639	20.529	− 57.281	2.504
5100	4.974	52.432	47.700	24.136	20.527	− 58.837	2.521
5200	4.973	52.529	47.792	24.633	20.524	− 60.393	2.538
5300	4.973	52.624	47.882	25.131	20.521	− 61.950	2.554
5400	4.973	52.717	47.971	25.628	20.515	− 63.509	2.570
5500	4.973	52.808	48.058	26.125	20.510	− 65.062	2.585
5600	4.973	52.898	48.144	26.623	20.503	− 66.619	2.600
5700	4.973	52.986	48.228	27.120	20.494	− 68.175	2.614
5800	4.972	53.072	48.310	27.617	20.486	− 69.728	2.627
5900	4.972	53.157	48.392	28.114	20.475	− 71.284	2.640
6000	4.972	53.241	48.472	28.612	20.464	− 72.842	2.653

HYDROGEN FLUORIDE (HF)
(Ideal Gas)(mol. wt. = 20.01)Dec. 31, 1963.

T, °K.	C_p°	S°	$-(F^\circ - H^\circ_{298})/T$	$H^\circ - H^\circ_{298}$	ΔH_f°	ΔF_f°	$Log_{10}K_P$
	cal mole⁻¹ deg⁻¹			kcal mole⁻¹			
0	.000	.000	INFINITE	− 2.055	− 64.788	− 64.788	INFINITE
100	6.962	33.904	47.700	− 1.380	− 64.840	− 64.976	141.342
200	6.962	38.729	42.146	− .683	− 64.795	− 65.134	71.172
298	6.964	41.509	41.509	.000	− 64.800	− 65.301	47.865
300	6.964	41.552	41.509	.013	− 64.801	− 65.305	47.572
400	6.967	43.556	41.782	.709	− 64.836	− 65.468	35.768
500	6.973	45.111	42.298	1.406	− 64.892	− 65.620	28.681
600	6.986	46.384	42.877	2.104	− 64.960	− 65.759	23.952
700	7.015	47.463	43.457	2.804	− 65.037	− 65.886	20.570
800	7.063	48.402	44.017	3.508	− 65.119	− 66.002	18.030
900	7.129	49.238	44.552	4.217	− 65.204	− 66.107	16.052
1000	7.211	49.993	45.059	4.934	− 65.288	− 66.204	14.468
1100	7.303	50.684	45.539	5.660	− 65.372	− 66.291	13.170
1200	7.402	51.324	45.995	6.395	− 65.452	− 66.371	12.087
1300	7.504	51.921	46.428	7.140	− 65.530	− 66.444	11.170
1400	7.606	52.480	46.841	7.896	− 65.605	− 66.512	10.382
1500	7.705	53.009	47.234	8.662	− 65.678	− 66.573	9.699
1600	7.800	53.509	47.611	9.437	− 65.749	− 66.632	9.101
1700	7.891	53.985	47.972	10.221	− 65.817	− 66.683	8.572
1800	7.977	54.438	48.319	11.015	− 65.883	− 66.734	8.102
1900	8.058	54.872	48.652	11.817	− 65.946	− 66.779	7.681
2000	8.133	55.287	48.974	12.626	− 66.008	− 66.820	7.301
2100	8.204	55.685	49.284	13.443	− 66.068	− 66.860	6.958
2200	8.270	56.069	49.584	14.267	− 66.127	− 66.897	6.645
2300	8.331	56.438	49.874	15.097	− 66.184	− 66.930	6.359
2400	8.389	56.793	50.155	15.933	− 66.240	− 66.961	6.097
2500	8.442	57.137	50.427	16.775	− 66.296	− 66.991	5.856
2600	8.493	57.469	50.692	17.621	− 66.351	− 67.017	5.633
2700	8.540	57.790	50.949	18.473	− 66.406	− 67.041	5.426
2800	8.584	58.102	51.199	19.329	− 66.459	− 67.065	5.234
2900	8.625	58.404	51.442	20.190	− 66.513	− 67.083	5.055
3000	8.664	58.697	51.679	21.054	− 66.567	− 67.105	4.888
3100	8.701	58.981	51.910	21.922	− 66.620	− 67.122	4.732
3200	8.735	59.258	52.135	22.794	− 66.673	− 67.137	4.585
3300	8.768	59.528	52.355	23.669	− 66.727	− 67.150	4.447
3400	8.799	59.790	52.570	24.548	− 66.781	− 67.163	4.317
3500	8.829	60.045	52.780	25.429	− 66.836	− 67.171	4.194
3600	8.857	60.294	52.985	26.313	− 66.891	− 67.180	4.078
3700	8.884	60.537	53.186	27.201	− 66.947	− 67.187	3.968
3800	8.910	60.775	53.383	28.090	− 67.003	− 67.192	3.864
3900	8.934	61.006	53.575	28.982	− 67.062	− 67.198	3.765
4000	8.958	61.233	53.764	29.877	− 67.120	− 67.200	3.671

HYDROGEN FLUORIDE (HF) (Continued)

	cal mole⁻¹ deg⁻¹			kcal mole⁻¹			
T, °K.	C_p°	S°	$-(F^\circ - H_{298}^\circ)/T$	$H^\circ - H_{298}^\circ$	ΔH_f°	ΔF_f°	$\mathrm{Log}_{10}K_p$
4100	8.980	61.454	53.949	30.774	− 67.180	− 67.202	3.582
4200	9.002	61.671	54.130	31.673	− 67.240	− 67.201	3.497
4300	9.023	61.883	54.308	32.574	− 67.302	− 67.199	3.415
4400	9.043	62.091	54.482	33.478	− 67.365	− 67.197	3.338
4500	9.063	62.294	54.654	34.383	− 67.429	− 67.194	3.263
4600	9.081	62.494	54.822	35.290	− 67.494	− 67.186	3.192
4700	9.100	62.689	54.987	36.199	− 67.561	− 67.179	3.124
4800	9.117	62.881	55.150	37.110	− 67.628	− 67.169	3.058
4900	9.135	63.069	55.309	38.023	− 67.697	− 67.158	2.995
5000	9.151	63.254	55.466	38.937	− 67.767	− 67.149	2.935
5100	9.168	63.435	55.621	39.853	− 67.839	− 67.136	2.877
5200	9.184	63.613	55.773	40.770	− 67.911	− 67.120	2.821
5300	9.199	63.788	55.922	41.690	− 67.985	− 67.102	2.767
5400	9.214	63.961	56.070	42.610	− 68.061	− 67.089	2.715
5500	9.229	64.130	56.215	43.532	− 68.137	− 67.069	2.665
5600	9.244	64.296	56.358	44.456	− 68.215	− 67.049	2.617
5700	9.258	64.460	56.498	45.381	− 68.295	− 67.026	2.570
5800	9.272	64.621	56.637	46.308	− 68.375	− 67.004	2.525
5900	9.286	64.780	56.774	47.236	− 68.457	− 66.981	2.481
6000	9.299	64.936	56.908	48.165	− 68.541	− 66.956	2.439

APPENDIX G

HEAT OF FORMATION OF SOME COMMON PURE FUELS AND OXIDIZERS
AT 25°C (298.16°K)(IDEAL GAS)
(Kcal/gm mole)*
These values are compatible with the JANNAF tables and Appendix F.

Fuel	Formula	ΔH_f^o	Fuel	Formula	ΔH_f^o
Methane	CH_4	−17.89	Benzene	C_6H_6	19.82
Ethane	H_2H_6	−20.24	n-Heptane	C_7H_{16}	−44.88
Ethylene	C_2H_4	12.50	Toluene	C_7H_8	11.95
Acetylene	C_2H_2	54.19	n-Octane	C_8H_{18}	−49.82
Propane	C_3H_8	−24.85	Ethylene oxide	C_2H_4O	−12.58
Cyclopropane	C_3H_6	12.74	Propylene oxide	C_3H_6O	−22.17
Propylene (propene)	C_3H_6	4.88	Methyl alcohol	CH_3OH	−48.08
Propene (methylacetylene)	C_3H_4	44.32	Ethyl alcohol	C_2H_5OH	−56.12
			Acetone	C_3H_6O	−52.00
n-Butane	C_4H_{10}	−30.15	Dimethyl ether	C_2H_6O	−43.99
isoButane	C_4H_{10}	−32.15	Diethyl ether	$C_4H_{10}O$	−60.28
1,2 Butadiene	C_4H_6	38.77	Nitro methane	CH_3NO_2	−17.86
1,3 Butadiene	$C4H_6$	26.33	Methyl nitrate	CH_3NO_3	−28.80
n-Pentane	C_5H_{12}	−35.00	Hydrazine	N_2H_4	22.75
n-Hexane	C_6H_{14}	−39.96	Hydrogen peroxide	H_2O_2	−32.53
Cyclohexane	C_6H_{12}	−29.43	Ozone	O_3	34.00

*Taken from D. R. Stull, E. F. Westrum, and G. C. Sinke, *The Chemical Thermodynamics of Organic Compounds*, (New York: Wiley, 1969), with permission.

Index